Die Grundlehren der mathematischen Wissenschaften

in Einzeldarstellungen
mit besonderer Berücksichtigung
der Anwendungsgebiete

Band 217

Bo Stenström

Rings of Quotients

An Introduction to Methods of Ring Theory

Springer-Verlag
New York Heidelberg Berlin 1975

Bo Stenström
Matematiska Institutionen, Stockholms Universitet

AMS Subject Classifications (1970):
16-01, 16-02, 16 A 08, 18-01, 18 E 15, 18 E 40

ISBN 0-387-07117-2 Springer-Verlag New York Heidelberg Berlin
ISBN 3-540-07117-2 Springer-Verlag Berlin Heidelberg New York

Typesetting, printing, and binding: Universitätsdruckerei H. Stürtz AG, Würzburg.

Library of Congress Cataloging in Publication Data. Stenström, Bo.T. Rings of quotients. (Die Grundlehren der mathematischen Wissenschaften in Einzeldarstellungen: Bd. 217.) Bibliography: p. Includes index. 1. Quotient rings. 2. Torsion theory (Algebra). 3. Categories (Mathematics). I. Title. II. Series. QA251.5.S8. 512'.55. 75-1003.

Contents

Introduction

The theory of rings of quotients has its origin in the work of Ø. Ore and K. Asano on the construction of the total ring of fractions, in the 1930's and 40's. But the subject did not really develop until the end of the 1950's, when a number of important papers appeared (by R.E. Johnson, Y. Utumi, A.W. Goldie, P. Gabriel, J. Lambek, and others). Since then the progress has been rapid, and the subject has by now attained a stage of maturity, where it is possible to make a systematic account of it (which is the purpose of this book).

The most immediate example of a ring of quotients is the field of fractions Q of a commutative integral domain A. It may be characterized by the two properties:

(i) For every $q \in Q$ there exists a non-zero $s \in A$ such that $qs \in A$.
(ii) Q is the maximal over-ring of A satisfying condition (i).

The well-known construction of Q can be immediately extended to the case when A is an arbitrary commutative ring and S is a multiplicatively closed set of non-zero-divisors of A. In that case one defines the *ring of fractions* $Q = A[S^{-1}]$ as consisting of pairs (a, s) with $a \in A$ and $s \in S$, with the declaration that $(a, s) = (b, t)$ if there exists $u \in S$ such that $uta = usb$. The resulting ring Q satisfies (i), with the extra requirement that $s \in S$, and (ii).

If A is a non-commutative ring and S is a multiplicatively closed set of non-zero-divisors of A, then a *right ring of fractions* $A[S^{-1}]$ can be defined in the same way as in the commutative case, but for this to work one has to assume that S satisfies the following condition:

(∗) For each $a \in A$ and $s \in S$, there exist $b \in A$ and $t \in S$ such that $at = sb$.

Every element in S then becomes invertible in $A[S^{-1}]$, and the elements of $A[S^{-1}]$ may be written as as^{-1} with $a \in A$ and $s \in S$. In particular, when S consists of all the non-zero-divisors of A, the condition (∗) is called the *Ore condition* and $A[S^{-1}]$ is called the *classical right ring of quotients* of A. We devote Chap. II to a study of rings of fractions with respect to multiplicatively closed sets S (which may contain zero-divisors also). This chapter may be considered as a preparation for the general theory of rings of quotients, and the concepts and results developed there will serve as illustrations in the later chapters. We return to the subject of classical rings of quotients in the final chapter (Chap. XV).

There are examples of rings A with over-rings B of A such that B strictly contains the classical ring of quotients of A, but still B may be viewed as a kind of ring of quotients of A. (As an example one may take a boolean algebra A, which is its own classical ring of quotients, and its boolean completion B.) In these examples, the previous condition (i) is valid in a modified formulation:

(i)′ For every $q \in B$ there exists a right ideal $\mathfrak{a}$ of A such that $q\mathfrak{a} \subset A$ and $(\mathfrak{a}:a)$ has no non-zero left annihilators for arbitrary $a \in A$.

The condition (i)′ together with the correspondingly modified (ii) characterize the *maximal right ring of quotients* of A.

On the other hand, let $\mathfrak{F}$ be a filter of right ideals of A with the properties:

1) If $\mathfrak{a} \in \mathfrak{F}$ and $a \in A$, then $(\mathfrak{a}:a) \in \mathfrak{F}$.

2) If $\mathfrak{a} \in \mathfrak{F}$ and $\mathfrak{b}$ is a right ideal such that $(\mathfrak{b}:a) \in \mathfrak{F}$ for all $a \in \mathfrak{a}$, then $\mathfrak{b} \in \mathfrak{F}$.

Such a filter $\mathfrak{F}$ is called a *Gabriel topology* on A. One may construct a *ring of quotients* $A_{\mathfrak{F}}$ by considering the set of homomorphisms $\varphi\colon \mathfrak{a} \to A$ for $\mathfrak{a} \in \mathfrak{F}$, identifying φ with $\psi\colon \mathfrak{b} \to A$ if φ equals ψ on some right ideal $\mathfrak{c} \in \mathfrak{F}$ with $\mathfrak{c} \subset \mathfrak{a} \cap \mathfrak{b}$. More precisely, one defines

$$A_{\mathfrak{F}} = \varinjlim \operatorname{Hom}(\mathfrak{a}, A/t(A)),$$

where $t(A) = \{a \in A \mid a\mathfrak{b} = 0 \text{ for some } \mathfrak{b} \in \mathfrak{F}\}$ and the direct limit is taken over $\mathfrak{a} \in \mathfrak{F}$. As a special case, the ring of fractions $A[S^{-1}]$ is obtained by considering the set of homomorphisms $sA \to A$ for $s \in S$. These general rings of quotients are introduced in Chap. IX, which may be considered as the central chapter of the book.

For a Gabriel topology $\mathfrak{F}$ and every A-module M, one may construct the *module of quotients* $M_{\mathfrak{F}}$ by considering homomorphisms $\varphi\colon \mathfrak{a} \to M$ for $\mathfrak{a} \in \mathfrak{F}$, in the same way as for $A_{\mathfrak{F}}$. The modules $M_{\mathfrak{F}}$ form a full subcategory **Mod**-$(A, \mathfrak{F})$ of the category **Mod**-$A_{\mathfrak{F}}$ of all $A_{\mathfrak{F}}$-modules, and it turns out that this *quotient category* has several excellent properties: it is an abelian category with exact direct limits and a generator (i.e. a *Grothendieck category*). Moreover, the "localization" functor a: **Mod**-$A \to$ **Mod**-$(A, \mathfrak{F})$ given by $a(M) = M_{\mathfrak{F}}$ is exact and is a left adjoint of the inclusion functor i: **Mod**-$(A, \mathfrak{F}) \to$ **Mod**-A. However, the functor i is in general not exact. These categorial aspects of the theory are discussed in Chap. X. There we will also prove a converse result, namely that every Grothendieck category **C** is the quotient category **Mod**-$(A, \mathfrak{F})$ for a certain Gabriel topology $\mathfrak{F}$ on the endomorphism ring A of a generator for **C**. In this way the theory of Grothendieck categories is intimately related to the theory of rings of quotients. We therefore devote Chap. V to a study of the properties of Grothendieck categories; the necessary background material on abelian categories is given in Chap. IV.

In many important cases, e.g. for modules of fractions, the functor a has the simple form $a(M) = M \otimes_A A_{\mathfrak{F}}$. Then **Mod**-$(A, \mathfrak{F})$ coincides with the module category **Mod**-$A_{\mathfrak{F}}$, and also several other properties of modules of fractions carry over to this situation. We will study this case in Chap. XI.

It is natural to attempt to classify all Gabriel topologies on a given ring A. This has been successfully done for certain noetherian rings (including all commutative noetherian rings) in terms of their prime ideal structure (Chap. VII), and for rings with suitable minimum conditions on right ideals (Chap. VIII). To every Gabriel topology $\mathfrak{F}$ there corresponds a notion of torsion: if M is a right A-module, then $x \in M$ is an $\mathfrak{F}$-*torsion element* if $x\mathfrak{a} = 0$ for some $\mathfrak{a} \in \mathfrak{F}$. In Chap. VI we study this torsion concept and determine the relations between the notions of Gabriel topology and torsion.

The purpose of this book is not only to describe the theory of rings of quotients, but also to give an introduction to the basic methods and results of ring theory at large. We devote Chap. I to a detailed account of the theory of modules, with applications to some important classes of rings (semi-simple, regular, coherent). Noetherian rings are discussed in Chap. VII and rings with various minimum conditions are treated in Chap. VIII. Chaps. III–V deal with modular lattices and abelian categories, which are fundamental tools in the general theory of rings and particularly useful in the theory of rings of quotients.

A very useful tool for ring theory is provided by homological algebra. However, an introduction to homological algebra in this book would have carried us too far away from the subject of rings of quotients. We have therefore preferred to avoid the use of homological methods (except for a few examples and applications).

There are a number of topics in the theory of rings of quotients which are not mentioned at all in this book. The most notable omissions concern the applications of the theory of classical rings of quotients:

(1) Rings with polynomial identities (Posner [1]).

(2) Structure theory of hereditary noetherian prime rings (Michler [3], Jategaonkar [4]).

(3) Maximal orders and Asano orders (Jacobson [1], Michler [4], Robson [2]).

(4) Principal right ideal rings (Goldie [3], Jategaonkar [3]).

(5) The Faith-Utumi theorem (Faith and Utumi [1], Lambek [4]).

The only prerequisites for a reading of this book are a knowledge of elementary "abstract" algebra (to the extent covered e.g. by the first chapters of Herstein's "Topics in algebra" (Herstein [3])), and of the basic notions of set theory (e.g. Zorn's lemma) and general topology.

A short preliminary version of this book was published in the Springer series "Lecture Notes in Mathematics" under the title "Rings and modules of quotients" (Vol 237, 1971).

Notations and Conventions

$\subset$	inclusion (*not* necessarily proper)
$\mathbb{N}$	natural integers
$\mathbb{Z}$	rational integers
$\mathbb{Q}$	rational numbers
$\mathbb{R}$	real numbers
$\mathbb{C}$	complex numbers
$M_n(A)$	the ring of $n \times n$-matrices with entries in A.
$UT_n(A)$	the ring of upper triangular $n \times n$-matrices with entries in A.

If A is a ring and $S \subset A$, then $r(S) = \{a \in A \mid Sa = 0\}$,

$$l(S) = \{a \in A \mid aS = 0\}.$$

All modules are right A-modules, unless otherwise is stated. Every ring has an identity element.

Chapter I. Modules

This chapter introduces the basic conceptual machinery to be used in the book, namely the theory of modules. In the following chapters this theory will be further developed on a more general and abstract level, in the form of lattice theory and the theory of abelian categories.

§ 1. Basic Definitions

Let A be a ring (with 1 as always).

Definition. A *right A-module* is an abelian group M together with a map $M \times A \to M$, written $(x, a) \mapsto x a$, such that

$M1$. $(x+y) a = x a + y a$,
$M2$. $x(a+b) = x a + x b$,
$M3$. $x(a b) = (x a) b$,
$M4$. $x 1 = x$,

for all $x, y \in M$ and $a, b \in A$.

Similarly we define *left A-modules* by letting A operate to the left of M. We write M_A (${}_A M$) to indicate that M is a right (left) A-module. In this book we shall mainly deal with right modules, so "*module*" *will mean a right module, unless otherwise is stated.* When A is commutative, there is no need to distinguish between right and left A-modules.

We leave it as an exercise to show that

$$0 a = 0, \quad x 0 = 0 \quad \text{and} \quad (-x) a = x(-a) = -(x a).$$

Let M be an A-module. A subset L of M is a *submodule* of M if L is an additive subgroup and $x \in L$, $a \in A$ implies $x a \in L$, i.e. L is a module under operations inherited from M. When L is a submodule of M, we can define the *quotient module* M/L to be the additive quotient group M/L together with the operation $M/L \times A \to M/L$ given by $(\bar{x}, a) \mapsto \overline{x a}$. This operation is well-defined, for if $\bar{x} = \bar{y}$, then $x - y \in L$ and hence $x a - y a = (x - y) a \in L$. One easily verifies that this makes M/L into an A-module.

Let M and N be A-modules. A map α: $M \to N$ is a *homomorphism* (or *A-linear map*) if

$$\alpha(x+y) = \alpha(x) + \alpha(y) \quad \text{and} \quad \alpha(x a) = \alpha(x) a \qquad \text{for } x, y \in M, \ a \in A.$$

If α: $M \to N$ and β: $N \to P$ are homomorphisms, then the composed map $\beta\alpha$: $M \to P$ is also a homomorphism. For each module M there is an *identity* homomorphism 1_M. If α: $M \to N$ is A-linear, we define its *kernel* as $\operatorname{Ker}\alpha = \{x \in M \mid \alpha(x) = 0\}$ and its *image* as $\operatorname{Im}\alpha = \{\alpha(x) \mid x \in M\}$. $\operatorname{Ker}\alpha$ is a submodule of M, and $\operatorname{Im}\alpha$ is a submodule of N. α is called a *monomorphism* if $\alpha(x) = \alpha(y)$ implies $x = y$ for $x, y \in M$, and this is known to be equivalent to $\operatorname{Ker}\alpha = 0$. α is an *epimorphism* if $\operatorname{Im}\alpha = N$. E.g. if L is a submodule of M, then there is a canonical monomorphism $L \to M$ and epimorphism $M \to M/L$. α is an *isomorphism* if there exists a homomorphism β: $N \to M$ such that $\alpha\beta = 1_N$ and $\beta\alpha = 1_M$. β is unique and is written as α^{-1}. M and N are then *isomorphic* modules, and we write $M \simeq N$. One easily verifies:

Lemma 1.1. *A homomorphism is an isomorphism if and only if it is both a monomorphism and an epimorphism.*

Proposition 1.2. *A homomorphism α: $M \to N$ induces an isomorphism*

$$M/\operatorname{Ker}\alpha \cong \operatorname{Im}\alpha.$$

Proof. The induced map $\bar{x} \mapsto \alpha(x)$ is a homomorphism, and it clearly is a monomorphism and an epimorphism. □

A sequence of homomorphisms of A-modules

$$\cdots \to M_{i-1} \xrightarrow{\alpha_{i-1}} M_i \xrightarrow{\alpha_i} M_{i+1} \xrightarrow{\alpha_{i+1}} \cdots$$

is said to be *exact at* M_i if $\operatorname{Ker}\alpha_i = \operatorname{Im}\alpha_{i-1}$, and is *exact* if it is exact at each M_i (except at the ends). E.g. if L is a submodule of M, then the sequence of canonical homomorphisms

$$0 \to L \to M \to M/L \to 0$$

is exact. Conversely, if a sequence $0 \to M' \xrightarrow{\alpha} M \xrightarrow{\beta} M'' \to 0$ is exact (in which case it is called a *short exact sequence*), then α is a monomorphism, β is an epimorphism and $M'' \cong M/\operatorname{Im}\alpha$. The following result ("the short 5-lemma") is useful:

Proposition 1.3. *Let the diagram*

$$\begin{array}{ccccccccc}
0 & \longrightarrow & L & \xrightarrow{\alpha} & M & \xrightarrow{\beta} & N & \longrightarrow & 0 \\
 & & \downarrow{\scriptstyle\lambda} & & \downarrow{\scriptstyle\mu} & & \downarrow{\scriptstyle\eta} & & \\
0 & \longrightarrow & L' & \xrightarrow[\alpha']{} & M' & \xrightarrow[\beta']{} & N' & \longrightarrow & 0
\end{array}$$

be commutative with exact rows. If λ and η are monomorphisms (epimorphisms, resp. isomorphisms), then so is μ.

Proof. Suppose λ and η are monomorphisms, and let $x \in M$ with $\mu(x) = 0$. Then $\eta\beta(x) = \beta'\mu(x) = 0$, and η monomorphism implies $x \in \operatorname{Ker}\beta = \operatorname{Im}\alpha$. If we put $x = \alpha(y)$, then $\alpha'\lambda(y) = \mu\alpha(y) = \mu(x) = 0$, and $\alpha'\lambda$ a monomorphism implies $y = 0$. Hence $x = 0$. The proof for the other two cases is left to the reader. □

With the help of Prop. 1.2 we can easily prove the two Noether isomorphism theorems. Let $L \subset M \subset N$ be submodules. There is then induced a monomorphism

$M/L \to N/L$, and clearly the sequence of canonical homomorphisms

$$0 \to M/L \to N/L \to N/M \to 0$$

is exact. Hence we have:

Proposition 1.4. *If* $L \subset M \subset N$ *are submodules, then* $(N/L)/(M/L) \cong N/M$.

Let L and M be two submodules of N. Define $L+M$ to be the submodule $\{x+y \mid x \in L, y \in M\}$ of N. There is a homomorphism $\alpha\colon L \to (L+M)/M$ given by $x \mapsto \bar{x}$. It is clearly an epimorphism, and $\operatorname{Ker}\alpha = \{x \in L \mid \bar{x} = 0\} = L \cap M$. Hence:

Proposition 1.5. *If* L *and* M *are submodules of* N, *then* $(L+M)/M \cong L/(L \cap M)$.

Let M and N be A-modules. The set of all homomorphisms $M \to N$ is denoted by $\operatorname{Hom}_A(M, N)$. This is an abelian group, for if α and β are homomorphisms, then one can define $\alpha+\beta$ by the rule $(\alpha+\beta)(x) = \alpha(x)+\beta(x)$; one verifies that $\alpha+\beta$ is a homomorphism, and that this makes $\operatorname{Hom}_A(M, N)$ into an abelian group.

Examples

1. $\mathbb{Z}$-modules. Every abelian group may be considered as a $\mathbb{Z}$-module, by defining $xn = x + \cdots + x$ (n times) for $n > 0$. Conversely, if M is a $\mathbb{Z}$-module, then $xn = x(1 + \cdots + 1) = x \cdot 1 + \cdots + x \cdot 1 = x + \cdots + x$, so a $\mathbb{Z}$-module is nothing more than an abelian group.

2. Ideals. The ring A may be considered as a right (and as a left) A-module by using the ring multiplication for the operation of A on itself. The submodules of A_A are precisely the right ideals of A.

3. Bimodules. Let A and B be two rings. M is an *A-B-bimodule* if M is a left A-module and a right B-module (with the same underlying additive group), such that $(ax)b = a(xb)$ for $a \in A$, $x \in M$, $b \in B$.

4. Endomorphism rings. Let M be a right A-module. A homomorphism $\alpha\colon M \to M$ is called an *endomorphism* (and is an *automorphism* if it is an isomorphism). The abelian group $\operatorname{Hom}_A(M, M)$ becomes a ring if one uses the composition of maps as multiplication. This ring is called the *endomorphism ring* of M, and it is sometimes denoted by $\operatorname{End}_A(M)$. If we let it operate to the left on M as $\alpha x = \alpha(x)$, then M becomes an $\operatorname{End}_A(M)$-A-bimodule. The endomorphism ring of M, when M is considered as a left $\operatorname{End}_A(M)$-module, is called the *bicommutator* of M and is denoted by $\operatorname{Bic}_A(M)$. There is a ring homomorphism $\eta\colon A \to \operatorname{Bic}_A(M)$, given by $\eta(a)\colon x \mapsto xa$ for $x \in M$.

5. Restriction of scalars. If $\varphi\colon A \to B$ is a homomorphism of rings, then every B-module M may be considered as an A-module by "restriction of scalars", i.e. by defining $xa = x\varphi(a)$ for $x \in M$, $a \in A$.

6. The opposite ring. If A is a ring, we define the *opposite* ring A^{op} to have the same underlying abelian group as A, but with a new product $*$ defined as $a * b = b \cdot a$, where $\cdot$ denotes the product in A. A left A-module is the same thing as a right A^{op}-module.

§2. Sums and Products of Modules

Throughout this section we let A be a ring, and all modules are right A-modules. Let I be a set, and assume we are given a family $(M_i)_I$ of modules M_i (the family is thus indexed by the set I). The cartesian product $\prod_I M_i = \{(x_i)_I \mid x_i \in M_i\}$ can be made into a module by defining the operations component-wise:

$$(x_i)_I + (y_i)_I = (x_i + y_i)_I, \qquad (x_i)_I\, a = (x_i\, a)_I.$$

This module is called the *direct product* of the modules M_i. It contains a submodule

$$\bigoplus_I M_i = \{(x_i)_I \mid x_i = 0 \text{ for almost all } i \in I\},$$

which is called the *(external) direct sum* of the modules M_i (we will use the term "almost all" to mean "all but a finite number"). There are canonical homomorphisms $\prod_I M_i \to M_j$, called *projections*, and $M_j \to \prod_I M_j$, called *injections*, for each $j \in I$. The same is true for $\bigoplus_I M_i$. If $I = \{1, \ldots, n\}$, then $\prod_I M_i = \bigoplus_I M_i$, and this is usually written as $M_1 \times \cdots \times M_n$ or $M_1 \oplus \cdots \oplus M_n$.

Let M be a module and $(M_i)_I$ a family of submodules of M. We define

$$\sum_I M_i = \{\textstyle\sum_i x_i \in M \mid x_i \in M_i \text{ with } x_i = 0 \text{ for almost all } i \in I\},$$

which clearly is the smallest submodule of M containing all M_i. We can also define a homomorphism $\alpha\colon \bigoplus_I M_i \to M$ by $\alpha((x_i)_I) = \sum x_i$. The image of α is precisely $\sum_I M_i$. If α happens to be a monomorphism, so that $\sum_I M_i \cong \bigoplus_I M_i$, then we say that the module $\sum_I M_i$ is the *(internal) direct sum* of its submodules M_i. The symbol $\oplus$ is usually employed also for internal direct sums. It is worth noting that the property of directness is of finite character, i.e. a sum $\sum_I M_i$ is direct if and only if the sums $\sum_J M_i$ are direct for all finite subsets J of I.

When all $M_i = M$, $i \in I$, we use the special notations

$$\prod_I M_i = M^I, \qquad \bigoplus_I M_i = M^{(I)} \quad \text{and} \quad M \times \cdots \times M = M^n.$$

If L and N are two modules, there is an obvious exact sequence

$$0 \to L \to L \oplus N \to N \to 0,$$

which is said to be "split". Slightly more generally we say that an exact sequence $0 \to L \to M \to N \to 0$ is *split* if there is an isomorphism $M \cong L \oplus N$ such that the diagram

$$\begin{array}{ccccccccc} 0 & \longrightarrow & L & \longrightarrow & M & \longrightarrow & N & \longrightarrow & 0 \\ & & \| & & \downarrow{\scriptstyle\cong} & & \| & & \\ 0 & \longrightarrow & L & \longrightarrow & L \oplus N & \longrightarrow & N & \longrightarrow & 0 \end{array} \tag{1}$$

commutes.

Proposition 2.1. *The following properties of an exact sequence*

$$0 \to L \xrightarrow{\alpha} M \xrightarrow{\beta} N \to 0$$

are equivalent:

(a) *The sequence splits.*

(b) *There exists a homomorphism $\varphi: M \to L$ such that $\varphi\alpha = 1_L$.*

(c) *There exists a homomorphism $\psi: N \to M$ such that $\beta\psi = 1_N$.*

Proof. It is clear that (a) implies (b) and (c). Suppose (b) is satisfied. The maps $\varphi: M \to L$ and $\beta: M \to N$ can be used to define $\mu: M \to L \oplus N$ so that the diagram (1) commutes. μ is an isomorphism by Prop. 1.3. Hence the sequence splits. The proof of (c) $\Rightarrow$ (a) goes dually. $\square$

A module M is said to be *generated* by a family $(x_i)_I$ of elements of M if each $x \in M$ can be written $x = \sum_I x_i a_i$ with all but a finite number of a_i equal to 0. If it furthermore is true that the coefficients a_i are uniquely determined by x, then the family $(x_i)_I$ is a *basis* for M. A module is called *free* if there exists a basis for it.

Proposition 2.2. *A module M is free if and only if $M \cong A^{(I)}$ for some I.*

Proof. The module $A^{(I)}$ is free on a basis $(e_i)_I$, where e_i has j-th component 1 if $j = i$ and 0 if $j \neq i$. Conversely, if M is free on a basis $(x_i)_I$, then one defines $A^{(I)} \to M$ as $(a_i)_I \mapsto \sum x_i a_i$, which is an isomorphism as a consequence of the definition of a basis. $\square$

Note that if M is a free module on a basis $(x_i)_I$, then every homomorphism $\alpha: M \to N$ is uniquely determined by its values on $(x_i)_I$, and the elements $\alpha(x_i)$ can be chosen arbitrarily and independently of each other in N.

Proposition 2.3. *If A is a skew-field, then every right A-module is free.*

We will prove a somewhat more precise statement. For this we define a family $(x_i)_I$ of elements of M to be *linearly independent* if $\sum_I x_i a_i = 0$ (with almost all $a_i = 0$) implies that all $a_i = 0$. Then:

Proposition 2.4. *Let M be a module over a skew-field A. If S is a set of generators for M, and L is a linearly independent set with $L \subset S$, then there exists a basis B for M such that $L \subset B \subset S$.*

Proof. We will apply Zorn's lemma to the set $\mathcal{M} = \{L' \mid L \subset L' \subset S$ and L' is linearly independent$\}$, which is partially ordered under inclusion, and is non-empty since $L \in \mathcal{M}$. Suppose $\mathcal{T}$ is a totally ordered chain contained in $\mathcal{M}$. Let L'' be the union of all L' in $\mathcal{T}$. L'' will be an upper bound for $\mathcal{T}$ if we can show that $L'' \in \mathcal{M}$. But it is clear that $L \subset L'' \subset S$, and L'' is linearly independent, because every finite subset of L'' lies in some $L' \in \mathcal{T}$ and is therefore linearly independent. Hence we can apply Zorn's lemma to find a maximal element $B \in \mathcal{M}$. Since B is linearly independent, it remains to show that B generates M. Since S generates M, it is enough to show that every $x \in S$ is a linear combination of elements from B. Now if $x \in S$ but $x \notin B$, then $B \cup \{x\}$ is not linearly independent, so one has $xa + \sum x_i a_i = 0$ with $a \neq 0$ and $x_i \in B$. This gives $x = -\sum x_i a_i a^{-1}$, as desired. $\square$

Prop. 2.3 is now obtained by taking e.g. $L=\emptyset$ and $S=M$. A module over a skew-field is usually called a *vector space.*

Proposition 2.5. *Every module is a quotient module of a free module.*

Proof. Let $(x_i)_I$ be generators for the module M. Define $\alpha\colon A^{(I)} \to M$ as $\alpha((a_i)_I) = \sum x_i a_i$. α is an epimorphism, so $M \cong A^{(I)}/\mathrm{Ker}\,\alpha$. ◻

Examples

1. Direct summands and idempotents. If L is a submodule of a module M, then L is a *direct summand* of M if there exists a submodule L' of M such that $M = L \oplus L'$. The module M is *indecomposable* if it has no direct summands $\neq 0$, M. Note that M is the direct sum of two submodules L and L' if and only if $L + L' = M$ and $L \cap L' = 0$.

The direct summands of A_A correspond to *idempotent* elements of A, i.e. $e \in A$ such that $e^2 = e$. For if $e^2 = e$, then one gets a direct sum decomposition $A = eA \oplus (1-e)A$. Conversely, if $A = \mathfrak{a} \oplus \mathfrak{b}$ with $\mathfrak{a}$ and $\mathfrak{b}$ right ideals, then we can write $1 = e + f$ with $e \in \mathfrak{a}$, $f \in \mathfrak{b}$. Multiplication with e gives $e - e^2 = fe \in \mathfrak{a} \cap \mathfrak{b} = 0$, so $e = e^2$. For each $a \in \mathfrak{a}$ one similarly gets $a = ea \in eA$, so $\mathfrak{a} = eA$ and $\mathfrak{b} = (1-e)A$.

2. The free module on a set. For each set S we define the free right A-module $F_A(S)$ on the set S to be the set of all formal linear combinations $\sum_{s \in S} s \cdot a_s$ with $a_s \in A$ and $a_s = 0$ for almost all s, with module operations induced from A. Then $F_A(S)$ is a free module on the basis S ($s \cdot 1$ is identified with $s \in S$).

3. Group rings. Let A be a commutative ring and G a group. We define the *group ring* $A[G]$ to be the free A-module on the set G, with multiplication induced from the multiplication in the group G. Thus

$$A[G] = \{ \sum_{g \in G} a_g g \mid a_g \in A \text{ and } a_g = 0 \text{ for almost all } g \},$$

and

$$\sum_g a_g g \cdot \sum_h b_h h = \sum_k c_k k, \quad \text{where} \quad c_k = \sum_{gh=k} a_g b_h.$$

4. Subspaces of vector spaces. Let M be a vector space and L a subspace of M. Choose a basis $(x_i)_I$ for L. The family $(x_i)_I$ is linearly independent in M, and by Prop. 2.4 therefore contained in a basis $(x_j)_J$ for M, with $I \subset J$. Let N be the subspace of M generated by $(x_j)_{j \notin I}$. Then $M = L \oplus N$. So every subspace of a vector space is a direct summand.

§ 3. Finitely Generated Modules and Noetherian Modules

A module M is *finitely generated* if there exists a finite set of generators for M, or in other words, if there is an epimorphism $A^n \to M$ for some n. In particular, M is *cyclic* if it can be generated by a single element, or equivalently, if there is an epimorphism $A \to M$. It follows that M is cyclic if and only if $M \cong A/\mathfrak{a}$ for a right ideal $\mathfrak{a}$ of A.

Lemma 3.1. *Let L be a submodule of a module M. Then:*
(i) *If M is finitely generated, then so is M/L.*
(ii) *If L and M/L are finitely generated, then so is M.*

Proof. (i): If $x_1, \ldots, x_n$ generate M, then $\bar{x}_1, \ldots, \bar{x}_n$ generate M/L.
(ii): Suppose L is generated by $x_1, \ldots, x_m$ and M/L is generated by $\bar{y}_1, \ldots, \bar{y}_n$, where $y_i \in M$. If $x \in M$, then $\bar{x} = \sum \bar{y}_i a_i$ and hence $x - \sum y_i a_i \in L$, so $x - \sum y_i a_i = \sum x_j b_j$. Thus M is generated by $x_1, \ldots, x_m, y_1, \ldots, y_n$. □

The module M is *finitely presented* if there exists an exact sequence

$$A^m \to A^n \to M \to 0,$$

which means that M is not only finitely generated, but also the module of "relations between the generators of M" is finitely generated.

Proposition 3.2. *If M is finitely presented and $0 \to K \to L \to M \to 0$ is an exact sequence with L finitely generated, then K is finitely generated.*

Proof. Since M is finitely presented, there exists an exact sequence $A^m \to A^n \to M \to 0$. We get a commutative diagram

$$\begin{array}{ccccccc} & & A^m & \xrightarrow{\eta} & A^n & \xrightarrow{\mu} & M & \to 0 \\ & & \downarrow{\scriptstyle\kappa} & & \downarrow{\scriptstyle\lambda} & & \| & \\ 0 & \to & K & \xrightarrow[\alpha]{} & L & \xrightarrow[\beta]{} & M & \to 0, \end{array}$$

where $\lambda \colon A^n \to L$ is obtained by mapping the basis element $e_i \in A^n$ to an element $x_i \in L$ such that $\beta(x_i) = \mu(e_i)$; then $\beta\lambda\eta = \mu\eta = 0$, so $\operatorname{Im}\lambda\eta \subset \operatorname{Ker}\beta = \operatorname{Im}\alpha$, and therefore $\lambda\eta$ can be factored over α by means of a homomorphism $\kappa \colon A^m \to K$. The map $\alpha \colon K \to L$ induces $\bar{\alpha} \colon K/\operatorname{Im}\kappa \to L/\operatorname{Im}\lambda$, because

$$\alpha(\operatorname{Im}\kappa) = \operatorname{Im}\alpha\kappa = \operatorname{Im}\lambda\eta \subset \operatorname{Im}\lambda.$$

We have thus got a bigger commutative diagram

$$\begin{array}{ccccccc} & & A^m & \longrightarrow & A^n & \longrightarrow & M & \to 0 \\ & & \downarrow{\scriptstyle\kappa} & & \downarrow{\scriptstyle\lambda} & & \| & \\ 0 & \to & K & \xrightarrow[\alpha]{} & L & \xrightarrow[\beta]{} & M & \to 0 \\ & & \downarrow & & \downarrow & & & \\ & & K/\operatorname{Im}\kappa & \xrightarrow[\bar{\alpha}]{} & L/\operatorname{Im}\lambda. & & & \end{array}$$

We assert that $\bar{\alpha}$ is an isomorphism. Suppose $\bar{\alpha}(\bar{x}) = 0$ for some $x \in K$. Then $\alpha(x) \in \operatorname{Im}\lambda$, so $\alpha(x) = \lambda(y)$ for some $y \in A^n$. But $\mu(y) = \beta\lambda(y) = \beta\alpha(x) = 0$, so $y \in \operatorname{Im}\eta$. Writing $y = \eta(z)$ for $z \in A^m$, we get $\alpha\kappa(z) = \lambda\eta(z) = \lambda(y) = \alpha(x)$ and hence $x = \kappa(z)$, since α is a monomorphism. It follows that $\bar{x} = 0$, so $\bar{\alpha}$ is a monomorphism.

Next we show that $\bar{x} \in \operatorname{Im}\bar{\alpha}$ for every $x \in L$. We have $\beta(x) = \mu(y)$ for some $y \in A^n$. Then $x - \lambda(y) \in \operatorname{Ker}\beta$, so $x - \lambda(y) = \alpha(z)$ for some $z \in K$. But then $\bar{x} = \overline{\alpha(z)} = \bar{\alpha}(\bar{z})$.

The module $L/\operatorname{Im}\lambda$ is finitely generated by 3.1 (i), and so is then also the module $K/\operatorname{Im}\kappa$, since $\bar{\alpha}$ is an isomorphism. Also $\operatorname{Im}\kappa$ is finitely generated by 3.1 (i), and 3.1 (ii) then shows that K is finitely generated. □

It is of course in general not true that submodules of a finitely generated module are finitely generated. We call a module M *noetherian* if every submodule of M is finitely generated.

Proposition 3.3. *A module is noetherian if and only if every strictly ascending chain of submodules is finite.*

Proof. Let M be noetherian and $M_1 \subset M_2 \subset \cdots$ an ascending chain of submodules. The submodule $\bigcup_i M_i$ has a finite number of generators, and all of them must lie in some M_{i_0}. It follows that the chain gets stationary at M_{i_0}. Conversely, it is easy to see that the ascending chain condition for submodules implies that every submodule has a finite number of generators. □

The "ascending chain condition", i.e. finiteness of all strictly ascending chains, is usually abbreviated as ACC.

Proposition 3.4. *Let L be a submodule of M. Then M is noetherian if and only if both L and M/L are noetherian.*

Proof. M noetherian obviously implies that L is noetherian. It also implies that M/L is noetherian, because the submodules of M/L can be written as M'/L, where $L \subset M' \subset M$. Suppose conversely that L and M/L are noetherian. If M' is a submodule of M, then $L \cap M'$ is finitely generated as a submodule of L, and $M'/(L \cap M') \cong (L+M')/L$ is finitely generated as a submodule of M/L. It follows from Lemma 3.1 (ii) that M' is finitely generated. Hence M is noetherian. □

The ring A is *right noetherian* if A_A is a noetherian module, i.e. every right ideal of A is finitely generated.

Proposition 3.5. *If A is right noetherian, then every finitely generated module is noetherian.*

Proof. If A_A is noetherian, then every finitely generated free module is noetherian by Prop. 3.4, and therefore every finitely generated module is a quotient of a noetherian module and hence noetherian by Prop. 3.4. □

Corollary 3.6. *If A is right noetherian, then every finitely generated module is finitely presented.*

A large number of examples of noetherian rings is obtained from "Hilbert's basis theorem":

Proposition 3.7. *If A is a right noetherian ring, then the polynomial ring $A[X]$ is ring noetherian.*

Proof. Let $\mathfrak{a}$ be a right ideal of $A[X]$. For each n we let $\mathfrak{a}_n$ be the set of $a \in A$ for which there exists a polynomial in $\mathfrak{a}$ with leading term aX^n. It is clear that $\mathfrak{a}_n$ is a right ideal of A, and that $\mathfrak{a}_n \subset \mathfrak{a}_{n+1}$. The ascending chain $\mathfrak{a}_0 \subset \mathfrak{a}_1 \subset \cdots$

becomes stationary at some n_0. Let $\{a_{ij}\}$ be a finite family of generators for the right ideal $\mathfrak{a}_i$ of A ($i \leq n_0$), and let f_{ij} be a polynomial in $\mathfrak{a}$ with leading term $a_{ij} X^i$. We assert that the polynomials f_{ij} generate $\mathfrak{a}$.

If there are polynomials in $\mathfrak{a}$ which cannot be written as linear combinations of f_{ij}'s, then we choose one such polynomial of smallest degree, say $g(X) = b X^m + \cdots$. Then $b \in \mathfrak{a}_m$, and if $m \leq n_0$, then we can write $b = \sum_j a_{mj} b_j$. The polynomial

$$g(X) - \sum_j f_{mj}(X) b_j$$

is an element of $\mathfrak{a}$ of degree $< m$. It is therefore a linear combination of f_{ij}'s, and so is then also $g(X)$, a contradiction. If $m > n_0$, then we have $b = \sum_j a_{n_0 j} b_j$, and we look instead at the polynomial

$$g(X) - \sum_j f_{n_0 j}(X) X^{n - m_0} b_j,$$

and argue as before. □

Corollary 3.8. *If K is a field, then the ring $K[X_1, \ldots, X_n]$ is noetherian.*

If A is right noetherian and $\mathfrak{a}$ is a two-sided ideal of A, then the ring $A/\mathfrak{a}$ is also right noetherian.

Examples

1. Cyclic submodules. Let M be a module and $x \in M$. The element x generates a cyclic submodule xA of M. There is an epimorphism $\alpha: A \to xA$ given by $\alpha(a) = xa$, and $\operatorname{Ker} \alpha = \{a \mid xa = 0\} = \operatorname{Ann}(x)$, the *annihilator* of x. Hence $xA \cong A/\operatorname{Ann}(x)$.

2. Simple modules. A module M is *simple* (or *irreducible*) if $M \neq 0$ and the only submodules of M are 0 and M. Every simple module M is cyclic, in fact it is generated by any non-zero $x \in M$. It is clear that M is simple if and only if $M \cong A/\mathfrak{m}$, where $\mathfrak{m}$ is a maximal right ideal of A.

3. Artinian modules and rings. A module M is called *artinian* if every strictly descending chain of submodules is finite. The ring A is *right artinian* if it is artinian as a right A-module. Artinian rings will be studied in Chap. VIII, where we will prove the remarkable fact that every right artinian ring is right noetherian.

§4. Categories and Functors

Although category theory will be developed first in Chap. IV, we will have some use for the functorial language already in this chapter, e.g. for the study of tensor products. It is therefore convenient to introduce the basic concepts of category theory at this point.

In order to be able to speak about e.g. the category of *all* sets, we will use a set theory which distinguishes between sets and classes (thus one can form the class of all sets; for the set-theoretical foundations we refer to Cohn [3] and MacLane [1]).

A *category* **C** is defined to consist of three ingredients:

(i) a class Ob(**C**) of *objects* of **C**,

(ii) a set $\mathrm{Hom}_{\mathbf{C}}(C, C')$, whose elements are called *morphisms* from C to C', for each ordered pair (C, C') of objects,

(iii) a *composition* $\mathrm{Hom}(C', C'') \times \mathrm{Hom}(C, C') \to \mathrm{Hom}(C, C'')$ for each ordered triple (C, C', C'') of objects.

Before stating the axioms for categories we introduce a useful notation. To indicate that $\alpha \in \mathrm{Hom}(C, C')$ we write $\alpha: C \to C'$. The composition of $\alpha: C \to C'$ and $\beta: C' \to C''$ is denoted by $\beta\alpha$. The axioms for categories can now be given:

$C1$. $\mathrm{Hom}(C, C')$ and $\mathrm{Hom}(D, D')$ are disjoint sets if $(C, C') \neq (D, D')$,

$C2$. if $\alpha: C \to C'$, $\beta: C' \to C''$ and $\gamma: C'' \to C'''$ are morphisms, then $\gamma(\beta\alpha) = (\gamma\beta)\alpha$,

$C3$. for each object C there exists $1_C \in \mathrm{Hom}(C, C)$ such that $1_C\alpha = \alpha$ and $\beta 1_C = \beta$ for all $\alpha: C' \to C$ and $\beta: C \to C''$.

The *identity* morphism 1_C is uniquely determined by C, for if also $1'_C$ satisfies $C3$, then $1_C = 1_C \cdot 1'_C = 1'_C$. Some important categories for us are:

1. The category **Sets**: the class Ob(**C**) here is the class of all sets, and the morphisms are the maps between sets, with the usual composition.

2. The category **Ab**: the objects are the abelian groups and the morphisms are the group homomorphisms.

3. The category **Mod**-A for a given ring A: the objects are the right A-modules, and the morphisms are the module homomorphisms. Similarly we write A-**Mod** for the category of left A-modules.

4. Every partially ordered set P defines a category with P as the class of objects, and with a unique morphism $p \to q$ whenever $p \leq q$, while $\mathrm{Hom}(p, q) = \emptyset$ if $p \nleq q$. This is an example of a *small* category, i.e. a category where the class of objects actually is a set.

For each category **C** there is a *dual* category $\mathbf{C}^{\mathrm{op}}$, whose underlying set is the same as that of **C** but with

$$\mathrm{Hom}_{\mathbf{C}^{\mathrm{op}}}(C, C') = \mathrm{Hom}_{\mathbf{C}}(C', C)$$

and $\alpha * \beta = \beta \cdot \alpha$, where $*$ denotes composition in $\mathbf{C}^{\mathrm{op}}$ and $\cdot$ composition in **C** (so $\mathbf{C}^{\mathrm{op}}$ is obtained by "reversing the arrows" of **C**). Every definition or theorem for **C** dualizes to a corresponding definition or theorem for $\mathbf{C}^{\mathrm{op}}$.

A morphism $\alpha: C \to C'$ in a category **C** is an *isomorphism* if there exists $\beta: C' \to C$ such that $\alpha\beta = 1_{C'}$ and $\beta\alpha = 1_C$. C and C' are then *isomorphic* objects, which is indicated as $C \cong C'$.

Let **B** and **C** be two categories.

Definition. A *functor* T: $\mathbf{B} \to \mathbf{C}$ assigns to each object B in **B** an object $T(B)$ in **C**, and assigns to each morphism $\alpha: B \to B'$ in **B** a morphism $T(\alpha): T(B) \to T(B')$ in **C**, in such a way that:

$F1$. $T(\beta\alpha) = T(\beta)\,T(\alpha)$ for $\alpha: B \to B'$, $\beta: B' \to B''$ in **B**,

$F2$. $T(1_B) = 1_{T(B)}$.

A functor T: $\mathbf{B} \to \mathbf{C}$ thus defines a map

$$\mathrm{Hom}_{\mathbf{B}}(B, B') \to \mathrm{Hom}_{\mathbf{C}}(T(B), T(B')) \tag{1}$$

for each pair B, B' of objects in $\mathbf{B}$. T is called *faithful* if these maps are injective, and is called *full* if they are surjective. Functors $T\colon \mathbf{B}\to\mathbf{C}$ and $U\colon \mathbf{C}\to\mathbf{D}$ may in an obvious way be composed to a functor $UT\colon \mathbf{B}\to\mathbf{D}$. A functor $T\colon \mathbf{B}^{\mathrm{op}}\to\mathbf{C}$ is by tradition considered as a *contravariant functor* from $\mathbf{B}$ to $\mathbf{C}$.

One often wishes to compare two functors with each other and therefore makes the

Definition. Let S and T be functors $\mathbf{B}\to\mathbf{C}$. A *natural transformation* $\eta\colon S\to T$ is obtained by taking for each object B in $\mathbf{B}$ a morphism $\eta_B\colon S(B)\to T(B)$ in $\mathbf{C}$, so that for every morphism $\alpha\colon B\to B'$ in $\mathbf{B}$ one gets commutativity in the diagram

$$\begin{array}{ccc} S(B) & \xrightarrow{\eta_B} & T(B) \\ {\scriptstyle S(\alpha)}\downarrow & & \downarrow{\scriptstyle T(\alpha)} \\ S(B') & \xrightarrow[\eta_{B'}]{} & T(B'). \end{array}$$

η is a *natural equivalence* if each η_B is an isomorphism in $\mathbf{C}$.

In this book we shall be concerned only with categories having some kind of additive structure, like the category **Mod**-A. The additive structure is most conveniently introduced on the Hom-sets:

Definition. A category $\mathbf{C}$ is *preadditive* if each set $\mathrm{Hom}_{\mathbf{C}}(C, C')$ is an abelian group and the composition maps $\mathrm{Hom}(C', C'')\times\mathrm{Hom}(C, C')\to\mathrm{Hom}(C, C'')$ are bilinear.

If $\mathbf{B}$ and $\mathbf{C}$ are preadditive categories, then a functor $T\colon \mathbf{B}\to\mathbf{C}$ is *additive* if it satisfies:

$F3$. $T(\alpha+\alpha')=T(\alpha)+T(\alpha')$ for $\alpha, \alpha'\colon C\to C'$.

Thus T is additive if and only if the maps (1) are group homomorphisms.

Examples

1. Monoids and rings. A category with only one object $*$ is uniquely determined by the set $S=\mathrm{Hom}(*, *)$, which is a *monoid* (i.e. a set with an associative binary operation, having an identity element). Often one identifies the category with the monoid S. If both $\mathbf{B}$ and $\mathbf{C}$ are categories of this kind, then a functor $\mathbf{B}\to\mathbf{C}$ is just a homomorphism of monoids.

The monoid S is a preadditive category if there exists a second binary operation $(\alpha, \beta)\mapsto\alpha+\beta$ under which it is an abelian group, and such that

$$\alpha(\beta+\beta')=\alpha\beta+\alpha\beta', \qquad (\alpha+\alpha')\beta=\alpha\beta+\alpha'\beta.$$

So a preadditive category with only one object is the same as a ring. An additive functor between two such categories is a ring homomorphism.

2. Subcategories. If $\mathbf{B}$ and $\mathbf{C}$ are categories, then $\mathbf{B}$ is a *subcategory* of $\mathbf{C}$ if $\mathrm{Ob}(\mathbf{B})$ is a subclass of $\mathrm{Ob}(\mathbf{C})$, $\mathrm{Hom}_{\mathbf{B}}(B, B')$ is a subset of $\mathrm{Hom}_{\mathbf{C}}(B, B')$ for all B and B' in $\mathrm{Ob}(\mathbf{B})$, and the composition in $\mathbf{B}$ is the same as in $\mathbf{C}$. In that case there is a faithful functor $\mathbf{B}\to\mathbf{C}$. $\mathbf{B}$ is a *full* subcategory of $\mathbf{C}$ if this functor is full.

If **C** is a preadditive category and **B** is a full subcategory of **C**, then also **B** is preadditive and the inclusion functor $\mathbf{B} \to \mathbf{C}$ is additive.

3. Product categories. Let I be a set and $(\mathbf{C}_i)_I$ a family of categories. The *product category* $\prod_I \mathbf{C}_i$ is defined as:

$$\mathrm{Ob}(\prod_I \mathbf{C}_i) = \prod_I \mathrm{Ob}(\mathbf{C}_i),$$

$$\mathrm{Hom}((C_i)_I, (D_i)_I) = \prod_I \mathrm{Hom}_{\mathbf{C}_i}(C_i, D_i)$$

with composition defined component-wise. There is a functor $\prod_I \mathbf{C}_i \to \mathbf{C}_i$ for each $i \in I$. For finite I one uses the notation $\mathbf{C}_1 \times \cdots \times \mathbf{C}_n$.

4. The Hom functor. Let **C** be a category. We will define a functor $\mathrm{Hom}: \mathbf{C}^{\mathrm{op}} \times \mathbf{C} \to \mathbf{Sets}$. To the object (C, D) of $\mathbf{C}^{\mathrm{op}} \times \mathbf{C}$ we assign the set $\mathrm{Hom}_{\mathbf{C}}(C, D)$. To the pair $\alpha: C' \to C$, $\beta: D \to D'$ of morphisms in **C** we assign the map $\mathrm{Hom}(\alpha, \beta): \mathrm{Hom}(C, D) \to \mathrm{Hom}(C', D')$ defined as $\varphi \mapsto \beta \varphi \alpha$. It is easily verified that this really is a functor.

If **C** is a preadditive category, then $\mathrm{Hom}(\alpha, \beta)$ is evidently a group homomorphism, so Hom is a functor $\mathbf{C}^{\mathrm{op}} \times \mathbf{C} \to \mathbf{Ab}$. It is additive when considered as a functor separately in each one of the two variables.

5. The free module on a set. Let A be a ring and S a set. Let $F_A(S)$ denote the free right A-module on the set S (Example 2.2). Every map $\sigma: S \to S'$ of sets induces an A-linear map $F_A(S) \to F_A(S')$ by linear extension of σ. In this way we obtain a functor F_A: **Sets** $\to$ **Mod**-A. There is also a functor G: **Mod**-$A \to$ **Sets** wich just "forgets" the module structure. Furthermore there are natural transformations $\zeta: 1 \to GF_A$, where $\zeta_S: S \to GF_A(S)$ imbeds the basis S in the free module on S, and $\xi: F_A G \to 1$, where $\xi_M: F_A G(M) \to M$ is the canonical map from the free module on the set M to M.

6. Restriction of scalars. Let $\varphi: A \to B$ be a homomorphism of rings. Each right B-module M can be considered as a right A-module by $xa = x\varphi(a)$ for $x \in M$, $a \in A$. In this way we get an additive and faithful functor φ_*: **Mod**-$B \to$ **Mod**-A.

§5. Exactness of Functors between Module Categories

Let A and B be rings and T: **Mod**-$A \to$ **Mod**-B an additive functor. T is an *exact* functor if it carries exact sequences in **Mod**-A into exact sequences in **Mod**-B. T is *left exact* if it has the weaker property that for an exact sequence $0 \to M' \to M \to M'' \to 0$ in **Mod**-A, it gives exactness for the sequence $0 \to T(M') \to T(M) \to T(M'')$ in **Mod**-B. Similarly one defines *right exactness.*

Lemma 5.1. *T is exact if and only if it is both left and right exact.*

Proof. We must show that T is exact when it preserves exactness of short exact sequences. Consider an exact sequence $L \xrightarrow{\alpha} M \xrightarrow{\beta} N$ in **Mod**-A. It gives rise to three short exact sequences

$$0 \to \operatorname{Ker}\alpha \to L \to \operatorname{Im}\alpha \to 0$$

$$0 \to \operatorname{Ker}\beta \to M \to \operatorname{Im}\beta \to 0$$

$$0 \to \operatorname{Im}\beta \to N \to N/\operatorname{Im}\beta \to 0,$$

with $\operatorname{Im}\alpha = \operatorname{Ker}\beta$. T preserves the exactness of each one of these sequences, and from this it is easy to check that $\operatorname{Im} T(\alpha) = T(\operatorname{Im}\alpha)$ and $\operatorname{Ker} T(\beta) = T(\operatorname{Ker}\beta)$. It follows that $\operatorname{Im} T(\alpha) = \operatorname{Ker} T(\beta)$, so $T(L) \to T(M) \to T(N)$ is exact. □

In a similar way one shows that T is left exact if and only if it carries an exact sequence $0 \to M' \to M \to M''$ into an exact sequence $0 \to T(M') \to T(M) \to T(M'')$. If T is a contravariant functor **Mod**-$A \to$ **Mod**-B, then T is called *left exact* if an exact sequence $0 \to M' \to M \to M'' \to 0$ in **Mod**-A induces an exact sequence $0 \to T(M'') \to T(M) \to T(M')$.

Proposition 5.2. *The functor* Hom: $(\mathbf{Mod}\text{-}A)^{\mathrm{op}} \times \mathbf{Mod}\text{-}A \to \mathbf{Ab}$ *is left exact in each one of the variables.*

Proof. Let $0 \to M' \xrightarrow{\alpha} M \xrightarrow{\beta} M'' \to 0$ be an exact sequence of A-modules and let N be another A-module. We assert that the induced sequence

$$0 \to \operatorname{Hom}(M'', N) \xrightarrow{\beta^*} \operatorname{Hom}(M, N) \xrightarrow{\alpha^*} \operatorname{Hom}(M', N)$$

is exact. β^* is a monomorphism, for if $\varphi\colon M'' \to N$ has the property that $0 = \beta^*(\varphi) = \varphi\beta$, then clearly $\varphi = 0$. $\operatorname{Ker}\alpha^*$ consists of all $\varphi\colon M \to N$ such that $\varphi\alpha = 0$. But if $\varphi\alpha = 0$, then one can define $\bar{\varphi}\colon M'' \to N$ by the rule $\bar{\varphi}(\beta(x)) = \varphi(x)$, and hence $\varphi = \beta^*(\bar{\varphi})$. Conversely, if $\varphi \in \operatorname{Im}\beta^*$, then $\varphi = \psi\beta$ for some $\psi\colon M'' \to N$, and it follows that $\varphi\alpha = \psi\beta\alpha = 0$. Thus $\operatorname{Ker}\alpha^* = \operatorname{Im}\beta^*$, and we have established the asserted exactness.

The left exactness of $\operatorname{Hom}(N, \cdot)$ is proved in an analogous way, and the details are left to the reader. □

We devote the rest of this section to a discussion of the direct limit functor. A partially ordered set I is called *directed* if for every couple $i, j \in I$ there exists $k \in I$ such that $i \le k$ and $j \le k$. Assume that I is directed. A *direct system* of A-modules over I consists of

(i) a module M_i for each $i \in I$,

(ii) a homomorphism $\alpha_{ij}\colon M_i \to M_j$ for each $i \le j$, satisfying:

$$\alpha_{ii} = 1, \quad \text{and} \quad i \le j \le k \quad \text{implies } \alpha_{jk}\alpha_{ij} = \alpha_{ik}.$$

Given such a direct system (M_i, α_{ij}), we define its *direct limit* $\varinjlim_I M_i$, usually abbreviated as $\varinjlim M_i$, as follows: form the disjoint union $\coprod_I M_i$ of the sets M_i, and define a binary relation $\sim$ on $\coprod M_i$ by putting $x \sim y$ if $x \in M_i$, $y \in M_j$ and there exists $k \in I$ such that $i \le k$, $j \le k$ and $\alpha_{ik}(x) = \alpha_{jk}(y) \in M_k$. $\sim$ is an equivalence relation (when verifying its transitivity, one uses the fact that I is directed). We now set $\varinjlim M_i = \coprod M_i/\sim$. Loosely speaking, this means that the direct limit is obtained by identifying elements which become equal for big $i \in I$. For each $i \in I$ there is a canonical map $\iota_i\colon M_i \to \varinjlim M_i$, and $i \le j$ implies $\iota_j\alpha_{ij} = \iota_i$. Each

$x \in \varinjlim M_i$ has the form $x = \iota_i(x_i)$ for some $i \in I$ and $x_i \in M_i$; we say that x is "represented" by x_i.

We have to make $\varinjlim M_i$ an A-module. Suppose $x, y \in \varinjlim M_i$ are represented by $x_i \in M_i$ and $y_j \in M_j$. Choose $k \in I$ with $i \le k$, $j \le k$. We define $x + y$ as the element of $\varinjlim M_i$ represented by $\alpha_{ik}(x_i) + \alpha_{jk}(y_j) \in M_k$. One may verify that this is independent of the choice of representing elements. If $a \in A$, we define xa to be represented by $x_i a \in M_i$. One then verifies the module axioms for $\varinjlim M_i \ldots$. The canonical maps ι_i: $M_i \to \varinjlim M_i$ become module homomorphisms. Note in particular that $x_i \in M_i$ becomes zero in $\varinjlim M_i$ if and only if $\alpha_{ij}(x_i) = 0$ for some $j \ge i$.

We will now show that the direct limit is functorial. Let the directed set I be fixed. Let $\mathbf{M} = (M_i, \alpha_{ij})$ and $\mathbf{N} = (N_i, \beta_{ij})$ be direct systems of A-modules over I. A morphism φ: $\mathbf{M} \to \mathbf{N}$ is obtained by giving a homomorphism φ_i: $M_i \to N_i$ for each $i \in I$, so that

$$\begin{array}{ccc} M_i & \xrightarrow{\varphi_i} & N_i \\ {\scriptstyle \alpha_{ij}} \big\downarrow & & \big\downarrow {\scriptstyle \beta_{ij}} \\ M_j & \xrightarrow[\varphi_j]{} & N_j \end{array}$$

commutes when $i \le j$. In this way we get the category $\mathbf{Dir}(I, \mathbf{Mod}\text{-}A)$ of direct systems of A-modules over I. A sequence $\mathbf{M} \xrightarrow{\varphi} \mathbf{N} \xrightarrow{\psi} \mathbf{P}$ in this category is called *exact* if each corresponding sequence $M_i \xrightarrow{\varphi_i} N_i \xrightarrow{\psi_i} P_i$, $i \in I$, is exact in $\mathbf{Mod}\text{-}A$. Exact sequences in $\mathbf{Dir}(I, \mathbf{Mod}\text{-}A)$ can be handled in essentially the same way as in $\mathbf{Mod}\text{-}A$, because everything is done "pointwise" at each $i \in I$.

Proposition 5.3. *The direct limit is an exact functor*

$$\varinjlim: \mathbf{Dir}(I, \mathbf{Mod}\text{-}A) \to \mathbf{Mod}\text{-}A.$$

Proof. Let φ: $\mathbf{M} \to \mathbf{N}$ be a morphism of direct systems. We define $\varinjlim \varphi_i$: $\varinjlim M_i \to \varinjlim N_i$ as follows: $x \in \varinjlim M_i$ is represented by $x_i \in M_i$, and we let $(\varinjlim \varphi_i)(x)$ be represented by $\varphi_i(x_i) \in N_i$. One verifies easily that this definition is independent of the choice of representing element x_i. In this way we have now obtained a functor $\varinjlim$, and we are going to show that it is exact. Let

$$0 \to \mathbf{L} \xrightarrow{\varphi} \mathbf{M} \xrightarrow{\psi} \mathbf{N} \to 0$$

be an exact sequence in $\mathbf{Dir}(I, \mathbf{Mod}\text{-}A)$. It gives rise to a sequence

$$0 \to \varinjlim L_i \xrightarrow{\bar{\varphi}} \varinjlim M_i \xrightarrow{\bar{\psi}} \varinjlim N_i \to 0$$

in $\mathbf{Mod}\text{-}A$. $\bar{\varphi}$ is a monomorphism: if $x \in \varinjlim L_i$ is represented by $x_i \in L_i$ and $\bar{\varphi}(x) = 0$, then $\varphi_i(x_i)$ becomes zero in $\varinjlim N_i$ and hence $\beta_{ij}\varphi_i(x_i) = 0$ for some $j \ge i$; but then $\varphi_j \alpha_{ij}(x_i) = \beta_{ij}\varphi_i(x_i) = 0$ implies that $\alpha_{ij}(x_i) = 0$ and hence $x = 0$.

$\bar{\psi}$ is an epimorphism: if $y \in \varinjlim N_i$ is represented by $y_i \in N_i$, then there exists $x_i \in M_i$ with $\psi_i(x_i) = y_i$, and x_i represents an element of $\varinjlim M_i$ which maps to y under $\bar{\psi}$.

It is clear that $\operatorname{Im}\bar{\varphi} \subset \operatorname{Ker}\bar{\psi}$. Suppose $x \in \operatorname{Ker}\bar{\psi}$, and let x be represented by $x_i \in M_i$. Then $\psi_i(x_i)$ goes to zero in the limit, so $\beta_{ij}\psi_i(x_i) = 0$ for some $j \ge i$. We can equally well assume $j = i$. Then $x_i \in \operatorname{Ker}\psi_i = \operatorname{Im}\varphi_i$ and thus $x_i = \varphi_i(z_i)$ for some $z_i \in L_i$. It follows that $x \in \operatorname{Im}\bar{\varphi}$. Hence we have $\operatorname{Im}\bar{\varphi} = \operatorname{Ker}\bar{\psi}$, and exactness is established. $\square$

Finally we note that if (M_i, α_{ij}) is a direct system and N is an arbitrary module, then a homomorphism $\varphi\colon \varinjlim M_i \to N$ induces a family of homomorphisms $\varphi_i = \varphi\,\iota_i\colon M_i \to N$ $(i \in I)$, and this family is *compatible* in the sense that $\varphi_j \alpha_{ij} = \varphi_i$ when $i \leq j$. Conversely, every compatible family of homomorphisms $\varphi_i\colon M_i \to N$ induces a unique $\varphi\colon \varinjlim M_i \to N$ such that $\varphi\,\iota_i = \varphi_i$ for all $i \in I$.

Examples

1. Direct unions. Let M be a module and let $(M_i, \alpha_{ij})_I$ be a direct system of submodules M_i of M such that $\alpha_{ij}\colon M_i \to M_j$ are inclusion maps. Then $\varinjlim M_i = \sum_I M_i$, and this sum coincides with the union of the sets M_i. The direct limit of a system of submodules is therefore called a *direct union*. So e.g. is every module a direct union of finitely generated submodules.

2. Finite presentation of modules. If M is an arbitrary finitely generated module, then there is an exact sequence

$$0 \to K \to F \to M \to 0$$

with F finitely generated free. K is the direct union of its finitely generated submodules K_i $(i \in I)$. An inclusion $K_i \subset K_j$ induces a homomorphism $F/K_i \to F/K_j$. The finitely presented modules F/K_i $(i \in I)$ then make up a direct system, and $\varinjlim F/K_i = F/\sum K_i = M$ by the exactness of direct limits. Every finitely generated module is thus a direct limit of finitely presented modules. With a little further work one can show that an arbitrary module is a direct limit of finitely presented modules (D. Lazard [1]).

§6. Projective and Injective Modules

Let A be a ring. A right A-module P is *projective* if the functor $\mathrm{Hom}_A(P, \cdot)$ is exact, which means that for every epimorphism $\beta\colon M \to N$ and every $\varphi\colon P \to N$, there exists $\varphi'\colon P \to M$ such that $\beta\varphi' = \varphi$.

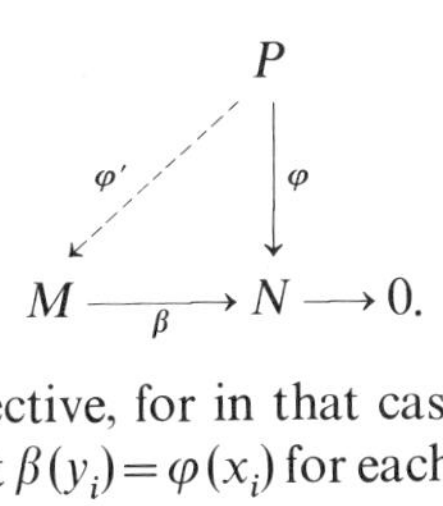

Every free module P is projective, for in that case we can choose a basis $(x_i)_I$ for P, then choose $y_i \in M$ such that $\beta(y_i) = \varphi(x_i)$ for each $i \in I$, and finally put $\varphi'(\sum x_i a_i) = \sum y_i a_i$.

Proposition 6.1. *The following properties of a module P are equivalent:*

(a) *P is projective.*

(b) *P is a direct summand of a free module.*

(c) *Every exact sequence $0 \to L \xrightarrow{\alpha} M \xrightarrow{\beta} P \to 0$ splits.*

Proof. (a) $\Rightarrow$ (c): Letting $1_P: P \to P$ correspond to the map φ in the definition above, there exists $\varphi': P \to M$ such that $\beta\varphi' = 1_P$. The sequence then splits by Prop. 2.1.

(c) $\Rightarrow$ (b): The module P can be written as the quotient of a free module F (Prop. 2.5), and the resulting exact sequence $0 \to K \to F \to P \to 0$ splits.

(b) $\Rightarrow$ (a): This follows from a more general result:

Lemma 6.2. *A direct sum $\bigoplus_I P_i$ is a projective module if and only if each P_i is projective.*

Proof. Suppose $P = \bigoplus_I P_i$ is projective. Consider an epimorphism $\beta: M \to N$ and a homomorphism $\varphi: P_i \to N$ for some $i \in I$. φ can be extended to $\bar{\varphi}: P \to N$ by putting $\bar{\varphi}(P_j) = 0$ for $j \neq i$. $\bar{\varphi}$ can be lifted to $\varphi': P \to M$, and the restriction of φ' to P_i lifts φ.

Suppose conversely that each P_i is projective. If $\beta: M \to N$ is an epimorphism and $\varphi: P \to N$ is given, then φ induces $\varphi_i: P_i \to N$ by restriction. Each φ_i can be lifted to $\varphi_i': P_i \to M$ so that $\beta\varphi_i' = \varphi_i$. Define $\varphi': P \to M$ as $\varphi'((x_i)_I) = \sum \varphi_i'(x_i)$. Then $\beta\varphi' = \varphi$. (A quicker way to formulate this proof would be to refer to Exercise 5.) ∎

The concept of "basis" for free modules can be extended to a weaker notion of "projective coordinates" for projective modules:

Proposition 6.3. *A module P is projective if and only if there exist a family $(x_i)_I$ of elements of P and homomorphisms $\varphi_i: P \to A$, such that for each $x \in P$ one has*

$$x = \sum_i x_i \varphi_i(x)$$

where $\varphi_i(x) = 0$ for all but a finite number of $i \in I$.

Proof. Let $\beta: F \to P$ be an epimorphism of a free module F onto P, let $(e_i)_I$ be a basis for F and put $x_i = \beta(e_i)$ for each $i \in I$. If P is projective, then β splits and there exists $\varphi: P \to F$ with $\beta\varphi = 1_P$. φ induces homomorphisms $\varphi_i: P \to A$ with the stated properties. Conversely, if the family (φ_i, x_i) exists and one defines β: $F \to P$ as before, then the maps φ_i induce $\varphi: P \to F$ with $\beta\varphi = 1_P$, and so P is a direct summand of F. ∎

Note that the family $(x_i)_I$ generates P, and that it can be chosen to be finite if P is finitely generated.

We now turn to the dual theory of injective modules. A right A-module E is *injective* if the functor $\mathrm{Hom}_A(\cdot, E)$ is exact, which means that for every monomorphism $\alpha: L \to M$ and every $\varphi: L \to E$, there exists $\varphi': M \to E$ such that $\varphi'\alpha = \varphi$.

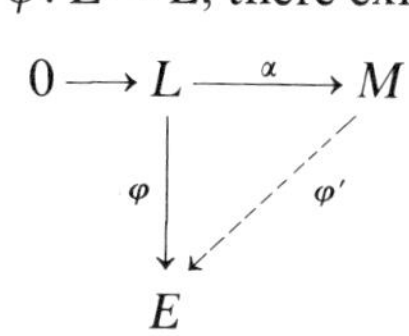

Dually to Lemma 6.2 one shows:

Lemma 6.4. *A direct product $\prod_I E_i$ is an injective module if and only if each E_i is injective.*

When determining whether a module is injective, it suffices to consider extensions over a very restricted class of monomorphisms:

Proposition 6.5. *A module E is injective if and only if for every right ideal $\mathfrak{a}$ of A and homomorphism $\varphi\colon \mathfrak{a}\to E$ there exists $y\in E$ such that $\varphi(a)=ya$ for all $a\in\mathfrak{a}$.*

Proof. The stated condition means simply that every $\varphi\colon \mathfrak{a}\to E$ can be extended to $\varphi'\colon A\to E$, so it is of course a necessary condition. Assume that E satisfies the condition. Let $\alpha\colon L\to M$ be a monomorphism and $\varphi\colon L\to E$ an arbitrary homomorphism. Consider the set $\mathscr{M}=\{\varphi'\colon L'\to E \mid L\subset L'\subset M$ and φ' extends $\varphi\}$. $\mathscr{M}$ can be partially ordered by declaring $\varphi'\le\varphi''$ if φ'' further extends φ'. If $\mathscr{T}$ is a totally ordered subset of $\mathscr{M}$, then we define $\tilde{L}$ as the sum of all L' in $\mathscr{T}$, and define $\tilde{\varphi}\colon \tilde{L}\to E$ so that it extends all $\varphi'\in\mathscr{T}$. $\tilde{\varphi}$ is then an upper bound for $\mathscr{T}$. The set $\mathscr{M}$ is thus inductive, and we can apply Zorn's lemma to obtain a maximal $\varphi_0\colon L_0\to E$ in $\mathscr{M}$. We must show that $L_0=M$. Suppose there exists $x\in M$ such that $x\notin L_0$. We will show that it is possible to extend φ_0 to $\psi\colon L_0+xA\to E$, which will give the desired contradiction. Put $\mathfrak{a}=\{a\in A \mid xa\in L_0\}$, which is a right ideal of A. There is a homomorphism $\alpha\colon \mathfrak{a}\to E$ given by $\alpha(a)=\varphi_0(xa)$, and by hypothesis there exists $y\in E$ such that $\alpha(a)=ya$ for all $a\in\mathfrak{a}$. Now define $\psi\colon L_0+xA\to E$ as $\psi(z+xa)=\varphi_0(z)+ya$ for $z\in L_0$. ψ is well-defined, for if $z+xa=z'+xa'$, then $z-z'=x(a'-a)$ implies $a'-a\in\mathfrak{a}$, and hence $\varphi_0(z)-\varphi_0(z')=\varphi_0(x(a'-a))=y(a'-a)=ya'-ya$, i.e. $\varphi_0(z)+ya=\varphi_0(z')+ya'$. Furthermore ψ is linear, and it extends φ_0. □

An injective module E is an *injective cogenerator* if $\operatorname{Hom}_A(M,E)\neq 0$ for every non-zero module M; since E is injective it actually suffices to require this for cyclic modules $M\neq 0$. The injective cogenerators are interesting because their existence (which will be established in § 9) guarantees that every module can be imbedded in an injective module:

Proposition 6.6.. *If E is an injective cogenerator, then for every module M there exists a monomorphism $M\to E^I$ for some set I.*

Proof. Put $I=\operatorname{Hom}_A(M,E)$, and define $\mu\colon M\to E^I$ as $\mu(x)=(\varphi(x))_{\varphi\in I}$. μ is evidently a homomorphism. If $x\neq 0$ in M, then there exists a non-zero homomorphism $\alpha\colon xA\to E$, and since E is injective, α can be extended to $\varphi\colon M\to E$ with $\varphi(x)\neq 0$. It follows that $\mu(x)\neq 0$, so μ is a monomorphism. □

Proposition 6.7. *An injective module E is a cogenerator if and only if it contains an isomorphic copy of each simple module.*

Proof. If E is a cogenerator and S is a simple module, then there exists a non-zero homomorphism $S\to E$, which is a monomorphism since its kernel must be zero.

Suppose on the other hand that E is an injective module containing every simple module. To show that E is a cogenerator, it suffices by a previous remark to prove that there is a non-zero homomorphism $C\to E$ for every cyclic module $C\neq 0$. But this is a consequence of the fact that C has a simple quotient module, or somewhat more generally:

Lemma 6.8. *Every non-zero finitely generated module M has a maximal proper submodule.*

Proof. Let $\mathscr{M}$ be the set of all proper submodules of M, partially ordered under inclusion. If $\mathscr{T}$ is a totally ordered subset of $\mathscr{M}$, let $\bar{L}$ be the sum of all $L \in \mathscr{T}$. If we can show that $\bar{L} \neq M$, then $\bar{L}$ will be an upper bound for $\mathscr{T}$ in $\mathscr{M}$, and Zorn's lemma can be applied to give a maximal proper submodule of M. If $\bar{L} = M$, then $\bar{L}$ contains a finite generating set $x_1, \ldots, x_n$ for M. Each x_i lies in a module belonging to $\mathscr{T}$, and then $\{x_1, \ldots, x_n\} \subset L$ for some $L \in \mathscr{T}$, which implies $L = M$, a contradiction. □

This Lemma establishes in particular the existence of maximal right ideals in every ring A.

Examples

1. Non-free projective modules. Let $A = \mathbb{Z}/6\mathbb{Z}$, which can be decomposed as $A = (\bar{2}) \oplus (\bar{3})$. The ideals $(\bar{2})$ and $(\bar{3})$ are projective modules which are not free.

2. Hereditary rings. The ring A is *right hereditary* if every right ideal of A is a projective module. E.g. the ring $\mathbb{Z}$ is hereditary. More generally, every *right principal domain*, i.e. a ring without zero-divisors in which every right ideal is principal, is right hereditary. For a less trivial example of a hereditary ring, see Exercise 24.

3. Semi-hereditary rings. The ring A is *right semi-hereditary* if every finitely generated right ideal of A is a projective module.

Proposition 6.9. *If A is right semi-hereditary, then every finitely generated submodule of a free module is isomorphic to a direct sum of finitely generated right ideals of A.*

Proof. Let M be a finitely generated submodule of a free module F. We may assume that F is finitely generated on a basis $x_1, \ldots, x_n$. We proceed by induction on n. If $n = 1$, then M is isomorphic to a finitely generated right ideal of A and, by hypothesis, projective. Let F' be the free submodule of F generated by $x_1, \ldots, x_{n-1}$, and let α: $F \to A$ be the projection $\alpha(\sum x_i a_i) = a_n$. The image of M under α is a finitely generated right ideal $\mathfrak{a}$, and we obtain a commutative diagram

$$\begin{array}{ccccccccc} 0 & \longrightarrow & M' & \longrightarrow & M & \longrightarrow & \mathfrak{a} & \longrightarrow & 0 \\ & & \downarrow & & \downarrow & & \downarrow & & \\ 0 & \longrightarrow & F' & \longrightarrow & F & \longrightarrow & A & \longrightarrow & 0 \end{array}$$

where the upper row splits since $\mathfrak{a}$ is projective. But then also M' is finitely generated, and thus isomorphic to a direct sum of finitely generated right ideals by the induction hypothesis. So is then also $M \cong M' \oplus \mathfrak{a}$. □

4. Divisible modules. We call a module M_A *divisible* if for each non-zero-divisor s of A and $x \in M$, there exists $y \in M$ such that $x = y s$. An injective module E must be divisible, for we can define φ: $s A \to E$ as $\varphi(s a) = x a$ since $s A$ is free, and the definition of injectivity gives $y \in E$ such that $x = y s$. But divisible modules are in general not injective (Exercise 25).

5. Injective modules over a principal domain

Proposition 6.10. *If A is a right principal domain, then a module is injective if and only if it is divisible.*

Proof. We know that an injective module is divisible. Suppose E is divisible. To show that E is injective, it suffices by Prop. 6.5 to consider a homomorphism $\varphi\colon aA \to E$. By the divisibility there exists $y \in E$ such that $\varphi(a) = ya$, and this is the element required by Prop. 6.5. ▯

This result applies in particular to the ring $A = \mathbb{Z}$. Thus e.g. $\mathbb{Q}$ and $\mathbb{Q}/\mathbb{Z}$ are injective $\mathbb{Z}$-modules. The module $\mathbb{Q}/\mathbb{Z}$ is also an example of an injective cogenerator in **Mod-**$\mathbb{Z}$. The existence of injective cogenerators for arbitrary rings will be proved in §9.

§7. Semi-Simple Modules and Rings

Recall that a non-zero module S is *simple* if it has no submodules other than 0 and S. More generally, a module is *semi-simple* if it is a sum of simple submodules. (Also the module (0) is semi-simple, as an empty sum of simple modules.) The following result shows that we can even take the sum to be direct:

Proposition 7.1. *Let S be a sum of simple submodules S_i $(i \in I)$ and let L be an arbitrary submodule of S. Then there exists $J \subset I$ such that $S = L \oplus (\bigoplus_J S_i)$.*

Proof. The direct sum is here to be interpreted as an internal direct sum of submodules. By an easy application of Zorn's lemma one finds a maximal subset J of I such that the sum $M = L + (\sum_J S_i)$ is direct. We must show that $M = S$, and to do this it suffices to show that M contains every S_i. But if $S_{i_0} \not\subset M$, then $M \cap S_{i_0} = 0$ by the simplicity of S_{i_0}, so the sum $M + S_{i_0}$ is direct, which contradicts the maximality of J. ▯

As a consequence we get the following alternative descriptions of semi-simple modules:

Proposition 7.2. *The following properties of a module S are equivalent:*
(a) *S is semi-simple.*
(b) *S is a direct sum of simple modules.*
(c) *Every submodule of S is a direct summand.*

Proof. (a) $\Leftrightarrow$ (b) follows from Prop. 7.1 (with $L = 0$), and also (a) $\Rightarrow$ (c) is an immediate consequence of Prop. 7.1.

(c) $\Rightarrow$ (a): The sum of all simple submodules of S is a direct summand of S, and in order to show that the complementary summand is zero, it is enough to show that every non-zero submodule L of S contains a simple submodule. The module L may as well be assumed cyclic, so by Lemma 6.8 it contains a maximal proper submodule M. The submodule M splits S as $S = M \oplus K$, and then $L = M \oplus (K \cap L)$. It follows that $K \cap L \cong L/M$ is a simple submodule of L. ▯

Corollary 7.3. *Let S be a sum of simple modules S_i ($i \in I$). If L is a submodule of S, then $L \cong \bigoplus_J S_j$ for some $J \subset I$.*

Proof. We have $S = L \oplus K$ for some $K \subset S$, and $S = K \oplus (\bigoplus_J S_j)$ with $J \subset I$. Then $L \cong S/K \cong \bigoplus_J S_j$. □

Corollary 7.4. *Every submodule and every quotient module of a semi-simple module is semi-simple.*

Let Ω denote the set of isomorphism classes of simple right A-modules. For every $\omega \in \Omega$ and every semi-simple module S we let S_ω denote the sum of all simple submodules of S of isomorphism class ω (S_ω is the *ω-isotypic component* of S).

Proposition 7.5. *If S is semi-simple, then $S = \bigoplus_\Omega S_\omega$.*

Proof. For each ω we have $S_\omega \cap (\sum_{\omega' \neq \omega} S_{\omega'}) = 0$, because Cor. 7.3 implies that each simple submodule of this module belongs both to ω and some $\omega' \neq \omega$. The sum $\sum_\Omega S_\omega$ is therefore direct (Exercise 6), and it is of course equal to S. □

If α: $S \to S'$ is a homomorphism between semi-simple modules S and S', then clearly $\alpha(S_\omega) \subset S'_\omega$. The endomorphism ring of a finite sum of simple modules can be described rather nicely. First we note:

Lemma 7.6 (Schur). *If S is a simple module, then $\mathrm{End}_A(S)$ is a skew-field.*

Proof. If α: $S \to S$ is a non-zero homomorphism, then $\mathrm{Ker}\,\alpha = 0$ and $\mathrm{Im}\,\alpha = S$, so α is an isomorphism. □

Let

$$S = S_{11} \oplus \cdots \oplus S_{1n_1} \oplus \cdots \oplus S_{k1} \oplus \cdots \oplus S_{kn_k},$$

where S_{ij} are simple modules such that $S_{ij} \cong S_{i'j'}$ if and only if $i = i'$. The endomorphism ring of S is the product of the endomorphism rings of each sum $S_{i1} \oplus \cdots \oplus S_{in_i}$. If we put $D_i = \mathrm{End}_A(S_{i1})$, which is a skew-field, then

$$\mathrm{End}_A(S) \cong M_{n_1}(D_1) \times \cdots \times M_{n_k}(D_k). \tag{1}$$

Definition. The ring A is *semi-simple* if A is semi-simple as a right A-module.

A right ideal of A which is simple as an A-module is called a *minimal* right ideal. A semi-simple ring is thus a direct sum of minimal right ideals, and every simple module is isomorphic to a minimal right ideal of A.

Proposition 7.7 *The following properties of a ring A are equivalent:*
(a) *A is a semi-simple ring.*
(b) *All right A-modules are semi-simple.*
(c) *All right A-modules are projective.*
(d) *All right A-modules are injective.*
(e) *Every right ideal of A is a direct summand of A.*
(f) *$A \cong M_{n_1}(D_1) \times \cdots \times M_{n_k}(D_k)$ for some skew-fields $D_1, \ldots, D_k$.*

Proof. (a) ⇒ (b): Every free module is semi-simple, and so is then also an arbitrary module by Prop. 2.5 and Cor. 7.4.

(a) ⇒ (e) by Prop. 7.2, and (e) ⇒ (d) with the help of Prop. 6.5.

(d) ⇒ (c): If $0 \to L \to M \to N \to 0$ is an arbitrary exact sequence, then L injective implies that the sequence splits (by the dual of Prop. 6.1 (c)). But the condition that all exact sequences split is equivalent to (c) by Prop. 6.1, and it is also equivalent to (b) by Prop. 7.2. Finally (b) ⇒ (a) trivially, so it now remains to show the equivalence of (a) and (f). That (a) implies (f) is immediate by applying the formula (1) to the semi-simple module A_A.

(f) ⇒ (a): It clearly suffices to show that the ring $M_n(D)$ is semi-simple when D is a skew-field. We can write $M_n(D)$ as a direct sum of right ideals $\mathfrak{a}_1, \ldots, \mathfrak{a}_n$, where $\mathfrak{a}_i$ consists of all matrices with only the i-th row different from zero. We will show that $\mathfrak{a}_i$ is a minimal right ideal of $M_n(D)$. One may consider $\mathfrak{a}_i$ as a left D-module, and as such one has $\mathfrak{a}_i \cong D^n$. The operation of $M_n(D)$ to the right of $\mathfrak{a}_i$ may be interpreted as the endomorphisms of the D-module $\mathfrak{a}_i$. Let $x \neq 0$ be arbitrary in $\mathfrak{a}_i$. By Prop. 2.4 there exists a basis B for the vector space $\mathfrak{a}_i$ such that $x \in B$. For any $y \in \mathfrak{a}_i$ one can use this basis to construct a D-linear map $\alpha \colon \mathfrak{a}_i \to \mathfrak{a}_i$ such that $\alpha(x) = y$. This shows that $x M_n(D) = \mathfrak{a}_i$ for each $x \neq 0$ in $\mathfrak{a}_i$, so $\mathfrak{a}_i$ is simple. □

Since the condition (f) is left-right symmetric, it follows that "right" can be replaced by "left" everywhere in the conditions of Prop. 7.7.

Consider a decomposition of the semi-simple ring A as given in condition (f). Each factor $M_{n_i}(D_i)$ is a two-sided ideal in A. We assert that there are no other two-sided ideals in A than the products of these factors. Since each ideal in A is of the form $\mathfrak{a}_1 \times \cdots \times \mathfrak{a}_k$, with $\mathfrak{a}_i$ an ideal in $M_{n_i}(D_i)$, this will follow from:

Proposition 7.8. *The following properties of a ring A are equivalent:*

(a) *A is a semi-simple ring and has no two-sided ideals except* 0 *and A.*

(b) *A is a semi-simple ring and there is only one isomorphism class of simple modules.*

(c) *$A \cong M_n(D)$ for some skew-field D.*

Proof. (a) ⇒ (b): The existence of two non-isomorphic simple modules would imply the existence of non-isomorphic minimal right ideals, and this would give rise to two-sided ideals in A by formula (1).

(b) ⇒ (c) is clear from the formula (1).

(c) ⇒ (a): Suppose $\mathfrak{a}$ is a two-sided ideal $\neq 0$ in $M_n(D)$. As a means to prove that $\mathfrak{a} = M_n(D)$, we show that $\mathfrak{a}$ contains the minimal right ideal $\mathfrak{a}_i$ consisting of matrices with only the i-th row different from zero. Choose a matrix $\beta \in \mathfrak{a}$ with some non-zero entry β_{st}. Let $\alpha \in \mathfrak{a}_i$ be the matrix with $\alpha_{is} = 1$ and all other entries zero. Then $0 \neq \alpha\beta \in \mathfrak{a}_i\,\mathfrak{a}$, so $0 \neq \mathfrak{a}_i\,\mathfrak{a} \subset \mathfrak{a}_i \cap \mathfrak{a}$, and $\mathfrak{a}_i$ minimal implies that $\mathfrak{a}_i \subset \mathfrak{a}$. □

A is a *simple ring* if it satisfies the equivalent conditions of Prop. 7.8.

Examples

1. Commutative rings. Every commutative semi-simple ring is a finite product of fields.

2. Socle. If M is a module, the sum of all simple submodules of M is called the *socle* of M and is denoted by $s(M)$. If $x \in s(M)$, then xA is a direct sum of a finite number of simple modules. From this it is easy to see that $s(M) = \{x \in M \mid \operatorname{Ann}(x)$ is a finite intersection of maximal right ideals$\}$.

3. Group rings. The semi-simplicity of certain group rings is a basic result for the theory of group representations. It is based on the following "averaging" principle. Suppose A is a commutative ring and G is a finite group. Let M and N be $A[G]$-modules, and suppose $\alpha\colon M \to N$ is an A-linear map. Define $\tilde{\alpha}\colon M \to N$ as

$$\tilde{\alpha}(x) = \sum_{g \in G} \alpha(x g^{-1})\, g .$$

Then $\tilde{\alpha}$ is of course A-linear, but it is even $A[G]$-linear. For we have

$$\tilde{\alpha}(x g_0) = \sum_g \alpha(x g_0 g^{-1})\, g = \sum_h \alpha(x h^{-1})\, h g_0 = \tilde{\alpha}(x)\, g_0 .$$

Proposition 7.9 (Maschke) *Let G be a finite group and let K be a field such that the characteristic of K does not divide the order of G. Then $K[G]$ is a semi-simple ring.*

Proof. Let M be a $K[G]$-module and consider a submodule L. It suffices to show that L is a direct summand of M. Since L is a subspace of the vector space M over K, there exists a K-linear map $\pi\colon M \to L$ such that $\pi | L = 1_L$. Put $\varphi = \frac{1}{n} \tilde{\pi}$, where n is the order of the group G and $1/n \in K$ by hypothesis. Then $\varphi\colon M \to L$ is $K[G]$-linear, and for $x \in L$ we have

$$\varphi(x) = \frac{1}{n} \sum_g \pi(x g^{-1})\, g = \frac{1}{n} \sum_g x g^{-1} \cdot g = x,$$

so $\varphi | L = 1_L$. $\square$

§8. Tensor Products

Let A be a ring, and let there be given modules L_A and ${}_A M$. We shall define the tensor product of L and M as a kind of linearization of the product $L \times M$. In a strive to make the definition more transparent, we will use an axiomatic formulation of a kind known as "universal mapping property". Let G be an abelian group. A map $\varphi\colon L \times M \to G$ will be called *bilinear* if

$$\varphi(x + x', y) = \varphi(x, y) + \varphi(x', y),$$
$$\varphi(x, y + y') = \varphi(x, y) + \varphi(x, y'),$$
$$\varphi(x a, y) = \varphi(x, a y) \quad \text{for} \quad a \in A .$$

Definition. A *tensor product* of L_A and ${}_A M$ is an abelian group T together with a bilinear map $\tau\colon L \times M \to T$, such that for every abelian group G and bilinear map $\varphi\colon L \times M \to G$ there exists a unique homomorphism $\alpha\colon T \to G$ satisfying $\alpha \tau = \varphi$.

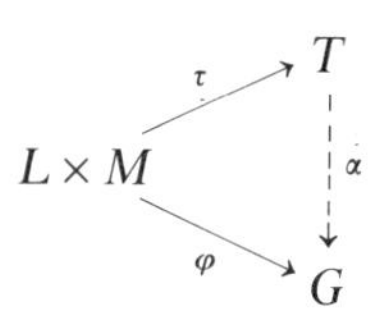

Proposition 8.1 (Unicity). *If (T, τ) and (T', τ') are tensor products of L_A and ${}_AM$, then there exists an isomorphism $\alpha\colon T \to T'$ such that $\alpha\tau = \tau'$.*

Proof. Since T is a tensor product and τ' is bilinear, there exists $\alpha\colon T \to T'$ with $\alpha\tau = \tau'$. Symmetrically there exists $\alpha'\colon T' \to T$ with $\alpha'\tau' = \tau$. Then $\alpha'\alpha\tau = \alpha'\tau' = \tau = 1_T \cdot \tau$, and the unicity of factorization over τ implies that $\alpha'\alpha = 1_T$. Similarly $\alpha\alpha' = 1_{T'}$, so α is an isomorphism. □

Proposition 8.2 (Existence). *A tensor product of L_A and ${}_AM$ exists.*

Proof. Let F be the free $\mathbb{Z}$-module on the set $L \times M$, i.e. $F = \{\sum n_i(x_i, y_i) \mid n_i \in \mathbb{Z}$ and summation is finite$\}$. Let R be the subgroup of F generated by all elements of the form

$$(x + x', y) - (x, y) - (x', y),$$

$$(x, y + y') - (x, y) - (x, y'),$$

$$(x a, y) - (x, a y).$$

Put $T = F/R$ and define $\tau\colon L \times M \to T$ as $\tau(x, y) = \overline{(x, y)}$, i.e. τ is the composition of the canonical maps

$$L \times M \xrightarrow{\iota} F \xrightarrow{\gamma} F/R.$$

τ is bilinear, as a consequence of the way we have defined R. Let $\varphi\colon L \times M \to G$ be a bilinear map into an arbitrary abelian group G. Define $\beta\colon F \to G$ by linear extension of φ. The bilinearity of φ implies that β is zero on the generators of the subgroup R, so $\beta(R) = 0$ and β can be factored over $\gamma\colon F \to F/R = T$ by a homomorphism $\alpha\colon T \to G$. Then $\alpha\tau = \alpha\gamma\iota = \beta\iota = \varphi$. The homomorphism α is uniquely determined by φ, because α is determined by $\alpha\gamma$ since γ is an epimorphism, and $\alpha\gamma$ is determined by its values on the basis $L \times M$ of F, i.e. by φ. □

Because of its unicity (up to isomorphism), we will speak of *the* tensor product of L and M. We denote it by $L \otimes_A M$, with generators $\tau(x, y) = x \otimes y$. Every element of $L \otimes_A M$ looks like $\sum n_i(x_i \otimes y_i)$ with finite summation, with relations

$$(x + x') \otimes y = x \otimes y + x' \otimes y,$$

$$x \otimes (y + y') = x \otimes y + x \otimes y',$$

$$x a \otimes y = x \otimes a y.$$

Proposition 8.3. *$A \otimes_A M \cong M$ for every ${}_AM$.*

Proof. Define $\tau\colon A \times M \to M$ as $\tau(a, x) = a x$. The map τ is bilinear, and every bilinear map $\varphi\colon A \times M \to G$ can be uniquely factored over τ by $\alpha\colon M \to G$ defined as $\alpha(x) = \varphi(1, x)$. The couple (M, τ) is thus a tensor product of A and M. □

If $A \otimes_A M$ is an arbitrary tensor product of A and M, then the isomorphism $A \otimes_A M \to M$ is given by $a \otimes x \mapsto a x$. The isomorphism is natural in the sense that there is a natural equivalence between the functor $A \otimes_A \cdot$ and the identity functor. In the future we will use the term "natural isomorphism" to indicate the existence of a natural equivalence between functors in the occurring variables.

Proposition 8.4. *If L_i $(i \in I)$ are right A-modules and M is a left A-module, then there is a natural isomorphism*

$$(\bigoplus_I L_i) \otimes_A M \cong \bigoplus_I (L_i \otimes_A M).$$

Proof. With the help of a suitable bilinear map one gets a homomorphism α: $(\bigoplus_I L_i) \otimes M \to \bigoplus_I (L_i \otimes M)$ such that $\alpha((x_i)_I \otimes y) = (x_i \otimes y)_I$. Similarly one obtains β: $\bigoplus_I (L_i \otimes M) \to (\bigoplus_I L_i) \otimes M$ such that $\beta((x_i \otimes y_i)_I) = \sum_i (\lambda_i(x_i) \otimes y_i)$, where λ_i: $L_i \to \bigoplus_I L_i$ is the inclusion map. It is clear that β is an inverse of α, so α is an isomorphism. The verification of the naturality is left to the reader. ▯

Corollary 8.5. *If F_A is free on a basis $(x_i)_I$, then $F \otimes_A M \cong M^{(I)}$ as $\mathbb{Z}$-modules.*

Proof. $F \otimes_A M \cong A^{(I)} \otimes_A M \cong (A \otimes_A M)^{(I)} \cong M^{(I)}$. (Note that $F \otimes_A M$ is in general not an A-module.) ▯

In particular, if F_A is free on a basis $(x_i)_I$ and ${}_AM$ is free on a basis $(y_j)_J$, then $F \otimes_A M$ is isomorphic to $A^{(I \times J)}$ as $\mathbb{Z}$-module. This observation determines all tensor products of vector spaces.

We will now look at the tensor product from a functorial point of view, and we will show that the tensor product may be considered as a functor **Mod**-$A \times A$-**Mod** $\to$ **Ab**. Let λ: $L \to L'$ and μ: $M \to M'$ be homomorphisms of right, resp. left A-modules. We assert that there is induced a homomorphism α: $L \otimes_A M \to L' \otimes_A M'$. Define φ: $L \times M \to L' \otimes_A M'$ as $\varphi(x, y) = \lambda(x) \otimes \mu(y)$. One verifies that φ is a bilinear map, so there exists a unique α: $L \otimes_A M \to L' \otimes_A M'$ such that $\alpha(x \otimes y) = \lambda(x) \otimes \mu(y)$. The homomorphism α is denoted by $\lambda \otimes \mu$ (warning: $\lambda \otimes \mu$ is here an element of $\mathrm{Hom}_{\mathbb{Z}}(L \otimes_A M, L' \otimes_A M')$ and *not* an element of $\mathrm{Hom}_A(L, L') \otimes_{\mathbb{Z}} \mathrm{Hom}_A(M, M')$, as the notation would suggest). So one has

$$(\lambda \otimes \mu)(x \otimes y) = \lambda(x) \otimes \mu(y).$$

If there are homomorphisms

$$L \xrightarrow{\lambda} L' \xrightarrow{\lambda'} L'' \quad \text{and} \quad M \xrightarrow{\mu} M' \xrightarrow{\mu'} M'',$$

then one has $(\lambda' \otimes \mu')(\lambda \otimes \mu) = \lambda' \lambda \otimes \mu' \mu$, because the two sides obviously have the same values on the generators $x \otimes y$ of $L \otimes M$. It follows that the tensor product really is a functor

$$\otimes\colon \mathbf{Mod}\text{-}A \times A\text{-}\mathbf{Mod} \to \mathbf{Ab}.$$

The functor is additive in each one of the two variables, i.e. one has $(\lambda + \lambda_1) \otimes \mu = \lambda \otimes \mu + \lambda_1 \otimes \mu$, etc.

One has to be somewhat careful when dealing with the tensor product, because it does not preserve submodules. Consider e.g. the $\mathbb{Z}$-submodule $\mathbb{Z}$ of $\mathbb{Q}$; tensoring with $\mathbb{Z}/2\mathbb{Z}$ gives

$$\mathbb{Z}/2\mathbb{Z} \cong \mathbb{Z} \otimes_{\mathbb{Z}} \mathbb{Z}/2\mathbb{Z} \to \mathbb{Q} \otimes_{\mathbb{Z}} \mathbb{Z}/2\mathbb{Z}$$

where the last tensor product is zero since

$$\frac{m}{n}\otimes\bar{1}=\frac{2m}{2n}\otimes\bar{1}=\frac{m}{2n}\otimes 2\cdot\bar{1}=\frac{m}{2n}\otimes 0=0.$$

The tensor product is thus not an exact functor in general. It is however right exact:

Proposition 8.6. *The tensor product is a right exact functor in each one of the two variables.*

Proof. By symmetry it suffices to consider one of the variables, so let us show that $L\otimes_A\cdot$ is right exact. Let

$$0\to M'\xrightarrow[\alpha]{} M\xrightarrow[\beta]{} M''\to 0$$

be an exact sequence of left A-modules. Is the sequence

$$L\otimes M'\xrightarrow[1\otimes\alpha]{} L\otimes M\xrightarrow[1\otimes\beta]{} L\otimes M''\to 0$$

exact? To show that $1\otimes\beta$ is an epimorphism, it suffices to show that each generator $x\otimes y''\in L\otimes M''$ is in the image of $1\otimes\beta$. But this is clear, for $y''=\beta(y)$ for some $y\in M$, and hence $x\otimes y''=(1\otimes\beta)(x\otimes y)$.

It is obvious that $\operatorname{Im}(1\otimes\alpha)\subset\operatorname{Ker}(1\otimes\beta)$. The reverse inclusion requires a little work. There is a commutative diagram

$$\begin{array}{ccc} L\otimes M & \xrightarrow{1\otimes\beta} & L\otimes M'' \\ \downarrow & \nearrow_{\xi} & \\ L\otimes M/\operatorname{Im}(1\otimes\alpha) & & \end{array}$$

since $\operatorname{Im}(1\otimes\alpha)\subset\operatorname{Ker}(1\otimes\beta)$. We will construct an inverse η of ξ. For this purpose we define $\varphi\colon L\times M''\to L\otimes M/\operatorname{Im}(1\otimes\alpha)$ as $\varphi(x,y'')=\overline{x\otimes y}$, where y is given by $y''=\beta(y)$. This map is well-defined, for if also $y''=\beta(y_1)$, then $y-y_1\in\operatorname{Ker}\beta=\operatorname{Im}\alpha$ and hence $x\otimes(y-y_1)\in\operatorname{Im}(1\otimes\alpha)$, giving $\overline{x\otimes y}=\overline{x\otimes y_1}$. Since φ is easily seen to be bilinear, there exists $\eta\colon L\otimes M''\to L\otimes M/\operatorname{Im}(1\otimes\alpha)$. Then

$$\xi\eta(x\otimes y'')=\xi(\overline{x\otimes y})=x\otimes\beta(y)=x\otimes y'',$$

and

$$\eta\xi(\overline{x\otimes y})=\eta(x\otimes\beta(y))=\overline{x\otimes y}.$$

It follows that ξ is an isomorphism, and hence $\operatorname{Im}(1\otimes\alpha)=\operatorname{Ker}(1\otimes\beta)$. □

We have proved in Prop. 8.4 that the tensor product "commutes" with direct sums. It also commutes with direct limits:

Proposition 8.7. *If $(L_i)_I$ is a direct system of right A-modules and M is a left A-module, then there is a natural isomorphism*

$$(\varinjlim L_i)\otimes_A M\cong\varinjlim(L_i\otimes_A M).$$

Proof. First it is clear that if (L_i,α_{ij}) is the given direct system of modules, then $(L_i\otimes M,\alpha_{ij}\otimes 1)$ is a direct system of abelian groups. The compatible family of canonical homomorphisms $L_i\to\varinjlim L_i$ gives a compatible family of homomorphisms $L_i\otimes M\to(\varinjlim L_i)\otimes M$, which induces $\alpha\colon\varinjlim(L_i\otimes M)\to(\varinjlim L_i)\otimes M$

such that $\alpha(\widetilde{x_i \otimes y}) = \tilde{x}_i \otimes y$ (here $x_i \in L_i$, and $\tilde{x}_i$ denotes the image of x_i in $\varinjlim L_i$). On the other hand, with the help of a suitable bilinear map

$$(\varinjlim L_i) \times M \to \varinjlim (L_i \otimes M)$$

one gets β: $(\varinjlim L_i) \otimes M \to \varinjlim (L_i \otimes M)$ such that $\beta(\tilde{x}_i \otimes y) = \widetilde{x_i \otimes y}$. β is then an inverse of α, so α is an isomorphism. □

We conclude this section with a very useful result which analyses what it is that makes an element of a tensor product zero.

Proposition 8.8. *Let $(y_i)_I$ be a family of generators for ${}_AM$, and let $(x_i)_I$ be a family of elements of L_A with almost all $x_i = 0$. Then $\sum x_i \otimes y_i = 0$ in $L \otimes_A M$ if and only if there exist a finite family $(u_j)_J$ of elements of L and a family $(a_{ji})_{J \times I}$ of elements of A such that:*

(i) $a_{ji} = 0$ *for almost all* (j, i),

(ii) $\sum_I a_{ji} y_i = 0$ *for each* $j \in J$,

(iii) $x_i = \sum_J u_j a_{ji}$ *for each* $i \in I$.

Proof. It is clear that the given conditions are sufficient to make $\sum x_i \otimes y_i = 0$, because they give

$$\sum_i x_i \otimes y_i = \sum_{j,i} u_j a_{ji} \otimes y_i = \sum_j u_j \otimes (\sum_i a_{ji} y_i) = \sum_j u_j \otimes 0 = 0.$$

Suppose on the other hand that $\sum_i x_i \otimes y_i = 0$. Since y_i $(i \in I)$ generate M, there is an exact sequence

$$0 \to K \xrightarrow{\alpha} A^{(I)} \xrightarrow{\beta} M \to 0$$

where β maps the canonical basis vector $e_i \in A^{(I)}$ to y_i. By tensoring with L we get an exact sequence

$$L \otimes K \xrightarrow{1 \otimes \alpha} L \otimes A^{(I)} \xrightarrow{1 \otimes \beta} L \otimes M \longrightarrow 0.$$

The hypothesis that $\sum x_i \otimes y_i = 0$ implies $\sum x_i \otimes e_i \in \operatorname{Ker}(1 \otimes \beta) = \operatorname{Im}(1 \otimes \alpha)$, so $\sum_i x_i \otimes e_i = \sum_j u_j \otimes \alpha(z_j)$ for some $u_j \in L$, $z_j \in K$. Each $\alpha(z_j) \in A^{(I)}$ can be expressed in the canonical basis as $\alpha(z_j) = \sum_i a_{ji} e_i$. Then $\beta\alpha = 0$ gives $0 = \beta(\sum_i a_{ji} e_i) = \sum_i a_{ji} y_i$, which is the desired condition (ii). We also have

$$\sum_i x_i \otimes e_i = \sum_j u_j \otimes \alpha(z_j) = \sum_{j,i} u_j \otimes a_{ji} e_i$$

in $L \otimes A^{(I)}$. Under the isomorphism $L \otimes A^{(I)} \cong L^{(I)}$, this gives $x_i = \sum_i u_j a_{ji}$ for each $i \in I$, i.e. condition (iii). □

Examples

1. Tensoring with cyclic modules. Let $\mathfrak{a}$ be a right ideal of A and M a left A-module. From the exact sequence $0 \to \mathfrak{a} \to A \to A/\mathfrak{a} \to 0$ we get an exact sequence

$$\mathfrak{a} \otimes M \xrightarrow{\alpha} M \to A/\mathfrak{a} \otimes M \to 0,$$

where $\operatorname{Im} \alpha = \{\sum a_i x_i \mid a_i \in \mathfrak{a},\ x_i \in M\} = \mathfrak{a} M$. It follows that $A/\mathfrak{a} \otimes_A M \cong M/\mathfrak{a} M$.

2. Direct unions. If L_A is a direct union of submodules L_i $(i \in I)$, then $L \otimes M = \varinjlim (L_i \otimes M)$ by Prop. 8.7, but it should be noted that $L_i \otimes M$ is not in general a submodule of $L \otimes M$.

3. Split exact sequences. If $0 \to M' \to M \to M'' \to 0$ is a split exact sequence of left A-modules, then it follows from Prop. 8.4 that a sequence $0 \to L \otimes M' \to L \otimes M \to L \otimes M'' \to 0$ also is split exact (one should verify that the maps are the right ones).

§9. Bimodules

Let A, B and C be rings. Recall that M is a *B-A-bimodule* if M is a left B-module and a right A-module, such that $(b\,x)\,a = b(x\,a)$ for $b \in B$, $x \in M$, $a \in A$. We write ${}_BM_A$ to indicate this situation.

The B-module structure of a bimodule ${}_BM_A$ can be described as a ring homomorphism $\varphi\colon B \to \mathrm{End}_A(M)$. For if ${}_BM_A$ is a bimodule, then we define φ as $\varphi(b)\colon x \mapsto b\,x$ for $x \in M$; conversely, if φ is given, then M becomes a B-A-bimodule by putting $b\,x = \varphi(b)(x)$.

Let T: **Mod**-$A \to$ **Mod**-C be an additive functor. For a bimodule ${}_BM_A$ we get a map

$$B \to \mathrm{End}_A(M) \to \mathrm{End}_C(T(M)) \tag{1}$$

where the second map, induced by T, is a ring homomorphism since T is an additive functor. By the previous remark it follows that $T(M)$ is a B-C-bimodule. If the functor T is contravariant, the second map in (1) is an antihomomorphism (i.e. it reverses the order of multiplication), and we get instead a ring homomorphism $B^{\mathrm{op}} \to \mathrm{End}_C(T(M))$, so $T(M)$ is a B^{op}-C-bimodule (or in other words, a B-C-bimodule with both B and C operating to the right).

We now apply these considerations to the functors Hom and $\otimes$. Let L be a right A-module, and consider the functor $\mathrm{Hom}_A(L, \cdot)$: **Mod**-$A \to$ **Mod**-$\mathbb{Z}$. If M is a B-A-bimodule, then $\mathrm{Hom}_A(L, M)$ becomes a left B-module. When one examines what (1) means in this case, it turns out that the B-module structure is defined as follows: let $b \in B$ and $\varphi \in \mathrm{Hom}_A(L, M)$; then $b\varphi\colon L \to M$ maps $x \in L$ to $b\varphi(x) \in M$.

Next we look at the contravariant functor $\mathrm{Hom}_A(\cdot, M)$: **Mod**-$A \to$ **Mod**-$\mathbb{Z}$. If L is a B-A-bimodule, then $\mathrm{Hom}_A(L, M)$ is a right B-module. If $b \in B$ and $\varphi \in \mathrm{Hom}_A(L, M)$, then $\varphi b\colon L \to M$ maps $x \in L$ to $\varphi(b\,x) \in M$. More generally, given two bimodules ${}_BL_A$ and ${}_CM_A$, one gets a C-B-bimodule $\mathrm{Hom}_A(L, M)$.

Finally we consider the tensor product. Given bimodules ${}_BL_A$ and ${}_AM_C$, one gets a B-C-bimodule $L \otimes_A M$ with $b(x \otimes y)c = (b\,x) \otimes (y\,c)$ for $b \in B$, $x \in L$, $y \in M$ and $c \in C$.

The associativity rule for tensor products can now be stated:

Proposition 9.1. *Given modules L_A, ${}_AM_B$ and ${}_BN$, there is a natural isomorphism*

$$(L \otimes_A M) \otimes_B N \cong L \otimes_A (M \otimes_B N).$$

Proof. We try to define the isomorphism on the generators of $(L \otimes_A M) \otimes_B N$ as $(x \otimes y) \otimes z \mapsto x \otimes (y \otimes z)$. The only problem is to show that this gives a well-defined homomorphism, for then by symmetry there is a similar map in the other direction, and the two maps are inverse of each other.

For each $z \in N$ we define $\alpha_z: M \to M \otimes_B N$ as $\alpha_z(y) = y \otimes z$, and then get a homomorphism $\beta_z = 1 \otimes \alpha_z: L \otimes_A M \to L \otimes_A (M \otimes_B N)$. We then define

$$\varphi: (L \otimes_A M) \times N \to L \otimes_A (M \otimes_B N)$$

as $\varphi(u, z) = \beta_z(u)$. φ is easily verified to be a bilinear map, and it induces the desired homomorphism $(L \otimes_A M) \otimes_B N \to L \otimes_A (M \otimes_B N)$. □

The next result is a very useful formula relating the tensor product to the Hom-functor. It is actually a rather immediate consequence of our axiomatic definition of the tensor product, but with an extra bimodule structure imposed. Let L_A, ${}_AM_B$ and N_B be given modules. We define $\operatorname{Bil}_B(L \times M, N)$ to consist of maps $\varphi: L \times M \to N$ which are additive in each variable and such that

$$\varphi(xa, y) = \varphi(x, ay) \quad \text{for } a \in A,$$

$$\varphi(x, yb) = \varphi(x, y)b \quad \text{for } b \in B.$$

There are natural isomorphisms

$$\operatorname{Hom}_B(L \otimes_A M, N) \cong \operatorname{Bil}_B(L \times M, N) \cong \operatorname{Hom}_A(L, \operatorname{Hom}_B(M, N)),$$

where the first isomorphism comes from the definition of tensor products (with the B-module structure taken into account), and the second isomorphism makes $\varphi: L \times M \to N$ correspond to $\tilde{\varphi}: L \to \operatorname{Hom}_B(M, N)$ defined as $\tilde{\varphi}(x): y \mapsto \varphi(x, y)$. (This is just a variant of the set theoretical bijection $(N^M)^L \cong N^{L \times M}$.) We have thus obtained:

Proposition 9.2. *Given modules L_A, ${}_AM_B$ and N_B, there is a natural isomorphism*

$$\operatorname{Hom}_B(L \otimes_A M, N) \cong \operatorname{Hom}_A(L, \operatorname{Hom}_B(M, N)).$$

With the help of this formula we can prove the important fact that every module can be imbedded in an injective module.

Proposition 9.3. *The module* $\operatorname{Hom}_{\mathbb{Z}}(A, \mathbb{Q}/\mathbb{Z})$ *is an injective cogenerator for* **Mod**-A.

Proof. Here we consider A as an A-$\mathbb{Z}$-bimodule. Then for every module L_A we get

$$\operatorname{Hom}_A(L, \operatorname{Hom}_{\mathbb{Z}}(A, \mathbb{Q}/\mathbb{Z})) \cong \operatorname{Hom}_{\mathbb{Z}}(L \otimes_A A, \mathbb{Q}/\mathbb{Z}) \cong \operatorname{Hom}_{\mathbb{Z}}(L, \mathbb{Q}/\mathbb{Z}).$$

Since $\mathbb{Q}/\mathbb{Z}$ is an injective cogenerator for **Mod**-$\mathbb{Z}$, it clearly follows that $\operatorname{Hom}_{\mathbb{Z}}(A, \mathbb{Q}/\mathbb{Z})$ is an injective cogenerator for **Mod**-A. □

Combining this result with Prop. 6.6 we obtain:

Corollary 9.4. *Every module is a submodule of an injective module.*

As an illustration of the usefulness of this result, we prove:

Proposition 9.5. *The following properties of a ring A are equivalent:*

(a) *A is right hereditary.*

(b) *Every submodule of a projective module is projective.*
(c) *Every quotient module of an injective module is injective.*

Proof. (a) $\Rightarrow$ (c): Let E be an injective module, and consider an epimorphism $\alpha: E \to M$. Suppose $\varphi: \mathfrak{a} \to M$ is a homomorphism from a right ideal $\mathfrak{a}$. Since $\mathfrak{a}$ is a projective module, φ can be lifted to a homomorphism $\varphi': \mathfrak{a} \to E$. But since E is injective, φ' can be extended to $\psi: A \to E$, and the composed map $\alpha\,\psi$ extends φ, as desired.

(c) $\Rightarrow$ (b): Let L be a submodule of a projective module P. Suppose there are given an exact sequence $0 \to K \to M \to N \to 0$ and a homomorphism $\varphi: L \to N$. If we consider M as a submodule of an injective module E, we obtain a commutative diagram with exact rows and columns

$$\begin{array}{ccccccc}
 & & & & 0 & & 0 & & \\
 & & & & \downarrow & & \downarrow & & \\
0 & \longrightarrow & K & \longrightarrow & M & \longrightarrow & N & \longrightarrow & 0 \\
 & & \Vert & & \downarrow{\scriptstyle\mu} & & \downarrow{\scriptstyle\eta} & & \\
0 & \longrightarrow & K & \longrightarrow & E & \longrightarrow & E/K & \longrightarrow & 0 \\
 & & & & \downarrow{\scriptstyle\mu'} & & \downarrow{\scriptstyle\eta'} & & \\
 & & & & E/M & = & E/M & & \\
 & & & & \downarrow & & \downarrow & & \\
 & & & & 0 & & 0 & &
\end{array}$$

Since E/K is an injective module, the map $\eta\,\varphi$ extends to $\varphi': P \to E/K$. But since P is projective, φ' can be lifted to $\psi: P \to E$. Then $\psi(L) \subset M$, because $\mu'\,\psi = \eta'\,\varphi'$ is zero on the submodule L. Thus ψ induces $L \to M$, which lifts the given map φ.

(b) $\Rightarrow$ (a) is trivial. $\square$

Example. Extension of scalars. Let $\varphi: A \to B$ be a homomorphism of rings. If M is a right A-module, then we may form $M \otimes_A B$, where B is considered as a left A-module by restriction of scalars, and $M \otimes_A B$ is a right B-module. E.g. if M is a real vector space, then $M \otimes_{\mathbb{R}} \mathbb{C}$ is the "complexification" of M.

§ 10. Flat Modules

Definition. *A* left *A*-module F is *flat* if the functor $\cdot \otimes_A F$ is exact.

Since the tensor product always is right exact, F is flat if and only if $\cdot \otimes F$ preserves monomorphisms of right A-modules.

Proposition 10.1. *If $(F_i)_I$ is a family of left A-modules, then $\bigoplus_I F_i$ flat if and only if each F_i is flat.*

Proof. Let $L \to M$ be a monomorphism of right A-modules. By Prop. 8.4 there is a commutative diagram

$$\begin{array}{ccc} L \otimes_A (\bigoplus_I F_i) & \longrightarrow & M \otimes_A (\bigoplus_I F_i) \\ \downarrow \cong & & \downarrow \cong \\ \bigoplus_I (L \otimes_A F_i) & \longrightarrow & \bigoplus_I (M \otimes_A F_i) \end{array}$$

The upper row is a monomorphism if and only if the lower one is, and this happens if and only if each $L \otimes F_i \to M \otimes F_i$ is a monomorphism. $\square$

Corollary 10.2. *Every projective module is flat.*

Proof. ${}_AA$ is flat since $L \otimes_A A \cong L$. Hence every free module is flat by Prop. 10.1, and it follows then from Props. 6.1 and 10.1 that projective modules are flat. $\square$

From Prop. 8.7 we obtain:

Proposition 10.3. *Every direct limit of flat modules is flat.*

The associativity relation between Hom and $\otimes$ establishes an intimate connection between flat modules and injective modules:

Proposition 10.4. *Let there be given modules ${}_AF_B$ and E_B, and assume that E_B is an injective cogenerator. Then ${}_AF$ is flat if and only if* $\mathrm{Hom}_B(F, E)$ *is an injective right A-module.*

Proof. Let α: $M \to N$ be an arbitrary monomorphism of right A-modules. It induces a commutative diagram

$$\begin{array}{ccc} \mathrm{Hom}_A(N, \mathrm{Hom}_B(F, E)) & \xrightarrow{\mathrm{Hom}(\alpha, 1)} & \mathrm{Hom}_A(M, \mathrm{Hom}_B(F, E)) \\ \downarrow \cong & & \downarrow \cong \\ \mathrm{Hom}_B(N \otimes_A F, E) & \xrightarrow[\mathrm{Hom}(\alpha \otimes 1, 1)]{} & \mathrm{Hom}_B(M \otimes_A F, E) \end{array}$$

with vertical isomorphisms furnished by Prop. 9.2. If ${}_AF$ is flat, then $\alpha \otimes 1: M \otimes_A F \to N \otimes_A F$ is a monomorphism, and E_B injective implies that $\mathrm{Hom}(\alpha \otimes 1, 1)$ is an epimorphism. By the commutativity it follows that $\mathrm{Hom}(\alpha, 1)$ is an epimorphism, and hence $\mathrm{Hom}_B(F, E)$ is an injective A-module.

Assume now instead that $\mathrm{Hom}_B(F, E)$ is an injective A-module. By reversing the previous argument we may conclude that $\mathrm{Hom}(\alpha \otimes 1, 1)$ is an epimorphism. Let x be any non-zero element of $M \otimes_A F$. Since E_B is an injective cogenerator, there exists a B-linear map φ: $M \otimes_A F \to E$ such that $\varphi(x) \neq 0$ (cf. the proof of Prop. 6.6). Since $\mathrm{Hom}(\alpha \otimes 1, 1)$ is an epimorphism, we have $\varphi = \psi(\alpha \otimes 1)$ for some ψ: $N \otimes_A F \to E$. From $\varphi(x) \neq 0$ must then follow that $(\alpha \otimes 1)(x) \neq 0$, so $\alpha \otimes 1$ is a monomorphism. Hence ${}_AF$ is flat. $\square$

If we take $B = \mathbb{Z}$ and $E = \mathbb{Q}/\mathbb{Z}$, we get in particular the remarkable:

Corollary 10.5. *A module ${}_AF$ is flat if and only if $\hat{F} = \mathrm{Hom}_{\mathbb{Z}}(F, \mathbb{Q}/\mathbb{Z})$ is an injective right A-module.*

We can use this result to get an analogue of Prop. 6.5 for flat modules.

Proposition 10.6. *A module ${}_AF$ is flat if and only if the canonical map $\mathfrak{a}\otimes_A F\to F$ is a monomorphism for every finitely generated right ideal $\mathfrak{a}$ of A.*

Proof. Suppose $\mathfrak{a}\otimes F\to F$ is a monomorphism for finitely generated right ideals $\mathfrak{a}$. Then it is so also for arbitrary right ideals $\mathfrak{a}$, for if $\sum a_i\otimes y_i\in\mathfrak{a}\otimes F$, then the a_i lie in a finitely generated right ideal $\mathfrak{b}\subset\mathfrak{a}$, and the composed map $\mathfrak{b}\otimes F\to\mathfrak{a}\otimes F\to F$ is a monomorphism. So if $\sum a_i\otimes y_i$ gets zero in F, then $\sum a_i\otimes y_i$ is zero in $\mathfrak{b}\otimes F$ and hence zero in $\mathfrak{a}\otimes F$.

The inclusion map $\mathfrak{a}\to A$ thus induces a monomorphism $\mathfrak{a}\otimes F\to F$, and there results a commutative diagram

$$\begin{array}{ccc}\mathrm{Hom}(A,\hat F) & \longrightarrow & \mathrm{Hom}(\mathfrak{a},\hat F)\\ \downarrow{\scriptstyle\cong} & & \downarrow{\scriptstyle\cong}\\ \hat F & \longrightarrow & \widehat{\mathfrak{a}\otimes F}\end{array}$$

where the lower map is an epimorphism. So is then also the upper map, and therefore $\hat F$ is injective by Prop. 6.5. But F is then flat by Cor. 10.5. □

The preceding result will now be used, together with Prop. 8.8, to prove a characterization of flat modules, which essentially states that a module F is flat if and only if all relations in F are due to relations existing already in the ring A.

Proposition 10.7. *A module ${}_AF$ is flat if and only if it satisfies: if $\sum_{i=1}^{n} b_i x_i=0$ for $b_i\in A$, $x_i\in F$, then there exist $u_1,\dots,u_m$ in F and $a_{ij}\in A$ $(i=1,\dots,n;\ j=1,\dots,m)$ such that $\sum_i b_i a_{ij}=0$ and $x_i=\sum_j a_{ij}u_j$.*

Proof. Suppose ${}_AF$ is flat and $\sum b_i x_i=0$ in F. Let $\mathfrak{a}$ be the right ideal generated by $b_1,\dots,b_n$. Then $\mathfrak{a}\otimes F\to F$ is a monomorphism, so we must have $\sum b_i\otimes x_i=0$ in $\mathfrak{a}\otimes F$, and we can apply Prop. 8.8 to obtain the stated condition.

Suppose conversely that F satisfies the condition in question. Let $\mathfrak{a}$ be an arbitrary right ideal and consider the map $\mathfrak{a}\otimes F\to F$. If $\sum b_i\otimes x_i\in\mathfrak{a}\otimes F$ goes to zero in F, then $\sum b_i x_i=0$, and there exist u_j and a_{ij} according to the condition. This gives immediately $\sum_i b_i\otimes x_i=\sum_{i,j} b_i\otimes a_{ij}u_j=\sum_{i,j} b_i a_{ij}\otimes u_j=\sum_i 0\otimes u_j=0$. The map $\mathfrak{a}\otimes F\to F$ is thus a monomorphism, and F is flat by Prop. 10.6. □

Examples

1. Flat modules are torsion-free. Suppose ${}_AF$ is flat and let s be a non-zero-divisor of A. Define $\alpha\colon A\to A$ as $\alpha(a)=sa$, which is a monomorphism of right A-modules. It gives rise to a commutative diagram

$$\begin{array}{ccc}A\otimes F & \xrightarrow{\alpha\otimes 1} & A\otimes F\\ \downarrow{\scriptstyle\cong} & & \downarrow{\scriptstyle\cong}\\ F & \xrightarrow[\beta]{} & F\end{array}$$

where $\beta(x)=sx$. Since $\alpha\otimes 1$ is a monomorphism, so is also β, and hence $x\neq 0$ implies $sx\neq 0$. The module F is thus *torsion-free* in the sense that $x\neq 0$ and s not a zero-divisor implies $sx\neq 0$.

2. Commutative rings of fractions. Let A be a commutative ring and let S be a subset of A which is closed under multiplication, with $1\in S$. We construct the *ring of fractions* $A[S^{-1}]$ by a generalization of the well-known construction of a field of fractions for an integral domain. Consider all pairs (a,s) with $a\in A$, $s\in S$. Put $(a,s)\sim(b,t)$ if there exists $u\in S$ such that $u(ta-sb)=0$. It is a routine matter to verify that $\sim$ is an equivalence relation (when A has zero-divisors, it is necessary to bring in the extra u to make the relation transitive). Let a/s denote the equivalence class containing (a,s). It is then possible to define

$$\frac{a}{s}+\frac{b}{t}=\frac{ta+sb}{st}, \qquad \frac{a}{s}\cdot\frac{b}{t}=\frac{ab}{st},$$

and this makes $A[S^{-1}]=\{a/s\}$ a commutative ring. There is a ring homomorphism $\varphi\colon A\to A[S^{-1}]$ such that $\varphi(a)=a/1$. Note that $\varphi(s)=s/1$ is invertible in $A[S^{-1}]$ for every $s\in S$. One has $a/s=0$ in $A[S^{-1}]$ if and only if $ua=0$ in A for some $u\in S$, as follows from the definition of $\sim$. In particular one gets

$$\operatorname{Ker}\varphi=\{a\in A\,|\,ua=0 \text{ for some } u\in S\}.$$

One may consider $A[S^{-1}]$ as an A-module by means of φ.

Proposition 10.8. *$A[S^{-1}]$ is a flat A-module.*

Proof. This follows by an easy application of Prop. 10.7. Another proof will be given in the next chapter. □

Two important special cases:

(i) S is the set of all non-zero-divisors of A. The ring $A[S^{-1}]$ is then called the *total ring of fractions* of A. It is the biggest ring of fractions of A containing A as a subring.

(ii) S is the set-theoretical complement of a prime ideal $\mathfrak{p}$ in A. The ring $A[S^{-1}]$ is then written $A_{\mathfrak{p}}$. It is a *local* ring, i.e. it has only one maximal ideal, namely $\mathfrak{p}A_{\mathfrak{p}}=\left\{\frac{a}{s}\,\middle|\,a\in\mathfrak{p}\right\}$.

§11. Pure Submodules

A short exact sequence $0\to M'\to M\to M''\to 0$ of left A-modules is called *pure* if $L\otimes_A M'\to L\otimes_A M$ is a monomorphism for every right A-module L. Since the tensor product commutes with direct limits, it suffices to require this for finitely generated modules L, and since every finitely generated module is a direct limit of finitely presented modules (Example 5.2), one may even assume L finitely presented. The following result relates flatness to purity:

Proposition 11.1. *The following properties of a module $_AF$ are equivalent:*
(a) *F is flat.*
(b) *Every exact sequence $0\to M'\to M\to F\to 0$ is pure.*
(c) *There is a pure exact sequence $0\to M'\to M\to F\to 0$ where M is a flat module.*

Proof. (a) ⇒ (b): If L is a right A-module, we choose an exact sequence $0\to K\to H\to L\to 0$ with H free. We get a commutative diagram

$$\begin{array}{ccccccccc} & & K\otimes M' & \longrightarrow & K\otimes M & \longrightarrow & K\otimes F & \longrightarrow & 0 \\ & & \downarrow & & \downarrow & & \downarrow{\scriptstyle\alpha} & & \\ 0 & \longrightarrow & H\otimes M' & \xrightarrow[\mu]{} & H\otimes M & \xrightarrow[\varphi]{} & H\otimes F & \longrightarrow & 0 \\ & & \downarrow & & \downarrow & & \downarrow & & \\ & & L\otimes M' & \xrightarrow[\beta]{} & L\otimes M & \longrightarrow & L\otimes F & \longrightarrow & 0 \\ & & \downarrow & & \downarrow & & \downarrow & & \\ & & 0 & & 0 & & 0 & & \end{array} \tag{1}$$

with exact rows and columns. By hypothesis we know that α is a monomorphism. We will show that it follows that β is a monomorphism, by making a diagram-chase. If $x\in\operatorname{Ker}\beta$, then x can be lifted to $y\in H\otimes M'$, and $\mu(y)\in H\otimes M$ goes to $\beta(x)=0$ in $L\otimes M$. Therefore $\mu(y)$ comes from some element $z\in K\otimes M$, and z maps to an element u in $K\otimes F$. But $\alpha(u)=\varphi\mu(y)=0$, so α a monomorphism implies that $u=0$. The element $z\in K\otimes M$ must then be the image of some $v\in K\otimes M'$, and v maps to y in $H\otimes M'$ since μ is a monomorphism. It follows that $x\in\operatorname{Ker}\beta$ is zero, so β is a monomorphism.

(b) ⇒ (c) is obvious (write F as the quotient of a free module M).

(c) ⇒ (a): We let $0\to M'\to M\to F\to 0$ be a pure exact sequence with M flat, and consider any exact sequence $0\to K\to H\to L\to 0$ of right A-modules. We then get a diagram (1), where now the middle column instead of the middle row is a short exact sequence. β is a monomorphism, so we are in a situation completely symmetrical to the one in (a) ⇒ (b), and it follows that α is a monomorphism. Hence F is flat. □

A submodule M' of M is a *pure* submodule if the exact sequence $0\to M'\to M\to M/M'\to 0$ is pure. The following characterization of pure submodules is an analogue of Prop. 10.7 for flatness:

Proposition 11.2. *A submodule M' of M is pure in M if and only if it satisfies: if $y_1,\dots,y_n$ are elements in M', (a_{ij}) is an $m\times n$-matrix of elements in A and the system of equations*

$$\sum_{i=1}^{m} X_i a_{ij} = y_j \qquad (j=1,\dots,n) \tag{2}$$

has a solution $(x_1,\dots,x_m)$ in M, then it has a solution in M'.

Proof. We assume the condition is satisfied, and we want to show that for every left module L we get a monomorphism $M'\otimes L\to M\otimes L$. Clearly we may assume L finitely generated, and since every finitely generated module is a direct limit of finitely presented modules (Example 5.2), we may even assume L finitely presented. Thus there is an exact sequence

$$A^m \xrightarrow{\alpha} A^n \to L \to 0,$$

where α is represented by a matrix (a_{ij}) by the rule α: $(x_i) \mapsto (\sum_i a_{ij} x_i)$. There results a commutative diagram

$$\begin{array}{ccccccc} M' \otimes A^m & \longrightarrow & M' \otimes A^n & \longrightarrow & M' \otimes L & \longrightarrow & 0 \\ \downarrow & & \downarrow & & \downarrow & & \\ M \otimes A^m & \longrightarrow & M \otimes A^n & \longrightarrow & M \otimes L & \longrightarrow & 0 \end{array} \tag{3}$$

where the first two vertical arrows are monomorphisms. We need the following lemma, whose proof is left to the reader:

Lemma 11.3. *Let the diagram*

$$\begin{array}{ccccccc} M' & \xrightarrow{\mu'} & M & \xrightarrow{\mu} & M'' & \longrightarrow & 0 \\ {\scriptstyle\alpha'}\downarrow & & {\scriptstyle\alpha}\downarrow & & {\scriptstyle\alpha''}\downarrow & & \\ N' & \xrightarrow[\eta']{} & N & \xrightarrow[\eta]{} & N'' & \longrightarrow & 0 \end{array}$$

be commutative with exact rows. Suppose α is a monomorphism. Then α'' is a monomorphism if and only if $\operatorname{Im}\eta' \cap \operatorname{Im}\alpha = \operatorname{Im}\alpha\mu'$.

We will apply the lemma to diagram (3). An element $(x_1, \ldots, x_m) \in M^m \cong M \otimes A^m$ maps to $u = \sum_i x_i \otimes (a_{ij}) \in M \otimes A^n$, which also is the image of $(y_1, \ldots, y_n) \in M'^n$ just when $\sum_i x_i a_{ij} = y_j$. The existence of a solution $(z_1, \ldots, z_m)$ in M' of this system means that u is the image of an element in $M' \otimes A^m = M'^m$, so the Lemma can be used to show that $M' \otimes L \to M \otimes L$ is a monomorphism.

Conversely, assume that M' is pure in M and that the system (3) has a solution in M. The matrix (a_{ij}) determines a homomorphism α: $A^m \to A^n$, and one puts $L = A^n/\operatorname{Im}\alpha$. One can now follow the preceding argument backwards to get the existence of a solution in M'. $\square$

Corollary 11.4. *Let M be a flat module and $0 \to K \to F \to M \to 0$ an exact sequence with F free. For any finite family $u_1, \ldots, u_m$ of elements of K there exists a homomorphism φ: $F \to K$ such that $\varphi(u_j) = u_j$ for $j = 1, \ldots, n$.*

Proof. We can write

$$u_j = \sum_{i=1}^{m} x_i a_{ij}, \qquad j = 1, \ldots, n, \tag{4}$$

where $x_1, \ldots, x_n$ is part of a basis for the free module F. Since M is flat, the given exact sequence is pure (Prop. 11.1), so the system (4) has a solution $z_1, \ldots, z_m$ in K. We can define φ: $F \to K$ so that $\varphi(x_i) = z_i$, and then get

$$\varphi(u_j) = \varphi(\sum_i x_i a_{ij}) = \sum_i z_i a_{ij} = u_j. \quad \square$$

A striking consequence of this Corollary is:

Corollary 11.5. *Every finitely presented flat module is projective.*

Proof. Let M be finitely presented and flat. Choose an exact sequence $0 \to K \to F \to M \to 0$ with F finitely generated free. K has a finite number of generators $u_1, \ldots, u_m$. If $\varphi: F \to K$ maps u_j to u_j, according to the corollary, then φ splits the sequence, and hence M is projective. ▯

Another very useful consequence of the corollary is:

Proposition 11.6. *Let A be a subring of a ring B. If M is a finitely generated and flat right A-module, and $M \otimes_A B$ is a projective B-module, then M is projective over A.*

Proof (Jøndrup [1]). Choose an exact sequence $0 \to K \to F \to M \to 0$ with F finitely generated free. Since M is flat, tensoring with B gives an exact sequence

$$0 \to K \otimes B \to F \otimes B \to M \otimes B \to 0$$

by Prop. 11.1. This sequence splits, because $M \otimes B$ was assumed to be B-projective. It follows that $K \otimes B$ is finitely generated over B, say with generators $k_1 \otimes 1, \ldots, k_n \otimes 1$. By Prop. 11.4 there exists a homomorphism $\varphi: F \to K$ such that $\varphi(k_j) = k_j$ for $j = 1, \ldots, n$. We will show that $\varphi(k) = k$ holds for all $k \in K$, so that φ splits and M is projective. If $k \in K$, we may write $k \otimes 1 = \sum (k_i \otimes 1) b_i$ with $b_i \in B$. Then

$$\varphi(k) \otimes 1 = (\varphi \otimes 1)(k \otimes 1) = \sum (\varphi \otimes 1)(k_i \otimes b_i) = \sum \varphi(k_i) \otimes b_i = \sum k_i \otimes b_i = k \otimes 1$$

in $K \otimes B$. But since F and M are flat, also K is flat (Exercise 36), and therefore $K \cong K \otimes A \to K \otimes B$ is a monomorphism. Hence $\varphi(k) = k$ for each $k \in K$. ▯

Example. Integral domains. If A is a commutative integral domain with field of fractions K, then every finitely generated flat A-module is projective by Prop. 11.6, for $M \otimes_A K$ is a vector space over K.

§12. Regular Rings

A ring A is called *regular* (in the sense of von Neumann) if for each $a \in A$ there exists $x \in A$ such that $a = axa$. The most important properties of such rings are summarized in the following statement (where "left" and "right" can be interchanged everywhere since the definition of regularity is left-right symmetric):

Proposition 12.1. *The following properties of a ring A are equivalent:*

(a) *A is regular.*

(b) *Every principal right ideal of A is generated by an idempotent element.*

(c) *Every finitely generated right ideal of A is generated by an idempotent element.*

(d) *Every left A-module is flat.*

Proof. (a) ⇒ (b): Given a principal right ideal aA, choose x such that $a = axa$. Then ax is idempotent and $aA = axA$.

(b) $\Rightarrow$ (a): Given $a \in A$, choose an idempotent e such that $eA = aA$. Then $e = ax$ for some x, and $a = ea = axa$.

(a) $\Rightarrow$ (c): It clearly suffices to show that if e and f are idempotents, then $eA + fA$ is a principal right ideal. We have $eA + fA = eA + (f - ef)A$, and if $x \in A$ is chosen so that $f - ef = (f - ef)x(f - ef)$, then $f' = (f - ef)x$ is an idempotent with $ef' = 0$, and $eA + fA = eA + f'A$. Now $eA + f'A = (e + f' - f'e)A$, because $e = (e + f' - f'e)e$ and $f' = (e + f' - f'e)f'$. Thus $eA + fA$ is principal.

(c) $\Rightarrow$ (d): The condition (c) means that every finitely generated right ideal is a direct summand of A, and Prop. 10.6 then immediately gives that all left A-modules are flat.

(d) $\Rightarrow$ (a): If $\mathfrak{a}$ is any right ideal and $\mathfrak{b}$ is any left ideal of A, then the flatness of $A/\mathfrak{b}$ implies that the canonical map $\mathfrak{a} \otimes A/\mathfrak{b} \to A/\mathfrak{b}$ is a monomorphism. But $\mathfrak{a} \otimes A/\mathfrak{b} \cong \mathfrak{a}/\mathfrak{a}\mathfrak{b}$ by Example 8.1, so this means that the canonical map $\mathfrak{a}/\mathfrak{a}\mathfrak{b} \to A/\mathfrak{b}$ is a monomorphism, i.e. $\mathfrak{a} \cap \mathfrak{b} = \mathfrak{a}\mathfrak{b}$. Choosing in particular $\mathfrak{a} = aA$ and $\mathfrak{b} = Aa$, we find that $aA \cap Aa = aAa$, and hence $a \in aAa$. $\square$

A particularly important class of regular rings is the one consisting of so-called "strongly regular" rings. Before introducing this class, we make a few definitions. An element $a \in A$ is called *nilpotent* if $a^n = 0$ for some n. The ring A is *reduced* if it has no nilpotent elements $\neq 0$. An idempotent element in A is *central* if it belongs to the center of A, i.e. commutes with every element of A. If e is a central idempotent, then the right ideal eA is actually a two-sided ideal, and A is decomposed as the direct sum of the two-sided ideals eA and $(1-e)A$. Conversely, if $A = \mathfrak{a} \oplus \mathfrak{b}$ where $\mathfrak{a}$ and $\mathfrak{b}$ are two-sided ideals, then there is a central idempotent e such that $\mathfrak{a} = eA$. We note:

Lemma 12.2. *Every idempotent in a reduced ring is central.*

Proof. If $e^2 = e$ and $a \in A$, then $(ea(1-e))^2 = ea(1-e)ea(1-e) = 0$, so $ea = eae$. Similarly $ae = eae$, so $ae = ea$. $\square$

Proposition 12.3. *The following properties of a ring A are equivalent:*

(a) *A is a reduced regular ring.*

(b) *Every principal right ideal is generated by a central idempotent.*

(c) *A is regular and every right ideal is two-sided.*

(d) *For every $a \in A$ there exists $x \in A$ such that $a = xa^2$.*

Proof. (a) $\Rightarrow$ (b) is clear from Lemma 12.2.

(b) implies that A is regular and that all principal right ideals are two-sided. But then every right ideal is two-sided, so we have (c).

(c) $\Rightarrow$ (d): For every $a \in A$ there exists x such that $a = axa$. Since xaA also is a left ideal, $axa = xab$ for some $b \in A$. Then $a = axa = xab = x \cdot axa \cdot b = xa \cdot a = xa^2$.

(d) $\Rightarrow$ (a): A is obviously reduced since $a = xa^2$ implies $a = x^{n-1}a^n$ for each n. We also have $(a - axa)^2 = a^2 + axa^2xa - a^2xa - axa^2 = 0$, so $a = axa$ and A is regular. $\square$

Definition. A is a *strongly regular* ring if it satisfies the equivalent conditions of Prop. 12.3.

Examples

1. Semi-simple rings. Every semi-simple is regular. Conversely, every right noetherian regular ring is semi-simple.

2. Endomorphism rings of semi-simple modules. Important examples of regular rings are obtained from the following result:

Proposition 12.4. *The endomorphism ring of a semi-simple module is regular.*

Proof. If M is a semi-simple module and $\alpha: M \to M$ is linear, then $M = \operatorname{Ker} \alpha \oplus K$ and $M = \operatorname{Im} \alpha \oplus N$ for submodules K and N. The map α induces an isomorphism $K \to \operatorname{Im} \alpha$, with inverse β: $\operatorname{Im} \alpha \to K$. β may be extended to ξ: $M \to M$ by putting $\xi(N) = 0$. Then $\alpha \xi \alpha = \alpha$, for if $x \in M$, then $x = y + z$ with $y \in \operatorname{Ker} \alpha$ and $z \in K$, which gives $\alpha \xi \alpha(x) = \alpha \xi \alpha(z) = \alpha(x)$. □

3. Products of skew-fields. If K_i $(i \in I)$ are skew-fields, then the product ring $\prod_I K_i$ is strongly regular. Conversely we have:

Proposition 12.5. *Every strongly regular ring is a subring of a product of skew-fields.*

Proof. Let $\mathfrak{m}_i$ $(i \in I)$ be the maximal right ideals of the strongly regular ring A. Since $\mathfrak{m}_i$ is a two-sided ideal, $A/\mathfrak{m}_i$ is a skew-field. Consider the canonical ring homomorphism $A \to \prod_I A/\mathfrak{m}_i$, which has the kernel $\mathfrak{a} = \bigcap_I \mathfrak{m}_i$. If $\mathfrak{a} \neq 0$, then $\mathfrak{a}$ contains an idempotent $e \neq 0$. Since e belongs to every maximal right ideal, $1 - e$ cannot belong to any maximal right ideal, so $(1-e)\,b = 1$ for some $b \in A$. But this is of course impossible. So $\mathfrak{a} = 0$, and A is a subring of $\prod A/\mathfrak{m}_i$. □

4. Boolean rings. A boolean ring is a commutative ring in which every element is idempotent. Such a ring is strongly regular.

§13. Coherent Rings

The coherent rings form a rather vast class of rings, comprising both noetherian rings and regular rings. In order to characterize these rings, we need some results on the commutation of tensor products and direct products. For each right A-module M and each family $(L_i)_I$ of left A-modules, there is a homomorphism

$$\varphi: M \otimes_A \big(\prod_I L_i\big) \to \prod_I (M \otimes_A L_i),$$

given by $\varphi(x \otimes (z_i)) = (x \otimes z_i)$. In particular, there is a homomorphism φ: $M \otimes A^I \to M^I$ given by $\varphi(x \otimes (a_i)) = (x a_i)$.

Lemma 13.1. *The following properties of the module M are equivalent:*

(a) *M is finitely generated.*

(b) *φ: $M \otimes_A (\prod_I L_i) \to \prod_I (M \otimes_A L_i)$ is an epimorphism for every family $(L_i)_I$ of left A-modules.*

(c) *φ: $M \otimes_A A^I \to M^I$ is an epimorphism for every set I.*

Proof. (a) $\Rightarrow$ (b): Suppose M has generators $x^{(1)}, \ldots, x^{(n)}$. If $(u_i)_I \in \prod_I (M \otimes L_i)$, then each u_i may be written as $u_i = \sum_{j=1}^{n} x^{(j)} \otimes z_{ji}$ with $z_{ji} \in L_i$. Therefore $(u_i)_I$ is the image of $\sum_j x^{(j)} \otimes (z_{ji})_I$, and so φ is an epimorphism.

(b) $\Rightarrow$ (c) is trivial.

(c) $\Rightarrow$ (a): Choose I to be the set underlying M. Consider the element $u \in M^M$ whose x-th component is x. Since $\varphi\colon M \otimes A^M \to M^M$ is an epimorphism, we have $u = \varphi(\sum_j x^{(j)} \otimes (a_{jx}))$ for certain $(a_{1x}), \ldots, (a_{nx}) \in A^M$ and $x^{(1)}, \ldots, x^{(n)} \in M$. Then clearly $x = \sum_j x^{(j)} a_{jx}$ for each $x \in M$, so $x^{(1)}, \ldots, x^{(n)}$ are generators for M. $\square$

Lemma 13.2. *The following properties of the module M are equivalent:*

(a) *M is finitely presented.*

(b) *$\varphi\colon M \otimes_A (\prod_I L_i) \to \prod_I (M \otimes_A L_i)$ is an isomorphism for every family $(L_i)_I$ of left A-modules.*

(c) *$\varphi\colon M \otimes_A A^I \to M^I$ is an isomorphism for every set I.*

Proof. (a) $\Rightarrow$ (b): Choose an exact sequence $0 \to K \to F \to M \to 0$ with F finitely generated free. There is induced a commutative diagram

$$\begin{array}{ccccccc}
K \otimes (\prod L_i) & \longrightarrow & F \otimes (\prod L_i) & \longrightarrow & M \otimes (\prod L_i) & \longrightarrow & 0 \\
\downarrow{\scriptstyle \varphi^K} & & \downarrow{\scriptstyle \varphi^F} & & \downarrow{\scriptstyle \varphi^M} & & \\
\prod (K \otimes L_i) & \longrightarrow & \prod (F \otimes L_i) & \longrightarrow & \prod (M \otimes L_i) & \longrightarrow & 0
\end{array}$$

with exact rows. Here φ^F is an isomorphism, since F is finitely generated free. When M is finitely presented, K is finitely generated and hence φ^K is an epimorphism by the preceding Lemma. An easy diagram chase shows that φ^M must then be a monomorphism, hence an isomorphism.

(b) $\Rightarrow$ (c) is trivial.

(c) $\Rightarrow$ (a): By the preceding Lemma we know that M is finitely generated. Choose again an exact sequence $0 \to K \to F \to M \to 0$ with F finitely generated free. We obtain a commutative diagram

$$\begin{array}{ccccccccc}
 & & K \otimes A^I & \longrightarrow & F \otimes A^I & \longrightarrow & M \otimes A^I & \longrightarrow & 0 \\
 & & \downarrow{\scriptstyle \varphi^K} & & \downarrow{\scriptstyle \varphi^F} & & \downarrow{\scriptstyle \varphi^M} & & \\
0 & \longrightarrow & K^I & \longrightarrow & F^I & \longrightarrow & M^I & \longrightarrow & 0
\end{array}$$

with exact rows, where φ^F and φ^M are isomorphisms. It follows by a diagram chase that φ^K is an epimorphism, so K is finitely generated, and hence M is finitely presented. $\square$

A module M is called *coherent* if it is finitely presented and also every finitely generated submodule of M is finitely presented.

Proposition 13.3 (Chase [1]). *The following properties of a ring A are equivalent:*

(a) *Every direct product of flat left A-modules is flat.*

(b) *A^I is a flat left A-module for every set I.*

(c) *Every finitely presented right A-module is coherent.*

(d) *A is coherent as a right A-module.*

(e) *$r(a)=\{b\in A \mid a\,b=0\}$ is a finitely generated right ideal for every $a\in A$, and the intersection of two finitely generated right ideals is finitely generated.*

Definition. A is said to be a *right coherent* ring if it satisfies these equivalent conditions.

Proof. (a) ⇒ (b) is trivial.

(b) ⇒ (c): Let L be a finitely generated submodule of a finitely presented module M. For each index set I we have a commutative diagram

$$\begin{array}{ccc} L\otimes A^I & \longrightarrow & M\otimes A^I \\ \Big\downarrow{\scriptstyle \varphi^L} & & \Big\downarrow{\scriptstyle \varphi^M} \\ L^I & \longrightarrow & M^I \end{array}$$

where the horizontal arrows are monomorphisms. Since φ^M is a monomorphism by Lemma 13.2, also φ^L is a monomorphism, and hence L is finitely presented.

(c) ⇒ (e): Since A is coherent, every finitely generated right ideal $\mathfrak{a}$ is finitely presented. For each $a\in A$ there is an exact sequence

$$0\to r(a)\to A\to aA\to 0, \qquad (1)$$

and aA finitely presented implies that $r(a)$ is finitely generated.

Let $\mathfrak{a}$ and $\mathfrak{b}$ be two finitely generated right ideals. There is an exact sequence

$$0\to \mathfrak{a}\cap\mathfrak{b}\to A\xrightarrow{\beta} A/\mathfrak{a}\oplus A/\mathfrak{b},$$

where $A/\mathfrak{a}\oplus A/b$ is finitely presented and therefore coherent. Since $\operatorname{Im}\beta$ is finitely generated, it follows that $\operatorname{Im}\beta$ is finitely presented, and hence $\mathfrak{a}\cap\mathfrak{b}$ is finitely generated.

(e) ⇒ (d): Consider a finitely generated right ideal $\mathfrak{a}$ with generators $a_1,\dots,a_n$. To show that $\mathfrak{a}$ is finitely presented, we use induction on n. When $n=1$, it follows from the exact sequence (1) that $\mathfrak{a}=a_1A$ is finitely presented. Assume that $\mathfrak{a}_1=a_1A+\dots+a_{n-1}A$ and $\mathfrak{a}_2=a_nA$ are finitely presented. The canonical map $\mathfrak{a}_1\oplus\mathfrak{a}_2\to\mathfrak{a}_1+\mathfrak{a}_2$ gives rise to an exact sequence

$$0\to\mathfrak{a}_1\cap\mathfrak{a}_2\to\mathfrak{a}_1\oplus\mathfrak{a}_2\to\mathfrak{a}\to 0,$$

where $\mathfrak{a}_1\oplus\mathfrak{a}_2$ is finitely presented and $\mathfrak{a}_1\cap\mathfrak{a}_2$ is finitely generated. It follows that $\mathfrak{a}$ is finitely presented (Exercise 11).

(d) ⇒ (a): Assume $(L_i)_I$ is a family of flat left A-modules. For each finitely generated right ideal $\mathfrak{a}$ we obtain monomorphisms

$$\mathfrak{a}\otimes(\prod L_i)\to\prod(\mathfrak{a}\otimes L_i)\to\prod L_i$$

by Lemma 13.2, and it follows from Prop. 10.6 that $\prod L_i$ is flat. ▯

Examples

1. Noetherian rings. Every right noetherian ring is right coherent.

2. Semi-hereditary rings. If A is right semi-hereditary, then every finitely generated right ideal is projective and hence finitely presented (Exercise 21), so A is right coherent. In particular, every regular ring is right and left coherent.

Exercises

A denotes a ring throughout the following exercises.

§1

1. Let R be a commutative ring. An (associative) *R-algebra* is a ring A which also is an R-module such that

$$(a\,r)\,b=a(b\,r)=(a\,b)\,r \quad \text{for} \quad r\in R \quad \text{and} \quad a, b\in A.$$

Show that:

(i) To give an R-algebra structure on a ring A is the same as to give a ring homomorphism $\varphi: R\to A$ such that $\varphi(R)$ is contained in the center of A.

(ii) The matrix ring $M_n(R)$ is an R-algebra.

(iii) If A is an R-algebra and L, M are A-modules, then $\operatorname{Hom}_A(L, M)$ may be viewed as an R-module.

2. Show that if L and M are A-modules, then $\operatorname{Hom}_A(L, M)$ may be considered as an $\operatorname{End}_A(M)$-$\operatorname{End}_A(L)$-bimodule.

3. Let M be a right A-module and put $\operatorname{Ann}(M)=\{a\in A\,|\,M\,a=0\}$. Call M *faithful* if $\operatorname{Ann}(M)=0$. Show that:

(i) $\operatorname{Ann}(M)$ is a two-sided ideal of A, and M is a faithful module over the ring $A/\operatorname{Ann}(M)$.

(ii) M is faithful as a left $\operatorname{End}_A(M)$-module.

(iii) M is faithful if and only if the canonical homomorphism $A\to\operatorname{Bic}(M)$ is a monomorphism.

§2

4. Let K and L be submodules of a module M. Show that there are exact sequences

$$0\to K\cap L\to K\oplus L\xrightarrow{\alpha} K+L\to 0,$$

$$0\to M/(K\cap L)\to M/K\oplus M/L\xrightarrow{\beta} M/(K+L)\to 0$$

where $\alpha(x, y)=x-y$ and $\beta(\bar{u}, \bar{v})=\overline{u-v}$.

5. If $(M_i)_I$ is a family of modules and N is any module, show that

$$\operatorname{Hom}(N, \prod_I M_i)\cong\prod_I \operatorname{Hom}(N, M_i),$$

$$\operatorname{Hom}(\bigoplus_I M_i, N)\cong\prod_I \operatorname{Hom}(M_i, N).$$

6. Let $(M_i)_I$ be a family of submodules of a module M. Show that M is the direct sum of the submodules M_i if and only if:

(1) $M=\sum_I M_i$;

(2) $M_{i_0}\cap(M_{i_1}+\cdots+M_{i_n})=0$ for every finite subset $J=\{i_1, \ldots, i_n\}$ of I and $i_0\notin J$.

7. Let M be a module and α_0 an ordinal. Suppose that for each ordinal $\alpha\leq\alpha_0$ there is a submodule M_α of M so that:

(1) $M_{\alpha+1}=M_\alpha\oplus N_\alpha$ for some submodule N_α of M;

(2) $M_\alpha=\sum_{\beta<\alpha} M_\beta$ when α is a limit ordinal;

(3) $M=\sum_{\alpha\leq\alpha_0} M_\alpha$.

Show that $M=\bigoplus_{\alpha<\alpha_0} N_\alpha$.

8. Show that if A is a commutative ring and G is a cyclic group of order $n<\infty$, then $A[G]\cong A[X]/(X^n-1)$.

9. Let M be an A-A-bimodule. Define a multiplication on the abelian group $A\times M$ as

$$(a, m)(b, m')=(a\,b, a\,m'+m\,b),$$

and show that:

(i) $A\times M$ becomes a ring (called the *trivial extension* of A by M), and there are ring homomorphisms $\alpha\colon A\to A\times M$ and $\varphi\colon A\times M\to A$ such that $\varphi\,\alpha=1_A$.

(ii) M becomes imbedded as a two-sided ideal $0\times M$ of $A\times M$, and $(0\times M)^2=0$.

10. Let A and B be rings and ${}_AM_B$ a bimodule. Consider M as a bimodule over the ring $A\times B$ by the rules $(a,b)\,x=a\,x$ and $x(a,b)=x\,b$. Show that:

(i) The trivial extension $(A\times B)\times M$ is isomorphic to the "generalized matrix ring" $\begin{pmatrix} A & M\\ 0 & B\end{pmatrix}$, constisting of elements

$$\begin{pmatrix} a & m\\ 0 & b\end{pmatrix} \quad \text{with } a\in A,\ b\in B \text{ and } m\in M,$$

with ordinary matrix addition and multiplication.

(ii) The right ideals of $\begin{pmatrix} A & M\\ 0 & B\end{pmatrix}$ are of the form

$$\left\{\begin{pmatrix} a & m\\ 0 & b\end{pmatrix}\middle|\, a\in\mathfrak{a},\ (m,b)\in L\right\},$$

where $\mathfrak{a}$ is a right ideal of A and L is a B-submodule of $M\oplus B$ such that $\mathfrak{a}M\subset L$. What is the corresponding result for left ideals?

§3

11. Let $0\to K\to L\to M\to 0$ be an exact sequence with L finitely presented and K finitely generated. Show that M is finitely presented.

12. Let A be an R-algebra (Exercise 1). Show that if R is a noetherian ring and A is a finitely generated R-module, then A is a right and left noetherian ring.

13. Let A be a commutative ring and M a noetherian A-module. Show that the quotient ring $A/\mathrm{Ann}(M)$ is noetherian.

14. Show that if A is a commutative noetherian ring and G is a finite group, then the ring $A[G]$ is right and left noetherian.

15. Show that the ring (cf. Exercise 10)

$$\begin{pmatrix} \mathbb{Z} & \mathbb{Q}\\ 0 & \mathbb{Q}\end{pmatrix}$$

is right noetherian, but not left noetherian.

§4

16. Show that the small categories may be considered as the objects of a category **Cat**, where the morphisms consist of the functors.

17. Let **C** be a category. A morphism $\alpha\colon C\to C'$ is called a *monomorphism* if for every couple $\xi,\eta\colon C''\to C$ of morphisms, $\alpha\,\xi=\alpha\,\eta$ implies $\xi=\eta$. *Epimorphisms* are defined dually: $\xi\,\alpha=\eta\,\alpha$ implies $\xi=\eta$ for $\xi,\eta\colon C'\to C''$. Show that:

(i) In the category of sets, α is a monomorphism (epimorphism) if and only if it is an injective (resp. surjective) map.

(ii) In the category **Mod**-A, these terms keep their usual meaning.

(iii) Let **C** be the category of commutative rings and ring homomorphisms. Show that $\mathbb{Z}\to\mathbb{Q}$ is both a monomorphism and an epimorphism in **C**.

18. Let A be a ring and let $1_{\mathbf{Mod}\text{-}A}$ denote the identity functor **Mod**-$A\to$ **Mod**-A. For each element a in the center of A we define a natural transformation $\varphi(a)\colon 1_{\mathbf{Mod}\text{-}A}\to 1_{\mathbf{Mod}\text{-}A}$ as $\varphi(a)_M\colon x\mapsto x\,a$ for $x\in M_A$. Show that this gives an isomorphism of rings

$$\varphi\colon \operatorname{cen} A\to \operatorname{Nat}(1_{\mathbf{Mod}\text{-}A}, 1_{\mathbf{Mod}\text{-}A}),$$

where $\operatorname{cen} A$ denotes the center of A and $\operatorname{Nat}(1_{\mathbf{Mod}\text{-}A}, 1_{\mathbf{Mod}\text{-}A})$ denotes the set of natural transformations $1_{\mathbf{Mod}\text{-}A}\to 1_{\mathbf{Mod}\text{-}A}$.

§5

19. For each abelian group G we let $t(G)$ denote its torsion subgroup.

(i) Show that every group homomorphism $G \to H$ induces $t(G) \to t(H)$ by restriction.

(ii) Deduce that t may be considered as a functor t: **Ab** $\to$ **Ab**, with a natural transformation $\eta\colon t \to 1_{\mathbf{Ab}}$ such that η_G is a monomorphism for every G.

(iii) Show that the functor t is left exact.

20. Let p a prime number and let $\mathbb{Z}(p^\infty)$ denote the subgroup of $\mathbb{Q}/\mathbb{Z}$ consisting of classes mod $\mathbb{Z}$ of elements a/p^n, $a \in \mathbb{Z}$.

(i) Show that $\mathbb{Z}(p^\infty)$ is a direct limit of the cyclic groups $\mathbb{Z}/p^n\mathbb{Z}$.

(ii) Define a direct system (M_n, α_{mn}), where $M_n = \mathbb{Z}(p^\infty)$ for every integer $n > 0$, and $\alpha_{mn}\colon M_m \to M_n$ is given by $\alpha_{mn}(x) = p^{n-m}x$ for $m \leq n$. Show that $\varinjlim M_n = 0$, although all α_{mn} are epimorphisms.

§6

21. Show that every finitely generated projective module is finitely presented.

22. Show that if E is an injective cogenerator, then $\beta\colon M \to N$ is an epimorphism if and only if $\beta^*\colon \mathrm{Hom}_A(N, E) \to \mathrm{Hom}_A(M, E)$ is a monomorphism.

23. Let R be a commutative ring, A an R-algebra (Exercise 1) and U an R-module. For each right (left) A-module M, put $M^* = \mathrm{Hom}_R(M, U)$. Show that:

(i) M^* is a left (resp. right) A-module.

(ii) The canonical homomorphism $\mu_M\colon M \to M^{**}$, defined as $\mu_M(m)\colon \alpha \mapsto \alpha(m)$ for $m \in M$ and $\alpha \in M^*$, is A-linear.

(iii) The two homomorphisms

$$M^* \underset{(\mu_M)^*}{\overset{\mu_{M^*}}{\rightleftarrows}} M^{***}$$

satisfy $(\mu_M)^* \cdot \mu_{M^*} = 1$; thus μ_{M^*} is a split monomorphism.

(iv) If the R-module U is an injective cogenerator, then $\mu_M\colon M \to M^{**}$ is a monomorphism for each A-module M.

24. Show that the ring (cf. Exercise 10)

$$\begin{pmatrix} \mathbb{Z} & \mathbb{Q} \\ 0 & \mathbb{Q} \end{pmatrix}$$

is right hereditary.

25. Let A be an integral domain with field of fractions K. Show that:

(i) K/A is a divisible A-module.

(ii) K/A is not an injective A-module, when $A = \mathbb{Z}[X]$.

§7

26. Let $\mathfrak{a}$ be a minimal right ideal of A and let S be a simple right A-module. Show that if $\mathfrak{a}$ and S are non-isomorphic, then $S\mathfrak{a} = 0$.

27. Show that if $\mathfrak{a}$ is a minimal right ideal of A, then either $\mathfrak{a}^2 = 0$ or $\mathfrak{a} = eA$ for some idempotent e.

28. Let A be a finite-dimensional algebra over an algebraically closed field K. Show that:

(i) If S is a simple A-module, then $\mathrm{End}_A(S) \cong K$. (Hint: show that if $\alpha \in \mathrm{End}_A(S)$, then $f(\alpha) = 0$ for some polynomial f over K.)

(ii) If A is semi-simple, then $A \cong M_{n_1}(K) \times \cdots \times M_{n_r}(K)$.

§8

29. Show that the homomorphism $\mathbb{C} \otimes_{\mathbb{R}} \mathbb{C} \to \mathbb{C}$, induced by multiplication, is not a monomorphism.

30. Let R be a commutative ring, and let A and B be two R-algebras (Exercise 1). Show that:

(i) $A \otimes_R B$ has the structure of an R-algebra.

(ii) A right $A \otimes_R B$-module is the same thing as an A^{op}-B-bimodule for which the induced R-module structures on the left and on the right coincide.

31. Let R be a commutative ring and A an R-algebra. Consider the R-algebra $A^{\mathrm{e}} = A^{\mathrm{op}} \otimes_R A$ and note that A is a right A^{e}-module (cf. preceding Exercise).

(i) Show that multiplication in A induces an A^{e}-linear map $\mu\colon A^{\mathrm{e}} \to A$.

(ii) Show that the kernel J of μ is the right ideal of A^{e} generated by all elements of the form $a \otimes 1 - 1 \otimes a$.

(iii) For each right A^e-module M we put $M^A=\{x\in M\,|\,x(a\otimes 1)=x(1\otimes a)$ for all $a\in A\}$. Show that $M^A=\{x\in M\,|\,xJ=0\}$, and that $M\mapsto M^A$ is a functor H: **Mod**-$A^e\to$ **Mod**-R.

(iv) Show that the functor H is naturally equivalent to the functor $\operatorname{Hom}_{A^e}(A,\cdot)$.

§9

32. Let modules L_A and ${}_AM_B$ be given. Show that if L_A and M_B are projective modules, then $L\otimes_A M$ is a projective B-module.

33. Let there be given a finitely presented module M_A, a bimodule ${}_BN_A$ and an injective module ${}_BE$. Show that there is an isomorphism

$$\operatorname{Hom}_B(\operatorname{Hom}_A(M,N),E)\cong M\otimes_A \operatorname{Hom}_B(N,E).$$

(Hint: first consider the case $M=A^n$; then use the right exactness in M of both sides.)

34. Let $A\times M$ be the trivial extension of A by a bimodule M (Exercise 9). Show that:

(i) A right module over the ring $A\times M$ is the same thing as a right A-module X together with an A-linear map α: $X\otimes_A M\to X$, such that the composed map

$$X\otimes_A M\otimes_A M\xrightarrow{\alpha\otimes 1} X\otimes_A M\xrightarrow{\alpha} X$$

is zero. (We use the notation (X,α) for this $A\times M$-module.)

(ii) If (X,α) and (Y,β) are $A\times M$-modules, then an $A\times M$-homomorphism $(X,\alpha)\to(Y,\beta)$ is the same thing as an A-homomorphism φ: $X\to Y$ such that the diagram

$$\begin{array}{ccc} X\otimes_A M & \xrightarrow{\alpha} & X \\ \downarrow{\scriptstyle \varphi\otimes 1} & & \downarrow{\scriptstyle \varphi} \\ Y\otimes_A M & \xrightarrow[\beta]{} & Y \end{array}$$

commutes.

35. Let A and B be rings and ${}_AM_B$ a bimodule. Consider M as an $A\times B$-bimodule as in Exerxise 10. Show that:

(i) $M\otimes_{A\times B}M=0$.

(ii) Let X be a right A-module and Y a right B-module, and suppose there is a B-linear map α: $X\otimes_A M\to Y$. Show that $X\oplus Y$ becomes a right module over the ring $\begin{pmatrix} A & M\\ 0 & B\end{pmatrix}$ by the rule

$$(x,y)\begin{pmatrix} a & m\\ 0 & b\end{pmatrix}=(x\,a,\alpha(x\otimes m)+y\,b).$$

(We use the notation $(X\;Y)_\alpha$ for this module.)

(iii) Every right module over the ring $\begin{pmatrix} A & M\\ 0 & B\end{pmatrix}$ is of the form $(X\;Y)_\alpha$.

(iv) A homomorphism $(X\;Y)_\alpha\to(X'\;Y')_\beta$ of $\begin{pmatrix} A & M\\ 0 & B\end{pmatrix}$-modules is a pair (φ,ψ), where φ: $X\to X'$ is A-linear, ψ: $Y\to Y'$ is B-linear and the diagram

$$\begin{array}{ccc} X\otimes_A M & \xrightarrow{\alpha} & Y \\ \downarrow{\scriptstyle \varphi\otimes 1} & & \downarrow{\scriptstyle \psi} \\ X'\otimes_A M & \xrightarrow[\beta]{} & Y' \end{array}$$

commutes.

§10

36. Let $0\to L\to M\to N\to 0$ be an exact sequence with N flat. Show that L is flat if and only if M is flat.

37. Let $0\to L\to M\to N\to 0$ be an exact sequence of left A-modules with M flat. Show that N is flat if and only if $\mathfrak{a}M\cap L=\mathfrak{a}L$ for every right ideal $\mathfrak{a}$ of A.

38. Use Exercises 33 and 22 to show that if M is a finitely presented flat module, then M is projective. (Cf. Cor. 11.5 for another proof.)

39. A flat module ${}_AF$ is *faithfully flat* if $M \otimes_A F = 0$ implies $M = 0$ for every module M_A. Show that:

(i) A flat module ${}_AF$ is faithfully flat if and only if $\mathfrak{m}F \neq F$ for every maximal right ideal $\mathfrak{m}$ of A.

(ii) Every non-zero free module is faithfully flat.

(iii) The $\mathbb{Z}$-module $\mathbb{Q}$ is not faithfully flat.

(iv) Suppose A is a subring of a ring B. Show that B/A is flat as a left A-module if and only if ${}_AB$ is faithfully flat.

(v) Suppose A is a commutative ring. Show that the A-module $\oplus A_{\mathfrak{m}}$, where the sum is taken over all maximal ideals $\mathfrak{m}$ of A, is faithfully flat.

§11

In the following exercises, let $\hat{M} = \mathrm{Hom}_{\mathbb{Z}}(M, \mathbb{Q}/\mathbb{Z})$.

40. Show that the following conditions are equivalent for an exact sequence $0 \to L \xrightarrow{\alpha} M \xrightarrow{\beta} N \to 0$ of right A-modules:

(a) The sequence $0 \to L \to M \to N \to 0$ is pure.

(b) For every finitely presented module F and homomorphism $\varphi\colon F \to N$, there exists $\psi\colon F \to M$ such that $\beta\psi = \varphi$.

(c) The sequence $0 \to \hat{N} \to \hat{M} \to \hat{L} \to 0$ of left A-modules is pure.

(d) The sequence $0 \to \hat{N} \to \hat{M} \to \hat{L} \to 0$ splits.

(Hint: Use Prop. 9.2 to show (a) $\Leftrightarrow$ (c) and (a) $\Leftrightarrow$ (d), and Exercise 33 to show (b) $\Leftrightarrow$ (c).)

41. Show that the canonical homomorphism $\mu_M\colon M \to \hat{\hat{M}}$ (Exercise 23) is a pure monomorphism. (Hint: use Exercises 23 and 40.)

42. A module Q is called *pure-injective* if for every pure sequence $0 \to L \xrightarrow{\alpha} M \xrightarrow{\beta} N \to 0$ and every homomorphism $\varphi\colon L \to Q$, there exists $\psi\colon M \to Q$ such that $\psi\alpha = \varphi$. Show that:

(i) The module $\hat{K}$ is pure-injective for every left module K.

(ii) If Q is pure-injective, then every pure sequence $0 \to Q \to M \to N \to 0$ splits.

(iii) Q is pure-injective if and only if Q is a direct summand of $\hat{K}$ for some left module K. (Hint: use Exercise 41.)

43. Show that the following properties of a module Q are equivalent:

(a) For every exact sequence $0 \to L \xrightarrow{\alpha} M \xrightarrow{\beta} N \to 0$ with N finitely presented, and every homomorphism $\varphi\colon L \to Q$, there exists $\psi\colon M \to Q$ such that $\psi\alpha = \varphi$.

(b) Every exact sequence $0 \to Q \to M \to N \to 0$ is pure.

(Hint: use Exercise 40(b).) A module Q with these properties is called *FP-injective* (or *absolutely pure*). When A is right noetherian, all FP-injective modules are injective.

44. Show that the following conditions are equivalent for a short exact sequence $0 \to L \to M \to N \to 0$ of abelian groups:

(a) L is pure in M.

(b) $L \cap nM = nL$ for every $n \in \mathbb{N}$.

(c) Each $y \in N$ is the image of an element in M having the same order as y.

45. Show that an abelian group is a flat $\mathbb{Z}$-module if and only if it is torsion-free.

§12

46. Let $\mathfrak{a}$ be a two-sided ideal of a regular ring A. Show that:

(i) The ring $A/\mathfrak{a}$ is regular.

(ii) $\mathfrak{a}$ is an intersection of maximal right ideals of A.

47. Let A be a regular ring. Show that:

(i) For each $a \in A$ there exists $\tilde{a} \in A$ such that $a\tilde{a}a = a$ and $\tilde{a}a\tilde{a} = \tilde{a}$.

(ii) A is strongly regular if and only if $\tilde{a}$ is uniquely determined by a, for every $a \in A$.

(iii) If A is strongly regular, C is a subring of A, and B is the intersection of all regular subrings of A containing C, then B is a (strongly) regular ring.

48. Show that the center of a regular ring is a regular ring.

49. Let A be a regular ring. Show that:

(i) If $\mathfrak{a}$ is a non-zero right or left ideal, then $\mathfrak{a}^n \neq 0$ for all n.

(ii) If $\mathfrak{a}$ is a two-sided ideal, then $r(\mathfrak{a}) = l(\mathfrak{a})$ and $\mathfrak{a} \cap r(\mathfrak{a}) = 0$.

50. Show that the endomorphism ring of a module M is regular if and only if every endomorphism α of M has the property that $\mathrm{Ker}\,\alpha$ and $\mathrm{Im}\,\alpha$ are direct summands of M.

51. Let M be a module with a regular endomorphism ring H. Show that the following conditions are equivalent:

(a) Each non-zero direct summand of M contains a non-zero indecomposable direct summand.

(b) Each non-zero right ideal of H contains a minimal non-zero right ideal.

§13

52. Show that if L is a submodule of a module M, and both L and M/L are coherent, then M is coherent.

53. Show that if A is a right coherent ring and $\mathfrak{a}$ is a two-sided ideal of A, which is finitely generated as a right ideal, then $A/\mathfrak{a}$ is a right coherent ring.

54. Show that the following properties of A are equivalent:

(a) A is right semi-hereditary.

(b) A is right coherent and every right ideal is a flat module.

(c) A is right coherent and every left ideal is a flat module.

55. Let A be a right coherent ring. Show that the following properties of a module M are equivalent:

(a) M is FP-injective (Exercise 43).

(b) For every finitely generated right ideal $\mathfrak{a}$ and every homomorphism $\varphi\colon \mathfrak{a} \to M$, there exists $x \in M$ such that $\varphi(a) = xa$ for all $a \in \mathfrak{a}$.

(c) $\hat{M} = \mathrm{Hom}_{\mathbb{Z}}(M, \mathbb{Q}/\mathbb{Z})$ is a flat left A-module.

(Hint: use Exercise 33.)

Chapter II. Rings of Fractions

In § 10 (Example 3) of the preceding chapter it was briefly indicated how the well-known construction of the field of fractions of a commutative integral domain could be generalized to arbitrary commutative rings. When trying to extend this construction to non-commutative rings, one finds that this is not always possible, but that one can give necessary and sufficient conditions for the existence of a ring of fractions. Such a condition was first found by Ø. Ore [1] around 1930 for the case of a skew-field of fractions of a domain. The existence of a total ring of fractions of an arbitrary ring was first considered by K. Asano [1]. General rings of fractions were studied by Elizarov [1], and a systematic theory of rings and modules of fractions was developed by P. Gabriel [1, 2] in connection with his theory of general rings of quotients.

§ 1. The Ring of Fractions

Let A be a ring and let S be a multiplicatively closed subset of A, i.e. $s, t \in S$ implies $st \in S$ and $1 \in S$. We define a *right ring of fractions* of A with respect to S as a ring $A[S^{-1}]$ together with a ring homomorphism $\varphi: A \to A[S^{-1}]$ satisfying:

$F1$. $\varphi(s)$ is invertible for every $s \in S$.

$F2$. Every element in $A[S^{-1}]$ has the form $\varphi(a)\varphi(s)^{-1}$ with $s \in S$.

$F3$. $\varphi(a)=0$ if and only if $as=0$ for some $s \in S$.

Similarly one defines a left ring of fractions $[S^{-1}]A$ of A with respect to S. It is not immediately clear that the axioms $F1$–3 determine $A[S^{-1}]$ uniquely, but that so is the case follows from the fact that $A[S^{-1}]$ is a solution of a universal problem. (We will use the term "universal problem" in an informal sense, because attempts towards a formalization of it tend to become a bit awkward. The unicity of solutions of universal problems can therefore not be proved formally here, but it can be done in each specific case by the same kind of argument as was employed in the typical case of tensor products (Prop. I.8.1).)

Proposition 1.1. *When $A[S^{-1}]$ exists, it has the following universal property: for every ring homomorphism $\psi: A \to B$ such that $\psi(s)$ is invertible in B for every $s \in S$, there exists a unique homomorphism $\sigma: A[S^{-1}] \to B$ such that $\sigma\varphi = \psi$.*

Proof. σ is defined as $\sigma(\varphi(a)\varphi(s)^{-1}) = \psi(a)\psi(s)^{-1}$. We have to verify that this is well-defined. So suppose $\varphi(a)\varphi(s)^{-1} = \varphi(b)\varphi(t)^{-1}$. Then $\varphi(a) = \varphi(b)\varphi(t)^{-1}\varphi(s)$

$=\varphi(b)\,\varphi(c)\,\varphi(u)^{-1}$ for some $c\in A$ and $u\in S$, by $F2$. This gives

$$\varphi(a)\,\varphi(u)=\varphi(b)\,\varphi(c)$$

and

$$\varphi(s)\,\varphi(u)=\varphi(t)\,\varphi(c),$$

from which we obtain with the use of $F3$ that $a\,u\,v=b\,c\,v$ and $s\,u\,v'=t\,c\,v'$ for some v and v' in S. Since $\psi(v)$ and $\psi(v')$ are invertible, this implies

$$\psi(a)\,\psi(u)=\psi(b)\,\psi(c)$$

and

$$\psi(s)\,\psi(u)=\psi(t)\,\psi(c),$$

and we can go backwards to find that $\psi(a)\,\psi(s)^{-1}=\psi(b)\,\psi(t)^{-1}$.

We leave to the reader to verify that σ is a homomorphism. It is clear that $\sigma\,\varphi=\psi$ and that σ is unique. □

Because of the unicity of the solution of a universal problem we may draw the following two conclusions:

Corollary 1.2. *$A[S^{-1}]$ is unique up to isomorphism.*

Corollary 1.3. *If both $A[S^{-1}]$ and $[S^{-1}]\,A$ exist, then they are naturally isomorphic.*

The existence of a solution of the universal problem stated in Prop. 1.1 does not imply that $A[S^{-1}]$ exists. This is clear since it may happen that $[S^{-1}]\,A$ exists but $A[S^{-1}]$ does not exist (see Example 5 below).

Note that it follows from $F3$ that the canonical homomorphism $\varphi\colon A\to A[S^{-1}]$ is a monomorphism if and only if S is contained in the set of non-zero-divisors. When so is the case, A is usually identified with $\varphi(A)$, and elements in $A[S^{-1}]$ may be written as $a\,s^{-1}$.

We now turn to the question of the existence of $A[S^{-1}]$. It is rather easy to show that the universal problem in Prop. 1.1 always has a solution (in a presentation of A by generators and relations, one adjoins a new generator $\bar{s}$ for each $s\in S$, and adds the new relations $s\,\bar{s}=\bar{s}\,s=1$). However, this solution is practically useless because in general it is difficult to tell what its elements look like, e.g. it is hard to determine the kernel of the canonical homomorphism from A into it.

Proposition 1.4. *Let S be a multiplicatively closed subset of A. $A[S^{-1}]$ exists if and only if S satisfies:*

$S1$. If $s\in S$ and $a\in A$, there exist $t\in S$ and $b\in A$ such that $s\,b=a\,t$.

$S2$. If $s\,a=0$ with $s\in S$, then $a\,t=0$ for some $t\in S$.

When $A[S^{-1}]$ exists, it has the form

$$A[S^{-1}]=A\times S/\sim$$

where $\sim$ is the equivalence relation defined as $(a,s)\sim(b,t)$ if there exist $c,d\in A$ such that $a\,c=b\,d$ and $s\,c=t\,d\in S$.

Proof. Suppose $A[S^{-1}]$ exists and let $\varphi\colon A\to A[S^{-1}]$ be the canonical map. We first verify $S1$. If $a\in A$ and $s\in S$, then $F2$ gives $\varphi(s)^{-1}\,\varphi(a)=\varphi(b)\,\varphi(t)^{-1}$ for some $b\in A$ and $t\in S$. Hence $\varphi(a\,t)=\varphi(s\,b)$, and by $F3$ this means that $a\,t\,u=s\,b\,u$

for some $u \in S$. Since $t\,u \in S$, this gives $S1$. If $s\,a=0$ with $s \in S$, then $\varphi(a)=0$ by $F1$, and $F3$ implies that $a\,t=0$ for some $t \in S$. Thus we also have $S2$.

Now suppose instead that $S1$ and $S2$ are satisfied. It is routine matter to verify that the stated relation $\sim$ is an equivalence relation. We form $A \times S/\sim$ and define addition and multiplication in the obvious way:

$$(a, s)+(b, t)=(a\,c+b\,d, u) \quad \text{where} \quad u=s\,c=t\,d \in S,$$

$$(a, s)\cdot(b, t)=(a\,c, t\,u) \quad \text{where} \quad s\,c=b\,u \quad \text{and} \quad u \in S.$$

One then has to verify that these definitions are independent of the choice of representing couples. This we leave to the industrious reader, as well as the verification that $A \times S/\sim$ is a ring and that $a \mapsto (a, 1)$ is a ring homomorphism $\varphi\colon A \to A \times S/\sim$. It is easy to see that we have arrived at a right ring of fractions: the inverse of $\varphi(s)$ is $(1, s)$ when $s \in S$, and $(a, s)=\varphi(a)\,\varphi(s)^{-1}$; finally, the kernel of φ is $\{a \in A \mid a\,s=0 \text{ for some } s \in S\}$. □

S is a *right denominator set* when it is multiplicatively closed and satisfies $S1$ and $S2$. Similarly one defines left and two-sided denominator sets. Sometimes S is called *right permutable* if it satisfies $S1$ and *right reversible* if it satisfies $S2$. We will now show that $S1$ implies $S2$ under certain circumstances. A right ideal of the form $r(S)=\{a \in A \mid S\,a=0\}$, for some subset S of A, is called a right *annihilator*.

Proposition 1.5. *Assume A satisfies* ACC *on right annihilators. If S is multiplicatively closed and satisfies $S1$, then S is a right denominator set.*

Proof. Suppose $s\,a=0$ for some $s \in S$. There exists an integer n such that $r(s^k)=r(s^n)$ for all $k \geq n$. By $S1$ one can find $t \in S$ and $b \in A$ such that $s^n\,b=a\,t$. Then $s^{n+1}\,b=s\,a\,t=0$, so $b \in r(s^{n+1})=r(s^n)$. Hence $a\,t=s^n\,b=0$, which establishes $S2$. □

Examples

1. Commutative rings. When A is a commutative ring, both $S1$ and $S2$ are automatically satisfied, so every multiplicatively closed set S is a denominator set.

2. Ore rings. The most important example of a multiplicatively closed set is the set S_{reg} of all *regular* elements (i.e. non-zero-divisors) of A. The ring of fractions $A[S_{\text{reg}}^{-1}]$ is usually called the *classical right ring of quotients* (or sometimes the *total right ring of fractions*) of A. We will denote it by $Q_{\text{cl}}^r(A)$, or more often simply by Q_{cl}. The corresponding left-hand notation is $Q_{\text{cl}}^l(A)$. Recall that when both $Q_{\text{cl}}^r(A)$ and $Q_{\text{cl}}^l(A)$ exist, they coincide. Since condition $S2$ is automatically satisfied, we have:

Proposition 1.6 (Ore [1]). *$Q_{\text{cl}}^r(A)$ exists if and only if A satisfies the right Ore condition, i.e. for a and s in A with s regular, there exist b and t in A with t regular, such that $a\,t=s\,b$.*

3. Ore domains. If A has no zero-divisors, the right Ore condition states that $a\,A \cap s\,A \neq 0$ for all non-zero elements a, s of A, and this is the same as saying that $\mathfrak{a} \cap \mathfrak{b} \neq 0$ for all non-zero right ideals $\mathfrak{a}$ and $\mathfrak{b}$. A ring without zero-divisors and satisfying the right Ore condition is called a *right Ore domain*, and its classical right ring of quotients is a skew-field. Numerous examples of Ore domains are obtained from:

Proposition 1.7. *Every right noetherian ring without zero-divisors is a right Ore domain.*

Proof. Let a and b be non-zero elements of A. Consider the right ideals $\mathfrak{a}_n = bA + abA + \cdots + a^n bA$. Since A is right noetherian, there exists a smallest n for which $\mathfrak{a}_n = \mathfrak{a}_{n+1}$, and then $a^{n+1} b = b c_0 + a b c_1 + \cdots + a^n b c_n$. This gives

$$b c_0 = a(a^n b - b c_1 - \cdots - a^{n-1} b c_n) \neq 0$$

where the minimality of n implies that the bracket is $\neq 0$. Thus $bA \cap aA \neq 0$, which is the right Ore condition. ▯

This result will be considerably improved in the next section.

4. Bezout domains. A *right Bezout domain* is a ring without zero-divisors in which every finitely generated right ideal is principal.

Proposition 1.8. *Every right Bezout domain is a right Ore domain.*

Proof. If A does not satisfy the Ore condition, then there exist non-zero elements a and b of A such that $aA \cap bA = 0$. Put $aA + bA = cA$ and write $b = cd$. Then $aA = aA/(aA \cap bA) \cong (aA + bA)/bA = cA/bA \cong A/dA$. Since $d \neq 0$, this is a contradiction. ▯

5. Skew polynomial rings. Our aim is to construct examples of rings which are Ore domains to the right but not to the left. Let K be a field and $\varphi\colon K \to K$ an endomorphism of K. The elements of the *skew polynomial ring* $K_\varphi[X]$ are formal polynomials

$$a_0 + X a_1 + \cdots + X^n a_n$$

with addition defined as usual, but with multiplication given by $aX = X\varphi(a)$, i.e.

$$X^m a \cdot X^n b = X^{m+n} \varphi^n(a) b.$$

This gives rise to a ring which contains K as a subring. Each element f of $K_\varphi[X]$ has a well-defined degree $\deg(f)$, and one has $\deg(fg) = \deg(f) + \deg(g)$. It follows in particular that $K_\varphi[X]$ has no zero-divisors. It is easily verified that $K_\varphi[X]$ is a right euclidean domain, i.e. for any f and g there exist q and r with $\deg r < \deg q$ such that $f = gq + r$. This implies that $K_\varphi[X]$ is a right principal domain and therefore a right Ore domain by Prop. 1.7 (or 1.8).

If φ is an automorphism of K, one can perform the euclidean division algorithm also to the left, and $K_\varphi[X]$ is a left Ore domain. But if there exists $a \in K$ which is not in the image of φ, then the left Ore condition fails for the couple (X, Xa), as one easily verifies. Thus $K_\varphi[X]$ is a left Ore domain if and only if φ is an automorphism of K.

6. Rings of quotients. A ring A is called a *ring of quotients* if every non-zero-divisor of A is invertible, i.e. A is its own classical right and left ring of quotients. E.g. every regular ring is a ring of quotients, as one easily sees. Also rings with minimum condition on right ideals are rings of quotients, in fact one has more generally:

Proposition 1.9. *If A satisfies* DCC *on principal right ideals, then A is a ring of quotients.*

Here DCC stands for the "descending chain condition", i.e. every strictly descending chain (of principal right ideals) is finite. The assertion is a consequence of:

Lemma 1.10. *Suppose A satisfies* DCC *on principal right ideals. The following properties of an element a are equivalent:*
(a) *a is invertible.*
(b) *a is regular.*
(c) $r(a)=0$.

Proof. Obviously (a) $\Rightarrow$ (b) $\Rightarrow$ (c). Assume the element a satisfies condition (c). Consider the descending chain $aA \supset a^2 A \supset \cdots$. By hypothesis, $a^n A = a^{n+1} A$ for some n. Thus $a^n = a^{n+1} b$ for some $b \in A$, and then $a^n(1-ab)=0$, and (c) implies $1=ab$. We then also have $aba=a$, and (c) implies $ba=1$. Hence (c) $\Rightarrow$ (a). ∎

Given a ring Q of quotients, one may be interested in investigating those subrings of Q for which Q is the classical right ring of quotients. Such subrings are called *right orders* in Q.

§2. Orders in a Semi-Simple Ring

In this section we prove a theorem by Goldie [2] which characterizes those rings which have a semi-simple classical ring of quotients. Before stating the theorem we have to make some definitions.

The ring A has *finite right rank* if there are no infinite direct sums of non-zero right ideals within A, i.e. A does not contain any right ideal of the form $\mathfrak{a} = \mathfrak{a}_1 \oplus \mathfrak{a}_2 \oplus \cdots$ with $\mathfrak{a}_i \neq 0$, $i \in \mathbb{N}$. Every right noetherian ring has of course finite right rank.

A right ideal $\mathfrak{a}$ is *nilpotent* if $\mathfrak{a}^n=0$ for some n, and is *nil* if every element of $\mathfrak{a}$ is nilpotent. The ring A is *prime* if there are no non-zero two-sided ideals $\mathfrak{a}$ and $\mathfrak{b}$ such that $\mathfrak{a}\mathfrak{b}=0$, and A is *semi-prime* if it has no non-zero nilpotent ideals. Finally, the right ideal $\mathfrak{a}$ is *essential* if $\mathfrak{a} \cap \mathfrak{b} \neq 0$ holds for every right ideal $\mathfrak{b} \neq 0$. We note:

Lemma 2.1. *The ring A is semi-simple if and only if it has no proper essential right ideals.*

Proof. If A is semi-simple, then every proper right ideal of A is a direct summand and can therefore not be essential. Conversely, suppose there are no essential right ideals $\neq A$. If $\mathfrak{a}$ is any right ideal of A, one can use Zorn's lemma to find a right ideal $\mathfrak{b}$ such that $\mathfrak{b}$ is maximal with the property that $\mathfrak{a} \cap \mathfrak{b} = 0$. But then $\mathfrak{a}+\mathfrak{b}$ is an essential right ideal of A, for if $\mathfrak{c}$ is a right ideal such that $(\mathfrak{a}+\mathfrak{b}) \cap \mathfrak{c} = 0$, then one has $\mathfrak{a} \cap (\mathfrak{b}+\mathfrak{c}) = 0$, which implies that $\mathfrak{c}=0$ by the maximality of $\mathfrak{b}$. It follows that $\mathfrak{a} \oplus \mathfrak{b} = A$, so every right ideal of A is a direct summand. Hence A is semi-simple. ∎

Theorem 2.2 (Goldie). *The following properties of the ring A are equivalent:*
(a) *A is a right order in a semi-simple ring.*
(b) *A has finite right rank, satisfies* ACC *on right annihilators and is a semi-prime ring.*
(c) *A right ideal of A is essential if and only if it contains a regular element.*

Proof. (a) ⇒ (b): Let Q be the semi-simple right ring of quotients of A. If $\mathfrak{a}$ and $\mathfrak{b}$ are right ideals of A such that $\mathfrak{a} \cap \mathfrak{b} = 0$, then it is easy to see that also $\mathfrak{a} Q \cap \mathfrak{b} Q = 0$. Since Q has finite right rank, this observation implies that also A has finite right rank. A satisfies ACC on right annihilators because $r_A(S) = r_Q(S) \cap A$ for any $S \subset A$, where the subscripts indicate in which ring the annihilator is taken.

Finally, let $\mathfrak{a}$ be a nilpotent two-sided ideal in A. $l(\mathfrak{a})$ is essential as a right ideal of A, because if $\mathfrak{b}$ is any non-zero right ideal of A, then there is a largest integer $n \geq 0$ such that $\mathfrak{b}\,\mathfrak{a}^n \neq 0$, and clearly $\mathfrak{b} \cap l(\mathfrak{a}) \supset \mathfrak{b}\,\mathfrak{a}^n \neq 0$. But if $l(\mathfrak{a})$ is essential in A, then $l(\mathfrak{a})Q$ is an essential right ideal in Q, as one easily verifies. Since Q is semi-simple, this means that $l(\mathfrak{a})\,Q = Q$, and we may write $1 = \sum a_i q_i$ with $a_i \in l(\mathfrak{a})$. There exists a regular element $s \in A$ such that all $q_i s \in A$. Then $s\,\mathfrak{a} = \sum a_i q_i s\,\mathfrak{a} \subset \sum a_i\,\mathfrak{a} = 0$, and it follows that $\mathfrak{a} = 0$.

(b) ⇒ (c): Every non-zero-divisor generates an essential right ideal:

Lemma 2.3. *If A has finite right rank and $r(s) = 0$, then sA is an essential right ideal.*

Proof. If $\mathfrak{a}$ is a non-zero right ideal such that $sA \cap \mathfrak{a} = 0$, then it is easy to see that one gets a direct sum $\mathfrak{a} \oplus s\,\mathfrak{a} \oplus \cdots \oplus s^n\,\mathfrak{a} \oplus \cdots$ in A. ☐

It requires more work to show that every essential right ideal contains a regular element. Here we follow the proof given by Goldie in [7]. We need:

Lemma 2.4. *If A is semi-prime and satisfies* ACC *on right annihilator ideals, then A has no right or left nil ideals $\neq 0$.*

Proof. Since Aa is nil if and only if aA is nil, it is enough to consider a left nil ideal Aa. Assume $Aa \neq 0$. Among the non-zero elements of Aa, choose one $b \in Aa$ with maximal right annihilator. For each $c \in A$, let k be the smallest integer such that $(c\,b)^k = 0$. Since $r(b) \subset r((c\,b)^{k-1})$, the choice of b implies that we actually have $r(b) = r((c\,b)^{k-1})$. It follows that $c\,b \in r(b)$ and hence $b\,A\,b = 0$ by the arbitrary choice of c. From this we obtain $(A\,b\,A)^2 = 0$, but every nilpotent ideal is zero, so $A\,b\,A = 0$. Hence $b = 0$, which is a contradiction, and we must have $Aa = 0$. ☐

Now let $\mathfrak{a}$ be an essential right ideal of A. We have to find a regular element in $\mathfrak{a}$. Since A has ACC on right annihilators and $\mathfrak{a}$ is not a nil ideal (by the preceding lemma), there exists $a_1 \neq 0$ in $\mathfrak{a}$ such that $r(a_1) = r(a_1^2)$. If $\mathfrak{a} \cap r(a_1) = 0$, we continue and choose $a_2 \in \mathfrak{a} \cap r(a_1)$ such that $a_2 \neq 0$ and $r(a_2) = r(a_2^2)$. If then $\mathfrak{a} \cap r(a_1) \cap r(a_2) \neq 0$, we go on and get $a_3 \in \mathfrak{a} \cap r(a_1) \cap r(a_2)$, and so on. At each step we obtain a direct sum $a_1 A \oplus \cdots \oplus a_k A$. This is proved by induction: suppose $a_1 A \oplus \cdots \oplus a_{k-1} A$ is direct and $a_k b = a_1 b_1 + \cdots + a_{k-1} b_{k-1}$; since for each $i < k$ we have $a_i a_k = 0$, we get $b_i \in r(a_i^2) = r(a_i)$ and hence $\sum_1^{k-1} a_i b_i = 0$. But A has finite right rank, so the process must stop at some stage, where we have $\mathfrak{a} \cap r(a_1) \cap \cdots \cap r(a_k) = 0$. Then $r(a_1) \cap \cdots \cap r(a_k) = 0$, so if $c = a_1 + \cdots + a_k \in \mathfrak{a}$, then $r(c) = 0$.

c is thus the kind of element we are looking for, but it remains to show that also $l(c) = 0$. Note that cA is an essential right ideal by Lemma 2.3. Define the *right singular ideal* of A as $Z = \{a \in A \mid r(a) \text{ is an essential right ideal of } A\}$. We leave as an exercise to show that Z is a two-sided ideal. Since $l(c) \subset Z$ and A contains no nilpotent ideals $\neq 0$, we may conclude that $l(c) = 0$ by using:

Lemma 2.5. *If A satisfies* ACC *on right annihilators, then the right singular ideal is nilpotent.*

Proof. We will show that the ascending chain $r(Z)\subset r(Z^2)\subset\cdots$ would be strictly ascending if Z were not nilpotent. If $Z^n\neq 0$, choose an element $a\in Z$ with $Z^{n-1}a\neq 0$ and largest possible right annihilator. For each $b\in Z$ we have $r(b)\cap aA\neq 0$ since $r(b)$ is essential in A. So there exists $c\in A$ such that $ac\neq 0$, but $bac=0$. This means that $r(ba)$ is strictly larger than $r(a)$, and by the choice of a we must therefore have $Z^{n-1}ba=0$. Since $b\in Z$ is arbitrary, we get $Z^n a=0$ and hence $r(Z^{n-1})$ is strictly contained in $r(Z^n)$. □

(c) ➡ (a): We first verify the right Ore condition. Let s be a regular element and a arbitrary in A. Since sA is essential, also $(sA:a)=\{c\,|\,ac\in sA\}$ is an essential right ideal, as is easily checked, and it therefore contains a regular element t. Thus $at=sb$ for some $b\in A$.

To prove that $Q=Q_{\mathrm{cl}}(A)$ is semi-simple, it suffices by Lemma 2.1 to show that Q has no proper essential right ideals. If $\mathfrak{a}$ is any right ideal of Q, then $(\mathfrak{a}\cap A)\,Q=\mathfrak{a}$ since every element $q\in\mathfrak{a}$ can be written as $q=as^{-1}$ with $a=qs\in\mathfrak{a}\cap A$. If furthermore $\mathfrak{a}$ is an essential right ideal of Q, then $\mathfrak{a}\cap A$ is an essential right ideal of A and thus contains a regular element t. Since t is invertible in Q, $\mathfrak{a}=(\mathfrak{a}\cap A)\,Q=Q$. This concludes the proof of the theorem. □

As a special case of Goldie's theorem we obtain a characterization of those rings A for which $Q_{\mathrm{cl}}(A)$ is a simple ring (Goldie [1], Lesieur and Croisot [2]):

Proposition 2.6. *The following properties of A are equivalent:*
(a) *A is a right order in a simple ring.*
(b) *A has finite right rank, satisfies* ACC *on right annihilators and is a prime ring.*

Proof. (a) ⇒ (b): It only remains to show that A is a prime ring. Let $\mathfrak{a}$ and $\mathfrak{b}$ be two-sided ideals of A with $\mathfrak{a}\,\mathfrak{b}=0$. Then $\mathfrak{b}Q\mathfrak{a}\cap A$ is a two-sided ideal of A and $(\mathfrak{b}Q\mathfrak{a}\cap A)^2=0$. Since A has no non-zero nilpotent ideals by the theorem, we have $\mathfrak{b}Q\mathfrak{a}\cap A=0$. But this clearly implies that $\mathfrak{b}Q\mathfrak{a}=0$. Since Q is a simple ring, it is prime, and therefore $\mathfrak{b}$ or $\mathfrak{a}$ is zero.

(b) ⇒ (a): Since every semi-simple prime ring is simple, it suffices to show that Q is a prime ring. Suppose $\mathfrak{a}$ and $\mathfrak{b}$ are two-sided ideals of Q such that $\mathfrak{a}\,\mathfrak{b}=0$. Then $(\mathfrak{a}\cap A)(\mathfrak{b}\cap A)=0$, and A prime implies that $\mathfrak{a}\cap A$ or $\mathfrak{b}\cap A$ is zero, which means that $\mathfrak{a}$ or $\mathfrak{b}$ is zero. □

Examples

1. Noetherian rings. Every semi-prime right noetherian ring has a semi-simple classical right ring of quotients.

2. Group rings. Let A be a commutative integral domain with field of fractions K, and let G be a finite group. Then $K[G]$ is an artinian ring, and is therefore a ring of quotients by Prop. 1.9. Clearly $A[G]$ is a two-sided order in $K[G]$. By Prop. I.7.9, $K[G]$ is a semi-simple ring if the characteristic of K does not divide the order of G.

§3. Modules of Fractions

Let S be a right denominator set in the ring A. For each right A-module M we define its *module of fractions* with respect to S as $M[S^{-1}] = M \otimes_A A[S^{-1}]$ with its canonical structure as a right $A[S^{-1}]$-module. This is a reasonable definition in view of the following two results:

Proposition 3.1. *The canonical A-linear map $\mu_M: M \to M[S^{-1}]$ has the following universal property: for each right $A[S^{-1}]$-module N and A-linear map $\alpha: M \to N$ there exists a unique $A[S^{-1}]$-linear map $\sigma: M[S^{-1}] \to N$ such that $\sigma\,\mu_M = \alpha$.*

Proof. $\mathrm{Hom}_{A[S^{-1}]}(M \otimes_A A[S^{-1}],\ N) \cong \mathrm{Hom}_A(M,\ \mathrm{Hom}_{A[S^{-1}]}(A[S^{-1}],\ N)) \cong \mathrm{Hom}_A(M, N)$. □

Proposition 3.2. *$M[S^{-1}] \cong M \times S/\sim$, where $\sim$ is the equivalence relation defined as $(x, s) \sim (y, t)$ if there exist $c, d \in A$ such that $x\,c = y\,d$ and $s\,c = t\,d \in S$.*

Proof. This is proved similarly to Prop. 1.4, by verifying that $M \times S/\sim$ solves the universal problem stated in Prop. 3.1. The operations in $M \times S/\sim$ are defined as:

$(x, s) + (y, t) = (x\,c + y\,d, u)$ where $u = s\,c = t\,d \in S$ (using $S1$),

$(x, s) \cdot (b, t) = (x\,c, t\,u)$ for $(b, t) \in A[S^{-1}]$, where $s\,c = b\,u$ and $u \in S$. □

Corollary 3.3. *The kernel of $\mu_M: M \to M[S^{-1}]$ consists of those $x \in M$ for which there exist $s \in S$ with $x\,s = 0$.*

From this corollary it follows that $t(M) = \{x \in M \mid x\,s = 0 \text{ for some } s \in S\}$ is a submodule of M, called the *S-torsion submodule*. M is an *S-torsion module* if $M = t(M)$, i.e. if $M[S^{-1}] = 0$, and is *S-torsion-free* if $t(M) = 0$.

Lemma 3.4. *$M/t(M)$ is S-torsion-free.*

Proof. If $x \in M$ and $\bar{x} \in t(M/t(M))$, then $x\,s \in t(M)$ for some $s \in S$, and this means that $x\,s\,t = 0$ for some $t \in S$. Hence $x \in t(M)$ and $\bar{x} = 0$. □

The assignment $M \mapsto M[S^{-1}]$ is a functor **Mod**-$A \to$ **Mod**-$A[S^{-1}]$. A basic fact is that this functor is exact, or in other words:

Proposition 3.5. *$A[S^{-1}]$ is flat as a left A-module.*

Proof. Let $\alpha: L \to M$ be a monomorphism in **Mod**-A. We must show that the induced homomorphism $L[S^{-1}] \to M[S^{-1}]$ is a monomorphism. Suppose $x\,s^{-1}$ (with $x \in L$, $s \in S$) has the property that $\alpha(x)\,s^{-1} = 0$ in $M[S^{-1}]$. By Prop. 3.2 this means that $\alpha(x)\,c = 0$ in M for some $c \in A$ such that $s\,c \in S$. Since α is a monomorphism, this implies $x\,c = 0$ in L, and then $x\,s^{-1} = 0$ in $L[S^{-1}]$. □

Obviously we have:

Proposition 3.6. *If M is a right $A[S^{-1}]$-module, then the canonical map $M \to M[S^{-1}]$ is an isomorphism.*

This result means that the category of modules of fractions, considered as a full subcategory of **Mod**-A, is "equivalent" (in a sense to be made precise in Chap. IV) to the category **Mod**-$A[S^{-1}]$.

We want to have an intrinsic characterization of those A-modules which are modules of fractions, i.e. may be considered as $A[S^{-1}]$-modules.

Definitions. (i) An A-module M is *S-injective* if for each right ideal $\mathfrak{a}$ of A such that $\mathfrak{a} \cap S \neq \emptyset$ and each homomorphism $\alpha: \mathfrak{a} \to M$, there exists $x \in M$ such that $\alpha(a) = x\,a$ for all $a \in \mathfrak{a}$.

(ii) M is *S-divisible* if $M = M\,s$ for each $s \in S$.

Proposition 3.7. *Assume S is a right denominator set in A. The following properties of a right A-module M are equivalent:*

(a) $M \cong M[S^{-1}]$.

(b) *M is S-torsion-free and S-injective.*

(c) *M is S-torsion-free and S-divisible.*

Proof. (a) ⇒ (c): The condition (a) means that M is an $A[S^{-1}]$-module. If now $x \in M$ and $x\,s = 0$ for some $s \in S$, then $x = x\,s \cdot s^{-1} = 0$. Hence M is S-torsion-free. For each $x \in M$ and $s \in S$ we can write $x = x\,s^{-1} \cdot s \in M\,s$, so M is S-divisible.

(c) ⇒ (a): M S-torsion-free implies that M is a submodule of $M[S^{-1}]$. For each $x \otimes s^{-1} \in M[S^{-1}]$ we can write $x = y\,s$ for some $y \in M$ by S-divisibility. Then $x \otimes s^{-1} = y \otimes s s^{-1} = y \otimes 1$, and so the map $M \to M[S^{-1}]$ is an epimorphism also.

(c) ⇒ (b): Suppose we are given $\alpha: \mathfrak{a} \to M$, where $\mathfrak{a}$ is a right ideal with $s \in \mathfrak{a} \cap S$. Write $\alpha(s) = x\,s$ for some $x \in M$. We want to show that $\alpha(a) = x\,a$ holds for all $a \in \mathfrak{a}$. By the condition $S1$ we may write $a\,t = s\,b$ for some $t \in S$ and $b \in A$. Then $\alpha(a)\,t = \alpha(s)\,b = x\,s\,b = x\,a\,t$, and M S-torsion-free implies that $\alpha(a) = x\,a$.

(b) ⇒ (c): Given $x \in M$ and $s \in S$, we define a homomorphism $\alpha: s\,A \to M$ by $\alpha(s\,a) = x\,a$. α is well-defined because if $s\,a = s\,b$, then $a\,t = b\,t$ for some $t \in S$ by $S2$, and $x\,a\,t = x\,b\,t$ implies $x\,a = x\,b$ since M is S-torsion-free. By hypothesis there exists $y \in M$ such that $\alpha(s) = y\,s$, and thus $x = y\,s$. This shows that $M = M\,s$. ▯

Examples

1. S_{reg}-torsion. The definition of S_{reg}-torsion elements was first proposed by Levy [1]. Assume S_{reg} is a right denominator set in A. Instead of "S_{reg}-torsion" and "S_{reg}-divisible", one uses the shorter terms "torsion" and "divisible". Note that every S_{reg}-injective module is divisible, because the homomorphism $\alpha: s\,A \to M$, given by $\alpha(s\,a) = x\,a$, is always well-defined when s is regular, and can be extended to A. Also, every flat module is torsion-free (by the argument used in Example I.10.1).

2. Orders in a semi-simple ring. Suppose A has a classical right ring of quotients Q. Then Q is semi-simple if and only if S_{reg}-injective modules are injective:

Proposition 3.8. *The following properties of a ring A with a classical right ring of quotients Q are equivalent:*

(a) *Every torsion-free and divisible right A-module is injective.*

(b) *Every right Q-module is injective over A.*

(c) *Every S_{reg}-injective right A-module is injective.*

(d) *Q is a semi-simple ring.*

Proof. (a) $\Leftrightarrow$ (b) by Prop. 3.7, and also (c) $\Rightarrow$ (b) follows from this proposition.

(b) $\Rightarrow$ (d): Let $\mathfrak{q}$ be any right ideal in Q. Since $\mathfrak{q}$ is injective over A, Q splits as an A-module as $Q=\mathfrak{q}\oplus K$. Now K is torsion-free and divisible, and is therefore a Q-module. Q thus splits as a right Q-module, so Q must be a semi-simple ring.

(d) $\Rightarrow$ (c): Let M be an S_{reg}-injective module. Consider any homomorphism $\alpha\colon \mathfrak{a}\to M$ from a right ideal $\mathfrak{a}$ of A. Let $\mathfrak{b}$ be a right ideal which is maximal with respect to $\mathfrak{a}\cap\mathfrak{b}=0$. α can trivially be extended to $\alpha'\colon \mathfrak{a}+\mathfrak{b}\to M$. But $\mathfrak{a}+\mathfrak{b}$ is an essential right ideal of A (see the proof of Lemma 2.1) and therefore contains a regular element s (Theorem 2.2). Since M is S_{reg}-injective, α' can be extended to the whole of A. □

§4. Invertible Ideals and Hereditary Orders

Let A be a right order in the ring Q. As usual we let S_{reg} denote the set of regular elements of A. For each right A-submodule I of Q we put $I^*=\{q\in Q \mid qI\subset A\}$.

Proposition 4.1. *If I is a right A-submodule of Q with $I\cap S_{\text{reg}}\neq\emptyset$, then $I^*\cong \operatorname{Hom}_A(I, A)$.*

Proof. If $q\in I^*$, then $x\mapsto qx$ defines a homomorphism $I\to A$. Conversely, suppose $\varphi\colon I\to A$ is a homomorphism. Let $s\in I\cap S_{\text{reg}}$. For each $x\in I$ there exists $t\in S_{\text{reg}}$ with $xt\in A$, and by the Ore condition there exists $b\in A$ and $u\in S_{\text{reg}}$ such that $sb=xtu$. Then

$$\varphi(x)\,tu=\varphi(xtu)=\varphi(sb)=\varphi(s)\,b=\varphi(s)\,s^{-1}sb=\varphi(s)\,s^{-1}xtu,$$

which implies $\varphi(x)=\varphi(s)\,s^{-1}x$ with $\varphi(s)\,s^{-1}\in I^*$. □

Definition. A right A-submodule I of Q is *right invertible* if there exist $a_1,\dots,a_n\in I$ and $q_1,\dots,q_n\in I^*$ such that $1=\sum a_i q_i$.

Note that under these circumstances we have for each $x\in I$ that $x=\sum a_i q_i x$ with $q_i x\in A$, so $a_1,\dots,a_n$ generate the module I. Thus:

Proposition 4.2. *Every right invertible A-submodule of Q is finitely generated.*

Proposition 4.3. *A right A-submodule I of Q is right invertible if and only if $I\cap S_{\text{reg}}\neq\emptyset$ and I is a projective module.*

Proof. Suppose I is right invertible, i.e. $1=\sum a_i q_i$ with $a_i\in I$ and $q_i I\subset A$. Define homomorphisms $\varphi_i\colon I\to A$ as $\varphi_i(x)=q_i x$. The family (φ_i, a_i) clearly satisfies the requirements of Prop. I.6.3, and therefore makes I into a finitely generated projective module. Choose $s\in S_{\text{reg}}$ such that all $q_i s\in A$. Then $s=\sum a_i q_i s\in I\cap S_{\text{reg}}$.

Assume conversely that I is a projective A-module with $I\cap S_{\text{reg}}\neq\emptyset$. Let $(\varphi_j, a_j)_J$ be a family of projective coordinates for I. Each $\varphi_j\colon I\to A$ is of the form $\varphi_j(x)=q_j x$ for some $q_j\in I^*$. If $s\in I\cap S_{\text{reg}}$, then $s=\sum a_j\varphi_j(s)=\sum a_j q_j s$, and s regular implies $1=\sum a_j q_j$. Hence I is right invertible. □

A divisible module has the following injectivity property with respect to invertible right ideals:

Proposition 4.4. *Suppose A is a two-sided order in Q. Let $\mathfrak{a}$ be a right invertible right ideal of A and let M be a divisible module. For each homomorphism $\alpha: \mathfrak{a} \to M$ there exists $x \in M$ such that $\alpha(a) = xa$ for all $a \in \mathfrak{a}$.*

Proof. Let $a_1, \ldots, a_n \in \mathfrak{a}$ and $q_1, \ldots, q_n \in I^*$ with $\sum a_i q_i = 1$. Since A is assumed to be also a left order in Q, there exists $s \in S_{\text{reg}}$ such that all $sq_i \in A$. The divisibility of M implies that we can write $\alpha(a_i) = x_i s$ with $x_1, \ldots, x_n \in M$. For every $a \in \mathfrak{a}$ we obtain

$$\alpha(a) = \alpha(\textstyle\sum a_i q_i a) = \sum \alpha(a_i)\, q_i a = \sum x_i s q_i a = xa$$

with $x = \sum x_i s q_i \in M$. ▯

This result is used to prove the following characterization of hereditary orders (due to Levy [1]):

Proposition 4.5. *Suppose A is a two-sided order in a semi-simple ring. The following properties of A are equivalent:*

(a) *A is right hereditary.*

(b) *Every right ideal containing a regular element is right invertible.*

(c) *Every divisible module is injective.*

Proof. (a) ⇒ (b) is a consequence of Prop. 4.3.

(b) ⇒ (c): Every divisible module is S_{reg}-injective by Prop. 4.4, and is therefore injective by Prop. 3.8.

(c) ⇒ (a): Every quotient module of an injective module is divisible, and is therefore injective by (c). Hence A is right hereditary by Prop. I.9.5. ▯

The following is another important property of orders in semi-simple rings:

Proposition 4.6. *If A is a two-sided order in a semi-simple ring, then every finitely generated torsion-free module is a submodule of a free module.*

Proof. Let M be a torsion-free module with generators $x_1, \ldots, x_m$. By Cor. 3.3 there is a monomorphism $M \to M \otimes_A Q$, where Q is the classical right ring of quotients of A. Now $M \otimes_A Q$ is a finitely generated Q-module, and since Q is semi-simple, this implies that $M \otimes_A Q$ is a submodule of a finitely generated free Q-module F. Let $v_1, \ldots, v_n$ be a basis for F over Q. Then $x_j = \sum_i v_i q_{ij}$ for certain $q_{ij} \in Q$. Choose $s \in S_{\text{reg}}$ such that all $sq_{ij} \in A$. Then $x_j = \sum_i v_i s^{-1} \cdot sq_{ij}$, so M is imbedded in the free A-submodule of F generated by $v_1 s^{-1}, \ldots, v_n s^{-1}$. ▯

Corollary 4.7. *Let A be a two-sided order in a semi-simple ring. The following properties of A equivalent:*

(a) *A is right semi-hereditary.*

(b) *Every finitely generated torsion-free module is projective.*

(c) *Every torsion-free module is flat.*

(d) *Every right ideal is flat.*

Proof. (a) ⇒ (b): Every finitely generated torsion-free module is a submodule of a free module (Prop. 4.6), and therefore projective (Prop. I.6.9).

(b) ⇒ (c): Every module is the direct limit of its finitely generated submodules, and the direct limit of projective modules is flat (Prop. I.10.3).

(c) $\Rightarrow$ (d) is trivial.
(d) $\Rightarrow$ (a) follows from Prop. I.11.6. □

Examples

1. Dedekind domains. When A is a commutative integral domain, Prop. 4.5 gives:

Proposition 4.8. *The following properties of a commutative integral domain A are equivalent:*
(a) *A is hereditary.*
(b) *Every non-zero ideal is invertible.*
(c) *Every divisible module is injective.*

A ring satisfying these conditions is called a *Dedekind domain.* Every Dedekind domain is a noetherian ring (by Prop. 4.2), and its invertible modules form an abelian group under multiplication.

2. Prüfer domains. A semi-hereditary commutative integral domain is called a *Prüfer domain.*

Exercises

§1

Let S be a right denominator set in A, and $\varphi\colon A \to A[S^{-1}]$ the canonical homomorphism.

1. Show that $\varphi\colon A \to A[S^{-1}]$ is an epimorphism in the category of rings (as defined in Exercise I.17).
2. Show that if $q_1, \ldots, q_n \in A[S^{-1}]$, then there exists $s \in S$ such that $q_1 s, \ldots, q_n s \in \varphi(A)$.
3. Let A be a right Ore domain. Show that also the polynomial ring $A[X]$ is a right Ore domain.
4. Consider S as a subset of $\mathrm{M}_n(A)$ in the obvious way (diagonal matrices). Show that:
 (i) S is a right denominator set of $\mathrm{M}_n(A)$.
 (ii) $\mathrm{M}_n(A)[S^{-1}] \cong \mathrm{M}_n(A[S^{-1}])$.
 (iii) $\mathrm{UT}_n(A)[S^{-1}] \cong \mathrm{UT}_n(A[S^{-1}])$.
5. Show that if A is injective both as a right and as a left A-module, then A is a ring of quotients.

§2

6. Let S and T be subsets of A. Show that:

$$r(l(r(S))) = r(S),$$

$$r(S \cup T) = r(S) \cap r(T),$$

$$S \subset T \quad \text{implies} \quad r(S) \supset r(T).$$

7. Show that A satisfies ACC on right annihilators if and only if A satisfies DCC on left annihilators.
8. Show that if A is a prime ring, then every two-sided ideal is essential in A.
9. Show that
 (i) $Q_{\mathrm{cl}}(\mathrm{M}_n(A)) \cong \mathrm{M}_n(Q_{\mathrm{cl}}(A))$.
 (ii) $Q_{\mathrm{cl}}(\mathrm{UT}_n(A)) \cong \mathrm{UT}_n(Q_{\mathrm{cl}}(A))$.
 (Cf. Exercise 4.)

§3

10. Let A be a right Ore domain with skew-field of fractions Q. Show that A is a left Ore domain if and only if Q is flat as a right A-module. (Hint: the Ore condition for a, b in A is obtained by noting that the relation $1 \cdot a = ab^{-1} \cdot b$ in Q comes from a relation in A by flatness.)

11. Let S be a right denominator set in A. Let M be a right A-module and $\mu\colon M \to M[S^{-1}]$ the canonical homomorphism. A submodule L of M is called *S-saturated* in M if it satisfies: if $x \in M$ and $xs \in L$ for some $s \in S$, then $x \in L$. Show that:

(i) There is a bijective correspondence between S-saturated submodules of M and $A[S^{-1}]$-submodules of $M[S^{-1}]$, given by $L \mapsto L[S^{-1}]$ and $N \mapsto \mu^{-1}(N)$.

(ii) M is S-torsion-free if and only if $\operatorname{Ann}(x)$ is S-saturated in A for every $x \in M$.

(iii) If A is a right noetherian ring, then so is also $A[S^{-1}]$.

12. Let S be a right denominator set in A. Show that:

(i) If M is a right A-module and N is a right $A[S^{-1}]$-module, then

$$\operatorname{Hom}_{A[S^{-1}]}(M[S^{-1}], N) \cong \operatorname{Hom}_A(M, N).$$

(ii) A right $A[S^{-1}]$-module is injective if and only if it is injective as a right A-module.

13. Suppose S_{reg} is a right denominator set in A. Show that the following conditions are equivalent:

(a) A is a ring of quotients.

(b) There are no non-zero torsion modules.

(c) Every direct product of torsion modules is a torsion module.

14. Let A be a right order in Q. Show that:

(i) Every finitely generated right ideal of Q is of the form $\mathfrak{a}Q$, where $\mathfrak{a}$ is a finitely generated right ideal of A.

(ii) If A is right semi-hereditary, then so is also Q.

§4

15. Let A be a right order in Q. Show that if every divisible right A-module is injective, then A is right hereditary and Q is semi-simple.

16. Let R be a commutative integral domain with field of fractions K. Let Q be a semi-simple K-algebra (Exercise I.1), which is finite-dimensional over K.

(i) Let I be an R-submodule of Q. Show that the following properties of I are equivalent:

(a) I contains a basis for Q as a vector space over K.

(b) $IK = Q$.

(c) For each $q \in Q$ there exists $0 \neq r \in R$ such that $rq \in I$.

(ii) Let A be an R-subalgebra of Q. Show that A is an order in Q if and only if $AK = Q$.

(iii) Suppose R is a noetherian ring, and let A be an R-subalgebra of Q such that A is finitely generated as an R-module and $AK = Q$ (A is then called an *R-order* in Q). Show that the following properties of a right A-submodule I of Q are equivalent:

(a) I is finitely generated as an R-module, and $IK = Q$.

(b) There exist non-zero $r, s \in R$ such that $r \in I$ and $sI \subset A$.

17. Show that if A is a right hereditary ring and is a right order in a semi-simple ring, then A is right noetherian.

Chapter III. Modular Lattices

We now return to the development of general tools to be used in the study of rings of quotients. The theory of modular lattices is one such tool, dealing with the abstract aspects of the relation of inclusion between submodules of a module. In the context of rings of quotients, this theory will be indispensable when we shall elucidate the connection between submodules of a given module M and the submodules of a module of quotients of M. It may also be remarked that one of the immediate examples of a ring of quotients, not obtainable as a ring of fractions in the sense of Chap. II, is furnished by the completion of boolean algebras, which is a lattice theoretical phenomenon.

§ 1. Lattices

Let P be a partially ordered set, with the partial ordering denoted by $\leq$, and S a subset of P. An *upper bound* for S in P is an element $x \in P$ such that $s \leq x$ for all $s \in S$. An element $s_0 \in S$ is a *greatest* element in S if $s \leq s_0$ for all $s \in S$. There can be at most one greatest element in S. Similarly one defines *lower bound* and *least* element. In particular, a *least upper bound* for S is a least element in the set of upper bounds for S, and similarly for *greatest lower bound* for S.

A *lattice* is a partially ordered set in which every couple of elements x, y has a least upper bound (called the *join* of x and y, and written $x \vee y$) and a greatest lower bound (called the *meet* of and x and y, and written $x \wedge y$). It follows then by induction that every non-empty finite set of elements has a join and a meet.

Suppose L is a lattice. If one reverses the partial ordering in L, i.e. replaces $\leq$ by $\geq$, and interchanges $\vee$ and $\wedge$, then one obtains a new lattice, called the *opposite* (or *dual*) lattice of L and denoted by L^{op}. As in the case of categories, this leads to a duality principle, by which every concept or result for lattices carries with it a dual concept or result.

If L and L' are lattices, then a *morphism* $\alpha: L \to L'$ is a map from L to L' satisfying

$$\alpha(x \vee y) = \alpha(x) \vee \alpha(y), \qquad \alpha(x \wedge y) = \alpha(x) \wedge \alpha(y) \quad \text{for } x, y \in L.$$

In this way one obtains the category **Lat** of lattices. It is a subcategory of the category of partially ordered sets, where the morphisms are the order-preserving maps (we call them *order morphisms*). Since the lattice operations are defined in terms of the ordering, one obtains:

Proposition 1.1. *If L and L′ are lattices, then every order isomorphism $L \to L'$ is a lattice isomorphism.*

A lattice morphism $L^{op} \to L'$ is often said to be an *anti-morphism* from L to L'.

If L is a lattice, then a *sublattice* of L is a subset L' of L such that $x, y \in L'$ implies $x \vee y \in L'$ and $x \wedge y \in L'$; L' is then itself a lattice.

A lattice L is *complete* if every subset S of L has a least upper bound, written $\sup S$ or $\bigvee_{s \in S} s$ and called the *join* of S, and a greatest lower bound, written $\inf S$ or $\bigwedge_{s \in S} s$ and called the *meet* of S. In a complete lattice there exists a greatest element $\sup L$, denoted by 1, and a smallest element $\inf L$, denoted by 0. Note that we also have $1 = \inf \emptyset$ and $0 = \sup \emptyset$.

Proposition 1.2. *If L is a partially ordered set and every subset of L has a least upper bound in L, then L is a complete lattice.*

Proof. For each subset S of L, consider the set B of all lower bounds of S in L. (B is non-empty because $0 = \sup \emptyset \in L$.) Let x be the least upper bound of B in L. If $s \in S$, then for each $b \in B$ we have $b \leq s$, so $s \geq \sup B = x$. If y is any element such that $s \geq y$ for all $s \in S$, then $y \in B$ and hence $y \leq \sup B = x$. It follows that x is a greatest lower bound for S. ▯

If L is a complete lattice, then a *complete sublattice* of L is a sublattice L' of L such that $S \subset L'$ implies that $\sup S$ and $\inf S$ (taken in L!) belong to L'; L' is then itself a complete lattice.

Examples

1. Intervals. Let L be a lattice and a, b elements of L with $a \leq b$. Then $[a, b] = \{x \in L \mid a \leq x \leq b\}$ is a sublattice of L, called the *interval* between a and b.

2. Ideals and filters. If L is a lattice, then a subset I of L is called an *ideal* if it satisfies:

(i) $a \in I$, $x \leq a$ implies $x \in I$;
(ii) $a, b \in I$ implies $a \vee b \in I$;
(iii) If L has 0, then $0 \in I$.

In particular, for any $a \in L$ one obtains the *principal ideal* $\{x \in L \mid x \leq a\}$. Dually one defines the notion of filter: a subset F of L is a *filter* if:

(i) $a \in F$, $a \leq x$ implies $x \in F$;
(ii) $a, b \in F$ implies $a \wedge b \in F$;
(iii) If L has 1, then $1 \in F$.

3. The lattice of submodules. If M is a module, then the submodules of M form a complete lattice, which we will denote by $\mathsf{L}(M)$. When $K \subset L \subset M$, the interval $[K, L]$ in $\mathsf{L}(M)$ is isomorphic to the lattice of submodules of L/K. The lattices of right, resp. left, ideals of the ring A will be denoted by $\mathsf{L}(A.)$ and $\mathsf{L}(.A)$.

4. Completions. If L is a lattice, then an *upper* (resp. *lower*) *completion* of L is a complete lattice $\hat{L}$, containing L as a sublattice and such that $x = \sup(L \cap [0, x])$ (resp. $x = \inf(L \cap [x, 1])$) for every $x \in \hat{L}$. If $\hat{L}$ is both an upper and a lower completion of L, then $\hat{L}$ is a *completion* of L. E.g. the chain of real numbers is a completion of the rational numbers. One may show that every lattice has a unique completion, the so-called Dedekind-MacNeille completion (for the existence, see §8; for the unicity, see Schmidt [1]).

5. Closure systems. Let L be a complete lattice. If C is a subset of L, and C is closed under arbitrary meets (i.e. $S \subset C$ implies $\inf S \in C$), then it follows from Prop. 1.2 that C is a complete lattice. Such a subset C of L is called a *closure system* in L. Note that the meet in C is the same as in L, but that the join in C is given by

$$\sup_C S = \inf\{\text{upper bounds of } S \text{ in } C\}.$$

§2. Modularity

Let L be a lattice. The operations $\vee$ and $\wedge$ are associative and commutative. Furthermore one clearly has for any $a \le b$ in L that

$$(x \wedge b) \vee a \le (x \vee a) \wedge b \quad \text{for all } x \in L.$$

The lattice is *modular* if the reverse inequality also holds, i.e.

$$(c \wedge b) \vee a = (c \vee a) \wedge b \quad \text{for all } a, b, c \text{ with } a \le b.$$

Note that every sublattice of a modular lattice is modular.

Proposition 2.1. *Let a and b be elements of a modular lattice. Then there is a lattice isomorphism $[a \wedge b, a] \to [b, a \vee b]$.*

Proof. Define α: $[a \wedge b, a] \to [b, a \vee b]$ as $\alpha(x) = x \vee b$, and define β: $[b, a \vee b] \to [a \wedge b, a]$ as $\beta(y) = y \wedge a$. Then $\beta\alpha(x) = (x \vee b) \wedge a = (a \wedge b) \vee x = x$ by modularity, since $x \le a$. Dually, $\alpha\beta(y) = (y \wedge a) \vee b = (b \vee a) \wedge y = y$. The map β is thus an inverse of α. Clearly α preserves joins, and it preserves meets because its inverse β does so. □

This result leads to a refinement of the notion of isomorphism between intervals of a modular lattice. Two intervals of L are *similar* if there exist elements a, b of L such that one of the intervals can be written as $[a \wedge b, a]$ and the other one as $[b, a \vee b]$. Two intervals I and J of L are *projective* if there exists a chain $I = I_0$, $I_1, \ldots, I_n = J$ of intervals such that I_{i-1} and I_i are similar. Projective intervals are of course isomorphic as lattices.

Let L be a lattice with 0 and 1. If $a \in L$, then a *complement* of a in L is an element $c \in L$ such that

$$a \wedge c = 0, \qquad a \vee c = 1.$$

The lattice L is *complemented* if every element of L has a complement in L.

Proposition 2.2. *If L is a complemented modular lattice, then every interval of L is complemented.*

Proof. Let $a \le b$ in L and $d \in [a, b]$, and suppose d has a complement c in L. Then one verifies that

$$a \vee (c \wedge b) = b \wedge (a \vee c)$$

is a complement of d in $[a, b]$. □

The modularity of L depends on a unicity property of relative complements:

Proposition 2.3. *The lattice L is modular if and only if every interval I of L has the following property: if $c \in I$ has two complements a, b in I with $a \leq b$, then $a = b$.*

Proof. If L is modular, then so is every interval, so for the necessity we may assume $I = L$. Then

$$b = b \wedge 1 = b \wedge (a \vee c) = a \vee (b \wedge c) = a \vee 0 = a.$$

Conversely, if a, b, c are elements of L with $a \leq b$, then we have the modular inequality

$$a_1 = (c \wedge b) \vee a \leq (c \vee a) \wedge b = a_2.$$

Then $a_1 \wedge c = \big((c \wedge b) \vee a\big) \wedge c \geq (c \wedge a) \vee (c \wedge b) = c \wedge b$, and $a_1 \leq b$ implies $a_1 \wedge c = c \wedge b$. Also $a_2 \wedge c = (c \vee a) \wedge b \wedge c = b \wedge c$. Further we have $a_1 \vee c = (c \wedge b) \vee a \vee c = a \vee c$, and finally $a_2 \vee c = \big((c \vee a) \wedge b\big) \vee c \leq (b \vee c) \wedge (c \vee a) = a \vee c$, and $a \leq a_2$ implies $a_2 \vee c = a \vee c$. Thus a_1 and a_2 are complements of c in $[b \wedge c, a \vee c]$, and by hypothesis they must be equal, which proves the modularity of L. ▯

Examples

1. The lattice of submodules. If M is a module, then the lattice $\mathsf{L}(M)$ of submodules is modular. For let K, L and L' be submodules of M such that $L' \subset L$. If $x \in (K + L') \cap L$, then $x = y + z$ with $y \in K$ and $z \in L'$. But $y = x - z \in K \cap L$, so $x \in (K \cap L) + L'$, which shows that $(K + L') \cap L = (K \cap L) + L'$.

If two intervals $[K', K]$ and $[L', L]$ in $\mathsf{L}(M)$ are projective, then it follows from the Noether isomorphism theorem (Prop. I.1.5), applied to each one of the similar pairs of intervals connecting $[K', K]$ and $[L', L]$, that the quotient modules K/K' and L/L' are isomorphic.

A member of $\mathsf{L}(M)$ has a complement if and only if it is a direct summand of M. The lattice $\mathsf{L}(M)$ is complemented if and only if M is a semi-simple module (Prop. I.7.2).

2. Coherent rings. If A is a right coherent ring, then the finitely generated right ideals form a sublattice of the lattice $\mathsf{L}(A.)$ (by Prop. I.13.3(e)). In particular, when A is a regular ring, the principal right ideals form a complemented modular sublattice of $\mathsf{L}(A.)$.

§3. Lattices with Chain Condition

Let L be a modular lattice with elements 0 and 1. Two chains

$$a = a_0 \leq a_1 \leq \cdots \leq a_m = b, \tag{1}$$

$$a = b_0 \leq b_1 \leq \cdots \leq b_n = b \tag{2}$$

between the same pair of elements of L are called *equivalent* if $m = n$ and there is a permutation π of $\{1, \ldots, n\}$ such that the intervals $[a_{i-1}, a_i]$ and $[b_{\pi(i)-1}, b_{\pi(i)}]$ are projective. A *refinement* of a chain (1) is obtained by inserting further elements in the chain. The following result is known as the "Schreier refinement theorem":

Proposition 3.1. *Any two finite chains between the same pair of elements in a modular lattice have equivalent refinements.*

Proof. Let (1) and (2) be the given chains. For each $i=1, \dots, m$ and $j=1, \dots, n$ we put

$$a_{ij}=(a_i \wedge b_j) \vee b_{j-1}=(a_i \vee b_{j-1}) \wedge b_j,$$

$$b_{ji}=(b_j \wedge a_i) \vee a_{i-1}=(b_j \vee a_{i-1}) \wedge a_i.$$

The interval

$$[(a_{i-1} \vee b_{j-1}) \wedge a_i \wedge b_j, a_i \wedge b_j] \tag{3}$$

is then similar to $[a_{i-1,j}, a_{ij}]$, because we have

$$a_{i-1,j} \wedge (a_i \wedge b_j)=(a_{i-1} \vee b_{j-1}) \wedge b_j \wedge a_i \wedge b_j=(a_{i-1} \vee b_{j-1}) \wedge a_i \wedge b_j,$$

while $a_{i-1} \wedge b_j \le a_i$ gives

$$\begin{aligned} a_{i-1,j} \vee (a_i \wedge b_j) &= (a_{i-1} \wedge b_j) \vee b_{j-1} \vee (a_i \wedge b_j) \\ &= a_i \wedge ((a_{i-1} \wedge b_j) \vee b_j) \vee b_{j-1}=(a_i \wedge b_j) \vee b_{j-1}=a_{ij}. \end{aligned}$$

By symmetry, the interval (3) is also similar to $[b_{j-1,i}, b_{ji}]$. So $[b_{j-1,i}, b_{ji}]$ is projective to $[a_{i-1,j}, a_{ij}]$, and it follows that the two chains

$$a=a_{01} \le a_{11} \le \dots \le a_{m1} \le a_{12} \le \dots \le a_{mn}=b,$$

$$a=b_{01} \le b_{11} \le \dots \le b_{n1} \le b_{12} \le \dots \le b_{nm}=b,$$

where $a_{mj}=b_j=a_{0,j+1}$ and $b_{ni}=a_i=a_{0,i+1}$, are equivalent. They are refinements of (2) and (1) respectively. □

A *composition chain* between a and b is a chain

$$a=a_0<a_1<\dots<a_m=b \tag{4}$$

which has no refinement, except by introducing repititions of the given elements a_i. The integer m is the *length* of the chain. From Prop. 3.1 we immediately get the "Jordan-Hölder theorem":

Corollary 3.2. *Any two composition chains between the same pair of elements in a modular lattice are equivalent.*

A modular lattice L is of *finite length* if there is a composition chain between 0 and 1, and by Cor. 3.2 we can define the *length* of L to be the length of such a composition chain. From Prop. 3.1 we may also conclude:

Proposition 3.3. *In a modular lattice of finite length, every chain* (4) *can be refined to a composition chain.*

A lattice L is *noetherian* (or *satisfies* ACC) if there is no infinite strictly ascending chain $a_0<a_1<\cdots$ in L, and is *artinian* (or *satisfies* DCC) if there is no infinite strictly descending chain $a_0>a_1>\cdots$ in L. These chain conditions can also be formulated as maximum (minimum) conditions. If S is a subset of L, then a *maximal* element of S is an element $a \in S$ such that $a \le x$ for $x \in S$ implies $x=a$. Dually one defines *minimal* elements of S.

Proposition 3.4. *A lattice L is noetherian (artinian) if and only if every non-empty subset of L has a maximal (minimal) element.*

Proof. It is quite clear that the maximum condition implies ACC, and the converse implication follows with the help of Zorn's lemma. ▯

Proposition 3.5. *A modular lattice is of finite length if and only if it is both noetherian and artinian.*

Proof. If L is of finite length m, then every strictly ascending (or strictly descending) chain in L consists of at most $m+1$ elements, so L is noetherian and artinian. Suppose conversely that L is noetherian. For every $a \neq 0$ in L there exists by Prop. 3.4 a maximal element b such that $b < a$. By repeated use of this observation we get a descending chain $1 > a_1 > a_2 \dots$. If now L also is artinian, this chain stops after a finite number of steps, and we have thus obtained a composition chain between 1 and 0. L is thus of finite length. ▯

Proposition 3.6. *Let a be an element of a modular lattice L. Then L is noetherian (artinian) if and only if both intervals* $[0, a]$ *and* $[a, 1]$ *are noetherian (artinian).*

Proof. If L is noetherian or artinian, then clearly every interval of L is like-wise. Suppose conversely that the intervals $[0, a]$ and $[a, 1]$ are noetherian, and let $b_1 < b_2 < \cdots$ be a strictly ascending chain in L. Then there exists an integer n such that

$$b_n \wedge a = b_{n+1} \wedge a = c,$$

$$b_n \vee a = b_{n+1} \vee a = d.$$

Applying Prop. 2.3 to the element a in $[c, d]$, we obtain $b_n = b_{n+1}$. Hence L must be noetherian. Similarly for the artinian case. ▯

Examples

1. Modules of finite length. Let M be a module. A composition chain between $\{0\}$ and M in $\mathsf{L}(M)$ is a chain of submodules

$$\{0\} = M_0 \subset M_1 \subset \cdots \subset M_n = M$$

such that each quotient module M_i/M_{i-1} is simple. In view of the meaning of projectivity between intervals in $\mathsf{L}(M)$ (Example 2.1), the Jordan-Hölder theorem takes the form:

Proposition 3.7. *Let*

$$0 = M_0 \subset M_1 \subset \cdots \subset M_n = M$$

and

$$0 = M'_0 \subset M'_1 \subset \cdots \subset M'_m = M$$

be two composition chains. Then $n = m$, *and there exists a permutation* π *of* $\{1, \dots, n\}$ *such that* $M_i/M_{i-1} \cong M_{\pi(i)}/M_{\pi(i)-1}$.

A module M is called *noetherian* (*artinian*, of *finite length*) when the lattice $\mathsf{L}(M)$ is noetherian (etc.). From Prop. 3.6 we obtain (cf. Prop. I.3.4):

Proposition 3.8. *Let $0 \to M' \to M \to M'' \to 0$ be an exact sequence of modules. Then M is noetherian (artinian resp. of finite length) if and only if both M' and M'' are so.*

2. Finite-dimensional vector spaces. A vector space V is of finite length if and only if it is finite-dimensional, and its length is the dimension of V.

3. Irreducible decomposition. An element a of the lattice L is *(meet-) irreducible* if $a = b \wedge c$ implies $a = b$ or $a = c$. An *irreducible decomposition* of a in L is obtained by writing a as the meet of a finite number of irreducible elements in L.

Proposition 3.9. *Every element in a noetherian lattice has an irreducible decomposition.*

Proof. If there are elements without an irreducible decomposition, we can choose a maximal such element a. Then a is reducible, i.e. $a = b \wedge c$ for some $b > a$ and $c > a$. The elements b and c then have irreducible decompositions, and so has then also $a = b \wedge c$, a contradiction. □

Every irreducible decomposition $a = a_1 \wedge \cdots \wedge a_n$ may be assumed to be *irredundant*, i.e. $a_i \geq \bigwedge_{j \neq i} a_j$ does not hold for any i. For a modular lattice it can be proved that two irredundant irreducible decompositions

$$a = a_1 \wedge \cdots \wedge a_n = b_1 \wedge \cdots \wedge b_m$$

must have $n = m$ (see e.g. Cohn [3], p. 76). In the case of a lattice of submodules, a stronger unicity theorem will be proved in Chap. V (Cor. V.5.11).

§4. Distributive Lattices

Proposition 4.1. *The following conditions on a lattice L are equivalent:*

(a) $(a \wedge b) \vee c = (a \vee c) \wedge (b \vee c)$ *for all a, b, c in L.*

(b) $(a \vee b) \wedge c = (a \wedge c) \vee (b \wedge c)$ *for all a, b, c in L.*

(c) $(a \vee c) \wedge b \leq (a \wedge b) \vee c$ *for all a, b, c in L.*

Proof. (a) ⇒ (c): Since always $b \leq c \vee b$, we get $(a \vee c) \wedge b \leq (a \vee c) \wedge (c \vee b) = (a \wedge b) \vee c$.

(c) ⇒ (a): $a \wedge b \leq a$ implies $(a \wedge b) \vee c \leq a \vee c$, and similarly we have $(a \wedge b) \vee c \leq b \vee c$, so $(a \wedge b) \vee c \leq (a \vee c) \wedge (b \vee c)$ always holds. From (c) we get $(a \wedge b) \vee c \geq ((a \vee c) \wedge b) \vee c \geq (a \vee c) \wedge (b \vee c)$.

(a) ⇔ (b) follows by the duality principle. □

A lattice is called *distributive* if it satisfies the equivalent conditions of this Proposition. From condition (c) we get in particular:

Corollary 4.2. *Every distributive lattice is modular.*

Proposition 4.3. *An element in a distributive lattice can have at most one complement.*

Proof. Suppose L is a distributive lattice with 0 and 1, and that $a \in L$ has the complements b and c. Then

$$c = c \wedge (a \vee b) = (a \wedge c) \vee (b \wedge c) = b \wedge c,$$

which implies $c \leq b$. By symmetry we also have $b \leq c$, so $b = c$. □

A complemented distributive lattice (with 0 and 1) is called a *boolean algebra.* For each element a in a boolean algebra L we let a^* denote the unique complement of a in L.

Proposition 4.4. *The following properties of a lattice L are equivalent:*
(a) *L is a boolean algebra.*
(b) *Each $a \in L$ has a unique complement a^*, and $a \wedge b = 0$ holds if and only if $b \leq a^*$.*

Proof. (a) ⇒ (b): If $a \wedge b = 0$, then

$$b = b \wedge 1 = b \wedge (a \vee a^*) = (b \wedge a) \vee (b \wedge a^*) = b \wedge a^*,$$

so $b \leq a^*$. The converse is clear.

(b) ⇒ (a): First note that $a = a^{**}$ for every $a \in L$, since both are complements of a^*. If a and b are elements of L, then $x \in L$ satisfies $a \wedge x \leq b$ if and only if $a \wedge x \wedge b^* = 0$. Therefore there exists a greatest element x satisfying $a \wedge x \leq b$, namely $x = (a \wedge b^*)^*$, which we denote by $b : a$.

Now let a, b, c be any three elements of L, and consider $d = (a \wedge c) \vee (b \wedge c)$. Since $a \wedge c \leq d$ and $b \wedge c \leq d$, we have $a \leq d:c$ and $b \leq d:c$. Hence $a \vee b \leq d:c$, which gives

$$c \wedge (a \vee b) \leq c \wedge (d : c) \leq d = (a \wedge c) \vee (b \wedge c).$$

This implies distributivity, since the converse inequality always holds. □

In a complete boolean algebra, arbitrary joins and meets are distributive:

Proposition 4.5. *If L is a complete boolean algebra, then*

$$\Big(\bigvee_I a_i\Big) \wedge c = \bigvee_I (a_i \wedge c),$$

$$\Big(\bigwedge_I a_i\Big) \vee c = \bigwedge_I (a_i \vee c)$$

hold for an arbitrary family $(a_i)_I$ and c in L.

Proof. The two equalities hold in any distributive lattice for a finite family $(a_i)_I$, by induction. In a complete boolean algebra it suffices by duality to prove the first equality. It is clear that

$$\bigvee_I (a_i \wedge c) \leq \Big(\bigvee_I a_i\Big) \wedge c$$

always holds. To obtain the converse inequality, we show that $\big(\bigvee_I a_i\big) \wedge c \leq u$ for every upper bound u of the set $\{a_i \wedge c\}$. From $a_i \wedge c \leq u$ we get

$$a_i = (a_i \wedge c) \vee (a_i \wedge c^*) \leq u \vee c^*,$$

and hence

$$\Big(\bigvee_I a_i\Big) \wedge c \leq (u \vee c^*) \wedge c = u \wedge c \leq u,$$

as desired. □

A *subalgebra* of a boolean algebra L is a sublattice L' of L such that $0 \in L'$ and such that $a \in L'$ implies $a^* \in L'$. A *homomorphism* between two boolean algebras A and B is a lattice morphism $\alpha\colon A \to B$ such that $\alpha(a^*) = \alpha(a)^*$ for all $a \in A$.

Examples

1. Boolean algebra of subsets. If S is a set, then the set $\mathsf{P}(S)$ of all subsets of S is a boolean algebra in an obvious way. Every subset of $\mathsf{P}(S)$ containing $\emptyset$ and closed under complements and finite unions is a subalgebra of $\mathsf{P}(S)$.

2. Arithmetical rings. A commutative ring with a distributive lattice of ideals $\mathsf{L}(A)$ is called *arithmetical*. It has been proved by C.U. Jensen [1] that A is arithmetical if and only if for every maximal ideal $\mathfrak{m}$ of A, the ideals of the local ring $A_{\mathfrak{m}}$ are totally ordered under inclusion.

3. Central idempotents. We recall that a central idempotent of a ring A is an element e belonging to the center of A such that $e^2 = e$.

Proposition 4.6. *The central idempotents of a ring A form a boolean algebra* $\mathsf{B}(A)$.

Proof. We define a partial ordering on the set $\mathsf{B}(A)$ of central idempotents by setting $e \leq f$ if $ef = e$. Then $\mathsf{B}(A)$ becomes a complemented lattice with

$$e \wedge f = ef, \qquad e \vee f = e + f - ef, \qquad e^* = 1 - e,$$

as one easily verifies. It follows from Prop 4.4 that $\mathsf{B}(A)$ is a boolean algebra. □

4. Boolean rings. Recall that a commutative ring is said to be boolean if every element is idempotent (Example I.12.3). Every boolean ring A becomes a boolean algebra by modifying the operations:

$$a \vee b = a + b - ab, \qquad a \wedge b = ab.$$

Since this is just a particular case of Prop. 4.6, we write $\mathsf{B}(A)$ when A is considered as a boolean algebra.

Conversely, every boolean algebra L becomes a boolean ring by defining

$$a + b = (a \wedge b^*) \vee (a^* \wedge b), \qquad ab = a \wedge b.$$

This gives a bijective correspondence between boolean rings and boolean algebras. It is easily verified that under this correspondence, the notions of homomorphism and ideal for boolean rings are carried over to the same notions for boolean algebras.

5. Strongly regular rings. If A is a strongly regular ring, then every idempotent is central (Lemma I.12.2), and we obtain:

Proposition 4.7. *If A is a strongly regular ring, then the principal right ideals of A form a boolean algebra which is isomorphic to* $\mathsf{B}(A)$.

Proof. It is enough to show that the map $e \mapsto eA$ is an isomorphism between $\mathsf{B}(A)$ and the lattice of principal right ideals. We first note that $eA \subset fA$ means that $e = fx$ for some $x \in A$, which gives $fe = f \cdot fx = fx = e$. Thus $eA \subset fA$ implies $e \leq f$. It follows that the map $e \mapsto eA$ is a bijection, and that it is an order isomorphism, because we also have that $e \leq f$ implies $eA \subset fA$. The map is then a lattice isomorphism by Prop. 1.1. □

The converse statement is also true:

Proposition 4.8. *If A is a regular ring such that the lattice of principal right ideals is distributive, then A is strongly regular.*

Proof. It suffices to show that every idempotent e is central. For arbitrary $a \in A$, consider the principal right ideal $\mathfrak{a} = (e + (1-e)\,a\,e)\,A$. Distributivity implies that

$$(e A \cap \mathfrak{a}) \oplus ((1-e)\,A \cap \mathfrak{a}) = \mathfrak{a},$$

and $e + (1-e)\,a\,e \in \mathfrak{a}$ then gives $e \in e A \cap \mathfrak{a}$, $(1-e)\,a\,e \in (1-e)\,A \cap \mathfrak{a}$. Hence $e = e\,x + (1-e)\,a\,e\,x$ for some $x \in A$, and multiplication from the left with e gives $e = e\,x$, and so $(1-e)\,a\,e = 0$. Thus $a\,e = e\,a\,e$, and replacing e by $1-e$, we also get $e\,a = e\,a\,e$. It follows that $a\,e = e\,a$. ▯

§5. Continuous Lattices

Let L denote a complete lattice throughout this section. The lattice L is *upper continuous* if for every directed subset D of L and every $a \in L$, one has

$$\Big(\bigvee_{d \in D} d\Big) \wedge a = \bigvee_{d \in D} (d \wedge a).$$

If the opposite lattice L^{op} is upper continuous, then L is said to be *lower continuous*. Finally, one calls L *continuous* if it is both upper and lower continuous. The following criterion for upper continuity of complemented lattices is of purely technical interest, but it will be needed in Chap. XII:

Proposition 5.1. *Suppose L is modular and complemented. Then L is upper continuous if and only if it satisfies the following condition:*

if D is a directed subset of L and there exists a non-zero $c \in L$ such that $c \wedge d = 0$ for all $d \in D$, then $\sup D < 1$.

Proof. If L is upper continuous, and $c \neq 0$ has the property that $c \wedge d = 0$ for all $d \in D$, then $\big(\bigvee_D d\big) \wedge c = \bigvee_D (d \wedge c) = 0$, which shows that $\bigvee_D d \neq 1$.

Conversely, assume that the condition is satisfied. Let $a \in L$ and D be a directed subset of L. Since $d \wedge a \leq \big(\bigvee_D d\big) \wedge a$ for every $d \in D$, one has at least $\bigvee_D (d \wedge a) \leq \big(\bigvee_D d\big) \wedge a$. By Prop. 2.2 there exists $c \in L$ such that

$$\bigvee_D (d \wedge a) \vee c = \Big(\bigvee_D d\Big) \wedge a$$

and

$$\Big(\bigvee_D (d \wedge a)\Big) \wedge c = 0.$$

Also let b be a complement of $\bigvee_D d$ in L. Then for each $d \in D$ we have

$$(d \vee b) \wedge c = (d \vee b) \wedge \Big(\bigvee_D d\Big) \wedge c = \Big(\Big(\Big(\bigvee_D d\Big) \wedge b\Big) \vee d\Big) \wedge c = d \wedge c$$

$$= d \wedge a \wedge c \leq \bigvee_D (d \wedge a) \wedge c = 0.$$

From the hypothesis now follows that if $c \neq 0$, then

$$\left(\bigvee_D d\right) \vee b = \bigvee_D (d \vee b) < 1,$$

which is a contradiction. Hence $c = 0$, as desired. □

An element c of L is *compact* if whenever $c \leq \bigvee_{d \in D} d$ for a directed subset D of L, there is a $d_0 \in D$ such that $a \leq d_0$. (This terminology is of course inspired from the Heine-Borel covering theorem for compact spaces.) If L is upper continuous, then c is compact if and only if it has the property that whenever $a = \bigvee_{d \in D} d$ for a directed subset D of L, there is a $d_0 \in D$ such that $a = d_0$. The lattice L is *compact* if 1 is a compact element, and is *compactly generated* if every element of L is a join of compact elements. This join may be taken over a directed set, because of:

Lemma 5.2. *If a and b are compact, then $a \vee b$ is compact.*

Proof. If D is directed and $a \vee b \leq \bigvee_D d$, then $a \leq d_0$ and $b \leq d_1$ for some d_0, d_1 in D. Since D is directed, there is a $d_2 \in D$ such that $a \vee b \leq d_0 \vee d_1 \leq d_2$. □

Proposition 5.3. *Every compactly generated lattice is upper continuous.*

Proof. Suppose L is compactly generated, and let D be a directed subset of L. For each $a \in L$ we trivially have $\bigvee_D (a \wedge d) \leq a \wedge \left(\bigvee_D d\right)$, so it remains to show the converse inequality. To do this, it suffices to show that $c \leq \bigvee_D (a \wedge d)$ holds for every compact c such that $c \leq a \wedge \left(\bigvee_D d\right)$. But $c \leq \bigvee_D d$ implies that $c \leq d_0$ for some $d_0 \in D$, and then $c \leq a \wedge d_0$, so one has $c \leq \bigvee_D (a \wedge d)$. □

A non-zero element a of L is an *atom* if $b < a$ implies $b = 0$, and L is *locally atomic* if every element of L is a join of atoms.

Proposition 5.4. *Every locally atomic, upper continuous and modular lattice L is complemented.*

Proof. Let a be an element of L. By applying Zorn's lemma and using the upper continuity of L, one shows that there exists a maximal family $(s_i)_I$ of atoms such that $a \wedge \left(\bigvee_I s_i\right) = 0$ (cf. the proof of Prop. I.7.1). Write $c = \bigvee_I s_i$. In order to prove that c is a complement of a, it suffices to show that $a \vee c$ contains all atoms of L. Suppose s is an atom such that $s \nleq a \vee c$. Then $s \wedge (a \vee c) = 0$. By the modularity of L one has

$$a \wedge (c \vee s) \leq (a \vee c) \wedge (c \vee s) = \big((a \vee c) \wedge s\big) \vee c = c,$$

and hence $a \wedge (c \vee s) \leq a \wedge c = 0$. This contradicts the maximality of the family $(s_i)_I$. □

In order to obtain a more complete analogy with the case of semi-simple modules (Prop. I.7.2), we have to assume that L is compactly generated:

Proposition 5.5. *The following properties of a modular lattice L are equivalent:*
(a) *L is compactly generated and locally atomic.*
(b) *L is compactly generated and complemented.*
(c) *L is upper continuous and locally atomic.*

Proof. (a) $\Rightarrow$ (b) is a consequence of Props. 5.3 and 5.4.

(b) $\Rightarrow$ (c): If L is compactly generated, then it is upper continuous by Prop. 5.3. If moreover L is complemented, then to prove that L is locally atomic, it suffices to show that every compact element contains an atom (cf. the proof of Prop. I.7.2). This is done in the same way as for modules.

(c) $\Rightarrow$ (a): It suffices to prove:

Proposition 5.6. *If L is upper continuous, then every atom is compact.*

Proof. Suppose a is an atom and D is a directed subset of L such that $a \leq \bigvee_{d \in D} d$. If $a \nleq d$ for each $d \in D$, then $a \wedge d = 0$, and it follows that $a = a \wedge (\bigvee_D d) = \bigvee_D (a \wedge d) = 0$, which is a contradiction. ▯

Examples

1. The lattice of submodules. If M is a module, then the lattice $\mathsf{L}(M)$ of submodules is upper continuous, but it is in general not continuous. The compact elements of $\mathsf{L}(M)$ are the finitely generated submodules of M, and $\mathsf{L}(M)$ is thus compactly generated. $\mathsf{L}(M)$ is locally atomic if and only if M is a semi-simple module.

2. Boolean algebras. Every complete boolean algebra is continuous (Prop. 4.5). E.g. if S is an arbitrary set, then the boolean algebra $\mathsf{P}(S)$ of all subsets of S is continuous and locally atomic.

§6. Pseudo-Complemented Lattices

Throughout this section we assume L to be a modular lattice with 0 and 1. We define a *pseudo-complement* of an element $a \in L$ to be an element c such that $a \wedge c = 0$ and c is maximal with this property. The lattice L is *pseudo-complemented* if for every interval $I \subset L$ and every $a \in I$, there is a pseudo-complement of a in I.

Lemma 6.1. *Every complement is a pseudo-complement.*

Proof. Let c be a complement of a, and assume $a \wedge c' = 0$ for some $c' \geq c$. Then $c = (a \wedge c') \vee c = (a \vee c) \wedge c' = 1 \wedge c' = c'$, so c is maximal with $a \wedge c = 0$. ▯

Corollary 6.2. *Every complemented modular lattice is pseudo-complemented.*

A very useful observation is:

Proposition 6.3. *Every upper continuous modular lattice is pseudo-complemented.*

Proof. Let d belong to an interval $[a, b]$ in L, and consider the set C of all c in $[a, b]$ with $c \wedge d = a$. Using the upper continuity of L, one sees that every totally ordered subset of C has an upper bound in C. Hence Zorn's lemma can be applied to give a maximal element in C, which will be a pseudo-complement of d in $[a, b]$. □

If c is a pseudo-complement of a in L, then $a \vee c$ turns out to be an element which is quite large in L, in the following sense:

Definitions. An element a is *essential* in L if $a \wedge c \neq 0$ holds for every $c \neq 0$ in L. More generally, if $a \leq b$ in L, then b is an *essential extension* of a if a is essential in $[0, b]$. Finally, a is *essentially closed* in L if a has no essential extension $\neq a$ in L.

Note that the essential elements form a filter in L, i.e. if a and b are essential, then $a \wedge b$ is essential, and every $c \geq a$ is essential.

Proposition 6.4. *If c is a pseudo-complement of a in L, then $a \vee c$ is essential in L.*

Proof. Suppose $d \in L$ and $(a \vee c) \wedge d = 0$. Then $a \wedge (c \vee d) \leq (a \vee c) \wedge (c \vee d) = ((a \vee c) \wedge d) \vee c = c$ with the help of modularity. Since $a \wedge c = 0$, this gives $a \wedge (c \vee d) = 0$, and the maximality of c implies that $c \vee d = c$ and hence $d = 0$. □

Proposition 6.5. *Suppose L is pseudo-complemented. If b is a pseudo-complement of a in L, then there exists a pseudo-complement c of b in L such that $a \leq c$. Furthermore, c is then a maximal essential extension of a in L.*

Proof. Let c be a pseudo-complement of $a \vee b$ in the lattice $[a, 1]$. Then $a \leq c$, and c is clearly a pseudo-complement of b in L. To show that a is essential in $[0, c]$, assume that $d \leq c$ and $d \wedge a = 0$. Then $a \wedge (b \vee d) = a \wedge c \wedge (b \vee d) = a \wedge (d \vee (b \wedge c)) = a \wedge d = 0$, and the maximality of b implies that $d = 0$. Finally we show that c is maximal as an essential extension of a in L. If c' is an element of L such that $c < c'$, then $c' \wedge b \neq 0$ since c is maximal with respect to $c \wedge b = 0$, but $c' \wedge b \wedge a \leq b \wedge a = 0$, so a is not essential in $[0, c']$. □

Proposition 6.6. *Suppose L is pseudo-complemented. An element a of L is a pseudo-complement of some element in L if and only if a is essentially closed in L.*

Proof. If a has no essential extension $\neq a$, then $a = c$ in the context of Prop 6.5, so a is a pseudo-complement of b. Suppose conversely that a is a pseudo-complement of some $b \in L$. Choose a pseudo-complement b' of a such that $b \leq b'$. Then a is clearly a pseudo-complement also of b', and Prop. 6.5 tells us that a is a maximal essential extension of itself in L. □

There is a connection between essential elements and atoms which is worth mentioning:

Proposition 6.7. *Assume L is modular and compactly generated. The join of all atoms of L is equal to the meet of all essential elements of L.*

Proof. If a is an atom and b is essential in L, then $a \wedge b \neq 0$ implies $a \leq b$, so the join s of all atoms is smaller than the meet e of all essential elements. If we now can show that the interval $[0, e]$ is complemented, then it will follow from Prop. 5.5 that e is a join of atoms, and hence $s = e$. Suppose $a \leq e$. Choose a pseudo-complement c of a in L. Then $a \vee c$ is essential in L by Prop. 6.4, so $e \leq a \vee c$. It follows that $c \wedge e$ is a complement of a in $[0, e]$, because $(c \wedge e) \vee a = (a \vee c) \wedge e = e$ by modularity. □

Examples

1. The lattice of submodules. If M is a module, then the lattice $\mathsf{L}(M)$ of submodules of M is pseudo-complemented by Prop. 6.3. A submodule L of M is *essential* in M if it is an essential element of $\mathsf{L}(M)$, and that is the case if and only if for every $x \neq 0$ in M there exists $a \in A$ such that $0 \neq x\,a \in L$. Prop. 6.7 states that the socle of M is equal to the intersection of all essential submodules of M.

2. Artinian lattices. Suppose L is an artinian lattice. The minimality condition implies that for each non-zero a in L there exists an atom s such that $s \leq a$. The join of all atoms of L is therefore an essential element of L, in fact it is the smallest essential element of L.

3. Distributive lattices. If L is a distributive lattice, then an element $a \in L$ can have at most one pseudo-complement in L, for if $a \wedge b = a \wedge c = 0$, then also $a \wedge (b \vee c) = 0$. A pseudo-complemented, distributive lattice is called a *Brouwer lattice* (Birkhoff [1]).

§7. Closure Operators

Let L be a complete lattice. A *closure operator* on L is a mapping $\Phi: L \to L$, written $\Phi(a) = a^c$, such that:

($Cl1$) $a \leq b$ implies $a^c \leq b^c$;

($Cl2$) $a \leq a^c$;

($Cl3$) $(a^c)^c = a^c$.

An element a is *closed* under Φ if $a = a^c$.

Proposition 7.1. *The closed elements form a complete lattice L^c.*

Proof. If (a_i) is a family of closed elements, then $\bigwedge a_i \leq a_i$ implies $(\bigwedge a_i)^c \leq a_i^c = a_i$, and hence $(\bigwedge a_i)^c \leq \bigwedge a_i$. It follows that $(\bigwedge a_i)^c = \bigwedge a_i$, so $\bigwedge$ is a meet operation in L^c. By Prop. 1.1 we may conclude that L^c is a complete lattice. □

In general, L^c is not a sublattice of L. The meet in L^c is the same as in L, but the join $\overline{\bigvee}$ in L^c is given by

$$\overline{\bigvee} a_i = (\bigvee a_i)^c$$

for an arbitrary family (a_i) of closed elements. The closed elements form a "closure system" in L, in the sense of Example 1.5. Conversely, if S is a closure system in L, then one may define a closure operator as

$$a^c = \inf\{s \in S \mid a \leq s\}.$$

In this way one establishes a bijective correspondence between closure operators and closure systems.

A closure operator may satisfy one of the additional axioms:

($Cl4$) $(a \vee b)^c = a^c \vee b^c$;

($Cl5$) $(a \wedge b)^c = a^c \wedge b^c$.

When $Cl4$ is satisfied, L^c is a sublattice of L. This will rarely be the case in the coming applications to ring theory, where instead $Cl5$ is of interest.

Proposition 7.2. *If L is a complete modular lattice with a closure operator satisfying Cl5, then also L^c is modular.*

Proof. Suppose a, b, c are closed elements with $a \leq b$. Then

$$b \wedge (c \,\bar{\vee}\, a) = b^c \wedge (c \vee a)^c = (b \wedge (c \vee a))^c = (a \vee (c \wedge b))^c = a \,\bar{\vee}\, (c \wedge b). \quad \square$$

A similar result holds for distributivity. A closure operator is called *finitary* if for every directed subset D of L one has

$$\bigvee_{d \in D} d^c = \Big(\bigvee_{d \in D} d\Big)^c,$$

or in other words, if every directed join of closed elements is a closed element. Since this in particular implies that directed joins in L^c are the same as in L, it follows:

Proposition 7.3. *If L is an upper continuous lattice with a finitary closure operator, then L^c is upper continuous.*

Proposition 7.4. *Let Φ be a finitary closure operator on a compactly generated lattice L. A closed element a is compact in L^c if and only if $a = b^c$ for a compact element b in L.*

Proof. Suppose a is compact in L^c. We can write $a = \bigvee_D d$, where D is a directed family of compact elements in L. Then $a = a^c = \bigvee d^c$, and since a is compact in L^c, we get $a = d^c$ for some $d \in D$. Conversely, if b is compact in L, and D is a directed set of closed elements with $b^c \leq \bigvee_D d$, then $b \leq d$ for some $d \in D$, and then also $b^c \leq d$. Hence b^c is compact in L^c. $\square$

Corollary 7.5. *If Φ is a finitary closure operator on a compactly generated lattice L, then also L^c is compactly generated.*

Examples

1. S-saturated submodules. Let S be a right denominator set in a ring A. For each module M and submodule L of M, put

$$L^c = \{x \in M \mid x\, s \in L \text{ for some } s \in S\}.$$

This defines a closure operator on $\mathsf{L}(M)$, which satisfies $Cl5$ and is finitary. The closed submodules are called *S-saturated;* they correspond to the $A[S^{-1}]$-submodules of $M[S^{-1}]$ (see Exercise II.11).

2. Topological closure. If X is a topological space, then the closed subsets form a closure system in the lattice $\mathsf{P}(X)$ of subsets of X. The corresponding closure operator satisfies $Cl4$.

§8. Galois Connections

Let L and L' be complete lattices. A *Galois connection* between L and L' is a pair of mappings $\sigma\colon L \to L'$ and $\tau\colon L' \to L$, satisfying:

(1) $x_1 \leq x_2$ implies $\sigma(x_1) \geq \sigma(x_2)$ for $x_1, x_2 \in L$.
(2) $y_1 \leq y_2$ implies $\tau(y_1) \geq \tau(y_2)$ for $y_1, y_2 \in L'$.
(3) $x \leq \tau\,\sigma(x)$ and $y \leq \sigma\,\tau(y)$ for $x \in L$, $y \in L'$.

From condition (3), applied to $y = \sigma(x)$, we obtain $\sigma(x) \leq \sigma\,\tau\,\sigma(x)$. But applying (1) to $x \leq \tau\,\sigma(x)$, we also get $\sigma(x) \geq \sigma\,\tau\,\sigma(x)$. Hence we have

$$\sigma\,\tau\,\sigma(x) = \sigma(x), \qquad \tau\,\sigma\,\tau(y) = \tau(y) \quad \text{for all } x \in L,\ y \in L'.$$

It follows that the maps $\tau\,\sigma$ and $\sigma\,\tau$ are closure operators on L and L', resp. The closed elements in L are those which are of the form $\tau(y)$ for some $y \in L'$. Furthermore, it is clear that σ and τ induce an anti-isomorphism between the corresponding lattices of closed elements.

There are many important closure operators which arise by Galois connections. Of special interest for us will be one particular example, namely that of annihilator ideals, which we will discuss in some detail. Let A be a ring. There is a Galois connection between right and left ideals of A, given by l: $\mathsf{L}(A.) \to \mathsf{L}(.A)$ and r: $\mathsf{L}(.A) \to \mathsf{L}(A.)$, defined as

$$l(\mathfrak{a}) = \{a \in A \mid a\,\mathfrak{a} = 0\}, \qquad r(\mathfrak{b}) = \{a \in A \mid \mathfrak{b}\,a = 0\}.$$

The right annihilator ideals $r(\mathfrak{b})$ thus form a complete lattice $\mathsf{Ann}(A.)$, which is anti-isomorphic to the lattice $\mathsf{Ann}(.A)$ of left annihilator ideals $l(\mathfrak{a})$.

If a right ideal $\mathfrak{a}$ is generated by an idempotent e, then $\mathfrak{a}$ is a right annihilator ideal, in fact $\mathfrak{a} = r(A(1-e))$. In many rings, all annihilator ideals arise in this way. We note:

Lemma 8.1. *The following two conditions are equivalent for a ring:*
(a) *Every right annihilator ideal is generated by an idempotent.*
(b) *Every left annihilator ideal is generated by an idempotent.*

Proof. If (a) holds, then for every subset S of the ring A we have $r\,l(S) = eA$ for some idempotent e, and then

$$l(S) = l\,r\,l(S) = l(eA) = A(1-e),$$

so (b) holds. Symmetrically, (b) implies (a). $\square$

A *Baer ring* is a ring in which every right (and left) annihilator ideal is generated by an idempotent. From the preceding considerations it follows:

Proposition 8.2. *If A is a Baer ring, then the set of principal right ideals generated by idempotents form a complete lattice, which is anti-isomorphic to the lattice of principal left ideals generated by idempotents.*

Let us now specialize our discussion further to regular rings. Assume that A is a regular ring. Then there is a Galois connection between the two lattices $\mathsf{Princ}(A.)$ and $\mathsf{Princ}(.A)$ of principal right, resp. left, ideals of A. This connection is given by l and r, and in terms of idempotents it can be described as

$$eA \mapsto A(1-e), \qquad Ae \mapsto (1-e)A.$$

The regular ring A is called *complete* when either (and hence both) of the lattices $\mathsf{Princ}(A.)$ and $\mathsf{Princ}(.A)$ are complete.

Proposition 8.3. *A regular ring is complete if and only if it is a Baer ring.*

Proof. Every regular Baer ring is complete by Prop. 8.2. Conversely, suppose A is a complete regular ring, and let $\mathfrak{a}$ be a right annihilator ideal. We assert that $\mathfrak{a}=e_0 A$, where $e_0 A$ is the meet in $\mathsf{Princ}(A.)$ of the right ideals $(1-e)A$ for all idempotents e in $l(\mathfrak{a})$. If $b\in l(\mathfrak{a})$, then $Ab=Af$ for some idempotent f, and also $f\in l(\mathfrak{a})$. Hence $e_0 A\subset(1-f)A$, which implies $be_0 A\subset Af(1-f)=0$, and thus $e_0 A\subset r(b)$. It follows that $e_0 A\subset r(l(\mathfrak{a}))=\mathfrak{a}$. On the other hand, if $a\in\mathfrak{a}$, then for every idempotent $e\in l(\mathfrak{a})$ we have $a=(1-e)a\in(1-e)A$, so $a\in e_0 A$. Thus $\mathfrak{a}=e_0 A$. $\square$

Examples

1. The Dedekind-MacNeille completion. Let L be a lattice. For each subset S of L we put

$$Ub(S)=\{\text{upper bounds of } S \text{ in } L\},$$

$$Lb(S)=\{\text{lower bounds of } S \text{ in } L\}.$$

Ub and Lb define a Galois connection of the set $\mathsf{P}(L)$ of subsets of L with itself. Note that $Ub(S)$ is a filter in L, while $Lb(S)$ is an ideal in L. For each subset S of L we write $\bar{S}=Lb(Ub(S))$. The ideals S which are closed in the sense that $S=\bar{S}$, form a complete lattice $\tilde{L}=\mathsf{P}(L)^c$. The join and meet of closed ideals I_α in $\tilde{L}$ are given by

$$\bigwedge I_\alpha=\bigcap I_\alpha, \qquad \bigvee I_\alpha=\overline{\bigcup I_\alpha}.$$

The lattice L is isomorphic to the sublattice of $\tilde{L}$ consisting of principal ideals. That $\tilde{L}$ is a completion of L follows from the observation that for each closed ideal S one has

$$S=\bigvee_{x\in S}[0,x]=\bigcap_{x\in Ub(S)}[0,x].$$

It is easily verified that $L\to\tilde{L}$ is not only a lattice morphism, but that it preserves all those joins and meets which happen to exist in L.

2. Completion of boolean algebras. Let L be a boolean algebra with complementation $*$. For each subset S of L we put

$$S^*=\{a\in L\,|\,a\wedge s=0 \text{ for all } s\in S\}.$$

Then $S\mapsto S^*$ is a symmetric Galois correspondence of $\mathsf{P}(L)$ with itself. Since $a\wedge s=0$ is equivalent to $a\le s^*$, we obtain

$$S^{**}=\{a\,|\,a\le b^* \text{ for all } b\in S^*\}=Lb\{b^*\,|\,b\in S^*\}.$$

But $b\in S^*$ is equivalent to $b\wedge s=0$ for all $s\in S$, i.e. $s\le b^*$ for all $s\in S$. Hence $S^{**}=Lb(Ub(S))$, and it follows that the ideals closed under the operator $S\mapsto S^{**}$ are the same as those ideals which define the Dedekind-MacNeille completion. We can now prove:

Proposition 8.4. *The Dedekind-MacNeille completion of a boolean algebra L is also a boolean algebra.*

Proof. If I and J are ideals, then

$$(I\cup J)^*=\{a\,|\,a\wedge b=0 \text{ for } b\in I\cup J\}=I^*\cap J^*.$$

Hence $(I \cup I^*)^{**} = (I^* \cap I^{**})^* = 0^* = L$. Therefore I^* is the complement of I in the lattice $\tilde{L}$ of closed ideals. If J is an ideal such that $I \cap J = 0$, then for all $a \in I$, $b \in J$ we have $a \wedge b \in I \cap J = 0$, so $J \subset I^*$. Hence L is a boolean algebra by Prop. 4.4. $\square$

3. A complete regular ring. The endomorphism ring of a vector space V over a skew-field is a complete regular ring. In fact, since regularity is clear, it remains to see that it is a Baer ring. Now if S is a family of endomorphisms of V, then

$$r(S) = \{\alpha \in \operatorname{End}(V) \mid \alpha(V) \subset \bigcap_{\sigma \in S} \operatorname{Ker} \sigma\},$$

which may be generated by a projection onto $\bigcap \operatorname{Ker} \sigma$.

4. Strongly regular rings. If A is a strongly regular ring, then it follows from Prop. 4.6 that A is complete if and only if the boolean algebra $\mathsf{B}(A)$ is complete. In particular, a boolean ring is complete as a regular ring if and only if it is complete as a boolean algebra.

Exercises

§1

1. Define a *semi-lattice* to be a set S with a binary operation $*$ which is commutative, associative and satisfies $s * s = s$ for all $s \in S$. Show that:

(i) A lattice is a semi-lattice under $\vee$ and also under $\wedge$.

(ii) A semi-lattice S becomes partially ordered if we put $s \leq t$ whenever there exists $u \in S$ such that $s * u = t$, and then $s \vee s' = s * s'$ for all s, s'.

2. Let L be a semi-lattice under the operation $\vee$. Define an *ideal* in L in the same way as for lattices. Show that:

(i) The ideals of L form a lattice $\hat{L}$ with

$$I \wedge J = I \cap J,$$
$$I \vee J = \{x \in L \mid x \leq a \vee b \text{ for some } a \in I,\ b \in J\}.$$

(ii) $\hat{L}$ is a complete lattice.

3. If L is a lattice, show that:

(i) L is isomorphic to a sublattice of $\hat{L}$, consisting of the principal ideals.

(ii) $\hat{L}$ is an upper completion of L (it is called the *ideal completion* of L).

4. The chain of real numbers $0 \leq x \leq 1$ is not isomorphic to its ideal completion.

5. If L is a complete lattice and $f: L \to L$ an order morphism, show that there exists $a \in L$ such that $f(a) = a$. (Hint: put $S = \{x \mid x \leq f(x)\}$ and $a = \sup S$; show that $a \in S$, and that $f(a) \leq f^2(a)$ implies $f(a) \in S$ and therefore also $f(a) \leq a$.)

§2

6. Show that the lattice of normal subgroups of an arbitrary group is modular.

7. Show that if M is a coherent module, then the finitely generated submodules of M form a sublattice of $\mathsf{L}(M)$.

8. Show that if L is a modular lattice, then the ideal completion (Exercise 3) of L is modular.

§3

9. Show that a complemented modular lattice is noetherian if and only if it is artinian.

10. Show that a lattice L is noetherian if and only if every ideal of L is principal.

11. Show that in the lattice of ideals of a commutative ring, every prime ideal is a meet-irreducible element.

§4

12. Show that if L is a distributive lattice with 0 and 1, then the elements of L which have a complement form a sublattice of L.

13. Let L be a distributive lattice and $\hat{L}$ its ideal completion (Exercise 3). Show that:

(i) $I \vee J = \{a \vee b \mid a \in I, b \in J\}$ in $\hat{L}$.

(ii) $\hat{L}$ is a distributive lattice.

14. Show that if a boolean algebra is a noetherian lattice, then it is finite.

15. Let X be a topological space, and let $\mathrm{Cl}(X)$ (resp. $\mathrm{Op}(X)$) denote the sets of closed (open) subsets of X. Show that:

(i) $\mathrm{Cl}(X)$ and $\mathrm{Op}(X)$ are complete, distributive lattices, and that they are anti-isomorphic to each other.

(ii) $\mathrm{Clop}(X) = \mathrm{Cl}(X) \cap \mathrm{Op}(X)$ is a boolean algebra.

16. Verify that the ring $\mathbb{Z}$ is arithmetical.

§5

17. Let $\hat{L}$ be the ideal completion of a lattice L (Exercise 3). Show that:

(i) An ideal I is a compact element of $\hat{L}$ if and only if I is a principal ideal.

(ii) $\hat{L}$ is compactly generated.

18. Let L be a compactly generated lattice, and let K be the semi-lattice of compact elements of L. Show that L is isomorphic to the ideal completion $\hat{K}$ of K.

§6

19. Show that a pseudo-complemented modular lattice L is complemented if and only if 1 is the only essential element of L.

20. Let L be a pseudo-complemented lattice. Show that two elements a and b have the same pseudo-complements if and only if they are both essential extensions of $a \wedge b$.

21. If M is a module and L is an essential submodule of M, show that $(L : x) = \{a \in A \mid x a \in L\}$ is an essential right ideal of A for every $x \in M$.

22. Let $L = \mathsf{P}(S)$ be the boolean algebra of subsets of an infinite set S. If I is the ideal consisting of all finite subsets of S, show that I is an essential element in the lattice $\hat{L}$ of all ideals of L. (Hence $\hat{L}$ is not a complemented lattice.)

23. Let L be a distributive lattice with 0. Show that the lattice $\hat{L}$ of all ideals of L is a Brouwer lattice (cf. Exercise 13).

24. Show that the open subsets of a topological space form a Brouwer lattice (cf. Exercise 15).

§7

25. Show that if L is a complete lattice, then the closure operators on L form a complete lattice.

26. Let L be a compactly generated lattice. Show that a closure operator on L is finitary if and only if it satisfies the following condition: an element $a \in L$ is closed if and only if a is a directed join in L of closures of compact elements.

§8

27. Let A be a Baer ring, and show that:

(i) If S is a subset of the center of A, then $r(S)$ is generated by a central idempotent. (Hint: $eA = r(S) = l(S) = Af$ for idempotents e, f. Show that this implies $e = f$, and that $e\,a = e\,a\,e = a\,e$.)

(ii) The center of A is a Baer ring.

(iii) The boolean algebra $\mathsf{B}(A)$ of central idempotents is complete.

28. Let L be a lattice, and let $\varphi : L \to \tilde{L}$ be the canonical morphism of L into its Dedekind-MacNeille completion. Verify that φ preserves all joins and meets existing in L.

29. Let X be a topological space and define a Galois connection between the lattice $\mathrm{Op}(X)$ of open subsets of X and the lattice $\mathrm{Cl}(X)$ of closed subsets of X as: if $S \in \mathrm{Op}(X)$, let $\bar{S}$ be the closure of S; if $S \in \mathrm{Cl}(X)$, let S° be the interior of S. An open set S is *regular* if $S = \bar{S}^\circ$. Show that the regular open sets form a complete boolean algebra, where S^* is the complement (in the sense of set theory) of $\bar{S}$.

Chapter IV. Abelian Categories

Most categories used in the context of ring theory are abelian categories, i.e. they are preadditive and one can define kernels, images etc. of their morphisms in such a way that these notions may be handled, on the whole, as in the theory of modules. In this chapter we develop the basic concepts of additive category theory.

§ 1. Equivalence of Categories

Two categories $\mathbf{C}$ and $\mathbf{D}$ are *isomorphic* if there exist functors S: $\mathbf{C} \to \mathbf{D}$ and T: $\mathbf{D} \to \mathbf{C}$ such that $TS = 1_{\mathbf{C}}$ and $ST = 1_{\mathbf{D}}$. In practice however, this notion of isomorphism turns out to be too restrictive. In fact, there are numerous examples of categories $\mathbf{C}$ and $\mathbf{D}$ which have essentially the same structure, but where there is a bijective correspondence between the isomorphism classes of objects rather than between the individual objects. We therefore make the

Definition. A functor S: $\mathbf{C} \to \mathbf{D}$ is an *equivalence* if there exist a functor T: $\mathbf{D} \to \mathbf{C}$ and natural equivalences $TS \to 1_{\mathbf{C}}$ and $ST \to 1_{\mathbf{D}}$.

Often it is the following description of equivalences that lies more close at hand:

Proposition 1.1. *A functor S: $\mathbf{C} \to \mathbf{D}$ is an equivalence if and only if it is full and faithful and every object of $\mathbf{D}$ is isomorphic to an object of the form $S(C)$.*

Proof. Suppose S is an equivalence with T given as in the definition above. The natural equivalence $1 \to TS$ produces for every morphism α: $C \to C'$ in $\mathbf{C}$ a commutative diagram

$$\begin{array}{ccc} C & \xrightarrow{\cong} & TS(C) \\ {\scriptstyle\alpha}\downarrow & & \downarrow{\scriptstyle TS(\alpha)} \\ C' & \xrightarrow[\cong]{} & TS(C'), \end{array} \tag{1}$$

which shows that S is faithful. By symmetry also T is faithful. A morphism β: $S(C) \to S(C')$ induces conversely α: $C \to C'$ by the requirement that (1) should commute, with $TS(\alpha)$ replaced by $T(\beta)$. But then $TS(\alpha) = T(\beta)$, and T faithful implies $\beta = S(\alpha)$. Thus S is full. Finally there exists for every object D of $\mathbf{D}$ an isomorphism $D \cong ST(D)$.

Suppose conversely that S is full and faithful, and that for each D in $\mathbf{D}$ we can find an object $T(D)$ in $\mathbf{C}$ with an isomorphism ξ_D: $ST(D) \to D$. A morphism

β: $D \to D'$ in **D** induces a morphism $\xi_{D'}^{-1} \beta \xi_D$: $ST(D) \to ST(D')$, and since S is full and faithful, there is a unique morphism $T(D) \to T(D')$, which we denote by $T(\beta)$, such that $\xi_{D'}^{-1} \beta \xi_D = ST(\beta)$. It is easy to verify that in this way we have obtained a functor T: $\mathbf{D} \to \mathbf{C}$ and a natural equivalence ξ: $ST \to 1$.

For each C in **C** there is in particular an isomorphism $\xi_{S(C)}$: $STS(C) \to S(C)$, and since S is full and faithful, we must have $\xi_{S(C)} = S(v_C)$ for a unique isomorphism v_C: $TS(C) \to C$. To prove that v_C is natural in C, consider the diagram (1) induced by α: $C \to C'$. Applying S to this diagram we get a diagram which is commutative by the definition of v and the naturality of ξ. The faithfulness of S then implies that (1) commutes. Hence also v: $TS \to 1$ is a natural equivalence. □

Examples

1. Finite products of module categories. Let $A_1, \ldots, A_n$ be rings and $A_1 \times \cdots \times A_n$ the direct product ring. Then there is an equivalence

$$\textbf{Mod-}A_1 \times \cdots \times \textbf{Mod-}A_n \to \textbf{Mod-}(A_1 \times \cdots \times A_n),$$

given by $(M_1, \ldots, M_n) \mapsto M_1 \times \cdots \times M_n$. The reader should verify that this equivalence does not extend to infinite products of rings.

2. Modules of fractions. Let S be a right denominator set in the ring A. It was proved in Chap. II, § 3, that the full subcategory of **Mod**-A, consisting of modules of fractions $M[S^{-1}]$, is equivalent to **Mod**-$A[S^{-1}]$. From Prop. II.3.7 also follows that **Mod**-$A[S^{-1}]$ is equivalent to the category of S-torsion-free and S-divisible right A-modules.

3. Pontryagin duality. Equivalences often appear in the form of a duality functor, i.e. an equivalence between $\mathbf{C}^{\mathrm{op}}$ and **D**. E.g. the Pontryagin duality theory establishes an equivalence between $\mathbf{Ab}^{\mathrm{op}}$ and the category of compact abelian groups and continuous group homomorphisms.

§ 2. Kernels and Cokernels

Throughout this section we assume **C** to be a preadditive category. A *zero object* is an object Z of **C** such that $\mathrm{Hom}(C, Z) = \{0\}$ and $\mathrm{Hom}(Z, C) = \{0\}$ for every object C of **C**. Any two zero objects are isomorphic, and they will therefore be identified to one single zero object 0 of **C**. If a zero object 0 exists, then the zero element of the group $\mathrm{Hom}_{\mathbf{C}}(C, C')$ is identical to the composed morphism $C \to 0 \to C'$, as follows from the bilinearity of the composition map

$$\mathrm{Hom}(0, C') \times \mathrm{Hom}(C, 0) \to \mathrm{Hom}(C, C').$$

A morphism α: $B \to C$ is a *monomorphism* if $\alpha\xi = 0$ implies $\xi = 0$ for every ξ: $X \to B$. Dually, α is an *epimorphism* if $\eta\alpha = 0$ implies $\eta = 0$ for every η: $C \to X$. Two monomorphisms α: $B \to C$ and α': $B' \to C$ are *equivalent* if there is an isomorphism γ: $B \to B'$ such that $\alpha'\gamma = \alpha$. An equivalence class of monomorphisms into C is called a *subobject* of C. In order to avoid notational complications, we will however make an abuse of notation (and language) by denoting a subobject by some representing monomorphism. If $\alpha: B \to C$ and $\alpha': B' \to C$ are two subobjects of C, then we write $B \subset B'$ if there is a morphism $\gamma: B \to B'$ such that $\alpha' \gamma = \alpha$ (note that γ will then be a monomorphism). In particular, if both $B \subset B'$ and $B' \subset B$ hold, then the two monomorphisms represent the same subobject of C. Dually we define *quotient objects.* Clearly:

Lemma 2.1. *Every isomorphism is both a monomorphism and an epimorphism.*

C will now be assumed to possess a zero object.

Definition. A *kernel* of a morphism $\alpha: B \to C$ is a morphism $\kappa: K \to B$ such that
(1) $\alpha \kappa = 0$;
(2) for every $\xi: X \to B$ with $\alpha \xi = 0$, there exists a unique $\gamma: X \to K$ such that $\xi = \kappa \gamma$.

It is easily seen that every kernel is a monomorphism. The kernel of α solves a universal problem, and it is therefore unique; more precisely we have:

Proposition 2.2. *Any two kernels of* α *represent the same subobject.*

Proof. Let $\kappa: K \to B$ and $\kappa': K' \to B$ be kernels of $\alpha: B \to C$. Since $\alpha \kappa' = 0$, there exists $\gamma: K' \to K$ such that $\kappa \gamma = \kappa'$. Similarly there exists $\beta: K \to K'$ such that $\kappa' \beta = \kappa$. Hence

$$\kappa \gamma \beta = \kappa' \beta = \kappa \cdot 1_K,$$

$$\kappa' \beta \gamma = \kappa \gamma = \kappa' \cdot 1_{K'}.$$

But κ and κ' are monomorphisms, so $\gamma \beta = 1_K$ and $\beta \gamma = 1_{K'}$, and thus β is an isomorphism. ▯

Because of this unicity we may speak of *the* kernel of α (when it exists!), and we write $\kappa = \ker \alpha$, $K = \operatorname{Ker} \alpha$ for a kernel $\kappa: K \to B$ of α. One shows without difficulty:

Proposition 2.3. *A morphism* α *is a monomorphism if and only if* $\operatorname{Ker} \alpha = 0$.

By dualization one obtains the notion of *cokernel*, denoted as $\operatorname{Coker} \alpha$. The following observation will be needed later on:

Proposition 2.4. *If* α *is the kernel of some morphism and if* $\operatorname{Coker} \alpha$ *exists, then* $\alpha = \ker(\operatorname{coker} \alpha)$.

Proof. Let $\alpha: B \to C$ be the kernel of $\gamma: C \to D$. Then $\gamma \alpha = 0$ implies that γ factors over $\operatorname{coker} \alpha$. If now $\xi: X \to C$ is any morphism such that $\operatorname{coker} \alpha \cdot \xi = 0$, then it follows that $\gamma \xi = 0$, and ξ factors uniquely over $\ker \gamma = \alpha$. Therefore α is a kernel of $\operatorname{coker} \alpha$. ▯

Examples

1. Module categories. In the category **Mod**-A, for a ring A, kernels have the same signification as in Chap. I, while the cokernel of $\alpha: B \to C$ is the canonical homomorphism $C \to C/\operatorname{Im} \alpha$. In this category, every monomorphism is a kernel of some morphism, and every epimorphism is a cokernel.

2. Rings. Let A be a ring, considered as a preadditive category with one object. An element $a \in A$ is a monomorphism (epimorphism) if and only if $r(a)=0$ (resp. $l(a)=0$).

§3. Products and Coproducts

We still assume **C** to be preadditive with a zero object.

Definition. A *product* of a family $(C_i)_{i\in I}$ of objects of **C** is an object C together with morphisms $\pi_i\colon C \to C_i$ $(i \in I)$, such that for each object X and morphisms $\xi_i\colon X \to C_i$ $(i \in I)$ there is a unique morphism $\xi\colon X \to C$ with $\pi_i \xi = \xi_i$ for all $i \in I$.

Here we are again facing a "universal problem", and so the product is unique up to isomorphism, and we denote it by $\prod_I C_i$. The canonical morphisms $\pi_i\colon \prod_I C_i \to C_i$ are called the *projections*. Note that the definition of $\prod_I C_i$ can be summarized by the formula

$$\operatorname{Hom}(X, \prod_I C_i) \cong \prod_I \operatorname{Hom}(X, C_i), \tag{1}$$

where the second product is taken in the category of abelian groups. If $\alpha_i\colon B_i \to C_i$ $(i \in I)$ is a family of morphisms, then the composed morphisms $\prod B_i \to B_i \to C_i$ induce a morphism $\alpha\colon \prod B_i \to \prod C_i$. This makes it possible to consider the product as a functor $\mathbf{C}^I \to \mathbf{C}$.

Proposition 3.1. *If $\alpha_i\colon B_i \to C_i$ $(i \in I)$ are monomorphisms, then the induced morphism $\alpha\colon \prod_I B_i \to \prod_I C_i$ is a monomorphism.*

Proof. Let $\xi, \eta\colon X \to \prod B_i$ be morphisms such that $\alpha\xi = \alpha\eta$. For each α_i we then have $\alpha_i \pi_i \xi = \alpha_i \pi_i \eta$, which implies that $\pi_i \xi = \pi_i \eta$ for all i. The unicity of the factorization over $\prod B_i$ gives $\xi = \eta$. □

The dual notion of product is *coproduct* (or *direct sum*), and it is denoted as $\bigoplus_I C_i$. It is characterized by the formula

$$\operatorname{Hom}(\bigoplus_I C_i, X) \cong \prod_I \operatorname{Hom}(C_i, X), \tag{2}$$

where the isomorphism is induced by the canonical *injections* $\iota_i\colon C_i \to \bigoplus_I C_i$. The dual of Prop. 3.1 states that every coproduct of epimorphisms is an epimorphism. When all C_i are equal to a certain object C, we use the notations C^I and $C^{(I)}$ for the product and the coproduct.

Let $(C_i)_I$ be a family of objects such that both $\prod_I C_i$ and $\bigoplus_I C_i$ exist. For each $j \in I$ we define morphisms $\delta_{ij}\colon C_j \to C_i$ as

$$\delta_{ij} = \begin{cases} 1 & \text{if } i=j \\ 0 & \text{if } i \neq j. \end{cases}$$

By the definition of products there exists a unique morphism $\iota'_j\colon C_j \to \prod_I C_i$ such that $\pi_i \iota'_j = \delta_{ij}$. From this it follows in particular that each projection π_i is an epimorphism. Dually, the morphisms δ_{ij} induce $\pi'_i\colon \bigoplus_I C_j \to C_i$ such that $\pi'_i \iota_j = \delta_{ij}$, and the injections ι_j are therefore monomorphisms.

Let us now assume that I is finite, say $I = \{1, \dots, n\}$. In that case we have $\iota'_1 \pi_1 + \dots + \iota'_n \pi_n = 1$, because for each C_i we get

$$\pi_i(\iota'_1 \pi_1 + \dots + \iota'_n \pi_n) = \pi_i \iota'_i \pi_i = \pi_i = \pi_i \cdot 1,$$

and the unicity of factorizations over the product implies $\iota'_1 \pi_1 + \dots + \iota'_n \pi_n = 1$. For a finite coproduct we similarly obtain $\iota_1 \pi'_1 + \dots + \iota_n \pi'_n = 1$. This symmetry between finite products and coproducts actually implies that the two notions coincide:

Proposition 3.2. *Suppose there are given morphisms $\iota_i\colon C_i \to C$ and $\pi_i\colon C \to C_i$ $(i = 1, \dots, n)$ such that*

$$\pi_i \iota_j = \delta_{ij} \quad \textit{and} \quad \iota_1 \pi_1 + \dots + \iota_n \pi_n = 1.$$

Then C is a product and a coproduct of the objects C_i, with projections π_i and injections ι_i.

Proof. By symmetry it suffices to show that C is a product. Suppose there are given morphisms $\xi_i\colon X \to C_i$. Define $\xi\colon X \to C$ as $\xi = \iota_1 \xi_1 + \dots + \iota_n \xi_n$. Then $\pi_j \xi = \pi_j \iota_1 \xi_1 + \dots + \pi_j \iota_n \xi_n = \xi_j$, as desired. There is only one choice for ξ, for if $\xi'\colon X \to C$ is such that $\pi_i \xi' = \xi_i$, then $\xi' = (\iota_1 \pi_1 + \dots + \iota_n \pi_n)\xi' = \iota_1 \pi_1 + \dots + \iota_n \xi_n = \xi$. □

Thus there is no need to distinguish between products and coproducts over finite index sets; we will alternatively use the notions $C_1 \times \dots \times C_n$ and $C_1 \oplus \dots \oplus C_n$.

It is convenient to use matrix notation for morphisms between finite products. A morphism $\beta\colon \prod_J B_j \to \prod_I C_i$, with I and J finite, is denoted by the matrix (β_{ij}), where $\beta_{ij} = \pi_i^C \beta \iota_j^B\colon B_j \to C_i$. With this convention for the writing of indices, composition of morphisms corresponds to matrix multiplication, as the reader may verify. As an application of this notation we show:

Proposition 3.3. *Let* **C** *and* **D** *be preadditive categories in which finite products exist. A functor $T\colon$* **C** $\to$ **D** *is additive if and only if it preserves finite products.*

Proof. The characterization of finite products given in Prop. 3.1 entails that every additive functor preserves finite products. For the converse, let α and β be morphisms $C \to C'$ in **C**. Then $\alpha + \beta$ is obtained as the composition

$$C \xrightarrow{\binom{1}{1}} C \oplus C \xrightarrow{\begin{pmatrix} \alpha & 0 \\ 0 & \beta \end{pmatrix}} C' \oplus C' \xrightarrow{(1\ 1)} C'.$$

From this it is easy to see that if T preserves finite products, then $T(\alpha + \beta) = T(\alpha) + T(\beta)$. □

Examples

1. Module categories. For a module category **Mod**-A, the results of this section are known from Chap. I (except for Prop. 3.3).

2. The zero object. A zero object of **C** is a product (and a coproduct) of the empty family of objects.

3. Torsion groups. Let **Tors** be the full subcategory of **Ab** consisting of the torsion groups. It is a preadditive category and it has kernels, cokernels and coproducts, which are all computed as in **Ab**. It also has products, but the product in **Tors** of a family of torsion groups is obtained by taking the torsion subgroup of the direct product in **Ab**.

§4. Abelian Categories

Let **C** be a preadditive category with the property that every morphism has a kernel and a cokernel. For a morphism $\alpha: B \to C$ there is a canonical factorization as indicated by the commutative diagram

$$\begin{array}{ccccccc} \mathrm{Ker}\,\alpha & \longrightarrow & B & \xrightarrow{\alpha} & C & \longrightarrow & \mathrm{Coker}\,\alpha \\ & & \downarrow{\scriptstyle\lambda} & & \uparrow{\scriptstyle\mu} & & \\ & & \mathrm{Coker}(\ker\alpha) & \xrightarrow[\bar\alpha]{} & \mathrm{Ker}(\mathrm{coker}\,\alpha) & & \end{array}$$

where $\bar{\alpha}$ is obtained as follows: $\mathrm{coker}\,\alpha \cdot \alpha = 0$ implies $\alpha = \mu\,\beta$ for some $\beta: B \to \mathrm{Ker}(\mathrm{coker}\,\alpha)$; then $\mu\,\beta \cdot \ker\alpha = \alpha \cdot \ker\alpha = 0$, which implies $\beta \cdot \ker\alpha = 0$ since μ is a monomorphism, and hence β factors as $\beta = \bar{\alpha}\,\lambda$.

Consider as an example the case $\mathbf{C} = \mathbf{Ab}$. Here $\mathrm{Coker}(\ker\alpha) \cong B/\mathrm{Ker}\,\alpha$ and $\mathrm{Ker}(\mathrm{coker}\,\alpha) \cong \mathrm{Im}\,\alpha$, and $\bar{\alpha}$ is an isomorphism. The following therefore appears to be a reasonable definition:

Definition. A category **C** is *abelian* if

*A*1. **C** is preadditive.

*A*2. Every finite family of objects has a product (and coproduct).

*A*3. Every morphism has a kernel and a cokernel.

*A*4. $\bar{\alpha}: \mathrm{Coker}(\ker\alpha) \to \mathrm{Ker}(\mathrm{coker}\,\alpha)$ is an isomorphism for every morphism α.

The axiom *A*4 can be replaced by the following axiom:

*A*4′. Every morphism α has a factorization $\alpha = \gamma\,\beta$, where β is a cokernel and γ is a kernel.

In fact, it is obvious that *A*4 implies *A*4′. Suppose on the other hand that **C** satisfies *A*1–3 and *A*4′. If α factors as in *A*4′, then it follows from Prop. 2.4 that $\beta = \mathrm{coker}(\ker\beta)$ and $\gamma = \ker(\mathrm{coker}\,\gamma)$. But $\ker\beta = \ker\alpha$ since γ is a monomorphism, and similarly $\mathrm{coker}\,\gamma = \mathrm{coker}\,\alpha$. We conclude that $\mathrm{Coker}(\ker\alpha) = \mathrm{Ker}(\mathrm{coker}\,\alpha)$.

Note that the axioms are self-dual, so **C** is abelian if and only if $\mathbf{C}^{\mathrm{op}}$ is. For the rest of this section we assume that **C** is abelian. The preceding discussion shows:

Proposition 4.1. (i) *If α is a monomorphism and an epimorphism, then α is an isomorphism.*

(ii) *If α is a monomorphism, then* $\alpha = \ker(\mathrm{coker}\,\alpha)$.

(iii) *If α is an epimorphism, then* $\alpha = \mathrm{coker}(\ker\alpha)$.

When A is a subobject of B, we will often write B/A for the quotient object $\operatorname{Coker}(A \to B)$ of B.

For every morphism $\alpha\colon A \to B$ we define the *image* of α as $\operatorname{Im}\alpha = \operatorname{Ker}(\operatorname{coker}\alpha)$. Thus α has a factorization

$$A \xrightarrow{\alpha'} \operatorname{Im}\alpha \xrightarrow{\gamma} B$$

where α' is an epimorphism and γ is a monomorphism. If A' is a subobject of A, we write $\alpha(A')$ for the image of the composed morphism $A' \to A \xrightarrow{\alpha} B$.

In an abelian category one can work with exact sequences in much the same way as one is used to in the case of modules. A sequence

$$\cdots \longrightarrow C_{i-1} \xrightarrow{\alpha_{i-1}} C_i \xrightarrow{\alpha_i} C_{i+1} \longrightarrow \cdots$$

is *exact at* C_i if $\operatorname{Im}\alpha_{i-1} = \operatorname{Ker}\alpha_i$ (equal as subobjects of C_i). The whole sequence is *exact* if it is exact at each C_i. The "short 5-lemma" (Prop. I.1.3) holds also in an abelian category, and we leave its proof to the reader as an exercise. *Split* exact sequences can be defined and characterized as for modules (Prop. I.2.1); the details are left to the reader.

Let C be an object of $\mathbf{C}$ with a family $(C_i)_I$ of subobjects, and assume $\bigoplus_I C_i$ exists. The monomorphisms $C_i \to C$ induce a morphism $\alpha\colon \bigoplus_I C_i \to C$. The image of α is called the *sum* of the subobjects C_i and is denoted by $\sum_I C_i$ (or $C_1 + \cdots + C_n$ in the finite case). The sum is *direct*, and the family $(C_i)_I$ of subobjects is said to be *independent*, if α is a monomorphism, in which case it induces an isomorphism $\bigoplus_I C_i \cong \sum_I C_i$.

Dually, the epimorphisms $C \to C/C_i$ induce $\beta\colon C \to \prod_I C/C_i$ (when the product exists). The kernel of β is the *intersection* of the subobjects C_i and is denoted by $\bigcap_I C_i$ (or $C_1 \cap \cdots \cap C_n$ in the finite case).

It comes natural to look at the concepts of sum and intersection from the point of view of lattice theory. However, we cannot be sure that the class of subobjects of C (with equivalent monomorphisms identified) actually is a set. But there is a partial ordering on the class $\mathsf{L}(C)$ of subobjects of C, and we can apply the results of lattice theory as long as we restrict our attention to subsets of $\mathsf{L}(C)$, e.g. when taking joins and meets.

Proposition 4.2. *If $(C_i)_I$ is a family of subobjects of C, then $\sum_I C_i$ is a least upper bound and $\bigcap_I C_i$ is a greatest lower bound for the family.*

Proof. It is obvious that $\sum_I C_i$ is an upper bound for the family $(C_i)_I$. Suppose D is a subobject of C such that every $C_i \subset D$. Then the composition of each $C_i \to C$ with $C \to C/D$ is the zero morphism, and it follows that the induced morphism $\bigoplus_I C_i \to C$ also gets zero when composed with $C \to C/D$. Hence $\sum_I C_i \subset D$, so the sum is a least upper bound for the family. The result for intersections follows by duality. □

It follows that $\mathsf{L}(C)$ is a "lattice" (one sometimes uses the term *big lattice* when it is not a set), which is complete when **C** has products and coproducts.

Let T: $\mathbf{C} \to \mathbf{D}$ be an additive functor between abelian categories **C** and **D**. One defines *(left, right) exactness* of T as in the module case (Chap. I, §5). Note that T always preserves split exactness by Prop. 3.3.

Proposition 4.3. *The functor* Hom: $\mathbf{C}^{\mathrm{op}} \times \mathbf{C} \to \mathbf{Ab}$ *is left exact in each one of the variables.*

The proof may be left to the reader (it follows almost immediately from the definition of kernels and cokernels).

Examples

1. Module categories. If A is a ring, then **Mod**-A is clearly abelian.

2. Abelian subcategories. If **C** is an abelian category and **B** is a subcategory which also is abelian, then **B** is said to be an *abelian subcategory* of **C** if the inclusion functor $\mathbf{B} \to \mathbf{C}$ is exact. The category **Tors** of abelian torsion groups is an abelian subcategory of **Ab**. If A is a ring, then the noetherian modules form a full abelian subcategory of **Mod**-A.

3. Projective and injective objects. Let **C** be an abelian category. An object C is *projective* if the functor $\mathrm{Hom}(C, \cdot)$: $\mathbf{C} \to \mathbf{Ab}$ is exact, and is *injective* if $\mathrm{Hom}(\cdot, C)$: $\mathbf{C}^{\mathrm{op}} \to \mathbf{Ab}$ is exact. With the help of the formulas (1) and (2) of §3 one obtains (cf. Lemmas I.6.2 and I.6.4):

Proposition 4.4. *Let* $(C_i)_I$ *be a family of objects. Then:*
(i) $\bigoplus_I C_i$ *is projective if and only if each* C_i *is projective.*
(ii) $\prod_I C_i$ *is injective if and only if each* C_i *is injective.*

The category **C** is said to have *enough projectives* if every object of **C** is a quotient object of a projective object, and to have *enough injectives* if every object is a subobject of an injective object. The category **Mod**-A has enough projectives and injectives by the results of Chap. I. The category **Tors** has enough injectives, but has no projective objects other than 0.

§5. Pullbacks and Pushouts

In this section we let **C** be an abelian category.

Definition. Let α_1: $C_1 \to C$ and α_2: $C_2 \to C$ be morphisms in **C**. A *pullback* for α_1 and α_2 is an object P together with morphisms π_1: $P \to C_1$, π_2: $P \to C_2$ such that:

(1) $\alpha_1 \pi_1 = \alpha_2 \pi_2$;

(2) For every object X and morphisms ξ_1: $X \to C_1$, ξ_2: $X \to C_2$ with $\alpha_1 \xi_1 = \alpha_2 \xi_2$, there exists a unique γ: $X \to P$ such that $\pi_1 \gamma = \xi_1$ and $\pi_2 \gamma = \xi_2$.

The diagram

$$\begin{array}{ccc} P & \xrightarrow{\pi_2} & C_2 \\ {\scriptstyle \pi_1}\downarrow & & \downarrow{\scriptstyle \alpha_2} \\ C_1 & \xrightarrow[\alpha_1]{} & C \end{array} \tag{1}$$

is then called a *pullback diagram*. The pullback is unique up to isomorphism, as one easily verifies.

If $\alpha\colon B \to C$ is a morphism, then a diagram

$$\begin{array}{ccc} P & \longrightarrow & 0 \\ {\scriptstyle \pi}\downarrow & & \downarrow \\ B & \xrightarrow[\alpha]{} & C \end{array}$$

is a pullback if and only if $P = \operatorname{Ker} \alpha$. Kernels are thus particular cases of pullbacks. Also finite products are instances of pullbacks, because

$$\begin{array}{ccc} C_1 \times C_2 & \xrightarrow{\pi_2} & C_2 \\ {\scriptstyle \pi_1}\downarrow & & \downarrow \\ C_1 & \longrightarrow & 0 \end{array}$$

is a pullback diagram.

Conversely, every pullback can be constructed with the use of kernels and products. For if $\alpha_1\colon C_1 \to C$ and $\alpha_2\colon C_2 \to C$ are given morphisms, then we put

$$P = \operatorname{Ker}(\alpha_1 p_1 - \alpha_2 p_2) \tag{2}$$

where $p_1\colon C_1 \times C_2 \to C_1$ and $p_2\colon C_1 \times C_2 \to C_2$ are the projections; this gives rise to a commutative diagram

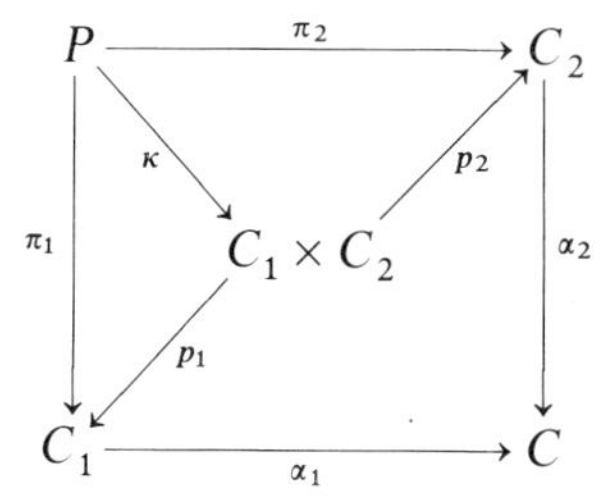

where the outer square is easily verified to be a pullback diagram.

Proposition 5.1. *Let* (1) *be a pullback diagram. Then:*

(i) *If* α_1 *is a monomorphism, then so is also* π_2.

(ii) *If* α_1 *is an epimorphism, then so is also* π_2.

(iii) *If* α_1 *is the kernel of a morphism* $\beta\colon C \to D$, *then* π_2 *is the kernel of* $\beta\alpha_2$.

Proof. (i): If $\xi\colon X \to P$ is a morphism such that $\pi_2 \xi = 0$, then $\alpha_1 \pi_1 \xi = \alpha_2 \pi_2 \xi = 0$, and α_1 a monomorphism gives that $\pi_1 \xi = 0$. The unicity of the factorization over P then implies $\xi = 0$.

(ii): Using (2) we get an exact sequence

$$0 \to P \xrightarrow{\kappa} C_1 \times C_2 \xrightarrow{\mu} C \to 0, \tag{3}$$

where $\mu = \alpha_1 p_1 - \alpha_2 p_2$ is an epimorphism because $\mu i_1 = \alpha_1$ is an epimorphism (where $i_1: C_1 \to C_1 \times C_2$ is the injection). Suppose $\xi: C_2 \to X$ is a morphism such that $\xi \pi_2 = 0$. Then $0 = \xi \pi_2 = \xi p_2 \kappa$ implies that there exists $\eta: C \to X$ with $\eta \mu = \xi p_2$. This gives $\eta \alpha_1 = \eta \mu i_1 = \xi p_2 i_1 = 0$, and α_1 an epimorphism implies $\eta = 0$. But then $\xi p_2 = 0$, and hence $\xi = 0$.

(iii): We certainly have $\beta \alpha_2 \pi_2 = 0$, and if $\xi: X \to C_2$ is a morphism such that $\beta \alpha_2 \xi = 0$, then ξ factors over $\alpha_1 = \ker \beta$ by a morphism $\eta: X \to C_1$ with $\alpha_1 \eta = \alpha_2 \xi$. Hence ξ factors over the pullback P, and does so uniquely since π_2 is a monomorphism. □

When α_1 is a monomorphism, one may think of the pullback object P as the "inverse image" of C_1 under α_2, and one often writes $\alpha_2^{-1}(C_1)$ for P. In this language, (iii) of Prop. 5.1 just states that $\alpha_2^{-1}(\operatorname{Ker} \beta) = \operatorname{Ker} \beta \alpha_2$. If both α_1 and α_2 are monomorphisms, then it is quite clear that P is the greatest lower bound of C_1 and C_2 in C, i.e. $P = C_1 \cap C_2$. The exact sequence (3) in this case becomes:

$$0 \to C_1 \cap C_2 \to C_1 \oplus C_2 \xrightarrow{\mu} C_1 + C_2 \to 0, \tag{4}$$

where μ is induced by the inclusions of C_1 and C_2 in C (up to a minus sign). It follows that the sum $C_1 + C_2$ is direct if and only if $C_1 \cap C_2 = 0$. After these remarks we can prove the Noether isomorphism theorem:

Proposition 5.2. *If C_1 and C_2 are subobjects of C, then*

$$(C_1 + C_2)/C_1 \cong C_2/(C_1 \cap C_2).$$

Proof. With the use of Prop. 5.1 (iii) we get a commutative diagram with exact rows

$$\begin{array}{ccccccccc}
0 & \longrightarrow & C_1 \cap C_2 & \longrightarrow & C_2 & \longrightarrow & (C_1 + C_2)/C_1 & \longrightarrow & 0 \\
 & & \downarrow & & \downarrow & & \| & & \\
0 & \longrightarrow & C_1 & \longrightarrow & C_1 + C_2 & \longrightarrow & (C_1 + C_2)/C_1 & \longrightarrow & 0
\end{array}$$

since the left square is a pullback. The exactness of the upper row immediately leads to the desired conclusion. □

Another consequence of Prop. 5.1 is that one can construct the pullback of a short exact sequence $0 \to C_0 \to C_1 \to C \to 0$ with respect to a morphism $C_2 \to C$. For this we consider the commutative diagram

$$\begin{array}{ccccccccc}
0 & \longrightarrow & C_0 & \xrightarrow{\beta_0} & P & \xrightarrow{\beta} & C_2 & \longrightarrow & 0 \\
 & & \| & & \downarrow{\scriptstyle \gamma} & & \downarrow{\scriptstyle \alpha_2} & & \\
0 & \longrightarrow & C_0 & \xrightarrow[\alpha_0]{} & C_1 & \xrightarrow[\alpha_1]{} & C & \longrightarrow & 0
\end{array}$$

where the right square is a pullback diagram and β_0 is induced by α_0 and the zero morphism $C_0 \to C_2$. It remains to see that β_0 is the kernel of β. Suppose $\xi: X \to P$

is such that $\beta\,\xi=0$. Then $\alpha_1\gamma\,\xi=0$, so $\gamma\,\xi=\alpha_0\,\lambda$ for some $\lambda\colon X\to C_0$. Now $\beta_0\,\lambda$ and ξ give the same results when composing them with γ or β, so $\xi=\beta_0\,\lambda$ by the universal property of pullbacks. Hence the upper row is exact.

When $\alpha_2\colon C_2\to C$ is a monomorphism, this construction shows that each subobject C_2 of C_1/C_0 is of the form P/C_0 for some subobject P of C_1.

With the preceding results at hand, we return to the lattice $\mathsf{L}(C)$ of subobjects of an object C, and prove:

Proposition 5.3. *The lattice* $\mathsf{L}(C)$ *is modular.*

Proof. We make use of Prop. III.2.3 to show the modularity. Since an interval $[B_1, B_2]$ of $\mathsf{L}(C)$ is isomorphic to the lattice $\mathsf{L}(B_2/B_1)$, it suffices to consider subobjects $B_1\subset B_2$ of C with a common complement C' in C. But in that case the monomorphism $B_1\to B_2$ induces a commutative diagram

$$\begin{array}{ccc} B_1\oplus C' & & \\ \downarrow & \searrow^{\cong} & \\ & & C \\ & \nearrow_{\cong} & \\ B_2\oplus C' & & \end{array}$$

which shows that $B_1=B_2$. $\square$

The results of this section may of course be dualized; the dual notion of pullback is *pushout*. A pushout diagram for morphisms $\beta_1\colon C\to C_1$ and $\beta_2\colon C\to C_2$ looks like

$$\begin{array}{ccc} C & \xrightarrow{\beta_1} & C_1 \\ {\scriptstyle\beta_2}\downarrow & & \downarrow{\scriptstyle\iota_2} \\ C_2 & \xrightarrow[\iota_1]{} & P \end{array}$$

with $P=\operatorname{Coker}(i_1\beta_1-i_2\beta_2\colon C\to C_1\oplus C_2)$.

Examples

1. Module categories. In the category **Mod**-A, for a ring A, the pullback formula (2) becomes

$$P=\{(x,y)\in C_1\times C_2\,|\,\alpha_1(x)=\alpha_2(y)\},$$

and the pushout formula becomes

$$P=(C_1\oplus C_2)/C' \quad\text{where}\quad C'=\{(\beta_1(x),\,-\beta_2(x))\,|\,x\in C\}.$$

2. Objects of finite length. Let C be an object of an abelian category. Since the lattice $\mathsf{L}(C)$ is modular, the Jordan-Hölder theorem holds, and it does so in the sharp form valid for modules (Prop. III.3.7), because Prop. 5.2 implies that projectivity of two intervals induces an isomorphism between corresponding quotient objects (cf. Example III.2.1). The object C is of *finite length* if the lattice $\mathsf{L}(C)$ is of finite length, i.e. there is a chain of subobjects

$$0=C_0\subset C_1\subset C_2\subset\cdots\subset C_n=C$$

such that each C_i/C_{i-1} is a simple object (i.e. a non-zero object with only two subobjects).

3. Pullback-pushout diagrams. Consider a commutative diagram in **C**

$$\begin{array}{ccccccccc} 0 & \longrightarrow & B & \longrightarrow & C_2 & \longrightarrow & D & \longrightarrow & 0 \\ & & \downarrow & & \downarrow & & \| & & \\ 0 & \longrightarrow & C_1 & \longrightarrow & C & \longrightarrow & D & \longrightarrow & 0 \end{array}$$

with exact rows. It is easy to verify that the left square is a pullback diagram (which is a converse of Prop. 5.1 (iii)), and that it also is a pushout diagram. A diagram of this kind appeared in the proof of Prop. 5.2.

§ 6. Generators and Cogenerators

Consider two abelian categories **C** and **D** and an additive functor T: $\mathbf{C} \to \mathbf{D}$. Recall that T is said to be faithful if $T(\alpha) \neq 0$ for every non-zero morphism α in **C** (Chap. I, §4).

Proposition 6.1. *An exact functor T is faithful if and only if $T(C) \neq 0$ for every non-zero object C.*

Proof. If T is faithful and $T(C)=0$, then $T(1_C)=0$ implies $1_C=0$ and hence $C=0$. Conversely, suppose $T(C)=0$ implies $C=0$. If $\alpha \neq 0$ in **C**, then $\operatorname{Im}\alpha \neq 0$ and so $T(\operatorname{Im}\alpha) \neq 0$. But the exactness of T implies that $T(\operatorname{Im}\alpha) = \operatorname{Im} T(\alpha)$, so $T(\alpha) \neq 0$. ▯

Definition. An object C of **C** is a *generator* for **C** if $\operatorname{Hom}(C, \cdot)$ is faithful, and is a *cogenerator* if $\operatorname{Hom}(\cdot, C)$ is faithful.

Proposition 6.2. *Suppose* **C** *has coproducts. If U is a generator, then for every object C there is an epimorphism $U^{(I)} \to C$ for some index set I.*

Proof. Put $I = \operatorname{Hom}(U, C)$, and define φ: $U^{(I)} = \bigoplus_{\alpha \in I} U_\alpha \to C$ to be the morphism induced by α: $U_\alpha \to C$ for each $\alpha \in I$. If γ: $C \to C'$ is any non-zero morphism, then $\operatorname{Hom}(U, \gamma) \neq 0$ implies the existence of α: $U \to C$ such that $\gamma\alpha \neq 0$, and hence $\gamma\varphi \neq 0$. This shows that φ is an epimorphism. ▯

Prop. 6.1 gives:

Proposition 6.3. *A projective object P is a generator if and only if there exists a non-zero morphism $P \to C$ for each $C \neq 0$.*

We put on record the duals of Prop. 6.2 and 6.3 (cf. Prop. I.6.6):

Proposition 6.4. *Suppose* **C** *has products. If V is a cogenerator, then for every object C there is a monomorphism $C \to V^I$ for some index set I.*

Proposition 6.5. *An injective object E is a cogenerator if and only if there exists a non-zero morphism $C \to E$ for each $C \neq 0$.*

A category is said to be *locally small* if the class of subobjects of any given object actually is a set (recall that subobjects are equivalence classes of monomorphisms). The existence of a generator may be conceived as a local smallness condition:

Proposition 6.6. *If* $\mathbf{C}$ *has a generator, then* $\mathbf{C}$ *is locally small.*

Proof. Let U be a generator and C any object. For each monomorphism $\beta: B \to C$, let $\langle\beta\rangle$ denote the set of all morphisms $U \to C$ which may be factored through β. Since $\mathrm{Hom}(U, C)$ is a set, it suffices to show that if $\beta: B \to C$ and $\beta': B' \to C$ represent different subobjects, then $\langle\beta\rangle \neq \langle\beta'\rangle$. Since β and β' are non-equivalent, it cannot happen that both monomorphisms $\gamma: B \cap B' \to B$ and $\gamma': B \cap B' \to B'$ are isomorphisms; say γ isn't. In that case γ is not an epimorphism, so $\mathrm{Coker}\,\gamma \neq 0$. Hence there exists $\alpha: U \to B$ such that $\mathrm{coker}\,\gamma \cdot \alpha \neq 0$. Then α cannot be factored over B', so $\langle\beta\rangle \neq \langle\beta'\rangle$. □

Examples

1. Generators for Mod-A. A module M is a generator for **Mod**-A if and only if A is a direct summand of some direct sum of copies of M, as follows from Prop. 6.2. This is in a remarkable way dual to the characterization of a projective module as a direct summand of a direct sum of copies of A. This duality has the following formal consequence:

Proposition 6.7. *Let* M *be a right* A*-module with endomorphism ring* $B = \mathrm{End}_A(M)$. *Then:*

(i) *If* M_A *is a generator, then* ${}_BM$ *is finitely generated projective.*
(ii) *If* M_A *is finitely generated projective, then* ${}_BM$ *is a generator.*

Proof. (i) If M_A is a generator, then $M^n \cong A \oplus K$ for some integer n and module K. Applying the functor $\mathrm{Hom}_A(\cdot, M)$ to this, we get $B^n \cong M \oplus \mathrm{Hom}_A(K, M)$ and hence M is finitely generated projective as a left B-module.
(ii) Similarly, from $A^n \cong M \oplus L$ we obtain $M^n \cong B \oplus \mathrm{Hom}_A(L, M)$. □

Every free module is a projective generator for **Mod**-A. Injective cogenerators for **Mod**-A were discussed in Chap. I (§6 and Prop. 9.3).

2. Generation and cogeneration of modules. If M is an A-module, then M is said to be *generated* by another module U if M may be written as a quotient module of a direct sum of copies of U. Dually, M is *cogenerated* by a module V if M may be imbedded as a submodule of a direct product of copies of V, i.e. for each $0 \neq x \in M$ there exists $\alpha: M \to V$ with $\alpha(x) \neq 0$ (cf. Prop. 6.4). E.g. every torsion-free abelian group is cogenerated by the group $\mathbb{Q}$ of rational numbers. A module is called *torsionless* if it is cogenerated by the module A_A.

3. Families of generators. A family $(U_i)_I$ of objects of the abelian category $\mathbf{C}$ is a *family of generators* for $\mathbf{C}$ if for each non-zero morphism $\alpha: B \to C$ in $\mathbf{C}$ there exists a morphism $\beta: U_i \to B$, for some $i \in I$, such that $\alpha\beta \neq 0$. If $\mathbf{C}$ has coproducts, this is clearly equivalent to saying that $\bigoplus_I U_i$ is a generator. The existence of a family of projective generators for $\mathbf{C}$ then implies that $\mathbf{C}$ has enough projectives.

§7. Functor Categories

If **I** is a small category and **C** is an arbitrary category, one can define the *functor category* **Fun**(**I**, **C**), where the objects are the functors $\mathbf{I} \to \mathbf{C}$ and the morphisms are the natural transformations between such functors. In fact, it is obvious how to compose two natural transformations $\xi\colon R \to S$, $\eta\colon S \to T$ between functors R, S, $T\colon \mathbf{I} \to \mathbf{C}$, to get a natural transformation $\eta\xi\colon R \to T$; the composition is clearly associative. For every functor $T\colon \mathbf{I} \to \mathbf{C}$ there is also an identity transformation $1_T\colon T \to T$. The set of all natural transformations $S \to T$ is usually written $\mathrm{Nat}(S, T)$; note that because of the smallness of **I** it actually is a set, for it may be considered as a subset of the product of the sets $\mathrm{Hom}(S(i), T(i))$ taken over all objects i of **I**.

It is a general rule that the functor category **Fun**(**I**, **C**) inherits all the nice properties of **C**. For example:

Proposition 7.1. *If* **I** *is a small category and* **C** *is preadditive (abelian), then* **Fun**(**I**, **C**) *is preadditive (abelian).*

Proof. Suppose **C** is preadditive. Let S and T be functors $\mathbf{I} \to \mathbf{C}$, and consider two natural transformations $\xi, \eta\colon S \to T$. A natural transformation $\xi + \eta$ is then defined as $(\xi+\eta)_i = \xi_i + \eta_i$ for every object i of **I**. This makes $\mathrm{Nat}(S, T)$ into an abelian group, and the composition of natural transformations becomes bilinear. The category **Fun**(**I**, **C**) is thus preadditive.

Now assume **C** is abelian. If $T_1, \ldots, T_n$ are functors $\mathbf{I} \to \mathbf{C}$, then a new functor T is defined as

$$T(i) = T_1(i) \times \cdots \times T_n(i),$$
$$T(\lambda) = T_1(\lambda) \times \cdots \times T_n(\lambda)$$

for objects i and morphisms λ of **I**. One verifies that $T = T_1 \times \cdots \times T_n$ in **Fun**(**I**, **C**), so this category has finite products. To show that it has kernels, consider a natural transformation $\eta\colon S \to T$. Define a functor $K\colon \mathbf{I} \to \mathbf{C}$ as

$$K(i) = \mathrm{Ker}\,\eta_i,$$
$$K(\lambda) = \text{induced from } S(\lambda) \text{ by restriction.}$$

$$\begin{array}{ccccc} K(i) & \longrightarrow & S(i) & \xrightarrow{\eta_i} & T(i) \\ {\scriptstyle K(\lambda)}\downarrow & & \downarrow{\scriptstyle S(\lambda)} & & \downarrow{\scriptstyle T(\lambda)} \\ K(j) & \longrightarrow & S(j) & \xrightarrow[\eta_j]{} & T(j). \end{array}$$

It is easy to see that $K \to S$ is a kernel for η. Similarly **Fun**(**I**, **C**) has cokernels. Note that kernels and cokernels (as well as products) in **Fun**(**I**, **C**) are constructed "point-wise" at the objects of **I**. From this it follows that the validity of the axiom $A\,4$ in **C** is inherited by **Fun**(**I**, **C**), which thus is an abelian category. □

Assume now that both **B** and **C** are preadditive categories and that **B** is small. Let **Hom**(**B**, **C**) denote the full subcategory of **Fun**(**B**, **C**) consisting of the additive functors. The category **Hom**(**B**, **C**) is also preadditive.

Proposition 7.2. *If* **I** *is a small preadditive category and* **C** *is abelian, then* **Hom**(**B**, **C**) *is an abelian subcategory of* **Fun**(**B**, **C**).

Proof. If $T_1, \ldots, T_n$ are additive functors $\mathbf{B} \to \mathbf{C}$, then also $T_1 \times \cdots \times T_n$ is additive. Similarly, if $\eta: S \to T$ is a natural transformation of additive functors S and T, then the functor K constructed in the proof of Prop. 7.1 is additive. The category $\mathbf{Hom}(\mathbf{B}, \mathbf{C})$ is thus abelian for the same reasons as $\mathbf{Fun}(\mathbf{B}, \mathbf{C})$. The inclusion functor $\mathbf{Hom}(\mathbf{B}, \mathbf{C}) \to \mathbf{Fun}(\mathbf{B}, \mathbf{C})$ is clearly exact. □

Of particular interest are the functor categories $\mathbf{Hom}(\mathbf{B}, \mathbf{Ab})$, because they generalize the module categories in a natural way (see Example 1). For each object B of $\mathbf{B}$ we let h_B denote the functor $\mathrm{Hom}(\cdot, B)$: $\mathbf{B}^{\mathrm{op}} \to \mathbf{Ab}$. The following assertion is known as the "Yoneda lemma".

Proposition 7.3. *Let* $\mathbf{B}$ *be a small preadditive category. For every object* B *of* $\mathbf{B}$ *and every additive functor* T: $\mathbf{B}^{\mathrm{op}} \to \mathbf{Ab}$ *there is an isomorphism*

$$\mathrm{Nat}(h_B, T) \cong T(B)$$

which is natural in B *and* T.

Proof. A natural transformation $\eta: h_B \to T$ determines in particular a homomorphism $\eta_B: h_B(B) = \mathrm{Hom}(B, B) \to T(B)$, and we define $\Theta_{B,T}: \mathrm{Nat}(h_B, T) \to T(B)$ as $\Theta_{B,T}(\eta) = \eta_B(1_B)$. Then $\Theta_{B,T}$ is a group homomorphism, as a direct consequence of the definition of addition of natural transformations. $\Theta_{B,T}$ is a monomorphism, because if $\eta: h_B \to T$ has the property that $\eta_B(1_B) = 0$, then for an arbitrary $\alpha: B' \to B$ in $\mathbf{B}$ we get $\eta_{B'}(\alpha) = T(\alpha)\eta_B(1_B) = 0$ and hence $\eta = 0$.

$$\begin{array}{ccc} \mathrm{Hom}(B', B) & \xrightarrow{\eta_{B'}} & T(B') \\ {\scriptstyle h_B(\alpha)}\uparrow & & \uparrow{\scriptstyle T(\alpha)} \\ \mathrm{Hom}(B, B) & \xrightarrow[\eta_B]{} & T(B) \end{array}$$

$\Theta_{B,T}$ is also an epimorphism, for if $x \in T(B)$ is given, then we define $\eta: h_B \to T$ at an object B' as $\eta_{B'}(\alpha) = T(\alpha)(x)$; it is easy to verify that η is a natural transformation, and $\Theta_{B,T}(\eta) = \eta_B(1_B) = T(1_B)(x) = x$.

It remains to prove that $\Theta_{B,T}$ is natural in B and T. This means that one should show that a morphism $\alpha: B \to B_1$ in $\mathbf{B}$ induces a commutative diagram

$$\begin{array}{ccc} \mathrm{Nat}(h_B, T) & \xrightarrow{\Theta_{B,T}} & T(B) \\ \uparrow & & \uparrow{\scriptstyle T(\alpha)} \\ \mathrm{Nat}(h_{B_1}, T) & \xrightarrow[\Theta_{B_1,T}]{} & T(B_1) \end{array}$$

where the left map is induced by $\mathrm{Hom}(\cdot, \alpha): h_B \to h_{B'}$, and that a natural transformation $\xi: S \to T$ induces a commutative diagram

$$\begin{array}{ccc} \mathrm{Nat}(h_B, S) & \xrightarrow{\Theta_{B,S}} & S(B) \\ \downarrow & & \downarrow{\scriptstyle \xi_B} \\ \mathrm{Nat}(h_B, T) & \xrightarrow[\Theta_{B,T}]{} & T(B). \end{array}$$

The detailed verifications of this may be left to the reader. □

If one applies the Yoneda lemma to $T=h_{B'}$, one gets a natural isomorphism $\mathrm{Nat}(h_B, h_{B'})\cong\mathrm{Hom}(B, B')$. The functor $B\mapsto h_B$ is therefore a full imbedding of $\mathbf{C}$ in $\mathbf{Hom}(\mathbf{C}^{\mathrm{op}}, \mathbf{Ab})$ (a faithful functor is an *imbedding* if it carries distinct objects into distinct objects). A functor $\mathbf{B}^{\mathrm{op}}\to\mathbf{Ab}$ which is naturally equivalent to a functor h_B, is called *representable*. Thus:

Corollary 7.4. *A small preadditive category* **B** *is equivalent to the full subcategory of* $\mathbf{Hom}(\mathbf{B}^{\mathrm{op}}, \mathbf{Ab})$ *consisting of the representable functors.*

Since a sequence $0\to T'\to T\to T''\to 0$ in $\mathbf{Hom}(\mathbf{B}^{\mathrm{op}}, \mathbf{Ab})$ is exact if and only if $0\to T'(B)\to T(B)\to T''(B)\to 0$ is exact for each B in **B**, we also get:

Corollary 7.5. *The family* $(h_B)_{B\in\mathbf{B}}$ *is a family of projective generators for* $\mathbf{Hom}(\mathbf{B}^{\mathrm{op}}, \mathbf{Ab})$.

Examples

1. Modules. Let A be a ring, considered as a preadditive category with only one object $*$, and let **C** be an arbitrary preadditive category. To give an additive functor $T\colon A\to\mathbf{C}$ is the same as to give an object $T(*)=C$ in **C** and a ring homomorphism

$$A=\mathrm{Hom}(*, *)\to\mathrm{Hom}_{\mathbf{C}}(T(*), T(*))=\mathrm{Hom}_{\mathbf{C}}(C, C).$$

An object C in **C** together with a ring homomorphism $A\to\mathrm{Hom}_{\mathbf{C}}(C, C)$ is called a *left A-object* in **C**. Dually, the *right A-objects* in **C** are given by a ring homomorphism $A^{\mathrm{op}}\to\mathrm{Hom}_{\mathbf{C}}(C, C)$.

Now choose $\mathbf{C}=\mathbf{Ab}$. The A-objects in **Ab** are clearly the A-modules. Thus we have category isomorphisms

$$A\text{-}\mathbf{Mod}\cong\mathbf{Hom}(A, \mathbf{Ab}),$$

$$\mathbf{Mod}\text{-}A\cong\mathbf{Hom}(A^{\mathrm{op}}, \mathbf{Ab}).$$

2. Presheaves. Let X be a topological space. An abelian *presheaf* F on X assigns to each open set U an abelian group $F(U)$, and assigns to each inclusion $U\subset V$ of open sets a group homomorphism $\rho_{V,U}\colon F(V)\to F(U)$, usually called the *restriction* from V to U, such that

1) $\rho_{U,U}=1_{F(U)}$,
2) if $U\subset V\subset W$, then $\rho_{V,U}\rho_{W,V}=\rho_{W,U}$.

If one defines a category $\mathbf{I}_X$ with open sets in X as objects and the inclusion maps as morphisms, then a presheaf is a functor $\mathbf{I}_X^{\mathrm{op}}\to\mathbf{Ab}$. It follows that the category of presheaves on X is abelian. As an example of a presheaf on X one may take the continuous functions, i.e. $F(U)$ is the abelian group of continuous real-valued functions on the open subset U of X, and $\rho_{V,U}$ is the map restricting a function from V to U, for $U\subset V$.

3. Diagram categories. As a very simple example, consider diagrams of the general form $C_0\to C_1\to C_2$ in **C**. A morphism between two such diagrams $C_0\to C_1\to C_2$ and $C'_0\to C'_1\to C'_2$ is defined to be a triple of morphisms $C_0\to C'_0$,

$C_1 \to C_1'$, $C_2 \to C_2'$ such that the diagram

$$\begin{array}{ccccc} C_0 & \longrightarrow & C_1 & \longrightarrow & C_2 \\ \downarrow & & \downarrow & & \downarrow \\ C_0' & \longrightarrow & C_1' & \longrightarrow & C_2' \end{array}$$

commutes. The class of diagrams of this form makes up a category.

A more systematic way to handle such diagram categories is to consider them as particular cases of functor categories. In the present example we let **I** be the category consisting of just three objects a, b, c and morphisms $\alpha: a \to b$, $\beta: b \to c$, $\beta\alpha: a \to c$ plus the identities. A diagram $C_0 \to C_1 \to C_2$ may be viewed as a functor $\mathbf{I} \to \mathbf{C}$. A morphism between diagrams is just a natural transformation of such functors. Hence the category of diagrams is isomorphic to the functor category **Fun**(**I**, **C**).

In a similar way one can treat more complicated diagrams, including diagrams with commutativity relations (but *not* with exactness conditions!). For example, a commutative diagram

$$\begin{array}{ccc} C_0 & \longrightarrow & C_1 \\ \downarrow & & \downarrow \\ C_2 & \longrightarrow & C_3 \end{array}$$

is a functor to **C** from the category

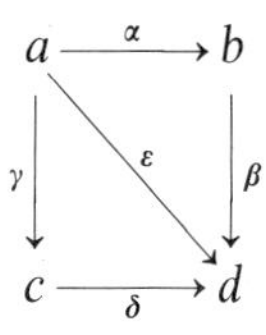

$\beta\alpha = \delta\gamma = \varepsilon$
$+$ identities.

On the other hand, every functor category **Fun**(**I**, **C**), where **I** is a small category, may be considered (though sometimes a bit artificially) as a diagram category. For this reason there is no need to develop any special theory for diagram categories.

4. Direct systems. Let I be a partially ordered set which is directed (Chap. I, § 5), and consider I as a category. If **C** is an arbitrary category, then a functor $I \to \mathbf{C}$ is called a *direct system* in **C**, while a functor $I^{\mathrm{op}} \to \mathbf{C}$ is an *inverse system* in **C**. Direct systems of modules were studied in Chap. I, § 5.

§ 8. Limits and Colimits

Let **C** be a preadditive category and let **I** be a small category. We propose to define the notion of "limit" of a functor $F: \mathbf{I} \to \mathbf{C}$. If X is an object of **C** and if we are given a morphism $\alpha_i: X \to F(i)$ for each $i \in \mathrm{Ob}(\mathbf{I})$, then the family (α_i) is called *compatible* if for every $\lambda: i \to j$ in **I** one has $\alpha_j = F(\lambda)\,\alpha_i$.

Definition. A *limit* (or *projective limit*) of the functor $F: \mathbf{I} \to \mathbf{C}$ is an object $\varprojlim F$ in $\mathbf{C}$ together with a compatible family of morphisms $\pi_i: \varprojlim F \to F(i)$, such that for each other compatible family $\xi_i: X \to F(i)$ there exists a unique $\xi: X \to \varprojlim F(i)$ with $\pi_i\, \xi = \xi_i$.

The limit of F solves a universal problem and is therefore unique up to isomorphism. The category $\mathbf{C}$ is called *complete* if the limit exists for every functor $F: \mathbf{I} \to \mathbf{C}$ when $\mathbf{I}$ is small.

Proposition 8.1. *If $\mathbf{C}$ is complete, then $\varprojlim$ may be considered as a functor $\mathbf{Fun}(\mathbf{I}, \mathbf{C}) \to \mathbf{C}$ for each small category $\mathbf{I}$.*

Proof. If F and G are functors $\mathbf{I} \to \mathbf{C}$, and $\eta: F \to G$ is a natural transformation, then one obtains a compatible family of morphisms $\varprojlim F \to F(i) \to G(i)$. By the definition of $\varprojlim G$, this family factors over a unique morphism $\varprojlim F \to \varprojlim G$, which we denote as $\varprojlim \eta$. It is now rather easy to check that $F \mapsto \varprojlim F$, $\eta \mapsto \varprojlim \eta$ gives a functor $\mathbf{Fun}(\mathbf{I}, \mathbf{C}) \to \mathbf{C}$. □

A *colimit* (or *inductive limit*) of $F: \mathbf{I} \to \mathbf{C}$ is defined in a dual fashion, and is denoted by $\varinjlim F$. In this case we have a compatible family $\psi_i: F(i) \to \varinjlim F$ which is universal for compatible families $\xi_i: F(i) \to X$. The category $\mathbf{C}$ is *cocomplete* if $\varinjlim F$ exists for every $F: \mathbf{I} \to \mathbf{C}$ when $\mathbf{I}$ is small.

Let us look at some particular cases.

Products and coproducts. Suppose $\mathbf{I}$ is a discrete category, i.e. the only morphisms are the identities. Then there are no compatibility conditions to fulfill, so $\varprojlim F = \prod F(i)$ and $\varinjlim F = \bigoplus F(i)$, both taken over $i \in \mathrm{Ob}(\mathbf{I})$.

Direct and inverse limits. Let $\mathbf{I}$ be a directed set. The colimit of a direct system $F: \mathbf{I} \to \mathbf{C}$ is called a *direct limit*, while the limit of an inverse system $F: \mathbf{I}^{\mathrm{op}} \to \mathbf{C}$ is an *inverse limit*.

Pullbacks and pushouts. Let $\mathbf{I}$ be the category

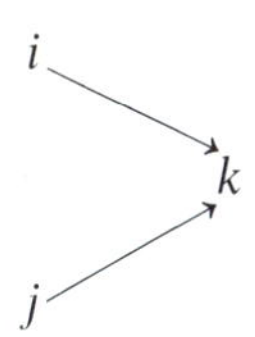

+identity morphisms,

and let $F: \mathbf{I} \to \mathbf{C}$ be a functor. The limit of F is embodied in a diagram

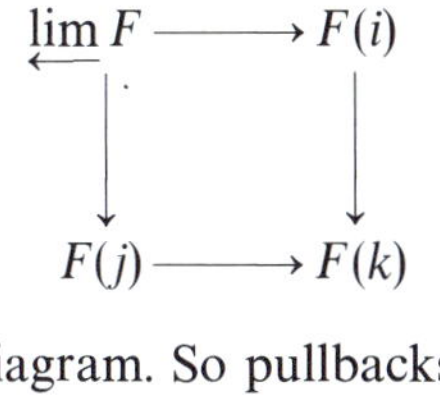

which actually is a pullback diagram. So pullbacks are particular cases of limits, and pushouts of colimits.

It is possible to build up the limit of $F: \mathbf{I} \to \mathbf{C}$ out of products and kernels, thereby generalizing the construction of pullbacks in §5. To accomplish this we need some notation. If $\lambda: i \to j$ is a morphism in $\mathbf{I}$, then we write $i = s(\lambda)$ (the *source* of λ) and $j=t(\lambda)$ (the *target* of λ). For each λ there is a morphism

$$\pi_\lambda = F(\lambda)\, p_{s(\lambda)} - p_{t(\lambda)}\colon \prod_i F(i) \to F(t(\lambda)),$$

where p_i denotes the projection to $F(i)$, and these π_λ induce a morphism

$$\pi\colon \prod_i F(i) \to \prod_\lambda F(t(\lambda)),$$

where the second product is taken over all morphisms λ in $\mathbf{I}$.

Proposition 8.2. *If* $\mathbf{C}$ *has products and kernels and* $F\colon \mathbf{I}\to\mathbf{C}$ *is a functor, then*

$$\varprojlim F = \operatorname{Ker}\Big(\prod_i F(i) \xrightarrow{\pi} \prod_\lambda F(t(\lambda))\Big).$$

Proof. Put $K=\operatorname{Ker}\pi$ and $u=\ker\pi\colon K\to\prod_i F(i)$. If $\varphi_i\colon K\to F(i)$ is defined as $\varphi_i = p_i u$, then (φ_i) is a compatible family of morphisms. If now $\xi_i\colon X\to F(i)$ is a compatible family, then there is induced a morphism $\xi\colon X\to\prod F(i)$ such that for each $\lambda\colon i\to j$ one has

$$\pi_\lambda\,\xi = \big(F(\lambda)\,p_{s(\lambda)} - p_{t(\lambda)}\big)\,\xi = F(\lambda)\,\xi_i - \xi_j = 0.$$

Therefore ξ factors uniquely over $K=\operatorname{Ker}\pi$, and hence (ξ_i) factors uniquely over (φ_i). This shows that $\operatorname{Ker}\pi$ is a limit for F, with $\varphi_i = p_i u$ as the canonical morphisms $\varprojlim F\to F(i)$. □

Corollary 8.3. *An abelian category is complete if and only if it has products.*

There is a dual result for the colimit:

Proposition 8.4. *If* $\mathbf{C}$ *has coproducts and cokernels and* $F\colon\mathbf{I}\to\mathbf{C}$ *is a functor, then*

$$\varinjlim F = \operatorname{Coker}\Big(\bigoplus_\lambda F(s(\lambda)) \xrightarrow{\iota} \bigoplus_i F(i)\Big)$$

with ι *induced by*

$$\iota_\lambda = u_{t(\lambda)}\,F(\lambda) - u_{s(\lambda)}\colon\ F(s(\lambda)) \to \bigoplus_i F(i).$$

Corollary 8.5. *Let* $\mathbf{C}$ *be a cocomplete abelian category and* $F\colon\mathbf{I}\to\mathbf{C}$ *a functor. If* $\alpha_i\colon F(i)\to C$ *is a compatible family of morphisms, inducing* $\alpha\colon \varinjlim F\to C$, *then* $\operatorname{Im}\alpha = \sum \operatorname{Im}\alpha_i$.

Proof. The morphisms α_i induce $\bar\alpha = \bigoplus_i F(i)\to C$, and $\bar\alpha$ can be factored as $\bigoplus F(i)\to\bigoplus\operatorname{Im}\alpha_i\to C$, where the first morphism is an epimorphism by the dual of Prop. 3.1. By the definition of sums we have

$$\sum \operatorname{Im}\alpha_i = \operatorname{Im}(\bigoplus \operatorname{Im}\alpha_i \to C) = \operatorname{Im}\big(\bigoplus F(i)\to C\big) = \operatorname{Im}\alpha,$$

because $\bar\alpha$ can also be factored by Prop. 8.4 as $\bigoplus F(i)\to\varinjlim F\xrightarrow{\alpha} C$, where the first morphism is an epimorphism. □

Let $\mathbf{C}$ and $\mathbf{D}$ be preadditive categories. A functor $T\colon\mathbf{C}\to\mathbf{D}$ is said to *preserve limits* if for every functor $F\colon\mathbf{I}\to\mathbf{C}$, where $\mathbf{I}$ is small and $\varprojlim F$ exists, one has $T(\varprojlim F) = \varprojlim TF$ and T carries the canonical morphisms $\varprojlim F\to F(i)$ into the canonical morphisms $\varprojlim TF\to TF(i)$. It follows from Prop. 8.2 that T preserves limits if and only if it preserves products and kernels.

Proposition 8.6. *Let $T: \mathbf{C} \to \mathbf{D}$ be an additive functor between two abelian categories $\mathbf{C}$ and $\mathbf{D}$. The following properties of T are equivalent:*

(a) *T is left exact.*

(b) *T preserves kernels.*

(c) *T preserves finite limits.*

Proof. By "finite limit" is meant the limit of a functor $\mathbf{I} \to \mathbf{C}$, where $\mathbf{I}$ has only finitely many objects and morphisms. The equivalence of (a) and (b) is proved precisely as in the case of modules, and so it may be left to the reader. The equivalence of (b) and (c) follows from the formula of Prop. 8.2. □

It is of particular interest to study limits and colimits in functor categories. Let $\mathbf{C}$ be a complete category and let $\mathbf{J}$ be a small category. We assert that $\mathbf{Fun}(\mathbf{J}, \mathbf{C})$ also is complete and that its limits may be computed "point-wise" in $\mathbf{C}$, as we already know is true for products and kernels. Suppose $\mathbf{I}$ is a small category and $F: \mathbf{I} \to \mathbf{Fun}(\mathbf{J}, \mathbf{C})$ is a functor. For each object j of $\mathbf{J}$ we define a functor $F_j: \mathbf{I} \to \mathbf{C}$ as

$$F_j(i) = F(i)(j).$$

$$F_j(\lambda) = F(\lambda)_j.$$

A morphism $\mu: j \to j'$ in $\mathbf{J}$ induces a natural transformation $\check{\mu}: F_j \to F_{j'}$, defined as

$$\check{\mu}_i = F(i)(\mu): \; F_j(i) \to F_{j'}(i).$$

$\varprojlim F$ is then the functor $\mathbf{J} \to \mathbf{C}$ given by

$$(\varprojlim F)(j) = \varprojlim F_j,$$

$$(\varprojlim F)(\mu) = \varprojlim \check{\mu}.$$

By this construction we may associate to the functor $F: \mathbf{I} \to \mathbf{Fun}(\mathbf{J}, \mathbf{C})$ a new functor $\check{F}: \mathbf{J} \to \mathbf{Fun}(\mathbf{I}, \mathbf{C})$, defined as

$$\check{F}(j) = F_j, \qquad \check{F}(\mu) = \check{\mu},$$

in terms of which we have arrived at the following conclusion:

Proposition 8.7. *If $\mathbf{J}$ is small and $\mathbf{C}$ is complete, then $\mathbf{Fun}(\mathbf{J}, \mathbf{C})$ is complete. If $F: \mathbf{I} \to \mathbf{Fun}(\mathbf{J}, \mathbf{C})$ is a functor, then $\varprojlim F$ is the composed functor*

$$\mathbf{J} \xrightarrow{\check{F}} \mathbf{Fun}(\mathbf{I}, \mathbf{C}) \xrightarrow{\varprojlim} \mathbf{C}. \tag{1}$$

The correspondence $F \to \check{F}$ is bijective, in fact there are category isomorphisms

$$\mathbf{Fun}(\mathbf{I}, \mathbf{Fun}(\mathbf{J}, \mathbf{C})) \cong \mathbf{Fun}(\mathbf{I} \times \mathbf{J}, \mathbf{C}) \cong \mathbf{Fun}(\mathbf{J}, \mathbf{Fun}(\mathbf{I}, \mathbf{C}))$$

as one may verify. If F corresponds to $\tilde{F}: \mathbf{I} \times \mathbf{J} \to \mathbf{C}$, then one also verifies (this is left to the reader) that

$$\varprojlim_{\mathbf{J}} (\varprojlim_{\mathbf{I}} F) = \varprojlim_{\mathbf{I} \times \mathbf{J}} \tilde{F} = \varprojlim_{\mathbf{I}} (\varprojlim_{\mathbf{J}} \check{F}). \tag{2}$$

An important consequence of all this is:

Proposition 8.8. *The functor* $\varprojlim$: $\mathbf{Fun}(\mathbf{J}, \mathbf{C}) \to \mathbf{C}$ *preserves limits.*

Proof. For every functor F: $\mathbf{I} \to \mathbf{Fun}(\mathbf{J}, \mathbf{C})$ we obtain

$$\varprojlim_{\mathbf{J}}(\varprojlim_{\mathbf{I}} F) = \varprojlim_{\mathbf{I}}(\varprojlim_{\mathbf{J}} \check{F}) = \varprojlim_{\mathbf{I}}(\varprojlim_{\mathbf{J}} \cdot F),$$

by using (2) and applying (1) to $\check{F}$ instead of F. □

An alternative proof of this important result will be given in §9. When also **J** is preadditive, similar results hold for **Hom**(**J**, **C**). There is of course a dual theory for colimits.

Examples

1. Limits of modules. Let F: $\mathbf{I} \to \mathbf{Mod}\text{-}A$ be a functor from a small category **I**. The formula of Prop. 8.2 becomes

$$\varprojlim F = \{(x_i) \in \prod_i F(i) \mid F(\lambda)\, x_i = x_j \text{ for every } \lambda\colon i \to j\}.$$

2. Direct limits of modules. The direct limit functor is always right exact by the dual of Prop. 8.8. In Chap. I, §5, it was proved that in the case of a module category, direct limits are even exact. More general abelian categories with this property will be the object of study in the next chapter.

3. Direct families of subobjects. Let **C** be a cocomplete abelian category, and let C be an object of **C**. A family $(C_i)_I$ of subobjects of C is said to be *direct* if I becomes a directed set when one defines $i \leq j$ whenever $C_i \subset C_j$. It should be noted that the direct limit $\varinjlim C_i$ is not necessarily a subobject of C, unless direct limits preserve monomorphisms. But there is a canonical morphism α: $\varinjlim C_i \to C$, and $\operatorname{Im} \alpha = \sum C_i$ by Cor. 8.5. We also make the following useful observation:

Proposition 8.9. *If* $(C_i)_I$ *is a direct family of subobjects of* C, *then the direct limit of the direct system* $(C/C_i)_I$ *of quotient objects is* $C/\sum C_i$.

Proof. The exact sequences $0 \to C_i \to C \to C/C_i \to 0$ induce in the direct limit an exact sequence

$$\varinjlim C_i \xrightarrow{\alpha} C \to \varinjlim C/C_i \to 0$$

by the dual of Prop. 8.8. Hence $\varinjlim C/C_i = C/\operatorname{Im} \alpha = C/\sum C_i$. □

4. Coproducts. If $(C_i)_I$ is an infinite family of objects, then the coproduct $\bigoplus_I C_i$ is the direct limit of the coproducts $\bigoplus_J C_j$ as J runs through all finite subsets of I.

5. The Hom functor. The functors $\operatorname{Hom}(C, \cdot)$: $\mathbf{C} \to \mathbf{Ab}$ and $\operatorname{Hom}(\cdot, C')$: $\mathbf{C}^{\mathrm{op}} \to \mathbf{Ab}$ both preserve limits. In fact, from the very definition of limits and colimits one has the formulas

$$\operatorname{Hom}_{\mathbf{C}}(C, \varprojlim D_i) \cong \varprojlim \operatorname{Hom}_{\mathbf{C}}(C, D_i),$$

$$\operatorname{Hom}_{\mathbf{C}}(\varinjlim C_i, D) \cong \varprojlim \operatorname{Hom}_{\mathbf{C}}(C_i, D),$$

where the right hand limits are taken in **Ab**.

§9. Adjoint Functors

Let $\mathbf{C}$ and $\mathbf{D}$ be two preadditive categories, and consider two additive functors $S\colon \mathbf{C}\to\mathbf{D}$ and $T\colon \mathbf{D}\to\mathbf{C}$. We say that T is a *right adjoint* of S (and symmetrically S is a *left adjoint* of T) if there is a natural equivalence

$$\eta\colon \operatorname{Hom}_{\mathbf{C}}(\cdot, T(\cdot))\to \operatorname{Hom}_{\mathbf{D}}(S(\cdot),\cdot)$$

of functors $\mathbf{C}^{\mathrm{op}}\times\mathbf{D}\to\mathbf{Ab}$, i.e. for each pair of objects $C\in\mathbf{C}$, $D\in\mathbf{D}$ there is an isomorphism

$$\eta_{C,D}\colon \operatorname{Hom}_{\mathbf{C}}(C, T(D))\to \operatorname{Hom}_{\mathbf{D}}(S(C), D) \tag{1}$$

which is natural in C and D. A right adjoint is uniquely determined by S:

Proposition 9.1. *If T and T' both are right adjoints of $S\colon \mathbf{C}\to\mathbf{D}$, then T and T' are naturally equivalent.*

Proof. (1) implies that there are isomorphisms

$$\lambda_{C,D}\colon \operatorname{Hom}(C, T(D))\cong \operatorname{Hom}(C, T'(D))$$

which are natural in C and D, and it is easy to show that if we put $\varphi_D=\lambda_{T(D),D}(1_{T(D)})\colon T(D)\to T'(D)$, then $\varphi\colon T\to T'$ is a natural equivalence. □

We will examine some of the consequences of an adjoint situation. By choosing $C=T(D)$ in (1) we get a morphism

$$\xi_D=\eta_{T(D),D}(1_{T(D)})\colon\ ST(D)\to D,$$

and the naturality in D implies that this produces a natural transformation $\xi\colon ST\to 1_{\mathbf{D}}$. Dually, there is a natural transformation $\zeta\colon 1_{\mathbf{C}}\to TS$ given by

$$\zeta_C=\eta^{-1}_{C,S(C)}(1_{S(C)})\colon\ C\to TS(C).$$

It is an important property of ξ and ζ that η may be reconstructed from each one of them. In fact we have

$$\eta_{C,D}(\alpha)=\xi_D\cdot S(\alpha)\quad\text{for}\quad \alpha\colon\ C\to T(D), \tag{2}$$

$$\eta^{-1}_{C,D}(\beta)=T(\beta)\cdot\zeta_C\quad\text{for}\quad \beta\colon\ S(C)\to D. \tag{3}$$

(2) follows from the commutativity of the diagram expressing the naturality of η with respect to $\alpha\colon C\to T(D)$:

$$\begin{array}{ccc}
\operatorname{Hom}(T(D), T(D)) & \xrightarrow{\eta_{T(D),D}} & \operatorname{Hom}(ST(D), D)\\
\downarrow & & \downarrow\\
\operatorname{Hom}(C, T(D)) & \xrightarrow[\eta_{C,D}]{} & \operatorname{Hom}(S(C), D),
\end{array}$$

and (3) follows dually. Also the naturality of η is a consequence of (2), or more precisely:

Proposition 9.2. *Let S: $\mathbf{C} \to \mathbf{D}$ and T: $\mathbf{D} \to \mathbf{C}$ be functors, and let there be given a natural transformation ξ: $ST \to 1_{\mathbf{D}}$. Define*

$$\eta_{C,D}: \operatorname{Hom}(C, T(D)) \to \operatorname{Hom}(S(C), D)$$

as $\eta_{C,D}(\alpha) = \xi_D \cdot S(\alpha)$. If $\eta_{C,D}$ is an isomorphism for all C and D, then η makes T into a right adjoint of S.

Proof. It only remains to show that η is natural in C and D. Let γ: $C \to C'$ be a morphism in $\mathbf{C}$. The diagram

$$\begin{array}{ccc} \operatorname{Hom}(C', T(D)) & \xrightarrow{\eta_{C',D}} & \operatorname{Hom}(S(C'), D) \\ \downarrow & & \downarrow \\ \operatorname{Hom}(C, T(D)) & \xrightarrow[\eta_{C,D}]{} & \operatorname{Hom}(S(C), D) \end{array}$$

is commutative, because on one hand it maps α: $C' \to T(D)$ to $\eta_{C,D}(\alpha\gamma) = \xi_D S(\alpha\gamma)$, and on the other hand it maps α to $\eta_{C',D}(\alpha) S(\gamma) = \xi_D S(\alpha) S(\gamma) = \xi_D S(\alpha\gamma)$. Thus η is natural in C.

A morphism β: $D \to D'$ in $\mathbf{D}$ induces a commutative diagram

$$\begin{array}{ccccc} \operatorname{Hom}(C, T(D)) & \xrightarrow{S(\cdot)} & \operatorname{Hom}(S(C), ST(D)) & \xrightarrow{\xi_D \cdot} & \operatorname{Hom}(S(C), D) \\ \downarrow & & \downarrow & & \downarrow \\ \operatorname{Hom}(C, T(D')) & \xrightarrow[S(\cdot)]{} & \operatorname{Hom}(S(C), ST(D')) & \xrightarrow[\xi_{D'} \cdot]{} & \operatorname{Hom}(S(C), D') \end{array}$$

which gives naturality in D. $\square$

By choosing $\alpha = \zeta_C$ in (2) and $\beta = \xi_D$ in (3), one obtains:

Proposition 9.3. *Let T be a right adjoint of S: $\mathbf{C} \to \mathbf{D}$. Then $\xi_{S(C)} \cdot S(\zeta_C) = 1_{S(C)}$ and $T(\xi_D) \cdot \zeta_{T(D)} = 1_{T(D)}$.*

The following is a basic property of adjoint functors:

Proposition 9.4. *A right adjoint functor preserves limits, while a left adjoint functor preserves colimits.*

Proof. Let T be a right adjoint of S: $\mathbf{C} \to \mathbf{D}$, and suppose F: $\mathbf{I} \to \mathbf{D}$ is a functor such that $\varprojlim F$ exists in $\mathbf{D}$. For an arbitrary object C in $\mathbf{C}$ we then have

$$\operatorname{Hom}(C, T(\varprojlim F)) \cong \operatorname{Hom}(S(C), \varprojlim F).$$

But a morphism $S(C) \to \varprojlim F$ corresponds to a compatible family of morphisms $S(C) \to F(i)$, which by the naturality of adjointness corresponds to a compatible family of morphisms $C \to TF(i)$. So $T(\varprojlim F)$ has the universal property defining $\varprojlim TF$, and hence $T(\varprojlim F) = \varprojlim TF$. $\square$

Another useful preservation property of adjoint functors is:

Proposition 9.5. *Let S: $\mathbf{C} \to \mathbf{D}$ be an exact functor between two abelian categories. If T: $\mathbf{D} \to \mathbf{C}$ is a right adjoint of S, and D is an injective object of $\mathbf{D}$, then $T(D)$ is injective in $\mathbf{C}$.*

Proof. There is a natural equivalence

$$\mathrm{Hom}_{\mathbf{C}}(\cdot, T(D)) \cong \mathrm{Hom}_{\mathbf{D}}(S(\cdot), D)$$

of functors $\mathbf{C} \to \mathbf{Ab}$. Since S is exact and D is injective, these functors must be exact, so $T(D)$ is injective. □

Examples

1. Limits and colimits. Assume that $\mathbf{C}$ is complete and let $\mathbf{I}$ be a small category. The functor $\varprojlim: \mathbf{Fun}(\mathbf{I}, \mathbf{C}) \to \mathbf{C}$ is the right adjoint of the functor $k: \mathbf{C} \to \mathbf{Fun}(\mathbf{I}, \mathbf{C})$, which to each C in $\mathbf{C}$ associates the constant functor $k_C: \mathbf{I} \to \mathbf{C}$ with $k_C(i) = C$, $k_C(\lambda) = 1_C$, and to each $\alpha: C \to C'$ in $\mathbf{C}$ associates the obvious natural transformation $k_C \to k_{C'}$. To see that this is true, we have to show that there are natural isomorphisms

$$\mathrm{Hom}_{\mathbf{C}}(C, \varprojlim F) \cong \mathrm{Nat}(k_C, F).$$

But this is clear, because a natural transformation $k_C \to F$ is just a compatible family of morphisms $C \to F(i)$.

From this adjointness follows that $\varprojlim$ preserves limits, a fact which was previously proved in Prop. 8.8.

The colimit functor $\varinjlim: \mathbf{Fun}(\mathbf{I}, \mathbf{C}) \to \mathbf{C}$ is a left adjoint of k.

2. Equivalences. Let $S: \mathbf{C} \to \mathbf{D}$ be an additive functor and assume that S is an equivalence, i.e. there exists $T: \mathbf{D} \to \mathbf{C}$ with natural equivalences $\xi: ST \to 1$ and $\zeta: 1 \to TS$. If we define $\eta_{C,D}: \mathrm{Hom}(C, T(D)) \to \mathrm{Hom}(S(C), D)$ as $\eta_{C,D}(\alpha) = \xi_D \cdot S(\alpha)$, then $\eta_{C,D}$ becomes an isomorphism because S is full and faithful and ξ_D is an isomorphism. It follows from Prop. 9.2 that η makes T into a right adjoint of S. By symmetry, T is also a left adjoint of S. In particular we may conclude that an equivalence preserves limits and colimits.

3. The tensor product. Let A and B be two rings and consider modules L_A, ${}_AM_B$ and N_B. By Prop. I.9.2 there is a natural equivalence of functors

$$\mathrm{Hom}_B(L \otimes_A M, N) \cong \mathrm{Hom}_A(L, \mathrm{Hom}_B(M, N)).$$

This means that $\cdot \otimes_A M: \mathbf{Mod}\text{-}A \to \mathbf{Mod}\text{-}B$ is a left adjoint of

$$\mathrm{Hom}_B(M, \cdot): \ \mathbf{Mod}\text{-}B \to \mathbf{Mod}\text{-}A.$$

4. Ring extensions. Let $\varphi: A \to B$ be a ring homomorphism, and define functors

$$\mathbf{Mod}\text{-}A \underset{\varphi^!}{\overset{\varphi^*}{\rightleftarrows}} \mathbf{Mod}\text{-}B \quad (\varphi^*, \varphi^!: \mathbf{Mod}\text{-}A \to \mathbf{Mod}\text{-}B,\ \varphi_*: \mathbf{Mod}\text{-}B \to \mathbf{Mod}\text{-}A)$$

as

$$\varphi^*(M) = M \otimes_A B \qquad \text{(extension of scalars)}$$

$$\varphi_*(N) = N \qquad \text{(restriction of scalars)}$$

$$\varphi^!(M) = \mathrm{Hom}_A(B, M).$$

φ^* is a left adjoint of φ_*, while $\varphi^!$ is a right adjoint of φ_*.

§ 10. Morita Equivalence

Let A and B be two rings. In this section we will examine the signification and implications of an equivalence between the module categories **Mod**-A and **Mod**-B. First of all we note the following result of Eilenberg [1] and Watts [1]:

Proposition 10.1. *The following assertions are equivalent for a functor* S: **Mod**-$A \to$ **Mod**-B:

(a) *S has a right adjoint.*

(b) *S is right exact and preserves direct sums.*

(c) *$S \cong \cdot \otimes_A P$ for some bimodule ${}_AP_B$, which is unique up to natural isomorphisms.*

Proof. The implications (a) ⇒ (b) and (c) ⇒ (a) are clear, so it remains to show (b) ⇒ (c). Assume S is right exact and preserves direct sums. Put $P = S(A)$. The functor S induces a ring homomorphism

$$A \cong \mathrm{Hom}_A(A, A) \to \mathrm{Hom}_B(S(A), S(A)),$$

which makes $S(A)$ into an A-B-bimodule. If M is an arbitrary A-module, there is an exact sequence

$$A^{(J)} \to A^{(I)} \to M \to 0.$$

When applying the functors $\cdot \otimes_A P$ and S to this sequence, we obtain the both rows of the commutative diagram

$$\begin{array}{ccccccc} P^{(J)} & \longrightarrow & P^{(I)} & \longrightarrow & M \otimes_A P & \longrightarrow & 0 \\ \| & & \| & & & & \\ S(A)^{(J)} & \longrightarrow & S(A)^{(I)} & \longrightarrow & S(M) & \longrightarrow & 0 \end{array}$$

Since both rows are exact, there is induced an isomorphism $M \otimes_A P \cong S(M)$. It is easily verified that this isomorphism is natural in M, so that we have a natural equivalence between S and $\cdot \otimes_A P$. □

Corollary 10.2. *The following assertions are equivalent for a functor* S: **Mod**-$B \to$ **Mod**-A:

(a) *S is an equivalence.*

(b) *There exist bimodules ${}_AP_B$ and ${}_BQ_A$ with bimodule isomorphisms $\alpha: P \otimes_B Q \to A$ and $\beta: Q \otimes_A P \to B$, and $S = \cdot \otimes_B Q$.*

One remarkable feature of condition (b) is its symmetry, which implies that **Mod**-B is equivalent to **Mod**-A if and only if B-**Mod** is equivalent to A-**Mod**. When such equivalences exist, we say that the rings A and B are *Morita equivalent.*

Proposition 10.3. *Let A and B be Morita equivalent rings with bimodules ${}_AP_B$ and ${}_BQ_A$ and bimodule isomorphisms $P \otimes_B Q \cong A$, $Q \otimes_A P \cong B$. Then:*

(i) *P_B, Q_A, ${}_BQ$ and ${}_AP$ are finitely generated projective generators.*

(ii) $\mathrm{Hom}_B(P, B) \cong Q$, $\mathrm{Hom}_A(Q, A) \cong P$, $\mathrm{Hom}_B(Q, B) \cong P$ *and* $\mathrm{Hom}_A(P, A) \cong Q$.

(iii) *The canonical ring homomorphisms*

$$\mathrm{Hom}_B(P, P) \leftarrow A \rightarrow \mathrm{Hom}_B(Q, Q),$$
$$\mathrm{Hom}_A(Q, Q) \leftarrow B \rightarrow \mathrm{Hom}_A(P, P)$$

are isomorphisms.

Proof. Since the functor $T = \cdot \otimes_A P$ is an equivalence, its right adjoint $\mathrm{Hom}_B(P, \cdot)$ is also an equivalence and is naturally equivalent to $S = \cdot \otimes_B Q$. It follows that $\mathrm{Hom}_B(P, \cdot)$ is right exact and preserves direct sums, and hence P_B is a finitely generated, projective module. Since $\mathrm{Hom}_B(P, \cdot)$ also is faithful, P_B is a generator. Applying the functor $\mathrm{Hom}_B(P, \cdot)$ to B, we obtain

$$\mathrm{Hom}_B(P, B) \cong S(B) = B \otimes_B Q \cong Q.$$

Finally,

$$\mathrm{Hom}_B(P, P) \cong S(P) = P \otimes_B Q \cong A$$

is the canonical ring isomorphism.

The other three cases follow by symmetry. □

A notable consequence of Prop. 10.3 is that Morita equivalence preserves the ideal structure:

Proposition 10.4. *Morita equivalent rings have isomorphic lattices of two-sided ideals.*

Proof. Let A and B be Morita equivalent, with a bimodule ${}_AP_B$ given according to Cor. 10.2. Since $\cdot \otimes_A P$: **Mod**-$A \rightarrow$ **Mod**-B is an equivalence, it induces a lattice isomorphism between right ideals of A and B-submodules of P. The two-sided ideals of A may be characterized as being the right ideals which are invariant under all endomorphisms of A in **Mod**-A, and therefore they correspond to those submodules of P_B which are invariant under all endomorphisms of P_B in **Mod**-B, and these are the A-B-submodules of ${}_AP_B$ by Prop. 10.3 (iii). By symmetry, the lattice of A-B-submodules of P is also isomorphic to the lattice of two-sided ideals of B, so A and B have isomorphic ideal lattices. □

We will now establish a sufficient condition for Morita equivalence, namely that if a module Q is a finitely generated projective generator for **Mod**-A, then it induces a Morita equivalence between A and $B = \mathrm{Hom}_A(Q, Q)$. We start by considering a more general situation. Let there be given a ring A and two right A-modules M and N. Put $B = \mathrm{Hom}_A(M, M)$ and $C = \mathrm{Hom}_A(N, N)$. There are homomorphisms

$$\alpha: \mathrm{Hom}_A(M, N) \otimes_B \mathrm{Hom}_A(N, M) \rightarrow C,$$
$$\beta: \mathrm{Hom}_A(N, M) \otimes_C \mathrm{Hom}_A(M, N) \rightarrow B,$$

defined as $\alpha(f \otimes g) = fg$, $\beta(g \otimes f) = gf$. The images of α and β are two-sided ideals of C resp. B.

We write $N \prec M$ if N is a direct summand of a finite direct sum of copies of M, and we write $N \sim M$ if $N \prec M$ and $M \prec N$.

Lemma 10.5. *The following properties of α are equivalent:*

(a) *α is an epimorphism.*

(b) *α is an isomorphism.*

(c) $N \prec M$.

Proof. (a)$\Leftrightarrow$(b): If α is an epimorphism, then the identity map of N can be written as $1=\sum f_i g_i$ for certain $f_i\colon M\to N$, $g_i\colon N\to M$ $(i=1,\ldots,n)$. If now $\alpha(\sum f_j'\otimes g_j')=0$, then by the associativity of composition of maps we obtain

$$\sum_j f_j'\otimes g_j'=\sum_{i,j} f_j'\otimes g_j' f_i g_i=\sum_{i,j} f_j' g_j' f_i\otimes g_i=(\sum_j f_j' g_j')\sum_i f_i\otimes g_i=0.$$

Hence α is an isomorphism, and the converse is trivial.

(a)$\Leftrightarrow$(c): N is a direct summand of M^n if and only if there exist $f_i\colon M\to N$ and $g_i\colon N\to M$ $(i=1,\ldots,n)$ such that $1=\sum f_i g_i$, i.e. if and only if the image of α is equal to C. □

There is of course a similar statement for β. Combining these results for α and β with Cor. 10.2, we obtain (Hirata [1]):

Proposition 10.6. *If M and N are right A-modules such that $M\sim N$, then* $\mathrm{Hom}_A(M,M)$ *and* $\mathrm{Hom}_A(N,N)$ *are Morita equivalent rings.*

Let us now in particular choose $N=A$. Then

$$\alpha\colon \mathrm{Hom}_A(M,A)\otimes_B M\to A,$$
$$\beta\colon M\otimes_A \mathrm{Hom}_A(M,A)\to B=\mathrm{Hom}_A(M,M),$$

are given by

$$\alpha(f\otimes x)=f(x),\qquad \beta(x\otimes f)=x f(\cdot).$$

The image of α is called the *trace ideal* of M in A, and is denoted by $\tau_A(M)$ or simply $\tau(M)$. We have $A\prec M$ if and only if M is a generator for **Mod**-A, and $M\prec A$ if and only if M is a finitely generated projective A-module.

The desired condition for Morita equivalence between A and $\mathrm{Hom}_A(M,M)$ is now obtained as a consequence of Prop. 10.3 and Prop. 10.6:

Proposition 10.7. *Let M be a right A-module and $B=\mathrm{Hom}_A(M,M)$. The functors $\cdot\otimes_A \mathrm{Hom}_A(M,A)$ and $\cdot\otimes_B M$ induce a Morita equivalence between A and B if and only if M is a finitely generated projective generator in* **Mod**-A.

Examples

1. Matrix rings. If A is any ring, then the matrix ring $M_n(A)$ is Morita equivalent to A. For we can choose M_A to be the free right A-module A^n. Then $\mathrm{Hom}_A(M,A)\cong {}_A A^n$, and

$$\alpha\colon \mathrm{Hom}_A(M,A)\otimes_{M_n(A)} M\to A \quad\text{and}\quad \beta\colon M\otimes_A \mathrm{Hom}_A(M,A)\to M_n(A)$$

take the form

$$\alpha((x_1,\ldots,x_n)\otimes(y_1,\ldots,y_n))=\sum x_i y_i,$$
$$\beta((y_1,\ldots,y_n)\otimes(x_1,\ldots,x_n))=(y_i x_j).$$

2. Commutative rings. Every generator M is a faithful module, i.e. $Ma=0$ implies $a=0$, but the converse does not always hold even for finitely generated projective modules (Exercise 33). However, when the ring A is commutative, we have:

Proposition 10.8. *If A is a commutative ring and P is a finitely generated, projective and faithful module, then P is a generator.*

Proof. Choose projective coordinates (x_i, φ_i) for P. Consider the system of equations

$$\sum x_i\, \varphi_i(x_j) = x_j,$$

and note that the determinant of the matrix $(\delta_{ij} - \varphi_i(x_j))$ may be written as $1-t$, with $t \in \tau_P(A)$. From linear algebra it is known that $x_i(1-t)=0$ for all x_i (see e.g. Lang [1], p. 355), and P faithful implies $1-t=0$, i.e. $1 \in \tau_P(A)$. □

3. Simple rings. If K is a skew-field, then it follows from Prop 10.4 and Example 1 that the matrix ring $M_n(K)$ has no two-sided ideals except the two obvious ones. This was proved already in Prop I.7.8 by a more direct method.

4. Morita invariance. A class **P** of rings is *Morita stable* if $A \in \mathbf{P}$ implies that all rings Morita equivalent to A also belong to **P** (thus e.g. all matrix rings $M_n(A)$ should belong to **P**). E.g. the class of semi-simple rings is Morita stable, because it is characterized by the property that all exact sequences split, and this property is of course preserved under category equivalences. From Prop. 10.4 follows that also the class of simple rings is Morita stable.

Exercises

§2

1. Show that for a preadditive category, the definitions of monomorphisms and epimorphisms made in Exercise I.17 agree with those presently used.

2. Show that in a preadditive category every idempotent endomorphism $e\colon C \to C$ has a kernel and a cokernel.

3. Let **F** be the full subcategory of **Ab** consisting of the torsion-free groups. Show that:

(i) The inclusion map $\mathbb{Z} \to \mathbb{Q}$ is both a monomorphism and an epimorphism in **F**.

(ii) If $\alpha\colon G \to H$ is a morphism in **F**, then Coker α exists and is equal to the intersection of all pure subgroups (Exercise I.44) of H containing $\alpha(G)$.

4. Consider the full subcategory of **Mod**-A consisting of the finitely generated modules. Show that this category has cokernels, but that it has kernels if and only if A is right noetherian.

§3

5. Our definition of products and coproducts may evidently be used in arbitrary categories. Show that in the category of sets, the cartesian products are products, while the disjoint unions are coproducts.

6. Let R be a commutative ring. Show that in the category of commutative R-algebras, the coproduct of A and B is $A \otimes_R B$.

7. Let $\alpha\colon B \to C$ be a morphism in a preadditive category, and assume that α has a kernel and that there exists $\beta\colon C \to B$ such that $\alpha\beta = 1_C$. Show that $B = C \oplus \operatorname{Ker} \alpha$.

§4

8. Show that the full subcategory of **Mod**-A consisting of the coherent modules (Exercise I.6) is a small abelian subcategory of **Mod**-A.

9. Let $\alpha\colon C \to C'$ be a morphism in an abelian category **C**. Consider a factorization of α as $C \xrightarrow{\xi} I \xrightarrow{\eta} C'$. Show that:

(i) If η is a monomorphism, then $\operatorname{Im} \alpha \subset I$.

(ii) If furthermore ξ is an epimorphism, then $\operatorname{Im} \alpha = I$ as subobjects of C'.

10. Let A be the endomorphism ring of an object C of an abelian category. Show that the following properties of $\alpha\in A$ are equivalent:

(a) $\alpha=\alpha\beta\alpha$ for some $\beta\in A$.

(b) $\operatorname{Ker}\alpha$ and $\operatorname{Im}\alpha$ are direct summands of C.

(c) The right ideal αA is a direct summand of A.

(d) The left ideal $A\alpha$ is a direct summand of A.

11. Let $\mathbf{C}$ be an abelian category and $F\colon \mathbf{C}\to\mathbf{C}$ an additive functor. Define a new category $\mathbf{C}\times F$ as follows. An object of $\mathbf{C}\times F$ is a couple (C,α), where C is an object of $\mathbf{C}$ and $\alpha\colon F(C)\to C$ is a morphism such that $\alpha\cdot F(\alpha)=0$. A morphism $\varphi\colon (C,\alpha)\to(C',\alpha')$ of $\mathbf{C}\times F$ is a morphism $\varphi\colon C\to C'$ in $\mathbf{C}$ such that the diagram

$$\begin{array}{ccc} F(C) & \xrightarrow{F(\varphi)} & F(C') \\ {\scriptstyle\alpha}\downarrow & & \downarrow{\scriptstyle\alpha'} \\ C & \xrightarrow[\varphi]{} & C' \end{array}$$

commutes. Show that:

(i) $\mathbf{C}\times F$ is a preadditive category with finite products and kernels.

(ii) If F is right exact, then $\mathbf{C}\times F$ is an abelian category and the forgetful functor $\mathbf{C}\times F\to\mathbf{C}$ is exact.

(iii) If $\mathbf{C}=\mathbf{Mod}\text{-}A$ and $F=\cdot\otimes_A M$ for a bimodule M, then $\mathbf{C}\times F$ is isomorphic to $\mathbf{Mod}\text{-}A\times M$, where $A\times M$ is the trivial extension of A by M (Exercise I.34).

§ 5

$\mathbf{C}$ is an abelian category in the following exercises.

12. Let

$$\begin{array}{ccccc} C_5 & \longrightarrow & C_3 & \longrightarrow & C_1 \\ \downarrow & & \downarrow & & \downarrow \\ C_4 & \longrightarrow & C_2 & \longrightarrow & C_0 \end{array}$$

be a commutative diagram in $\mathbf{C}$. Show that if each of the two squares is a pullback diagram, then the outer rectangle is a pullback.

13. Let $C_1,\dots,C_n$ be subobjects of C in $\mathbf{C}$. Show that the sum $C_1+\dots+C_n$ is direct if and only if $C_i\cap(\sum_{j\neq i} C_j)=0$ holds for every i, $i=1,\dots,n$.

14. Let C_1 and C_2 be subobjects of C in $\mathbf{C}$. Show that $C/(C_1+C_2)$ is a pushout of C/C_1 and C/C_2.

15. Let $\alpha\colon C\to C'$ be a morphism in $\mathbf{C}$. Consider the pullback diagram

$$\begin{array}{ccc} P & \xrightarrow{\pi_2} & C \\ {\scriptstyle\pi_1}\downarrow & & \downarrow{\scriptstyle\alpha} \\ C & \xrightarrow[\alpha]{} & C' \end{array}$$

Show that there is a canonical isomorphism $\operatorname{Coker}(\pi_1-\pi_2)\cong\operatorname{Im}\alpha$.

16. Let $0\to C_0\to P_0\to C\to 0$ and $0\to C_1\to P_1\to C\to 0$ be exact sequences in $\mathbf{C}$ with P_0 and P_1 projective objects. Show that $P_0\oplus C_1\cong P_1\oplus C_0$. (Hint: look at the pullback of $P_0\to C$ and $P_1\to C$.)

§ 6

17. Show that the following properties of a projective right A-module P are equivalent:

(a) P is a generator for $\mathbf{Mod}\text{-}A$.

(b) $P\otimes_A M=0$ implies $M=0$ for every ${}_AM$.

(c) $\operatorname{Hom}_A(P,N)=0$ implies $N=0$ for every N_A.

(d) $\operatorname{Hom}_A(P,S)\neq 0$ for every simple module S.

(e) $P\mathfrak{m}\neq P$ for every maximal left ideal $\mathfrak{m}$ of A.

18. Suppose A is a subring of a ring B. Show that B is a generator for $\mathbf{Mod}\text{-}A$ if and only if A is a direct summand of B as right A-modules.

19. Show that if $\mathfrak{a}$ is a two-sided ideal of A, then the cyclic module $A/\mathfrak{a}$ generates M if and only if $M\mathfrak{a}=0$.

20. Let M and N be A-modules. Show that:

(i) If $\alpha \in \mathrm{Bic}(M \oplus N)$, then $\alpha(M) \subset M$ and so there is induced a ring homomorphism

$$\varphi\colon \mathrm{Bic}(M \oplus N) \to \mathrm{Bic}(M).$$

(ii) If M generates N, then φ is a monomorphism.

21. Let M be an A-module with endomorphism ring $B = \mathrm{End}_A(M)$. Show that the following two conditions are equivalent:

(a) M_A is a generator.

(b) ${}_BM$ is finitely generated projective and $A = \mathrm{Bic}_A(M)$. (Hint: use Exercise 20 to determine the bicommutator of a generator.)

§ 7

22. Let $\mathbf{C}$ be an abelian category. The *endomorphism category* $\mathbf{End(C)}$ of $\mathbf{C}$ is defined as follows. An object of $\mathbf{End(C)}$ is a couple (C, α), where C is an object of $\mathbf{C}$ and α is an endomorphism of C. A morphism $\varphi\colon (C, \alpha) \to (C', \alpha')$ in $\mathbf{End(C)}$ is a morphism $\varphi\colon C \to C'$ such that the diagram

$$\begin{array}{ccc} C & \xrightarrow{\varphi} & C' \\ {\scriptstyle\alpha}\downarrow & & \downarrow{\scriptstyle\alpha'} \\ C & \xrightarrow[\varphi]{} & C' \end{array}$$

commutes. Similarly one defines the *automorphism category* $\mathbf{Aut(C)}$ as the full subcategory of $\mathbf{End(C)}$ consisting of (C, α) with α an automorphism. Show that:

(i) If $\mathbb{N}$ and $\mathbb{Z}$ are considered as categories with one object, then $\mathbf{End(C)} \cong \mathbf{Fun}(\mathbb{N}, \mathbf{C})$ and $\mathbf{Aut(C)} \cong \mathbf{Fun}(\mathbb{Z}, \mathbf{C})$.

(ii) If $\mathbf{C} = \mathbf{Mod}\text{-}A$, then $\mathbf{End(C)} \cong \mathbf{Mod}\text{-}A[X]$ and $\mathbf{Aut(C)} \cong \mathbf{Mod}\text{-}A[X, X^{-1}]$.

23. Let $\mathbf{C}$ be an abelian category, and let $\mathbf{Ses(C)}$ be the category whose objects are the short exact sequences of $\mathbf{C}$, and where a morphism $\mathbf{E} \to \mathbf{E}'$ is defined to be a triple $(\alpha_0, \alpha_1, \alpha_2)$ of morphisms in $\mathbf{C}$ yielding a commutative diagram

$$\begin{array}{rcccccccc} \mathbf{E}\colon & 0 \to & C_0 & \longrightarrow & C_1 & \longrightarrow & C_2 & \to 0 \\ & & {\scriptstyle\alpha_0}\downarrow & & {\scriptstyle\alpha_1}\downarrow & & {\scriptstyle\alpha_2}\downarrow & \\ \mathbf{E}'\colon & 0 \to & C'_0 & \longrightarrow & C'_1 & \longrightarrow & C'_2 & \to 0 \end{array}$$

Show that:

(i) $\mathbf{Ses(C)}$ is a preadditive category.

(ii) $(\alpha_0, \alpha_1, \alpha_2)$ is a monomorphism in $\mathbf{Ses(C)}$ if and only if α_0 and α_1 are monomorphisms in $\mathbf{C}$. Dualize.

(iii) If C is an arbitrary non-zero object of $\mathbf{C}$, then the morphism

$$\begin{array}{ccccccc} 0 \to & 0 & \longrightarrow & C & \longrightarrow & C & \to 0 \\ & {\scriptstyle 0}\downarrow & & {\scriptstyle 1}\downarrow & & {\scriptstyle 0}\downarrow & \\ 0 \to & C & \longrightarrow & C & \longrightarrow & 0 & \to 0 \end{array}$$

is both a monomorphism and an epimorphism in $\mathbf{Ses(C)}$, but it is not an isomorphism. (Hence $\mathbf{Ses(C)}$ is not an abelian category.)

(iv) Every morphism in $\mathbf{Ses(C)}$ has a kernel and a cokernel.

24. Let R be a commutative ring. The abelian category $\mathbf{C}$ is an *R-category* if each group $\mathrm{Hom}_{\mathbf{C}}(C, C')$ has an R-module structure such that composition of morphisms is R-bilinear. Show that:

(i) If $\mathbf{C}$ is an R-category, then $\mathrm{Hom}_{\mathbf{C}}(C, C)$ is an R-algebra (Exercise I.1) for each C in $\mathbf{C}$.

(ii) To give an R-category structure on $\mathbf{C}$ is the same as to give a ring homomorphism $R \to \mathrm{Nat}(1_{\mathbf{C}}, 1_{\mathbf{C}})$.

(iii) If A is an R-algebra, then $\mathbf{Mod}\text{-}A$ is an R-category (cf. Exercise I.18).

25. Let $\mathbf{C}$ be a small abelian category and T an additive functor $T\colon \mathbf{C} \to \mathbf{Ab}$. Show that if T is an injective object of $\mathbf{Hom(C, Ab)}$, then T is right exact.

§8

26. Let $\mathbf{C}$ be a complete abelian category and $F: \mathbf{C} \to \mathbf{C}$ an additive functor. Let $U: \mathbf{C} \times F \to \mathbf{C}$ be the forgetful functor (Exercise 11). Show that:

(i) $\mathbf{C} \times F$ is complete and $U(\varprojlim G) = \varprojlim UG$ for every $G: I \to \mathbf{C} \times F$.

(ii) If $\mathbf{C}$ is cocomplete and F preserves colimits, then $\mathbf{C} \times F$ is cocomplete and $U(\varinjlim G) = \varinjlim UG$ for every $G: I \to \mathbf{C} \times F$.

§9

In the following exercises, let $\mathbf{C}$ and $\mathbf{D}$ be preadditive categories, and let $S: \mathbf{C} \to \mathbf{D}$ be a left adjoint of $T: \mathbf{D} \to \mathbf{C}$.

27. Show that T is faithful if and only if $\xi_D: ST(D) \to D$ is an epimorphism for every D.

28. If $\mathbf{B}$ is a small preadditive category, then S and T induce functors

$$\mathbf{Hom}(\mathbf{B}, \mathbf{C}) \underset{T^*}{\overset{S^*}{\rightleftarrows}} \mathbf{Hom}(\mathbf{B}, \mathbf{D}).$$

Show that S^* is a left adjoint of T^*.

29. If $\mathbf{C}$ and $\mathbf{D}$ are small categories and $\mathbf{B}$ is an arbitrary preadditive category, then S and T induce functors

$$\mathbf{Hom}(\mathbf{D}, \mathbf{B}) \underset{T_*}{\overset{S_*}{\rightleftarrows}} \mathbf{Hom}(\mathbf{C}, \mathbf{B}).$$

Show that S_* is a right adjoint of T_*.

30. Suppose $\mathbf{C}$ and $\mathbf{D}$ are abelian categories. Show that if T carries injective objects into injective objects and if $\mathbf{D}$ has enough injective objects, then S is exact. (Hint: if $C' \to C$ is a monomorphism in $\mathbf{C}$, write $S(C')$ as a subobject of an injective object.)

31. Let P be an A-B-bimodule, assume that ${}_AP$ is finitely generated and projective, and put $P^* = \mathrm{Hom}_A(P, A)$. Show that:

(i) The functors $\mathbf{Mod}\text{-}A \to \mathbf{Mod}\text{-}B$, given by $\mathrm{Hom}_A(P^*, \cdot)$ and $\cdot \otimes_A P$, are naturally equivalent.

(ii) The functor $\cdot \otimes_A P$ is a right adjoint of the functor $\cdot \otimes_B P^*: \mathbf{Mod}\text{-}B \to \mathbf{Mod}\text{-}A$.

(iii) $M \otimes_A P$ is an injective B-module for every injective right A-module M if and only if P^* is a flat B-module.

§10

32. Let M be a generator or a cogenerator for $\mathbf{Mod}\text{-}A$. Show that M is a faithful module.

33. Consider the matrix ring $A = \begin{pmatrix} K & K \\ 0 & K \end{pmatrix}$, where K is a field. Put $e = \begin{pmatrix} 1 & 0 \\ 0 & 0 \end{pmatrix}$. Show that the right ideal eA is a finitely generated, projective and faithful A-module, but that it is not a generator for $\mathbf{Mod}\text{-}A$.

34. Suppose M and N are right R-modules with endomorphism rings $B = \mathrm{Hom}_R(M, M)$ and $A = \mathrm{Hom}_R(N, N)$. Show that if $N \prec M$, then:

(i) $\mathrm{Hom}_R(M, N)$ is a finitely generated projective right B-module and a generator as a left A-module.

(ii) $\mathrm{Hom}_R(N, M)$ is a finitely generated projective left B-module and a generator as a right A-module.

35. A *Morita context* $\langle A, B, P, Q, \alpha, \beta \rangle$ consists of two rings A and B, two bimodules ${}_AP_B$ and ${}_BQ_A$, and two bimodule homomorphisms $\alpha: P \otimes_B Q \to A$ and $\beta: Q \otimes_B P \to B$ such that the diagrams

$$\begin{array}{ccc} P \otimes_B Q \otimes_A P & \xrightarrow{\alpha \otimes 1} & A \otimes_A P \\ {\scriptstyle 1 \otimes \beta} \downarrow & & \downarrow {\scriptstyle \cong} \\ P \otimes_B B & \xrightarrow[\cong]{} & P \end{array} \qquad \begin{array}{ccc} Q \otimes_A P \otimes_B Q & \xrightarrow{\beta \otimes 1} & B \otimes_B Q \\ {\scriptstyle 1 \otimes \alpha} \downarrow & & \downarrow {\scriptstyle \cong} \\ Q \otimes_A A & \xrightarrow[\cong]{} & Q \end{array}$$

are commutative.

Let there be given a ring R and two right R-modules M and N. Show that by putting

$$B = \mathrm{Hom}_R(M, M), \quad A = \mathrm{Hom}_R(N, N), \quad P = \mathrm{Hom}_R(M, N), \quad Q = \mathrm{Hom}_R(N, M)$$

with the canonical pairings (as in the text), one obtains a Morita context (called the *derived context* of M and N).

36. Let $\langle A, B, P, Q, \alpha, \beta\rangle$ be a Morita context. The two-sided ideals $\tau_A = \operatorname{Im}\alpha$ and $\tau_B = \operatorname{Im}\beta$ are called the *trace ideals*. Show that:

(i) If $u \in \operatorname{Ker}\alpha$, then $u\tau_A = 0$.

(ii) If α is an epimorphism, then it is an isomorphism.

(iii) $\tau_A P = P\tau_B$ and $Q\tau_A = \tau_B Q$.

37. Let $\langle A, B, P, Q, \alpha, \beta\rangle$ be a Morita context. Define the generalized matrix ring

$$R = \begin{pmatrix} A & P \\ Q & B \end{pmatrix}_{\alpha,\beta}$$

by using ordinary matrix addition and multiplication (by means of α and β). (Cf. the special case of Exercise I.10.)

(i) Verify that one actually obtains a ring.

(ii) Put $e = \begin{pmatrix} 1 & 0 \\ 0 & 0 \end{pmatrix}$. Show that

$$A = eRe, \quad B = (1-e)R(1-e), \quad P = eR(1-e), \quad Q = (1-e)Re.$$

(iii) Show that if R is an arbitrary ring with an idempotent e, then

$$\langle eRe, (1-e)R(1-e), eR(1-e), (1-e)Re, \alpha, \beta\rangle$$

is a Morita context (for suitable α and β), and R is isomorphic to the corresponding generalized matrix ring.

38. Show that a Morita context may be described as a preadditive category with two objects.

Chapter V. Grothendieck Categories

A cocomplete abelian category $\mathbf{C}$ is called a *Grothendieck category* if direct limits are exact in $\mathbf{C}$ and $\mathbf{C}$ has a generator. Grothendieck categories play an important role in the theory of rings of quotients, being precisely the categories which appear as categories of modules of quotients. They constitute a natural extension of the class of module categories, with which they share a great number of important properties.

§ 1. Exactness of Direct Limits

Let $\mathbf{C}$ be a cocomplete abelian category. The colimit functor $\varinjlim: \mathbf{Fun}(\mathbf{I}, \mathbf{C}) \to \mathbf{C}$ is automatically right exact by the dual of Prop. IV.8.8. Thus direct limits are exact in $\mathbf{C}$ if and only if they preserve monomorphisms. Let C be an object of $\mathbf{C}$ and let $(C_i)_I$ be a direct family of subobjects of C. If direct limits in $\mathbf{C}$ are exact, then it follows from observations made in Chap. IV (§ 8, Example 3) that $\varinjlim C_i$ is a subobject of C which coincides with $\sum_I C_i$, and is called the *direct union* of the subobjects C_i. The exactness of direct limits implies in particular that direct unions preserve finite intersections, so for any subobject B of C one has

$$(\sum_I C_i) \cap B = \sum_I (C_i \cap B). \qquad (AB5)$$

In the terminology of Chap. III this means that the (big) lattice $\mathsf{L}(C)$ of subobjects of C is upper continuous. The designation "$AB5$" for the condition of upper continuity stems from Grothendieck's classical paper [1]. This condition is important because it implies exactness of direct limits:

Proposition 1.1. *The following properties of the cocomplete abelian category* $\mathbf{C}$ *are equivalent:*

(a) *Direct limits are exact in* $\mathbf{C}$.

(b) $\mathbf{C}$ *satisfies condition* $AB5$.

(c) *For every morphism* $\alpha: B \to C$ *and direct family* $(C_i)_I$ *of subobjects of* C, *one has*

$$\alpha^{-1}(\sum C_i) = \sum \alpha^{-1}(C_i).$$

Proof. (a) $\Rightarrow$ (b) is clear.

(b) ⇒ (c): The morphism α induces for each C_i a diagram

$$\begin{array}{ccc} \alpha^{-1}(C_i) & \longrightarrow & B \\ \downarrow & & \downarrow \\ C_i \cap \operatorname{Im}\alpha & \longrightarrow & \operatorname{Im}\alpha \end{array}$$

which is easily seen to be a pullback diagram. By Prop. IV.5.1 it gives an exact sequence

$$0 \to \alpha^{-1}(C_i) \to B \to \operatorname{Im}\alpha/(C_i \cap \operatorname{Im}\alpha) \to 0.$$

In the direct limit one obtains, with the use of Prop. IV.8.9 and after noting that $AB5$ implies $\sum(C_i \cap \operatorname{Im}\alpha) = (\sum C_i) \cap \operatorname{Im}\alpha$, a commutative diagram with exact rows

$$\begin{array}{ccccccccc} 0 \longrightarrow & \sum \alpha^{-1}(C_i) & \longrightarrow & B & \longrightarrow & \varinjlim \operatorname{Im}\alpha/(C_i \cap \operatorname{Im}\alpha) & \longrightarrow 0 \\ & \downarrow & & \downarrow & & \| & \\ 0 \longrightarrow & (\sum C_i) \cap \operatorname{Im}\alpha & \longrightarrow & \operatorname{Im}\alpha & \longrightarrow & \varinjlim \operatorname{Im}\alpha/(C_i \cap \operatorname{Im}\alpha) & \longrightarrow 0 \\ & \downarrow & & \downarrow & & \downarrow & \\ 0 \longrightarrow & \sum C_i & \longrightarrow & C & \longrightarrow & C/\sum C_i & \longrightarrow 0 \end{array}$$

The upper left square is a pullback diagram by Example IV.3.3 and the lower left square is obviously also a pullback diagram. It follows that the outer left square is a pullback diagram (Exercise IV.12), and hence $\sum \alpha^{-1}(C_i) = \alpha^{-1}(\sum C_i)$.

(c) ⇒ (b): If B is a subobject of C and $(C_i)_I$ is a direct family of subobjects of C, then one applies (c) to the monomorphism $\alpha\colon B \to C$, and obtains

$$B \cap (\sum C_i) = \alpha^{-1}(\sum C_i) = \sum \alpha^{-1}(C_i) = \sum (B \cap C_i).$$

It remains to prove that (b) and (c) together imply (a). We first show:

Lemma 1.2. *Suppose* **C** *satisfies $AB5$ and let $(C_i)_I$ be a direct system of objects in* **C**. *For each $k \in I$ one has*

$$\operatorname{Ker}(C_k \to \varinjlim C_i) = \sum_{j \geq k} \operatorname{Ker}(C_k \to C_j).$$

Proof. We will use the description of the colimit as a quotient object of a coproduct (Prop. IV.8.4). For each $k \in I$ there is a commutative diagram

$$\begin{array}{ccc} & & \bigoplus_I C_i \\ & \nearrow^{u_k} & \\ C_k & & \downarrow p \\ & \searrow_{\iota_k} & \\ & & \varinjlim C_i \end{array}$$

of canonical morphisms. Let R be the subset of $I \times I$ consisting of pairs (i, j) with $i \leq j$. For each $S \subset R$ we put

$$C_S = \sum_{(i,j) \in S} \operatorname{Im}(u_i - u_j \gamma_{ij}) \subset \bigoplus_I C_i,$$

where $\gamma_{ij}\colon C_i \to C_j$ is the canonical morphism for $i \leq j$. Then $\varinjlim C_i = \bigoplus_I C_i / C_R$. But C_R may be written as a direct union $C_R = \sum C_S$, where S runs through all finite subsets of R. Hence for each $k \in I$ one has

$$\operatorname{Ker} \iota_k = u_k^{-1}(C_R) = \sum u_k^{-1}(C_S)$$

by condition (c). Now in order to show that $\operatorname{Ker} \iota_k = \sum_{j \geq k} \operatorname{Ker} \gamma_{kj}$, it suffices to show that for every finite S one has $u_k^{-1}(C_S) \subset \operatorname{Ker} \gamma_{ks}$ for some $s \geq k$. Since S is finite, one can choose $s \in I$ such that $k \leq s$ and $(i, j) \in S$ implies $i \leq s$, $j \leq s$. Define $\beta\colon \bigoplus_I C_i \to C_S$ by requiring

$$\beta u_i = \begin{cases} \gamma_{is} & \text{when } i \leq s \\ 0 & \text{otherwise.} \end{cases}$$

For each $(i, j) \in S$ one then has

$$\beta(u_i - u_j \gamma_{ij}) = \gamma_{is} - \gamma_{js} \gamma_{ij} = 0,$$

so $C_S \subset \operatorname{Ker} \beta$. Hence

$$u_k^{-1}(C_S) \subset u_k^{-1}(\operatorname{Ker} \beta) = \operatorname{Ker} \beta u_k = \operatorname{Ker} \gamma_{ks},$$

as desired. □

We can now prove that (b) and (c) imply (a) in Prop. 1.1. Let $\alpha_i\colon B_i \to C_i$ $(i \in I)$ be a direct system of monomorphisms in **C**, with canonical morphisms $\beta_{ij}\colon B_i \to B_j$ and $\gamma_{ij}\colon C_i \to C_j$ when $i \leq j$. There are commutative diagrams

$$\begin{array}{ccc} B_i & \xrightarrow{\alpha_i} & C_i \\ \downarrow{\scriptstyle \iota'_i} & & \downarrow{\scriptstyle \iota_i} \\ \varinjlim B_i & \xrightarrow[\alpha]{} & \varinjlim C_i . \end{array}$$

Put $K = \operatorname{Ker} \alpha$. Since $\varinjlim B_i = \sum_I \operatorname{Im} \iota'_i$ by Prop. IV.8.4, we have $K = K \cap (\sum \operatorname{Im} \iota'_i) = \sum (K \cap \operatorname{Im} \iota'_i)$. It is therefore enough to show that $K \cap \operatorname{Im} \iota'_i = 0$ for every $i \in I$. Assume on the contrary that a certain $K \cap \operatorname{Im} \iota'_i \neq 0$. It follows that $\iota_i'^{-1}(K) \neq 0$ in B_i. With the use of Prop. IV.5.1 and Lemma 1.2 we get

$$\begin{aligned} 0 \neq \iota_i'^{-1}(K) &= \operatorname{Ker} \alpha \iota'_i = \operatorname{Ker} \iota_i \alpha_i = \alpha_i^{-1}(\operatorname{Ker} \iota_i) \\ &= \alpha_i^{-1}\Big(\sum_{j \geq i} \operatorname{Ker} \gamma_{ij}\Big) = \sum \alpha_i^{-1}(\operatorname{Ker} \gamma_{ij}) = \sum \operatorname{Ker} \gamma_{ij} \alpha_i \\ &= \sum \operatorname{Ker} \alpha_j \beta_{ij} = \sum \operatorname{Ker} \beta_{ij} = \operatorname{Ker} \iota'_i . \end{aligned}$$

This implies $\operatorname{Im} \iota'_i \cap K = \iota'_i(\iota_i'^{-1}(K)) = 0$, a contradiction which finishes the proof of Prop. 1.1. □

Examples

1. Module categories. If A is a ring, then **Mod**-A is a Grothendieck category by Prop. I.5.3.

2. Functor categories. If **B** is a small preadditive category, then the functor category $\mathbf{C}=\mathbf{Hom}(\mathbf{B}^{\mathrm{op}}, \mathbf{Ab})$ is a Grothendieck category. For colimits in **C** are computed "point-wise", so the exactness of direct limits in **Ab** implies their exactness in **C**. The objects in **B** form a family of projective generators for **C** (Cor. IV.7.5).

3. Presheaves. Let X be a topological space and **P** the category of abelian presheaves on X (Example IV.7.2). Direct limits in **P** are exact for the same reason as in the previous example. For each open set U of X we define a presheaf F_U as

$$F_U(V)=\begin{cases}\mathbb{Z} & \text{when} \quad V=U\\ 0 & \text{when} \quad V\neq U.\end{cases}$$

For an arbitrary presheaf G one then has an analogue of the Yoneda lemma, in that $\mathrm{Hom}_{\mathbf{P}}(F_U, G)\cong G(U)$. The F_U, for all open U, constitute a family of projective generators for **C**. Thus **P** is a Grothendieck category.

4. Direct sums. Let C be an object of a Grothendieck category and let $(C_i)_I$ be a family of subobjects of C. The sum $\sum_I C_i$ is direct if and only if this is true for every finite subset of I, and therefore one has that $\sum_I C_i$ is direct if and only if for every finite subset $\{i_0, i_1, \ldots, i_n\}$ of I one has $C_{i_0}\cap(C_{i_1}+\cdots+C_{i_n})=0$ (see Exercise IV.13).

5. Pseudo-complements. A locally small abelian category is called *pseudo-complemented* if the lattice $\mathsf{L}(C)$ of subobjects of any object C of **C** is pseudo-complemented. If **C** is cocomplete and satisfies $AB5$, then **C** is pseudo-complemented by Prop. III.6.3.

§2. Injective Envelopes

In this section we assume that **C** is a locally small abelian category and that **C** is pseudo-complemented. We propose to show that if an object C can be imbedded in an injective object, then this can be done in a certain minimal way.

As in Chap. III (§6) we say that a subobject B of C is *essential* if $B\cap C'\neq 0$ for every non-zero $C'\subset C$, i.e. B is an essential element of the lattice $\mathsf{L}(C)$. Slightly more generally, we say that a monomorphism $\alpha\colon B\to C$ is *essential* if $\operatorname{Im}\alpha$ is an essential subobject of C. The proof of the following fact is straight-forward:

Lemma 2.1. *If $\alpha\colon B\to C$ and $\beta\colon C\to D$ are monomorphisms, then $\beta\alpha$ is essential if and only if both α and β are essential.*

We also note that an injective extension encloses all essential extensions:

Lemma 2.2. *If $\alpha\colon C \to E$ is a monomorphism and E is an injective object, then for every essential monomorphism $\beta\colon C \to C'$ there exists a monomorphism $\gamma\colon C' \to E$ such that $\gamma\beta = \alpha$.*

Proof. Since E is injective, there exists a morphism $\gamma\colon C' \to E$ with $\gamma\beta = \alpha$. Then $\operatorname{Ker}\gamma \cap \operatorname{Im}\beta = \operatorname{Ker}\gamma\beta = \operatorname{Ker}\alpha = 0$, and since β is essential, this implies $\operatorname{Ker}\gamma = 0$. So γ is a monomorphism. □

Definition. An *injective envelope* of an object C is an essential monomorphism $C \to E$, where E is an injective object.

An injective envelope is uniquely determined by C:

Proposition 2.3. *If $\alpha\colon C \to E$ and $\alpha'\colon C \to E'$ are injective envelopes of C, then there is an isomorphism $\gamma\colon E \to E'$ such that $\gamma\alpha = \alpha'$.*

Proof. By Lemma 2.2 there exists a monomorphism $\gamma\colon E \to E'$ such that $\gamma\alpha = \alpha'$, and γ is essential by Lemma 2.1. But the injectivity of E causes $\operatorname{Im}\gamma$ to split off as a direct summand of E', so γ must be an isomorphism. □

Before considering the existence of injective envelopes we characterize injective objects in terms of essentiality:

Proposition 2.4. *An object E is injective if and only if every essential extension of E is an isomorphism.*

Proof. If E is injective, then every monomorphism $\alpha\colon E \to C$ splits, so if α is essential, then it must be an isomorphism. Assume conversely that all essential extensions of E are isomorphisms. If $\alpha\colon C \to C'$ is a monomorphism and $\varphi\colon C \to E$ is given, then form the pushout diagram

$$\begin{array}{ccc} C & \xrightarrow{\ \alpha\ } & C' \\ {\scriptstyle\varphi}\downarrow & & \downarrow{\scriptstyle\varphi'} \\ E & \xrightarrow[\ \beta\ ]{} & P. \end{array}$$

From the dual of Prop. IV.5.1 it is known that β is a monomorphism. Choose a pseudo-complement K of E in P, and let $\pi\colon P \to P/K$ be the canonical morphism. Since $K \cap E = 0$, the composed morphism $\pi\beta\colon E \to P \to P/K$ is a monomorphism. It is even an essential monomorphism, because every non-zero subobject of P/K can be written as L/K, where L strictly contains K, and since K is maximal with respect to $K \cap E = 0$, it follows that $L \cap E \neq 0$ and hence $\operatorname{Im}\pi\beta \cap L/K \neq 0$. By hypothesis, $\pi\beta$ is therefore an isomorphism. The morphism $(\pi\beta)^{-1}\pi\varphi'\colon C' \to E$ furnishes an extension of φ to C'. Hence E is an injective object. □

Proposition 2.5. *If C is a subobject of some injective object, then C has an injective envelope.*

Proof. Suppose C is a subobject of an injective object E. Choose a maximal essential extension C' of C within E, according to Prop. III.6.5. Every essential extension of C' can be imbedded in E by Lemma 2.2, and is therefore an isomorphism by the maximality of C'. It follows that C' is an injective object (Prop. 2.4), and so it is an injective envelope of C. □

In view of Prop. 2.3 it is legitimate to talk about *the* injective envelope of an object C (when it exists!), and we denote it by $E(C)$. However, one should bear in mind that $E(C)$ is determined only up to isomorphism. $E(C)$ can be imbedded in any injective object containing C, but the imbedding is usually not unique.

It is known that every finite direct sum of injective objects is injective. For injective envelopes one has the corresponding result:

Proposition 2.6. *The monomorphism* $C_1 \oplus \cdots \oplus C_n \to E(C_1) \oplus \cdots \oplus E(C_n)$ *induces an isomorphism*

$$E(C_1 \oplus \cdots \oplus C_n) \cong E(C_1) \oplus \cdots \oplus E(C_n).$$

This is an immediate consequence of:

Lemma 2.7. *If* $C_i \to C_i'$ $(i=1,\ldots,n)$ *are essential monomorphisms, then* $C_1 \oplus \cdots \oplus C_n \to C_1' \oplus \cdots \oplus C_n'$ *is essential.*

Proof. By induction it suffices to do the proof for $n=2$, so we show that each of the two monomorphisms $C_1 \oplus C_2 \to C_1' \oplus C_2 \to C_1' \oplus C_2'$ is essential. Consider e.g. the first one. Let π: $C_1' \otimes C_2 \to C_1'$ denote the canonical projection, and suppose β: $B \to C_1' \oplus C_2$ is an arbitrary non-zero monomorphism. Then either $\pi\beta=0$, in which case $\operatorname{Im}\beta \subset C_2$ and hence $\operatorname{Im}\beta \cap (C_1 \oplus C_2) \neq 0$, or $\pi\beta \neq 0$, in which case $\operatorname{Im}\pi\beta \cap C_1 \neq 0$ and hence $\operatorname{Im}\beta \cap (C_1 \oplus C_2) \neq 0$. In any case this shows that $C_1 \oplus C_2$ is essential in $C_1' \oplus C_2$. □

When direct limits in **C** are exact, one can extend Lemma 2.7 to arbitrary families of essential monomorphisms (see Exercise 6).

An object of **C** is called *indecomposable* if it cannot be written as the direct sum of two non-zero subobjects. A subobject B of an object C is *irreducible* in C if it cannot be written as the intersection of two strictly bigger subobjects of C (cf. Example III.3.3). It is clear that B is irreducible in C if and only if C/B is a *coirreducible* object, i.e. any two non-zero subobjects of C/B have non-zero intersection. With the use of Prop. 2.6 one obtains without difficulty:

Proposition 2.8. *The following properties of an injective object E are equivalent:*

(a) *E is indecomposable.*
(b) *Each subobject of E is coirreducible.*
(c) *E is the injective envelope of a coirreducible object.*
(d) *E is an injective envelope of each one of its non-zero subobjects.*

We know that an A-module E is injective if and only if homomorphisms $\mathfrak{a} \to E$, where $\mathfrak{a}$ is a right ideal of A, can be extended to A (Prop. I.6.5). This can be generalized to Grothendieck categories:

Proposition 2.9. *Let* **C** *be a Grothendieck category with a family of generators* $(U_i)_I$. *An object E is injective if and only if for every monomorphism* α: $C \to U_i$ *and morphism* φ: $C \to E$, *there exists* φ': $U_i \to E$ *such that* $\varphi'\alpha = \varphi$.

Proof. Consider an arbitrary monomorphism $\alpha\colon C \to C'$ and a morphism $\varphi\colon C \to E$. As in the proof of Prop. I.6.5 we can use Zorn's lemma to find a maximal extension of φ. The problem is then reduced to the extent that we may assume that φ cannot be extended any further within C', and we show that α is then an isomorphism. Suppose that α is not an isomorphism. Then there exists some morphism $\gamma\colon U_i \to C'$ such that $\operatorname{Im}\gamma$ is not contained in C. Construct the pullback diagram

$$\begin{array}{ccccccccc} 0 & \longrightarrow & K & \longrightarrow & \gamma^{-1}(C) & \longrightarrow & \operatorname{Im}\gamma \cap C & \longrightarrow & 0 \\ & & \| & & \downarrow & & \downarrow & & \\ 0 & \longrightarrow & K & \longrightarrow & U_i & \longrightarrow & \operatorname{Im}\gamma & \longrightarrow & 0 \end{array}$$

Since $\gamma^{-1}(C)$ is a subobject of U_i, the composed morphism

$$\gamma^{-1}(C) \to \operatorname{Im}\gamma \cap C \to C \to E$$

may be extended to $\beta\colon U_i \to E$. Since the restriction of β to K is zero, β can be factored over $\operatorname{Im}\gamma$ to give $\beta'\colon \operatorname{Im}\gamma \to E$. The diagram

$$\begin{array}{ccc} \operatorname{Im}\gamma \cap C & \longrightarrow & C \\ \downarrow & & \downarrow \varphi \\ \operatorname{Im}\gamma & \xrightarrow{\beta'} & E \end{array} \tag{1}$$

then commutes because $\gamma^{-1}(C) \to \operatorname{Im}\gamma \cap C$ is an epimorphism. By Chap. IV (§5, (4)) there is an exact sequence

$$0 \to \operatorname{Im}\gamma \cap C \to \operatorname{Im}\gamma \oplus C \to \operatorname{Im}\gamma + C \to 0.$$

Using the commutativity of (1), it follows from this sequence that $(-\beta', \varphi)$: $\operatorname{Im}\gamma \oplus C \to E$ induces a morphism $\operatorname{Im}\gamma + C \to E$ which strictly extends φ. This is the desired contradiction. $\square$

Examples

1. Grothendieck categories. Every Grothendieck category is pseudo-complemented, as was remarked already in §1. Later we will show (in Chap. X) that every Grothendieck category has enough injectives, and therefore has injective envelopes.

2. Projective covers. The definition of essential subobjects and pseudo-complements can of course be dualized. Stating the dualized concepts in terms of subobjects rather than quotient objects, we have for an arbitrary abelian category:

Definitions. 1. A subobject B of an object C is *small* in C if $B + C' \neq C$ for every subobject C' of C such that $C' \neq C$.

2. If B is a subobject of C, then a *supplement* of B in C is a subobject D of C such that $D + B = C$ and D is minimal with this property. The category $\mathbf{C}$ is *supplemented* if supplements exist for arbitrary subobjects.

3. A *projective cover* of an object C is an epimorphism $\alpha\colon P \to C$, where P is projective and $\operatorname{Ker}\alpha$ is small in P.

The dual of Prop. 2.5 takes the form:

Proposition 2.10. *If* **C** *is a supplemented category and* C *is a quotient object of some projective object, then* C *has a projective cover.*

3. Modules of fractions. Let S be a right denominator set in a ring A. Let M be an S-torsion-free module. Also the injective envelope $E(M)$ is then S-torsion-free, for otherwise the torsion submodule of $E(M)$ would intersect M non-trivially. The canonical imbedding $M \to M[S^{-1}]$ is essential, for if xs^{-1} is a non-zero element of $M[S^{-1}]$, then $0 \neq xs^{-1} \cdot s = x \in M$. Therefore $M[S^{-1}]$ may be considered as a submodule of an injective envelope $E(M)$ of M. More precisely we have:

Proposition 2.11. $M[S^{-1}] \cong \{u \in E(M) \mid us \in M$ *for some* $s \in S\}$.

Proof. If $us = x \in M$, then $us = xs^{-1} \cdot s$ implies $u = xs^{-1}$ since $E(M)$ is S-torsion-free. □

§3. Finitely Generated Objects

Let **C** be a Grothendieck category. An object C of **C** is called *finitely generated* if the lattice $\mathsf{L}(C)$ is compact, i.e. whenever $C = \sum C_i$ for a direct family of subobjects C_i of C, there exists an index i_0 such that $C = C_{i_0}$.

Lemma 3.1. *Let* $0 \to C' \to C \to C'' \to 0$ *be an exact sequence in* **C**. *Then:*
(i) *If* C *is finitely generated, then so is also* C''.
(ii) *If* C' *and* C'' *are finitely generated, then so is also* C.

Proof. (i): Suppose C'' is expressed as a direct union $C'' = \sum C_i''$. Each C_i'' can be written as $C_i'' = C_i/C'$ for a unique subobject C_i of C with $C' \subset C_i$. The family (C_i) is also direct, so $C = C_{i_0}$ for some i_0, and then $C'' = C_{i_0}/C' = C_{i_0}''$.

(ii): Suppose C is expressed as a direct union $C = \sum C_i$. Then C' finitely generated implies that $C' \subset C_{i_0}$ for some i_0. This gives $C'' = \sum_{i \geq i_0} (C_i/C')$, and C'' finitely generated implies that $C'' = C_{i_1}/C'$ for some i_1. It follows by the 5-lemma that $C = C_{i_1}$. □

In the proof we used the fact that if C is finitely generated and $C \subset \sum D_i$ for a direct family (D_i) of subobjects of a certain object, then $C \subset D_{i_0}$ for some i_0. More generally we have:

Proposition 3.2. *An object* C *is finitely generated if and only if the functor* $\mathrm{Hom}_{\mathbf{C}}(C, \cdot)$ *preserves direct unions.*

Proof. The precise formulation is that C is finitely generated if and only if the canonical homomorphism

$$\Phi\colon \varinjlim \mathrm{Hom}(C, D_i) \to \mathrm{Hom}\big(C, \textstyle\sum D_i\big)$$

is an isomorphism for every direct family $(D_i)_I$ of subobjects of a certain object D. It is of course automatically a monomorphism. If C is finitely generated and a morphism $\alpha\colon C \to \sum D_i$ is given, then $\mathrm{Im}\,\alpha$ is finitely generated by Lemma 3.1, so

$\operatorname{Im}\alpha \subset D_{i_0}$ for some i_0. This proves that Φ is an epimorphism, and hence an isomorphism.

The converse implication is equally obvious. □

For an arbitrary Grothendieck category one cannot be sure that there exist any non-zero finitely generated objects at all. We say that **C** is *locally finitely generated* if it has a family of finitely generated generators, or what amounts to the same thing, if the lattices $\mathsf{L}(C)$ are compactly generated for all $C \in \mathbf{C}$. The following technical lemma is sometimes useful, and easily proved:

Lemma 3.3. *Let* **C** *be locally finitely generated. If* $\alpha\colon B \to C$ *is an epimorphism and* C *is finitely generated, then there exists a finitely generated subobject* B' *of* B *such that* $\alpha(B') = C$.

An object C of **C** is *finitely presented* if it is finitely generated and every epimorphism $B \to C$, where B is finitely generated, has a finitely generated kernel. There is an analogue of Prop. 3.2 for finitely presented objects:

Proposition 3.4. *Suppose* **C** *is locally finitely generated. An object* C *is finitely presented if and only if the functor* $\operatorname{Hom}(C, \cdot)$ *preserves direct limits.*

Proof. We shall prove that C is finitely presented if and only if the canonical homomorphism

$$\Phi\colon \varinjlim \operatorname{Hom}(C, D_i) \to \operatorname{Hom}(C, \varinjlim D_i)$$

is an isomorphism for every direct system $(D_i)_I$ in **C**.

Suppose first that Φ always is an isomorphism. Then C is finitely generated by Prop. 3.2. Let $0 \to K \to B \to C \to 0$ be an exact sequence with B finitely generated. If $(K_i)_I$ is the family of finitely generated subobjects of K, then by Prop. IV.8.9 we have $C \cong \varinjlim B/K_i$. By hypothesis, this isomorphism factors over some B/K_{i_0}, which means that the sequence

$$0 \to K/K_{i_0} \to B/K_{i_0} \to C \to 0$$

is split exact. It follows that K/K_{i_0} is finitely generated, and then K is finitely generated by Lemma 3.1. Thus C is finitely presented.

Assume on the other hand that C is finitely presented. We first show that Φ is a monomorphism. Suppose $C \to D_i$ has the property that the composed morphism $C \to D_i \to \varinjlim D_i$ is zero. Since C is finitely generated, it follows from Lemma 1.2 and Prop. 3.2 that $C \to D_i \to D_j$ is zero for some $j \geq i$. Hence Φ is a monomorphism.

Next consider a morphism $\alpha\colon C \to \varinjlim D_i$. Let D_i' denote the image of D_i in $\varinjlim D_i$, so that $\varinjlim D_i = \sum D_i'$. Since C is finitely generated, α factors over some D_i' by a morphism $\alpha'\colon C \to D_i'$. Consider the pullback diagram

$$\begin{array}{ccccccccc}
0 & \longrightarrow & K & \longrightarrow & P & \longrightarrow & C & \longrightarrow & 0 \\
 & & \| & & \downarrow & & \downarrow{\scriptstyle \alpha'} & & \\
0 & \longrightarrow & K & \longrightarrow & D_i & \longrightarrow & D_i' & \longrightarrow & 0
\end{array}$$

By Lemma 3.3 there exists a finitely generated subobject P' of P such that the morphism $P' \to P \to C$ is an epimorphism. Its kernel $K \cap P'$ is finitely generated since C is finitely presented. By Lemma 1.2 we therefore have $K \cap P' \subset \operatorname{Ker}(D_i \to D_j)$

for some $j \geq i$. The composed morphism $P' \to P \to D_i \to D_j$ then factors over C to give the desired morphism $C \to D_j$, which shows that Φ is an epimorphism. □

Examples

1. Modules. Finitely presented modules were introduced in Chap. I, §3, with a different definition, which however agrees with the present one by Prop. I.3.2.

2. Projective objects. Every finitely generated projective object is finitely presented.

3. Functor categories. If $\mathbf{D}$ is a small preadditive category, then $\mathbf{C} = \mathbf{Hom}(\mathbf{D}^{\text{op}}, \mathbf{Ab})$ is locally finitely generated. In fact, the Yoneda lemma (Prop. IV.7.3) implies that $\mathbf{D}$ constitutes a family of finitely generated generators for $\mathbf{C}$. Conversely, if $\mathbf{C}$ is locally finitely generated and has enough projectives, then it may be shown that $\mathbf{C}$ is equivalent to a functor category $\mathbf{Hom}(\mathbf{D}^{\text{op}}, \mathbf{Ab})$ (see e.g. Mitchell [1], p. 104). In particular, if $\mathbf{C}$ has a finitely generated projective generator P, then $\mathbf{C}$ is equivalent to the category $\mathbf{Mod}\text{-}A$, where $A = \text{Hom}_{\mathbf{C}}(P, P)$. This will be proved in Chap. X in connection with the Popescu-Gabriel theorem.

§4. Locally Noetherian Categories

Let $\mathbf{C}$ be a Grothendieck category. An object C is called *noetherian* (*artinian*) if the lattice $\mathsf{L}(C)$ of subobjects is noetherian (artinian).

Proposition 4.1. *The object C is noetherian if and only if every subobject of C is finitely generated.*

Proof. An easy exercise. □

From Prop. III.3.6 we have:

Proposition 4.2. *Let $0 \to C' \to C \to C'' \to 0$ be an exact sequence in $\mathbf{C}$. Then C is noetherian if and only if both C' and C'' are noetherian.*

The category $\mathbf{C}$ is *locally noetherian* if it has a family of noetherian generators. In that case, every object is a direct union of noetherian subobjects, and every finitely generated object is noetherian. For locally noetherian categories one has a very satisfactory decomposition theory for injective objects.

Proposition 4.3. *Suppose $\mathbf{C}$ is locally finitely generated and has enough injective objects. Then $\mathbf{C}$ is locally noetherian if and only if every coproduct of injective objects is injective.*

Proof. Suppose $\mathbf{C}$ is locally noetherian, and let $(E_i)_I$ be a family of injective objects. To show that $\bigoplus_I E_i$ is injective, it suffices by Prop. 2.9 to consider a monomorphism $\alpha: B \to C$ of noetherian objects and extend every morphism $\varphi: B \to \bigoplus_I E_i$ to C. But since B is finitely generated, the image of φ is contained in a coproduct $\bigoplus_J E_i$ for a finite subset J of I (recall Example IV.8.4). Since $\bigoplus_J E_i$ certainly is injective, there is no problem of extending φ to $C \to \bigoplus_J E_i \to \bigoplus_I E_i$.

Assume conversely that every coproduct of injective objects is injective. We will show that every finitely generated object C is noetherian. Let $C_1 \subset C_2 \subset \cdots$ be an ascending chain of subobjects of C, and put $E_i = E(C/C_i)$. For each C_n and each $j \leq n$ we let $\varphi_{nj}\colon C_n \to E_j$ denote the composition of the canonical morphisms $C_n \to C \to C/C_j \to E_j$. For each n, these φ_{nj} induce a morphism $\varphi_n\colon C_n \to \bigoplus_1^\infty E_j$. The family (φ_n) is compatible and therefore induces $\varphi\colon \sum_n C_n \to \bigoplus_1^\infty E_j$. Since $\bigoplus_1^\infty E_j$ is assumed to be injective, φ extends to a morphism $\bar{\varphi}\colon C \to \bigoplus_1^\infty E_j$. But since C is finitely generated, $\bar{\varphi}$ factors through a finite coproduct $E_1 \oplus \cdots \oplus E_m$. It follows that $C_n = C_m$ for $n \geq m$. □

The argument used in the first half of the proof shows more generally:

Corollary 4.4. *If* **C** *is locally noetherian, then every direct union of injective objects is injective.*

Still more generally it can be shown that every direct limit of injective objects is injective in a locally noetherian category.

Proposition 4.5 (Matlis [1]). *If* **C** *is locally noetherian, then every injective object is a coproduct of indecomposable injective objects.*

Proof. Let E be an injective object. Consider all independent families $(E_i)_I$ of indecomposable injective subobjects of E. By Zorn's lemma (and the use of Example 1.4) one can find a maximal such family $(E_i)_I$. The sum $\sum_I E_i$ is an injective object by Prop. 4.3, so we can write $E = (\sum_I E_i) \oplus E'$. We want to show that $E' = 0$, and for this we only have to show that every non-zero injective object E' contains a non-zero indecomposable direct summand. Let C be a non-zero noetherian subobject of E'. Consider all injective subobjects E'' of E' such that $C \not\subset E''$. By Zorn's lemma, together with Prop. 3.2 and Cor. 4.4, there exists a maximal such subobject E''. Then $E' = E'' \oplus D$, and we assert that D is indecomposable. For if $D = D' \oplus D''$ with D' and D'' non-zero, then $(E'' + D') \cap (E'' + D'') = E''$ implies that one of the objects $E'' + D'$ or $E'' + D''$ does not contain C, which contradicts the maximality of E''. □

Examples

1. Module categories. Let A be a ring. The category **Mod**-A is locally finitely generated, and it is locally noetherian if and only if A is a right noetherian ring. Similarly, **Mod**-A is locally artinian (i.e. has an artinian generator) if and only if A is a right artinian ring (i.e. A is an artinian right A-module). It is known (see Chap. VIII) that every right artinian ring is right noetherian. However, it is not true in general that every locally artinian Grothendieck category is locally noetherian.

2. Commutative noetherian rings. For a commutative noetherian ring it is possible to classify all indecomposable injective modules, and thereby all injective modules.

Proposition 4.6 (Matlis [1]). *Let A be a commutative noetherian ring. There is a bijective correspondence between prime ideals of A and isomorphism classes of indecomposable injective modules, given by $\mathfrak{p} \mapsto E(A/\mathfrak{p})$.*

Proof. Every prime ideal is irreducible, so $E(A/\mathfrak{p})$ is indecomposable (Prop. 2.8). If $\mathfrak{p}$ and $\mathfrak{q}$ are prime ideals such that $E(A/\mathfrak{p}) \cong E(A/\mathfrak{q})$, then we may consider both $A/\mathfrak{p}$ and $A/\mathfrak{q}$ as submodules of $E(A/\mathfrak{p})$. Then $A/\mathfrak{p} \cap A/\mathfrak{q} \neq 0$, but $\operatorname{Ann}(x) = \mathfrak{p}$ for every non-zero $x \in A/\mathfrak{p}$, and $\operatorname{Ann}(y) = \mathfrak{q}$ for every non-zero $y \in A/\mathfrak{q}$, so we must have $\mathfrak{p} = \mathfrak{q}$. Finally we must show that if E is an indecomposable injective module, then there exists $x \in E$ such that $\operatorname{Ann}(x)$ is a prime ideal. The family of ideals $\operatorname{Ann}(y)$, for all $y \neq 0$ in E, has a maximal member $\operatorname{Ann}(x)$ since A is noetherian. We assert that $\operatorname{Ann}(x)$ is a prime ideal. Suppose $a, b \in A$ with $ab \in \operatorname{Ann}(x)$ but $b \notin \operatorname{Ann}(x)$. Then $bx \neq 0$ while $abx = 0$. So $a \in \operatorname{Ann}(bx)$. But $\operatorname{Ann}(x) \subset \operatorname{Ann}(bx)$, so by maximality we have $\operatorname{Ann}(x) = \operatorname{Ann}(bx)$, and hence $a \in \operatorname{Ann}(x)$. ▯

§5. The Krull-Remak-Schmidt-Azumaya Theorem

Let $\mathbf{C}$ be a Grothendieck category (with enough injectives). In this section we prove the unicity of direct sum decompositions $C_1 \oplus \cdots \oplus C_n$ when each C_i has a local endomorphism ring. First some words on local rings.

Definition. A ring is *local* if all the non-invertible elements form a proper ideal.

A local ring thus has precisely one maximal ideal, which also is the unique maximal right ideal.

Proposition 5.1. *If E is an indecomposable injective object of $\mathbf{C}$, then $\operatorname{Hom}_{\mathbf{C}}(E, E)$ is a local ring.*

Proof. It suffices to show that if α and β are non-invertible endomorphisms of E, then also $\alpha + \beta$ is non-invertible. If α or β is a monomorphism $E \to E$, then it splits E. Therefore we only need to show that if $\operatorname{Ker} \alpha \neq 0$ and $\operatorname{Ker} \beta \neq 0$, then $\operatorname{Ker}(\alpha + \beta) \neq 0$. But this is clear, since $\operatorname{Ker}(\alpha + \beta) \supset \operatorname{Ker} \alpha \cap \operatorname{Ker} \beta \neq 0$ by the co-irreducibility of E. ▯

Conversely, if C is an object such that $\operatorname{Hom}_{\mathbf{C}}(C, C)$ is local, then C must be indecomposable, for otherwise there would exist projections $\pi_i\colon C \to C_i$ and injections $\iota_i\colon C_i \to C$ $(i = 1, 2)$ such that $1 = \iota_1 \pi_1 + \iota_2 \pi_2$, where both $\iota_1 \pi_1$ and $\iota_2 \pi_2$ are non-invertible.

The crucial part of the unicity theorem is the following

Lemma 5.2. *Suppose $A \oplus B = C_1 \oplus \cdots \oplus C_n$, and that $\operatorname{Hom}_{\mathbf{C}}(A, A)$ is a local ring. Then there is an i such that $C_i \cong C_i' \oplus C_i''$ with $A \cong C_i''$ and $B \cong C_i' \oplus (\bigoplus_{j \neq i} C_j)$.*

Proof. The injections and projections associated to the direct summands are denoted by $\iota_A, \pi_A, \iota_B, \pi_B$ and ι_i, π_i $(i = 1, \ldots, n)$. In $\operatorname{Hom}_{\mathbf{C}}(A, A)$ we have

$$1 = \pi_A \iota_A = \pi_A \Big(\sum_i \iota_i \pi_i\Big) \iota_A = \sum_i \pi_A \iota_i \pi_i \iota_A,$$

and one of the terms $\pi_A \iota_i \pi_i \iota_A$ must be invertible since the endomorphism ring is local. Assume e.g. that $\pi_A \iota_1 \pi_1 \iota_A = \alpha$ is an automorphism of A. Put

$$\gamma = \alpha^{-1} \pi_A \iota_1 \colon C_1 \to A.$$

γ satisfies $\gamma \pi_1 \iota_A = 1_A$ and therefore gives rise to a split exact sequence

$$0 \to C_1' \to C_1 \to A \to 0.$$

Replacing C_1 by $A \oplus C_1'$, we have an isomorphism

$$A \oplus B \xrightarrow{\cong} A \oplus C_1' \oplus C_2 \oplus \cdots \oplus C_n$$

which on the first direct summand A reduces to $\gamma \pi_1 \iota_A = 1_A$. The conclusion now follows from the following

Lemma 5.3. *Let $\alpha = (\alpha_{ij})\colon A_1 \oplus A_2 \to B_1 \oplus B_2$ be a morphism such that both α and α_{11} are isomorphisms. Then $A_2 \cong B_2$.*

Proof. If $\alpha_{21} = 0$, then $(\alpha^{-1})_{22}$ is an inverse of α_{22}. The general case is reduced to the special one by replacing α by the isomorphism

$$\begin{pmatrix} 1 & 0 \\ -\alpha_{21}\alpha_{11}^{-1} & 1 \end{pmatrix} \alpha = \begin{pmatrix} \alpha_{11} & \alpha_{12} \\ 0 & \alpha_{22}' \end{pmatrix}. \quad \square$$

Proposition 5.4 (Krull-Remak-Schmidt-Azumaya). *If*

$$A_1 \oplus \cdots \oplus A_m \cong C_1 \oplus \cdots \oplus C_n,$$

where each A_i has a local endomorphism ring and each C_j is indecomposable, then $m = n$ and there exists a permutation π of $\{1, \ldots, n\}$ such that $A_i \cong C_{\pi(i)}$ for every i.

Proof. The case $m = 1$ is clear by the remark preceding Lemma 5.2. The Lemma 5.2 then makes it possible to proceed by induction on m. $\square$

The result can be extended to decompositions into infinitely many direct summands (see e.g. Pareigis [1]). By Prop. 5.1 we have:

Corollary 5.5. *If E is an injective object and*

$$E \cong E_1 \oplus \cdots \oplus E_m \cong F_1 \oplus \cdots \oplus F_n,$$

where each direct summand is indecomposable, then $m = n$ and there is a permutation π of $\{1, \ldots, n\}$ such that $E_i \cong F_{\pi(i)}$ for all i.

As an application of the Azumaya theorem we will develop a "rank" concept, which extends the notion of "dimension" for vector spaces. An object C is said to be of *finite rank* if there is no infinite independent family of subobjects of C. Obviously:

Lemma 5.6. *C is of finite rank if and only if $E(C)$ is of finite rank.*

Lemma 5.7. *If $C \neq 0$ is of finite rank, then C contains a non-zero coirreducible subobject.*

Proof. If C is not coirreducible, then $C \supset C_1 \oplus C_1'$. If C_1' is not coirreducible, then $C_1' \supset C_2 \oplus C_2'$ and hence

$$C \supset C_1 \oplus C_2 \oplus C_2'.$$

Continuing in this manner, one must finally reach a coirreducible C_n' since C is of finite rank. □

It follows from this lemma that an object of finite rank must be an essential extension of a finite coproduct of coirreducible subobjects. In particular we may conclude:

Proposition 5.8. *An object C is of finite rank if and only if $E(C) = E_1 \oplus \cdots \oplus E_n$ for a finite number n of indecomposable injective objects E_i.*

The number n is the *rank* of C, and it is equal to the maximal number of independent coirreducible subobjects of C. If C is not of finite rank, then we say that C has *rank* ∞.

Examples

1. Rank of integral domains. Let A be a ring without zero-divisors. Then A is a right Ore domain if and only if A is coirreducible as a right A-module, i.e. A is of rank 1.

Suppose A is of finite rank. Let $\mathfrak{a}$ be a non-zero coirreducible right ideal (Lemma 5.7), and choose $0 \neq a \in \mathfrak{a}$. If $\mathfrak{b}$ and $\mathfrak{c}$ are right ideals such that $\mathfrak{b} \cap \mathfrak{c} = 0$, then in particular $a\mathfrak{b} \cap a\mathfrak{c} = 0$, and since $\mathfrak{a}$ is coirreducible, this implies either $\mathfrak{b} = 0$ or $\mathfrak{c} = 0$. Hence A is of rank 1.

Thus A is of rank 1 or ∞, according as A is an Ore domain or not.

2. Irreducible decomposition. Let C be a noetherian object of $\mathbf{C}$. Each subobject B of C has an irreducible decomposition in C, say $B = C_1 \cap \cdots \cap C_n$, by Prop. III.3.9.

Proposition 5.9. *Suppose $B = C_1 \cap \cdots \cap C_n$ is an irredundant, irreducible decomposition of B in C. The canonical monomorphism β: $C/B \to C/C_1 \oplus \cdots \oplus C/C_n$ is essential.*

Proof. Let π_i denote the projection $C/C_1 \oplus \cdots \oplus C/C_n \to C/C_i$. The irredundancy means precisely that each $\pi_i \beta = 0$. Put $D_i = \operatorname{Im} \pi_i \beta$, which is an essential subobject of the coirreducible object C/C_i. Then $\bigoplus D_i$ is an essential subobject of $\bigoplus C/C_i$ by Lemma 2.7. But $\bigoplus D_i$ is contained in the image of α, so also α is essential. □

Corollary 5.10. *If $B = C_1 \cap \cdots \cap C_n$ is an irredundant, irreducible decomposition of B in C, then*

$$E(C/B) \cong E(C/C_1) \oplus \cdots \oplus E(C/C_n)$$

where each $E(C/C_i)$ is indecomposable.

The unicity theorem (Cor. 5.5) then implies:

Corollary 5.11. *If $B = C_1 \cap \cdots \cap C_n = C_1' \cap \cdots \cap C_m'$ are two irredundant, irreducible decompositions of B in C, then $m = n$ and there is a permutation π of $\{1, \ldots, n\}$ such that $E(B/C_i) \cong E(B/C_{\pi(i)}')$ for each i.*

3. Irreducible decomposition over a commutative noetherian ring. Let A be a commutative noetherian ring, and let L be a submodule of a noetherian module M. If $L = M_1 \cap \cdots \cap M_n$ is an irredundant, irreducible decomposition of L in M, then there are prime ideals $\mathfrak{p}_1, \ldots, \mathfrak{p}_n$ such that $E(M/M_i) \cong E(A/\mathfrak{p}_i)$. These prime ideals are said to be *associated* to L, and they are uniquely determined by L (Cor. 5.11 and Prop. 4.6).

§ 6. Spectral Categories

Definition. A Grothendieck category $\mathbf{C}$ is a *spectral category* if every short exact sequence in $\mathbf{C}$ splits.

In other words, a spectral category is a Grothendieck category in which every object is injective (and also projective). We note that the theory of spectral categories will involve regular rings:

Proposition 6.1. *If C is an object of a spectral category, then the endomorphism ring of C is regular.*

Proof. Use the proof of Prop. I.12.4 (cf. also Exercise IV.10). □

We will return to the connection between regular rings and spectral categories in Chap. XII. Let us now look at the least complicated kind of spectral categories, namely those in which every object is semi-simple. In the sequel we let $\mathbf{C}$ be a Grothendieck category. As in the case of modules (Chap. I, §7) we define:

Definition. An object S of $\mathbf{C}$ is *simple* if it is non-zero and has no other subobjects than 0 and S. S is *semi-simple* if it is a sum of simple subobjects. (The zero object is considered as semi-simple.)

Proposition 6.2. *Let S be a sum of simple subobjects S_i $(i \in I)$ and let C be an arbitrary subobject of S. Then there exists $J \subset I$ such that $S = C \oplus (\bigoplus_J S_i)$.*

Proof. Exactly as in the module case (Prop. I.7.1). □

Corollary 6.3. *Every semi-simple object is a coproduct of simple objects.*

Corollary 6.4. *If S is a semi-simple object, then every subobject of S is a direct summand.*

Proposition 6.5. *Every subobject and every quotient object of a semi-simple object is semi-simple.*

Proof. This is proved in the same way as Cor. I.7.4. □

Let Ω denote the set of isomorphism classes of simple objects in $\mathbf{C}$. For every $\omega \in \Omega$ and every semi-simple object S we let S_ω denote the sum of all simple subobjects of S of isomorphism class ω. Then we have (cf. Prop. I.7.5):

Proposition 6.6. *If S is semi-simple, then $S = \bigoplus_\Omega S_\omega$.*

If $\alpha: S \to S'$ is a morphism between semi-simple objects S and S', then $\alpha(S_\omega) \subset S'_\omega$. Since Schur's lemma holds for simple objects, it follows that the endomorphism ring of a finite sum of simple objects is a finite product of matrix rings over skew-fields.

From Cor. 6.4 we see that if every object of **C** is semi-simple, then **C** is a spectral category. A spectral category is called *discrete* if every object is semi-simple. These spectral categories can be described in quite an explicit fashion:

Proposition 6.7. *The following properties of a Grothendieck category* **C** *are equivalent:*

(a) **C** *is a discrete spectral category.*

(b) **C** *is a locally finitely generated spectral category.*

(c) *Every object of* **C** *is semi-simple.*

(d) **C** *has a family of simple generators.*

(e) **C** *is equivalent to a product category* $\prod_I \mathbf{Mod}\text{-}K_i$ *for a family of skew-fields* K_i $(i \in I)$.

Proof. (a) $\Leftrightarrow$ (c) $\Leftrightarrow$ (d) are clear.

(d) $\Rightarrow$ (e): For each $\omega \in \Omega$ we choose a simple object $S_\omega \in \omega$. Each endomorphism ring $K_\omega = \mathrm{Hom}_{\mathbf{C}}(S_\omega, S_\omega)$ is a skew-field by Schur's lemma. If C is an object of **C**, the abelian group $\mathrm{Hom}_{\mathbf{C}}(S_\omega, C)$ may be considered as a right K_ω-module in a natural way. Thus we have a functor $T: \mathbf{C} \to \prod_\Omega \mathbf{Mod}\text{-}K$ given by $T(C) = (\mathrm{Hom}_{\mathbf{C}}(S_\omega, C))_\Omega$. This functor is faithful because $(S_\omega)_\Omega$ is a family of generators for **C**. In order to see that T is full, we have to analyze it further. If $C_\omega \cong S_\omega^{(J)}$, then

$$\mathrm{Hom}_{\mathbf{C}}(S_\omega, C) = \mathrm{Hom}_{\mathbf{C}}(S_\omega, \bigoplus_\Omega C_{\omega'}) = \mathrm{Hom}_{\mathbf{C}}(S_\omega, C_\omega) = \mathrm{Hom}_{\mathbf{C}}(S_\omega, S_\omega^{(J)}) = K_\omega^{(J)},$$

where the last equality follows from the fact that S_ω is finitely generated. From this it is easy to see that the functor T is full, and still easier to see that every object of $\prod_\Omega \mathbf{Mod}\text{-}K_\omega$ is isomorphic to some $T(C)$.

(e) $\Rightarrow$ (b): Each $\mathbf{Mod}\text{-}K_i$ is a locally finitely generated spectral category, and the same is then true for a product category $\prod_I \mathbf{Mod}\text{-}K_i$.

(b) $\Rightarrow$ (c): This is a generalization of the implication (c) $\Rightarrow$ (a) in Prop. I.7.2. As in that special case it suffices to show that every non-zero object C contains a simple subobject. By hypothesis we may assume C finitely generated. But C then contains a maximal proper subobject (Exercise 10), so the proof may be concluded as for Prop. I.7.2. □

If **C** is an arbitrary spectral category, we let $\mathbf{C}_{\mathrm{dis}}$ (resp. $\mathbf{C}_{\mathrm{cont}}$) denote the full subcategory of **C** consisting of all semi-simple objects (resp. all objects without simple subobjects). $\mathbf{C}_{\mathrm{dis}}$ is a discrete spectral category, while $\mathbf{C}_{\mathrm{cont}}$ is a *continuous* spectral category, i.e. there are no simple objects in $\mathbf{C}_{\mathrm{cont}}$.

Proposition 6.8. *If* **C** *is a spectral category, then there is an equivalence*

$$\mathbf{C} \to \mathbf{C}_{\mathrm{dis}} \times \mathbf{C}_{\mathrm{cont}}.$$

Proof. For each C in **C** we form the sum $s(C)$ of all simple subobjects of C. The functor $C \mapsto (s(C), C/s(C))$ is clearly the alleged equivalence. □

Examples

1. Semi-simple rings. If A is a ring, then **Mod**-A is a spectral category if and only if A is a semi-simple ring, and in that case it is of course a discrete spectral category. The structure theorem for semi-simple rings (Prop. I.7.7) is a special case of Prop. 6.7, from which it can be obtained with the help of Morita theory.

2. $\mathbf{C}_{\text{dis}}$ for arbitrary C. For an arbitrary Grothendieck category $\mathbf{C}$ one can of course define $\mathbf{C}_{\text{dis}}$ as the full subcategory consisting of semi-simple objects. Then $\mathbf{C}_{\text{dis}}$ is a discrete spectral category, and it is an abelian subcategory of $\mathbf{C}$.

§7. The Spectral Category of a Grothendieck Category

Let $\mathbf{C}$ be a Grothendieck category. We will associate to $\mathbf{C}$ a spectral category **Spec-C** which in a certain sense describes the behavior of the injective objects of $\mathbf{C}$. The construction is due to Gabriel and Oberst [1].

Let the objects of **Spec-C** be the same as the objects of $\mathbf{C}$. For objects C and D of $\mathbf{C}$ we put

$$\mathrm{Hom}_{\mathbf{Spec\text{-}C}}(C,D)=\varinjlim \mathrm{Hom}_{\mathbf{C}}(C',D)$$

where the direct limit is taken over the down-wards directed family of essential subobjects C' of C. Note here that if C' and C'' are essential in C, then also $C'\cap C''$ is essential in C; more generally we have:

Lemma 7.1. *If*

$$\begin{array}{ccc} P & \xrightarrow{\alpha'} & B \\ \downarrow & & \downarrow \\ A & \xrightarrow[\alpha]{} & C \end{array}$$

is a pullback diagram with α an essential monomorphism, then also α' is an essential monomorphism.

Proof. α' is a monomorphism by Prop. IV.5.1. Suppose B' is a subobject of B such that $P\cap B'=0$. Factor $B'\to B\to C$ as an epimorphism $B'\to D$ followed by a monomorphism $D\to C$. Form successively the pullbacks

$$\begin{array}{ccc} P'' & \longrightarrow & B' \\ \downarrow & & \downarrow \\ P' & \longrightarrow & D \\ \downarrow & & \downarrow \\ A & \xrightarrow[\alpha]{} & C \end{array}$$

Since then also the outer square is a pullback, $P''=0$ by the unicity of pullbacks. $P''\to P'$ is an epimorphism by Prop. IV.5.1, and hence also $P'=0$. But α essential then implies $D=0$, and this gives $B'=0$ because the upper square is a pullback. □

It remains to define the composition in **Spec-C**. If $\varphi \in \mathrm{Hom}_{\mathbf{Spec\text{-}C}}(B, C)$ is represented by $\alpha\colon B' \to C$ and $\psi \in \mathrm{Hom}_{\mathbf{Spec\text{-}C}}(C, D)$ by $\beta\colon C' \to D$, then $\alpha^{-1}(C')$ is an essential subobject of B' (by the Lemma) and then also of B. We define $\psi\varphi$ to be the morphism in **Spec-C** represented by $\beta\alpha'$ as given by the diagram

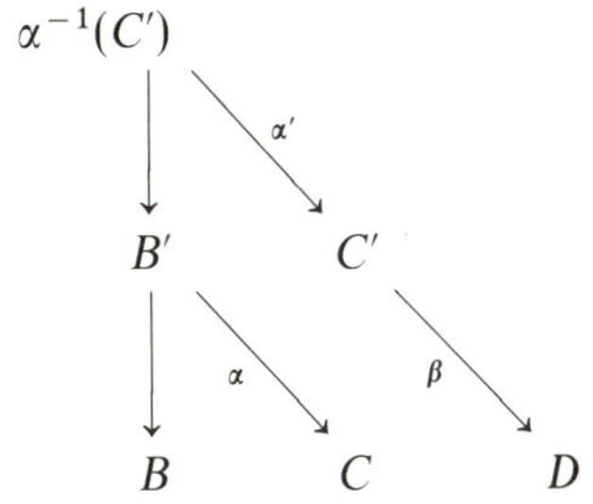

One may verify (although the computations become a bit awkward) that this composition is well-defined (independent of the choices of α and β) and that it is associative. This shows that **Spec-C** is a category.

There is a functor $P\colon \mathbf{C} \to \mathbf{Spec\text{-}C}$ which leaves objects unchanged and takes a morphism $C \to D$ to its canonical image in $\varinjlim \mathrm{Hom}(C', D)$. P carries each essential monomorphism in $\mathbf{C}$ into an isomorphism in **Spec-C**. P therefore identifies each object in $\mathbf{C}$ with its injective envelope (assuming that $\mathbf{C}$ has enough injectives).

Proposition 7.2. **Spec-C** *is a spectral category. P is a left exact functor which preserves coproducts.*

Proof. It is clear that **Spec-C** is a preadditive category and P is an additive functor. P preserves finite limits, because one has for every functor $F\colon \mathbf{I} \to \mathbf{C}$ with finite $\mathbf{I}$ that

$$\begin{aligned}\mathrm{Hom}(P(C), P(\varprojlim F)) &= \varinjlim \mathrm{Hom}(C', \varprojlim F) = \varinjlim \varprojlim \mathrm{Hom}(C', F(\cdot)) \\ &= \varprojlim \varinjlim \mathrm{Hom}(C', F(\cdot)) = \varprojlim \mathrm{Hom}(P(C), PF(\cdot)),\end{aligned}$$

using the exactness of direct limits in **Ab**. It follows in particular that **Spec-C** has finite products.

Let $\varphi\colon P(C) \to P(D)$ be an arbitrary morphism in **Spec-C**. It is represented by a morphism $\alpha\colon C' \to D$ for some essential subobject C' of C. Let C'' be a pseudo-complement of $\mathrm{Ker}\,\alpha$ in C'. Then $C'' + \mathrm{Ker}\,\alpha$ is an essential subobject of C by Prop. III.6.4, and α induces $\alpha'\colon C'' + \mathrm{Ker}\,\alpha \to D$. Also choose a pseudo-complement D' of $\mathrm{Im}\,\alpha'$ in D. There results a commutative diagram

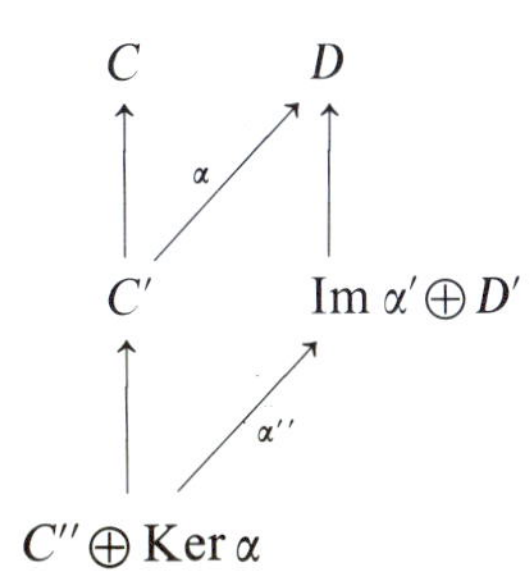

where the vertical arrows are essential monomorphisms. $P(\alpha'')$ then differs from α only by an isomorphism. The morphism α'' decomposes as

$$C''\oplus \operatorname{Ker}\alpha \to C'' \cong \operatorname{Im}\alpha' \to \operatorname{Im}\alpha' \oplus D',$$

i.e. as a right invertible morphism followed by a left invertible morphism. Also $P(\alpha'')$ has then a corresponding decomposition. It follows that $\varphi=\mu\lambda$ with $\lambda\lambda'=\mu'\mu=1$:

$$P(C) \underset{\lambda}{\overset{\lambda'}{\rightleftarrows}} P(C'') \underset{\mu}{\overset{\mu'}{\rightleftarrows}} P(D).$$

In order to verify the axioms $A3$ and $A4'$ for abelian categories, it suffices by symmetry to show that λ has a kernel and that λ is a cokernel. It is clear that λ is the cokernel of $1-\lambda'\lambda$: $P(C)\to P(C)$. One can also factor $1-\lambda'\lambda$ as an epimorphism followed by a monomorphism γ, and it is easy to see that γ is a kernel of λ. We have thus proved that **Spec-C** is an abelian category in which every morphism splits.

We have already verified that P is left exact. We now show that P preserves coproducts. Let $(C_i)_I$ be a family of objects in **C**. We have to show that $P(\bigoplus_I C_i)$ is a coproduct of the objects $P(C_i)$. Assume there are given morphisms φ_i: $P(C_i)\to P(D)$ for all $i\in I$. Each φ_i is represented by α_i: $C_i'\to D$ for some essential subobject C_i' of C. By Exercise 6, $\bigoplus_I C_i'$ is an essential subobject of $\bigoplus_I C_i$, so the α_i induce a morphism φ: $P(\bigoplus_I C_i)\to P(D)$ such that the diagrams

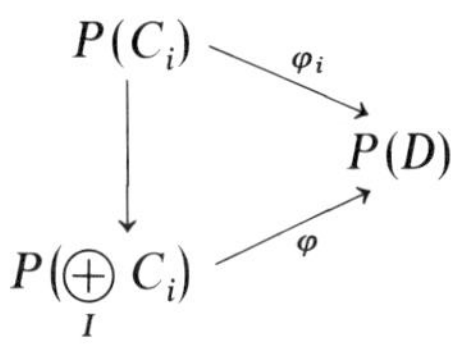

commute. Furthermore φ is uniquely determined by the given φ_i. Hence $P(\bigoplus_I C_i)=\bigoplus_I P(C_i)$. In particular this shows that **Spec-C** is cocomplete.

Next we show that P preserves sums of direct families of subobjects. Let $C\in\mathbf{C}$ be the direct union of subobjects C_i $(i\in I)$. We must show that $P(C)$ is the sum of the direct family of subobjects $P(C_i)$, which means that no proper subobject of $P(C)$ contains all $P(C_i)$. So what we shall prove is that if B is a subobject of C and $B\cap C_i$ is essential in C_i, for each $i\in I$, then B is essential in C. But this is clear, because if D is a non-zero subobject of C, then $D\cap C_i\neq 0$ for some $i\in I$, and hence $B\cap C_i\cap D\neq 0$.

Since P preserves both finite intersections and direct unions, the validity of axiom $AB5$ in **C** is transferred to **Spec-C**. To conclude the proof it now only remains to show that **Spec-C** has a family of generators. Let U be a generator for **C**, and let **U** be the set of (non-isomorphic) quotient objects of U. Every object C in **C** is an essential extension of a coproduct $\bigoplus_I U_i$ with $U_i\in\mathbf{U}$, as one easily sees. Then $P(C)=P(\bigoplus_I U_i)=\bigoplus_I P(U_i)$, so the objects $P(U')$, $U'\in\mathbf{U}$, form a family of generators for **Spec-C**. $\square$

Examples

1. Locally coirreducible categories. If C is an object in a Grothendieck category **C**, then $P(C)$ is simple if and only if every non-zero subobject of C is essential in C, i.e. C is coirreducible. Therefore one has:

Proposition 7.3. *The following properties of* **C** *are equivalent:*

(a) **Spec-C** *is discrete.*

(b) *Every non-zero object of* **C** *contains a non-zero coirreducible subobject.*

(c) *Every object of* **C** *is an essential extension of a coproduct of coirreducible objects.*

2. Objects of finite rank. If C is an object of finite rank in **C**, then $P(C)$ is a semi-simple object in **Spec-C**, and the rank of $P(C)$ equals the rank of C.

3. The spectral category of a ring. When A is a ring, we write **Spec**-A for **Spec**-(**Mod**-A). There is a bijective correspondence between the simple objects of **Spec**-A and the indecomposable injective A-modules. In particular, when A is a commutative noetherian ring, there is a bijective correspondence between the prime ideals of A and the isomorphism classes of simple objects of the discrete spectral category **Spec**-A.

Exercises

Let **C** be a Grothendieck category with enough injectives.

§ 1

1. Let $(C_i)_I$ be a family of objects of **C**. Show that the canonical morphism $\bigoplus_I C_i \to \prod_I C_i$ is a monomorphism.

2. Let $F: \mathbf{C} \to \mathbf{C}$ be a functor preserving colimits. Show that also $\mathbf{C} \times F$ is a Grothendieck category (cf. Exercises IV.11 and IV.26). (Hint: if U is a generator for **C**, then $F(U) \oplus U$ is the underlying object in **C** of a generator for $\mathbf{C} \times F$.)

§ 2

3. Show that a monomorphism $\alpha: C \to C'$ is essential if and only if every morphism $\beta: C' \to D$, such that $\beta\alpha$ is a monomorphism, is itself a monomorphism.

4. Let B be a subobject of C. Show that a subobject C' of C is a pseudo-complement of B in C if and only if the composed morphism $B \to C \to C/C'$ is an essential monomorphism.

5. Show that a subobject C' of C is essentially closed in C if and only if C' is the intersection of C with a direct summand of $E(C)$.

6. Let $C_i \to C_i'$ $(i \in I)$ be an arbitrary family of essential monomorphisms. Show that the induced monomorphism $\bigoplus_I C_i \to \bigoplus_I C_i'$ is essential.

7. Show that if M is a finitely generated module and $E(M)$ is a submodule of a free module, then $E(M)$ is a finitely generated projective module.

8. Show that if $\mathfrak{a}$ is a nilpotent two-sided ideal of A, and M is a module, then $M\mathfrak{a}$ is a small submodule of M.

9. Let $\mathfrak{a}$ be an irreducible right ideal of A, and let $b \notin \mathfrak{a}$. Show that $E(A/\mathfrak{a}) = E(A/(\mathfrak{a}:b))$ and that $(\mathfrak{a}:b)$ is irreducible.

§ 3

10. Show that every finitely generated object contains a maximal proper subobject.

11. Show that if C is the sum of finitely many finitely generated subobjects, then C is finitely generated.

12. Show that if $0 \to C' \to C \to C'' \to 0$ is exact with C finitely presented and C' finitely generated, then C'' is finitely presented.

13. Show that the following properties of an object C of $\mathbf{C}$ are equivalent:

(a) For every denumerable ascending chain of subobjects $C_1 \subset C_2 \subset \cdots$ of C such that $\sum C_i = C$, one has $C = C_n$ for some n.

(b) The functor $\mathrm{Hom}_{\mathbf{C}}(C, \cdot)$ preserves denumerable direct unions.

(c) The functor $\mathrm{Hom}_{\mathbf{C}}(C, \cdot)$ preserves coproducts.

(c′) The functor $\mathrm{Hom}_{\mathbf{C}}(C, \cdot)$ preserves denumerable coproducts.

The object C is called $\sum$-*generated* if these conditions are satisfied.

14. Show that every $\sum$-generated projective object is finitely generated.

15. Show that if $\mathbf{C}$ is locally finitely generated and every finitely generated object is injective, then every finitely presented object is projective.

§4

16. Let α be an endomorphism of an object C of $\mathbf{C}$. Show that:

(i) If C is noetherian and α is an epimorphism, then α is an isomorphism.

(ii) If C is artinian and α is a monomorphism, then α is an isomorphism.

(iii) If C is both noetherian and artinian, then $C = \mathrm{Im}\,\alpha^n \oplus \mathrm{Ker}\,\alpha^n$ for some n.

17. Show that if $\mathbf{C}$ is locally noetherian, then $E(\bigoplus_I C_i) = \bigoplus_I E(C_i)$ for an arbitrary family $(C_i)_I$ of objects.

18. Show that if C is a noetherian object, then $E(C)$ is a coproduct of finitely many indecomposable injective objects. (Hint: modify the proof of Prop. 3.7.)

19. Show that:

(i) A module E is FP-injective (Exercise I.43) if and only if for every finitely generated submodule L of a free module F and every homomorphism $\varphi: L \to E$, there exists a homomorphism $F \to E$ which extends φ.

(ii) Every direct sum of FP-injective modules is FP-injective.

20. (i) Show that an indecomposable injective $\mathbb{Z}$-module is isomorphic either to $\mathbb{Q}$ or to $\mathbb{Z}(p^\infty) = \mathbb{Q}/\mathbb{Z}_p$ for a prime p, where $\mathbb{Z}_p$ is the localization of $\mathbb{Z}$ at the prime ideal $p\mathbb{Z}$.

(ii) Decompose the $\mathbb{Z}$-module $\mathbb{Q}/\mathbb{Z}$ as a direct sum of indecomposable injective modules.

21. Let A be a commutative ring, $\mathfrak{p}$ a prime ideal of A and $S = A - \mathfrak{p}$. Show that:

(i) $E(A/\mathfrak{p})$ is S-torsion-free and S-divisible as an A-module.

(ii) $E(A/\mathfrak{p})$ is an injective envelope of the $A_{\mathfrak{p}}$-module $A_{\mathfrak{p}}/\mathfrak{p}A_{\mathfrak{p}}$.

§5

22. Show that if E is an indecomposable injective and projective module, then $E = eA$ for some idempotent e of A. (Hint: E is a submodule of a finitely generated free module by Exercise 7.)

23. Let C be an object of $\mathbf{C}$ of finite rank n. Show that a subobject D of C is essential in C if and only if D contains a coproduct of n coirreducible subobjects.

24. Show that the following properties of an object C are equivalent:

(a) C is of finite rank.

(b) C satisfies ACC on essentially closed subobjects.

(c) C satisfies DCC on essentially closed subobjects.

(Hint: use Exercise 5.) Deduce that the rank of C is equal to the maximal length of a chain of essentially closed subobjects of C.

§6

25. For each object C of $\mathbf{C}$, let $s(C)$ denote the sum of all simple subobjects of C. Show that:

(i) if C' is an essential subobject of C, then $s(C') = s(C)$.

(ii) the following three conditions for an object C are equivalent:

(a) There exist finitely many simple objects $S_1, \ldots, S_n$ such that $E(C) = E(S_1) \oplus \cdots \oplus E(S_n)$.

(b) $s(C)$ is finitely generated and is an essential subobject of C.

(c) C is an essential extension of an artinian object.

26. Deduce Prop. I.7.7 from Prop. 6.7 with the help of Morita theory.

§ 7

27. Show that the following properties of an exact sequence $0 \to C' \to C \to C'' \to 0$ in **C** are equivalent:

(a) The sequence $0 \to P(C') \to P(C) \to P(C'') \to 0$ is exact.

(b) C' is essentially closed in C.

(c) If B is an essential subobject of C and $B \subset C'$, then $B/C' \to C/C'$ is an essential monomorphism.

28. Show that the following properties of a ring A are equivalent:

(a) **Spec**-A is discrete.

(b) Every injective A-module is the injective envelope of a direct sum of indecomposable injective modules.

(c) Every proper right ideal $\mathfrak{a}$ of A can be written as $\mathfrak{a} = \mathfrak{b} \cap \mathfrak{c}$, where $\mathfrak{b}$ and $\mathfrak{c}$ are proper right ideals and at least one of them is irreducible.

(d) For every right ideal $\mathfrak{a}$ of A there exists $x \in A$ such that $(\mathfrak{a} : x)$ is irreducible.

Chapter VI. Torsion Theory

We have seen in Chap. II that to each ring of fractions of a ring A there is associated a notion of torsion for A-modules. The same will be true when we get to consider general rings of quotients of A, but here we will follow a converse course. We start by axiomatizing the concept of torsion, and then to each torsion theory we associate a ring of quotients. This chapter is devoted to a comprehensive study of the general aspects of torsion. The basic result will be that the particular notion of torsion, used in the theory of rings of quotients, can be desribed in three equivalent ways (Gabriel [2], Maranda [1]):

1) by the class of torsion modules,
2) by the right ideals which serve as annihilators of torsion elements,
3) by the functor assigning to each module its torsion submodule.

§ 1. Preradicals

One way of introducing a torsion concept for A-modules is to prescribe a functor on **Mod**-A which to each module associates a torsion submodule. In order to make formal duality arguments available, we will work in an abelian category, although all our applications deal with module categories.

Let **C** be an abelian category, which we assume to be complete, cocomplete and locally small. A *preradical* r of **C** assigns to each object C a subobject $r(C)$ in such a way that every morphism $C \to D$ induces $r(C) \to r(D)$ by restriction. In other words, a preradical is a subfunctor of the identity functor on **C**.

The class of all preradicals of **C** is a complete lattice, because there is a partial ordering in which $r_1 \leq r_2$ means that $r_1(C) \subset r_2(C)$ for all objects C, and any family (r_i) of preradicals has a least upper bound $\sum r_i$ and a greatest lower bound $\bigcap r_i$, defined in the obvious ways. If r_1 and r_2 are preradicals, one defines preradicals $r_1 r_2$ and $r_1 : r_2$ as

$$r_1 r_2(C) = r_1(r_2(C)),$$
$$(r_1 : r_2)(C)/r_1(C) = r_2(C/r_1(C)).$$

A preradical r is *idempotent* if $r r = r$ and is called a *radical* if $r : r = r$, i.e. if $r(C/r(C)) = 0$ for every object C.

Lemma 1.1. *If r is a radical and $D \subset r(C)$, then $r(C/D) = r(C)/D$.*

Proof. The canonical morphism $C \to C/D$ induces $r(C) \to r(C/D)$ with kernel D, so $r(C)/D \subset r(C/D)$. On the other hand, the canonical morphism $\alpha\colon C/D \to C/r(C)$ induces the zero morphism on $r(C/D)$, so $r(C/D) \subset \operatorname{Ker} \alpha = r(C)/D$. □

If r is a preradical of $\mathbf{C}$, then one can define a preradical r^{-1} of $\mathbf{C}^{\mathrm{op}}$ by setting $r^{-1}(M) = M/r(M)$. It is clear that r is idempotent (resp. a radical) of $\mathbf{C}$ if and only if r^{-1} is a radical (resp. an idempotent preradical) of $\mathbf{C}^{\mathrm{op}}$. This gives rise to a useful duality principle.

To a preradical r one can associate two classes of objects of $\mathbf{C}$, namely

$$\mathscr{T}_r = \{C \mid r(C) = C\},$$
$$\mathscr{F}_r = \{C \mid r(C) = 0\}.$$

Note that $\mathscr{F}_r$ in $\mathbf{C}$ is equal to $\mathscr{T}_{r^{-1}}$ in $\mathbf{C}^{\mathrm{op}}$.

Proposition 1.2. *$\mathscr{T}_r$ is closed under quotient objects and coproducts, while $\mathscr{F}_r$ is closed under subobjects and products.*

Proof. It is easy to see that $\mathscr{T}_r$ is closed under quotient objects. Let $(C_i)_I$ be an arbitrary family of objects in $\mathscr{T}_r$. Since $r(C_i) = C_i$, the image of each canonical monomorphism $C_i \to \bigoplus_I C_i$ is contained in $r(\bigoplus_I C_i)$, and it follows from the definition of coproducts that $r(\bigoplus_I C_i) = \bigoplus_I C_i$. The corresponding results for $\mathscr{F}_r$ follow by duality. □

Corollary 1.3. *If $C \in \mathscr{T}_r$ and $D \in \mathscr{F}_r$, then* $\operatorname{Hom}(C, D) = 0$.

A class $\mathscr{C}$ of objects is called a *pretorsion class* if it is closed under quotient objects and coproducts, and is a *pretorsion-free class* if it is closed under subobjects and products. Let $\mathscr{C}$ be a pretorsion class. If C is an arbitrary object of $\mathbf{C}$, and $t(C)$ denotes the sum of all subobjects of C belonging to $\mathscr{C}$, then clearly also $t(C) \in \mathscr{C}$. Hence every object C contains a largest subobject $t(C)$ belonging to $\mathscr{C}$. In this way $\mathscr{C}$ gives rise to a preradical t of $\mathbf{C}$, and t is clearly idempotent. Combining this procedure with the previous assignment $r \mapsto \mathscr{T}_r$, restricted to idempotent r, we obtain:

Proposition 1.4. *There is a bijective correspondence between idempotent preradicals of $\mathbf{C}$ and pretorsion classes of objects of $\mathbf{C}$. Dually, there is a bijective correspondence between radicals of $\mathbf{C}$ and pretorsion-free classes of objects of $\mathbf{C}$.*

In particular, if r is an arbitrary preradical of $\mathbf{C}$, and we let $\hat{r}$ denote the preradical corresponding to $\mathscr{T}_r$, i.e. $\hat{r}(C)$ is the largest subobject D of C such that $r(D) = D$, then it is easily verified that $\hat{r}$ is the largest idempotent preradical smaller than r (cf. Exercise 3). Thus:

Proposition 1.5. *For every preradical r there exists a largest idempotent preradical $\hat{r}$ smaller than r, and there exists a smallest radical $\bar{r}$ larger than r.*

The second half of the proposition is of course obtained by duality. Note that $\bar{r}(C)$ is the smallest subobject D of C such that $r(C/D) = 0$. It is possible to construct $\hat{r}$ and $\bar{r}$ by transfinite induction:

$\hat{r}$: if β is not a limit ordinal, then put $r^\beta = r r^{\beta-1}$, and if β is a limit ordinal, then put $r^\beta = \bigcap_{\alpha<\beta} r^\alpha$. This gives rise to a decreasing sequence of preradicals r^β, and $\hat{r} = \bigcap_\beta r^\beta$.

$\bar{r}$: if β is not a limit ordinal, then put $r_\beta = r_{\beta-1} : r$, i.e. r_β is given by

$$r_\beta(C)/r_{\beta-1}(C) = r(C/r_{\beta-1}(C)),$$

and for a limit ordinal β, put $r_\beta = \sum_{\alpha<\beta} r_\alpha$. This gives rise to an increasing sequence of preradicals r_β, and $\bar{r} = \sum_\beta r_\beta$.

The assignment $r \mapsto \bar{r}$ is a closure operator on the lattice of preradicals of **C**, and $r \mapsto \hat{r}$ is a closure operator on the opposite lattice. Hence the idempotent preradicals form a complete lattice, and so do also the radicals.

Proposition 1.6. (i) *If r is idempotent, then so is also $\bar{r}$.*
(ii) *If r is a radical, then so is also $\hat{r}$.*

Proof. By duality it suffices to prove e.g. (ii). Thus we should prove that if r is a radical, then $\hat{r}(C/\hat{r}(C)) = 0$ for every object C, i.e. $C/\hat{r}(C)$ has no subobject $\neq 0$ in $\mathcal{T}_r$. Suppose $D \supset \hat{r}(C)$ is a subobject of C such that $r(D/\hat{r}(C)) = D/\hat{r}(C)$. Then $\hat{r}(C) \subset r(D)$, and Lemma 1.1 implies $r(D) = D$, so $D \subset \hat{r}(C)$ and $D/\hat{r}(C) = 0$. □

Proposition 1.7. *The following assertions are equivalent for a preradical r:*
(a) *r is a left exact functor.*
(b) *If $D \subset C$, then $r(D) = r(C) \cap D$.*
(c) *r is idempotent and $\mathcal{T}_r$ is closed under subobjects.*

Proof. (a)$\Leftrightarrow$(b): Since the kernel of the morphism $r(C) \to r(D/C)$, induced from $C \to D/C$, is equal to $r(C) \cap D$, it is clear that (b) is equivalent to left exactness of r.

(b)$\Rightarrow$(c): By applying (b) to $r(C) \subset C$, one sees that r is idempotent. It is obvious that $\mathcal{T}_r$ is closed under subobjects.

(c)$\Rightarrow$(b): The inclusions $r(D) \subset r(C) \cap D \subset D$ are trivial. On the other hand, $r(C) \cap D$ belongs to $\mathcal{T}_r$ as a subobject of $r(C)$, and r idempotent implies $r(C) \cap D = r(D)$. □

A pretorsion class is called *hereditary* if it is closed under subobjects.

Corollary 1.8. *There is a bijective correspondence between left exact preradicals and hereditary pretorsion classes.*

Examples

1. Socle and Jacobson radical. Let **C** be a complete Grothendieck category. For every object C we let $s(C)$ denote the socle of C, i.e. the sum of all simple subobjects of C, and let $r(C)$ denote the Jacobson radical of C, i.e. the intersection of all maximal proper subobjects of C. Then s is a left exact preradical, while r is a radical.

2. S-torsion and S-divisibility. Let S be a right denominator set in a ring A. For every module M we let $t(M)$ be the S-torsion submodule of M, while $d(M)$ denotes the largest S-divisible submodule of M (which is easily shown to exist). Both t

and d are idempotent radicals. Furthermore t is left exact, which is not the case with d. It thus appears as if the preradicals connected to the customary torsion concepts should be left exact radicals (which are therefore often called "torsion radicals").

3. The pretorsion ideal. If r is a preradical of **Mod**-A, then $r(A)$ is a two-sided ideal, because it is invariant under all endomorphisms of A_A.

§2. Torsion Theories

Definition (Dickson [3]). A *torsion theory* for **C** is a pair $(\mathscr{T}, \mathscr{F})$ of classes of objects of **C** such that

(i) $\mathrm{Hom}(T, F)=0$ for all $T \in \mathscr{T}$, $F \in \mathscr{F}$.

(ii) If $\mathrm{Hom}(C, F)=0$ for all $F \in \mathscr{F}$, then $C \in \mathscr{T}$.

(iii) If $\mathrm{Hom}(T, C)=0$ for all $T \in \mathscr{T}$, then $C \in \mathscr{F}$.

$\mathscr{T}$ is called a *torsion class* and its objects are *torsion objects*, while $\mathscr{F}$ is a *torsion-free class* consisting of *torsion-free objects*.

Any given class $\mathscr{C}$ of objects *generates* a torsion theory in the following way:

$$\mathscr{F} = \{F \mid \mathrm{Hom}(C, F)=0 \text{ for all } C \in \mathscr{C}\},$$

$$\mathscr{T} = \{T \mid \mathrm{Hom}(T, F)=0 \text{ for all } F \in \mathscr{F}\}.$$

Clearly this pair $(\mathscr{T}, \mathscr{F})$ is a torsion theory, and $\mathscr{T}$ is the smallest torsion class containing $\mathscr{C}$. Dually, the class $\mathscr{C}$ *cogenerates* a torsion theory $(\mathscr{T}, \mathscr{F})$ such that $\mathscr{F}$ is the smallest torsion-free class containing $\mathscr{C}$.

Proposition 2.1. *The following properties of a class $\mathscr{T}$ of objects are equivalent:*

(a) *$\mathscr{T}$ is a torsion class for some torsion theory.*

(b) *$\mathscr{T}$ is closed under quotient objects, coproducts and extensions.*

Proof. A class $\mathscr{C}$ is said to be "closed under extensions" if for every exact sequence $0 \to C' \to C \to C'' \to 0$ with C' and C'' in $\mathscr{C}$, also $C \in \mathscr{C}$.

Suppose $(\mathscr{T}, \mathscr{F})$ is a torsion theory. $\mathscr{T}$ is obviously closed under quotient objects, and it is closed under coproducts because $\mathrm{Hom}(\bigoplus T_i, F) \cong \prod \mathrm{Hom}(T_i, F)$. Let $0 \to C' \to C \to C'' \to 0$ be exact with C' and C'' in $\mathscr{T}$. If F is torsion-free and there is a morphism $\alpha\colon C \to F$, then α is zero on C', so α factors over C''. But also $\mathrm{Hom}(C'', F)=0$, so $\alpha=0$. Hence $C \in \mathscr{T}$.

Conversely, assume that $\mathscr{T}$ is closed under quotient objects, coproducts and extensions. Let $(\mathscr{T}', \mathscr{F})$ be the torsion theory generated by $\mathscr{T}$. We want to show that $\mathscr{T} = \mathscr{T}'$, so suppose $\mathrm{Hom}(C, F)=0$ for all $F \in \mathscr{F}$. Since $\mathscr{T}$ is a pretorsion class, there is a largest subobject T of C belonging to $\mathscr{T}$. To show that $C=T$, it suffices to show that $C/T \in \mathscr{F}$. Suppose we have $\alpha\colon T'' \to C/T$ for some $T'' \in \mathscr{T}$. The image of α also belongs to $\mathscr{T}$, and if $\alpha \neq 0$, then we would get a subobject of C which strictly contains T and belongs to $\mathscr{T}$, since $\mathscr{T}$ is closed under extensions. This would contradict the maximality of T, and so we must have $\alpha=0$, and $C/T \in \mathscr{F}$. ▯

By duality one also has:

Proposition 2.2. *The following properties of a class $\mathscr{F}$ of objects are equivalent:*
(a) *$\mathscr{F}$ is a torsion-free class for some torsion theory.*
(b) *$\mathscr{F}$ is closed under subobjects, products and extensions.*

If $(\mathscr{T}, \mathscr{F})$ is a torsion theory, then $\mathscr{T}$ is in particular a pretorsion class, so every object C contains a largest subobject $t(C)$ belonging to $\mathscr{T}$, called the *torsion subobject* of C. An object C is torsion-free if and only if $t(C)=0$, because $C\in\mathscr{F}$ if and only if $\operatorname{Hom}(T, C)=0$ for all $T\in\mathscr{T}$. The idempotent preradical t is actually a radical, as is easily seen from the fact that $\mathscr{T}$ is closed under extensions.

Conversely, if t is an idempotent radical of $\mathbf{C}$, then one obtains a torsion theory $(\mathscr{T}_t, \mathscr{F}_t)$ with

$$\mathscr{T}_t=\{C \mid t(C)=C\}, \qquad \mathscr{F}_t=\{C \mid t(C)=0\}.$$

The Prop. 1.4 now specializes to:

Proposition 2.3. *There is a bijective correspondence between torsion theories and idempotent radicals.*

Corollary 2.4. *If r is an idempotent preradical, then $\bar{r}$ is the idempotent radical corresponding to the torsion theory generated by $\mathscr{T}_r$.*

Proof. Since $\bar{r}$ is the smallest idempotent radical containing r, it must correspond to the smallest torsion class containing $\mathscr{T}_r$. ▯

Proposition 2.5. *Let $\mathscr{C}$ be a class of objects closed under quotient objects. The torsion class generated by $\mathscr{C}$ consists of all objects C such that each non-zero quotient object of C has a non-zero subobject in $\mathscr{C}$.*

Proof. Let $(\mathscr{T}, \mathscr{F})$ be the torsion theory generated by $\mathscr{C}$. Since $\mathscr{C}$ is closed under quotient objects, an object belongs to $\mathscr{F}$ if and only if it has no non-zero subobjects in $\mathscr{C}$. Therefore the assertion is that an object C belongs to $\mathscr{T}$ if and only if C has no non-zero quotient objects in $\mathscr{F}$, and this is an obvious property of a torsion theory $(\mathscr{T}, \mathscr{F})$. ▯

Examples

1. Compact abelian groups. Pontryagin duality defines an equivalence between $\mathbf{Ab}^{\mathrm{op}}$ and the category of compact abelian groups. The standard torsion theory for $\mathbf{Ab}$ corresponds under this duality to the pair $(\mathscr{T}, \mathscr{F})$ where $\mathscr{T}$ is the class of connected compact groups and $\mathscr{F}$ is the class of totally disconnected compact groups (cf. Pontryagin [1], §33).

2. Splitting torsion theories. A torsion theory is *splitting* if the torsion subobject of C is a direct summand of C, for every object C. E.g. if e is a central idempotent of a ring A, and we put $t(M)=Me$ for every module M, then we get a splitting torsion theory $(\mathscr{T}, \mathscr{F})$ for $\mathbf{Mod}\text{-}A$ with

$$\mathscr{T}=\{M \mid Me=M\}, \qquad \mathscr{F}=\{M \mid Me=0\}.$$

A torsion theory obtained in this manner from a central idempotent e of a ring, is said to be *centrally splitting*.

3. Divisible objects. Let K be an object of $\mathbf{C}$. The torsion theory $(\mathscr{D}, \mathscr{R})$ generated by K may be described by means of Prop. 2.5 as:

$$\mathscr{D} = \{C \mid \mathrm{Hom}(K, C') \neq 0 \text{ for every non-zero quotient object } C' \text{ of } C\},$$

$$\mathscr{R} = \{C \mid \mathrm{Hom}(K, C) = 0\}.$$

The objects in $\mathscr{D}$ are called *K-divisible* and those in $\mathscr{R}$ are *K-reduced*. Each object C contains a largest K-divisible subobject $d(C)$. We may also define a preradical q by

$$q(C) = \sum \mathrm{Im}\,\alpha, \qquad \alpha \text{ ranging over } \mathrm{Hom}(K, C).$$

Clearly q is idempotent, and $d = \bar{q}$.

Of particular interest is the case when one chooses $K = E(A)$ in **Mod**-A. Every injective module is $E(A)$-divisible. In fact, the $E(A)$-divisible modules constitute the torsion class generated by the class of injective modules.

§3. Hereditary Torsion Theories

At this stage we have to impose on $\mathbf{C}$ conditions which are not of a self-dual nature, e.g. that $\mathbf{C}$ is a Grothendieck category with enough injectives. For the sake of simplicity we henceforth assume $\mathbf{C}$ to be a module category **Mod**-A for a ring A.

A torsion theory $(\mathscr{T}, \mathscr{F})$ is called *hereditary* if $\mathscr{T}$ is hereditary, i.e. $\mathscr{T}$ is closed under submodules. From Prop. 1.7 we recall that this occurs if and only if the associated radical t is left exact. Combining Cor. 1.8 and Prop. 2.3 we get

Proposition 3.1. *There is a bijective correspondence between hereditary torsion theories and left exact radicals.*

Proposition 3.2. *A torsion theory $(\mathscr{T}, \mathscr{F})$ is hereditary if and only if $\mathscr{F}$ is closed under injective envelopes.*

Proof. If t is left exact and $F \in \mathscr{F}$, then $t(E(F)) \cap F = t(F) = 0$, which implies $E(F) \in \mathscr{F}$ since F is essential in $E(F)$. Suppose conversely that $\mathscr{F}$ is closed under injective envelopes. If $T \in \mathscr{T}$ and $C \subset T$, then there is a homomorphism $\beta\colon T \to E(C/t(C))$ such that

$$\begin{array}{ccc} C & \longrightarrow & T \\ {\scriptstyle\alpha}\downarrow & & \downarrow{\scriptstyle\beta} \\ C/t(C) & \longrightarrow & E(C/t(C)) \end{array}$$

commutes. But $E(C/t(C))$ is torsion-free, so $\beta = 0$. This implies $\alpha = 0$ and hence $C = t(C) \in \mathscr{T}$. $\square$

Proposition 3.3. *Let $\mathscr{C}$ be a class of modules closed under submodules and quotient modules. The torsion theory generated by $\mathscr{C}$ is hereditary.*

Proof. We show that the class of torsion-free modules is closed under injective envelopes. Suppose F is torsion-free and there exists a non-zero $\alpha: C \to E(F)$ with $C \in \mathscr{C}$. Then $\operatorname{Im} \alpha \in \mathscr{C}$ and $F \cap \operatorname{Im} \alpha$ is a non-zero submodule of F belonging to $\mathscr{C}$, a contradiction. $\square$

Corollary 3.4. *If r is a left exact preradical, then also $\bar{r}$ is left exact.*

Proof. $\mathscr{T}_r$ is a hereditary pretorsion class, so Prop. 3.3 and Cor. 2.4 imply that $\bar{r}$ is left exact. $\square$

Corollary 3.5. *If r is a left exact preradical and M is a module, then $r(M)$ is an essential submodule of $\bar{r}(M)$.*

Proof. Suppose $L \subset \bar{r}(M)$ and $L \cap r(M) = 0$. Then $r(L) = 0$, which implies $\bar{r}(L) = 0$. But $\bar{r}(L) = \bar{r}(M) \cap L = L$, so $L = 0$. $\square$

Proposition 3.6. *A hereditary torsion theory is generated by the family of those cyclic modules $A/\mathfrak{a}$ which are torsion modules.*

Proof. A module M is a torsion module if and only if every cyclic submodule is a torsion module. From this remark the assertion easily follows. $\square$

A hereditary torsion theory is thus uniquely determined by the family of right ideals $\mathfrak{a}$ for which $A/\mathfrak{a}$ is a torsion module. These families of right ideals will be characterized in § 5. A kind of dual statement to Prop. 3.6 is:

Proposition 3.7. *A torsion theory is hereditary if and only if it can be cogenerated by an injective module.*

Proof. Let E be an injective module and put $\mathscr{T} = \{M \mid \operatorname{Hom}(M, E) = 0\}$. If $M \in \mathscr{T}$ and L is a submodule of M with a non-zero homomorphism $\alpha: L \to E$, then α extends to a homomorphism $M \to E$, which is impossible. Hence $L \in \mathscr{T}$, and the torsion theory cogenerated by E is hereditary.

Conversely, assume that $(\mathscr{T}, \mathscr{F})$ is a hereditary torsion theory. Put $E = \prod E(A/\mathfrak{a})$ with the product taken over all right ideals $\mathfrak{a}$ such that $A/\mathfrak{a} \in \mathscr{F}$. Then E is a torsion-free module, so $\operatorname{Hom}(M, E) = 0$ for every $M \in \mathscr{T}$. On the other hand, if $M \notin \mathscr{T}$, then there exists a cyclic submodule C of M with a non-zero homomorphism $\alpha: C \to F$ for some $F \in \mathscr{F}$. The image of α is cyclic torsion-free, so α induces a homomorphism $C \to E$ which can be extended to a non-zero map $M \to E$. We have thus shown that $M \in \mathscr{T}$ if and only if $\operatorname{Hom}(M, E) = 0$, and this means that E cogenerates the torsion theory. $\square$

When applying this result to an injective module of the form $E(M)$, it is useful to have available the following easily verified fact:

Lemma 3.8. *If L and M are modules, then* $\operatorname{Hom}(L, E(M)) = 0$ *if and only if* $\operatorname{Hom}(C, M) = 0$ *for every cyclic submodule C of L.*

We also note:

Proposition 3.9. *Consider a hereditary torsion theory cogenerated by the injective module E. A module M is torsion-free if and only if it is a submodule of a direct product of copies of E.*

Proof. Every submodule of a product E^I is of course torsion-free. Conversely, if M is torsion-free and $0 \neq x \in M$, then xA is not a torsion module, so $\operatorname{Hom}(xA, E) \neq 0$. Since E is injective, this means that for every non-zero $x \in M$ there exists $\mu\colon M \to E$ such that $\mu(x) \neq 0$. If one defines $\eta\colon M \to E^I$, where $I = \operatorname{Hom}(M, E)$, as $\eta(x) = (\mu(x))_{\mu \in I}$, then η is a monomorphism. □

Examples

1. Injective cogenerator. An injective cogenerator for **Mod**-A is the same as an injective module cogenerating the torsion theory $(0, \mathscr{F})$ with $\mathscr{F}$ consisting of all A-modules.

2. S-torsion. If S is a right denominator set in A, then the S-torsion and S-torsion-free modules, as defined in Chap. II, form a hereditary torsion theory.

3. Commutative localization. Let A be a commutative ring and $\mathfrak{p}$ a prime ideal of A. If $S = \{a \in A \mid a \notin \mathfrak{p}\}$, then the S-torsion theory is cogenerated by $E(A/\mathfrak{p})$. In fact, $0 \neq x \in A/\mathfrak{p}$ implies that $\operatorname{Ann}(x) = \mathfrak{p}$, so $A/\mathfrak{p}$ is S-torsion-free, and then also $E(A/\mathfrak{p})$ is S-torsion-free. On the other hand, if $\operatorname{Hom}(M, E(A/\mathfrak{p})) = 0$, then $\operatorname{Ann}(x) \not\subset \mathfrak{p}$ holds for every non-zero $x \in M$, so M is an S-torsion module.

§4. Linear Topologies

We have noticed that a hereditary torsion theory is characterized by the family of right ideals $\mathfrak{a}$ for which $A/\mathfrak{a}$ is a torsion module. It turns out that such a family of right ideals is the family of neighborhoods of 0 for a certain topology on A. For this reason we start with a general discussion of topological rings. It should be remarked in advance that these topological aspects are not crucial for the development of torsion theory, but in a way they make the picture more complete, and occasionally they may be of great value (e.g. in dealing with generalizations of the "Artin-Rees lemma").

An abelian group G is a *topological group* if it is equipped with a topology such that the group operations

$$(a, b) \mapsto a + b \quad \text{and} \quad a \mapsto -a$$

are continuous functions $G \times G \to G$ and $G \to G$. For a fixed $a \in G$, the translation map $x \mapsto a + x$ is then a homeomorphism, so U is a neighborhood of a if and only if $U - a$ is a neighborhood of 0. The topology of G is thus completely determined by the filter $\mathfrak{N}$ of neighborhoods of 0. This filter satisfies:

$N1$. For each $U \in \mathfrak{N}$, there exists $V \in \mathfrak{N}$ such that $V + V \subset U$.

$N2$. $U \in \mathfrak{N}$ implies $-U \in \mathfrak{N}$.

Conversely, if G is an abelian group with a filter $\mathfrak{N}$ of subsets, which all contain 0, and if $\mathfrak{N}$ satisfies $N1$–2, then there is a unique topology on G such that G is a topological group and $\mathfrak{N}$ is the system of neighborhoods of 0.

It is worth noting that if H is a subgroup of an abelian topological group G, then H is open if and only if it contains an interior point (for then all points are interior, by translation). So the open subgroups form a filter in the set of all subgroups.

A *topological ring* is a ring A with a topology making A into an additive topological group, such that the multiplication $(a,b)\mapsto ab$ is a continuous map $A\times A\to A$. Since one can write

$$ab-a_0 b_0=(a-a_0)(b-b_0)+(a-a_0)b_0+a_0(b-b_0),$$

the continuity of multiplication will follow if one only requires

1) for each $a\in A$, the maps $x\mapsto ax$ and $x\mapsto xa$ are continuous at 0;

2) the map $(a,b)\mapsto ab$ is continuous at $(0,0)$;

assuming of course that A is an additive topological group. Thus if A is a topological ring, the filter $\mathfrak{N}$ of neighborhoods of 0 satisfies, besides N1–2, also:

N3. For each $a\in A$ and $U\in\mathfrak{N}$, there exists $V\in\mathfrak{N}$ such that $aV\subset U$ and $Va\subset U$.

N4. For each $U\in\mathfrak{N}$ there exists $V\in\mathfrak{N}$ such that $V\cdot V\subset U$.

Conversely, if A is a ring with a filter $\mathfrak{N}$ of subsets which all contain 0, and if $\mathfrak{N}$ satisfies N1–4, then there is a unique topology on A, making A into a topological ring with $\mathfrak{N}$ as the system of neighborhoods of 0.

Suppose A is a topological ring. A *topological right A-module* is a right A-module M, equipped with a topology such that M is an additive topological group and the map $M\times A\to M$, given by $(x,a)\mapsto xa$, is continuous. The considerations made for topological rings can be repeated for modules, so the filter $\mathfrak{M}$ of neighborhoods of 0 is characterized by, besides N1–2, the axioms (let $\mathfrak{N}$ denote the filter of neighborhoods of 0 in A):

NM3. For each $x\in M$ and $U\in\mathfrak{M}$, there exists $V\in\mathfrak{N}$ such that $xV\subset U$.

NM4. For each $U\in\mathfrak{M}$ and $a\in A$, there exists $V\in\mathfrak{M}$ such that $Va\subset U$.

NM5. For each $U\in\mathfrak{M}$, there exists $V\in\mathfrak{M}$ and $W\in\mathfrak{N}$ such that $VW\subset U$.

In this book we shall be concerned exclusively with topologies defined by ideals or submodules. A topological ring A is *right linearly topological* if there is a fundamental system of neighborhoods of 0 consisting of right ideals. The set $\mathfrak{F}$ of all open right ideals then satisfies:

T1. If $\mathfrak{a}\in\mathfrak{F}$ and $\mathfrak{a}\subset\mathfrak{b}$, then $\mathfrak{b}\in\mathfrak{F}$.

T2. If $\mathfrak{a}$ and $\mathfrak{b}$ belong to $\mathfrak{F}$, then $\mathfrak{a}\cap\mathfrak{b}\in\mathfrak{F}$.

T3. If $\mathfrak{a}\in\mathfrak{F}$ and $a\in A$, then $(\mathfrak{a}:a)\in\mathfrak{F}$.

The two first axioms just state that $\mathfrak{F}$ is a filter, while T3 comes from N3. Conversely, if $\mathfrak{F}$ is a set of right ideals of A, satisfying T1–3, then there is a unique right linear topology on A with $\mathfrak{F}$ as a fundamental system of neighborhoods of 0. In fact, the axioms N1, 2 and 4, as well as one half of N3, are automatically satisfied when $\mathfrak{F}$ consists of right ideals, and the other half of N3 is covered by T3.

Suppose A is a right linearly topological ring with $\mathfrak{F}$ as the set of all open right ideals. A topological right A-module M is called a *linearly topological module* if it has a fundamental system of neighborhoods of 0 consisting of submodules. The open submodules of M satisfy:

TM1. If $L\subset L'$ are submodules and L is open, then L' is open.

TM2. If L and L' are open submodules, then $L\cap L'$ is open.

TM3. If L is an open submodule and $x\in M$, then $(L:x)\in\mathfrak{F}$.

Conversely, these axioms define uniquely a linear topology on M, in analogy with the case of rings.

In particular, for any right A-module M there is a strongest linear topology on M, namely the one for which the set of open submodules is

$$\mathfrak{F}(M)=\{L\subset M \mid (L:x)\in\mathfrak{F} \text{ for every } x\in M\}.$$

In fact, $\mathfrak{F}(M)$ satisfies $TM1$ and $TM2$ as a consequence of $T1$ and $T2$ for $\mathfrak{F}$, while $TM3$ is trivially satisfied. This topology is called the $\mathfrak{F}$-topology on M (the $\mathfrak{F}$-topology on A is the given topology!).

The module M is a linearly topological module under its discrete topology (in other words, the $\mathfrak{F}$-topology of M is discrete) if and only if the right annihilator ideals $\mathrm{Ann}(x)$ belong to $\mathfrak{F}$ for all $x\in M$. We will call a module M *$\mathfrak{F}$-discrete* if the $\mathfrak{F}$-topology of M is discrete.

Lemma 4.1. *The class of $\mathfrak{F}$-discrete modules is a hereditary pretorsion class.*

Proof. Since a module M is $\mathfrak{F}$-discrete if and only if $\mathrm{Ann}(x)\in\mathfrak{F}$ for every $x\in M$, it is clear that the class of $\mathfrak{F}$-discrete modules is closed under submodules. Closure under quotient modules follows from $T1$, and under direct sums from $T2$. ▯

By Cor. 1.8 there corresponds to $\mathfrak{F}$ a left exact preradical t, and one has $t(M)=\{x\in M \mid \mathrm{Ann}(x)\in\mathfrak{F}\}$. It will be convenient to call $t(M)$ the *$\mathfrak{F}$-pretorsion submodule* of M. We will also use alternatingly the terms "$\mathfrak{F}$-discrete module" and "$\mathfrak{F}$-pretorsion module". Cor. 1.8 can now be completed to:

Proposition 4.2. *There is a bijective correspondence between:*
(1) *Right linear topologies on A.*
(2) *Hereditary pretorsion classes of A-modules.*
(3) *Left exact preradicals of* **Mod**-*A.*

Proof. To each linear topology $\mathfrak{F}$ we have associated a pretorsion class $\mathscr{C}=\{M \mid \mathrm{Ann}(x)\in\mathfrak{F} \text{ for all } x\in M\}$. Conversely, if $\mathscr{C}$ is a hereditary pretorsion class, then we let $\mathfrak{F}$ be the family of right ideals $\mathfrak{a}$ for which $A/\mathfrak{a}\in\mathscr{C}$. This family $\mathfrak{F}$ satisfies $T1$ because $\mathscr{C}$ is closed under quotient modules, $T2$ because $A/(\mathfrak{a}\cap\mathfrak{b})$ is a submodule of $A/\mathfrak{a}\oplus A/\mathfrak{b}$, and satisfies $T3$ because if $\mathfrak{a}\in\mathfrak{F}$ and $a\in A$, then left multiplication by a induces an exact sequence

$$0\to(\mathfrak{a}:a)\to A\to A/\mathfrak{a},$$

which shows that $A/(\mathfrak{a}:a)\subset A/\mathfrak{a}$. Therefore $\mathfrak{F}$ defines a right linear topology on A.

It remains to verify that we have obtained a bijective correspondence (1)$\leftrightarrow$(2). Starting with a linear topology with the set $\mathfrak{F}$ of open right ideals, we get $\mathscr{C}=\{M \mid \mathrm{Ann}(x)\in\mathfrak{F} \text{ for all } x\in M\}$, and then get $\{\mathfrak{a} \mid A/\mathfrak{a}\in\mathscr{C}\}=\{\mathfrak{a} \mid (\mathfrak{a}:a)\in\mathfrak{F} \text{ for all } a\in A\}=\mathfrak{F}$ by $T1$. On the other hand, if we start with the class $\mathscr{C}$, we first get $\mathfrak{F}=\{\mathfrak{a} \mid A/\mathfrak{a}\in\mathscr{C}\}$ and then obtain $\{M \mid \mathrm{Ann}(x)\in\mathfrak{F} \text{ for all } x\in M\}=\{M \mid \text{each cyclic submodule}\in\mathscr{C}\}=\mathscr{C}$ because of the closure properties of $\mathscr{C}$. ▯

If A is an arbitrary ring and $\mathfrak{F}$ is a set of right ideals of A satisfying $T1$–3, we will by abuse of language call $\mathfrak{F}$ a *(right) topology.* The corresponding linear topology on A is called the *$\mathfrak{F}$-topology* on A, in accordance with terminology already introduced. By a *basis* for the topology $\mathfrak{F}$ we mean a subset $\mathfrak{B}$ of $\mathfrak{F}$ such that every right ideal in $\mathfrak{F}$ contains some $\mathfrak{b}\in\mathfrak{B}$.

Examples

1. Separated topologies. A topological group is called *separated* if every point is closed. A topology $\mathfrak{F}$ of right ideals is separated if and only if $\bigcap_{\mathfrak{a}\in\mathfrak{F}} \mathfrak{a}=(0)$.

2. The $\mathfrak{a}$-adic topology. Let $\mathfrak{a}$ be a two-sided ideal in a ring A. The powers $\mathfrak{a}^n$ form a basis for a linear topology on A, which is usually called the $\mathfrak{a}$-*adic* topology. The $\mathfrak{a}$-adic topology is separated if and only if $\bigcap_{n=1}^{\infty} \mathfrak{a}^n=0$.

§ 5. Gabriel Topologies

A hereditary torsion theory corresponds to a linear topology for which the class of discrete modules is closed unter extensions. In order to characterize such topologies we introduce a further axiom:

$T4$. If $\mathfrak{a}$ is a right ideal and there exists $\mathfrak{b}\in\mathfrak{F}$ such that $(\mathfrak{a}:b)\in\mathfrak{F}$ for every $b\in\mathfrak{b}$, then $\mathfrak{a}\in\mathfrak{F}$.

A family $\mathfrak{F}$ of right ideals of A satisfying axioms $T1$–4 is a *(right) Gabriel topology* on A. We can now state the main result of this chapter (which is due to Gabriel [2] and Maranda [1]):

Theorem 5.1. *There is a bijective correspondence between:*
(1) *Right Gabriel topologies on* A.
(2) *Hereditary torsion theories for* A.
(3) *Left exact radicals of* **Mod**-A.

Proof. We have already established the bijective correspondence (2)↔(3) in Prop. 3.1. Suppose $\mathfrak{F}$ is a Gabriel topology, and let $0\to L\to M\to N\to 0$ be an exact sequence of modules such that L and N are $\mathfrak{F}$-discrete. For each $x\in M$, put $\mathfrak{b}=\operatorname{Ann}(\bar{x})$, where $\bar{x}$ is the image of x in N. Then $\mathfrak{b}\in\mathfrak{F}$, and for each $b\in\mathfrak{b}$ we have $x\,b\in L$, so $\operatorname{Ann}(x\,b)\in\mathfrak{F}$. Since $\operatorname{Ann}(x\,b)=(\operatorname{Ann}(x):b)$, the axiom $T4$ then implies that $\operatorname{Ann}(x)\in\mathfrak{F}$. Hence M is $\mathfrak{F}$-discrete. The class of $\mathfrak{F}$-discrete modules is thus closed under extensions, and consequently it is a hereditary torsion class.

On the other hand, if $\mathscr{T}$ is a hereditary torsion class, then the corresponding topology $\mathfrak{F}=\{\mathfrak{a}\,|\,A/\mathfrak{a}\in\mathscr{T}\}$ satisfies $T4$. For if $\mathfrak{a}$ is a right ideal such that $(\mathfrak{a}:b)\in\mathfrak{F}$ for all $b\in\mathfrak{b}$ for some $\mathfrak{b}\in\mathfrak{F}$, we consider the exact sequence

$$0\to\mathfrak{b}/\mathfrak{a}\cap\mathfrak{b}\to A/\mathfrak{a}\to A/\mathfrak{a}+\mathfrak{b}\to 0$$

where $A/\mathfrak{a}+\mathfrak{b}\in\mathscr{T}$ since it is a quotient module of $A/\mathfrak{b}\in\mathscr{T}$, and also $\mathfrak{b}/\mathfrak{a}\cap\mathfrak{b}\in\mathscr{T}$ since $b\in\mathfrak{b}$ implies $((\mathfrak{a}\cap\mathfrak{b}):b)=(\mathfrak{a}:b)\in\mathfrak{F}$. Since $\mathscr{T}$ is closed under extensions, it follows that $A/\mathfrak{a}\in\mathscr{T}$ and hence $\mathfrak{a}\in\mathfrak{F}$. □

Thus if $\mathfrak{F}$ is a Gabriel topology on A, the corresponding hereditary torsion class consists of all modules which are discrete in their $\mathfrak{F}$-topology, or equivalently, for which all elements are annihilated by right ideals in $\mathfrak{F}$. These modules will be called $\mathfrak{F}$-*torsion* modules.

The task of checking axioms $T1$–4 is simplified by the fact that $T1$ and 2 actually follow from $T3$ and 4:

Lemma 5.2. *If $\mathfrak{F}$ is a non-empty set of right ideals of A satisfying $T3$ and 4, then $\mathfrak{F}$ also satisfies $T1$ and 2.*

Proof. $T1$: We first note that $T3$, together with the fact that $\mathfrak{F}$ is non-empty, implies that $A \in \mathfrak{F}$. Then suppose $\mathfrak{a} \in \mathfrak{F}$ and $\mathfrak{b} \supset \mathfrak{a}$. For each $a \in \mathfrak{a}$ we have $(\mathfrak{b} : a) = A \in \mathfrak{F}$, so $\mathfrak{b} \in \mathfrak{F}$ by $T4$.

$T2$: Suppose $\mathfrak{a}$ and $\mathfrak{b}$ belong to $\mathfrak{F}$. If $b \in \mathfrak{b}$, then $((\mathfrak{a} \cap \mathfrak{b}) : b) = (\mathfrak{a} : b) \cap (\mathfrak{b} : b) = (\mathfrak{a} : b) \in \mathfrak{F}$ by $T3$, so $\mathfrak{a} \cap \mathfrak{b} \in \mathfrak{F}$ by $T4$. □

We also remark that a Gabriel topology is closed under products:

Lemma 5.3. *Let $\mathfrak{F}$ be a Gabriel topology. If $\mathfrak{a}$ and $\mathfrak{b}$ belong to $\mathfrak{F}$, then $\mathfrak{a}\mathfrak{b} \in \mathfrak{F}$.*

Proof. For each $a \in \mathfrak{a}$ we have $(\mathfrak{a}\mathfrak{b} : a) \supset \mathfrak{b}$, so $\mathfrak{a}\mathfrak{b} \in \mathfrak{F}$ by $T1$ and 4. □

If $\mathfrak{F}_1$ and $\mathfrak{F}_2$ are topologies on A, we say that $\mathfrak{F}_1$ is *weaker* than $\mathfrak{F}_2$ (and $\mathfrak{F}_2$ is *stronger* than $\mathfrak{F}_1$) if $\mathfrak{F}_1 \subset \mathfrak{F}_2$. Since it is clear that any intersection of topologies is a topology, the topologies on A form a complete lattice $\mathsf{Top}(A)$. Since also every intersection of Gabriel topologies is a Gabriel topology, there is a closure operator J on $\mathsf{Top}(A)$ which to each topology $\mathfrak{E}$ associates the weakest Gabriel topology $J(\mathfrak{E})$ stronger than $\mathfrak{E}$. Prop. 3.1 establishes a lattice isomorphism between $\mathsf{Top}(A)$ and the lattice of pretorsion classes of A-modules. It follows that if $\mathfrak{E}$ is a topology, then $J(\mathfrak{E})$ is the Gabriel topology corresponding to the hereditary torsion theory generated by the class of $\mathfrak{E}$-discrete modules.

Proposition 5.4. *If $\mathfrak{E}$ is a topology, then $J(\mathfrak{E}) = \{\mathfrak{a} \mid$ for every $\mathfrak{b} \supset \mathfrak{a}$, $\mathfrak{b} \neq A$, there exists $a \notin \mathfrak{b}$ such that $(\mathfrak{b} : a) \in \mathfrak{E}\}$.*

Proof. Apply Prop. 2.5 with $\mathscr{C} = \{A/\mathfrak{a} \mid \mathfrak{a} \in \mathfrak{E}\}$. Note that the torsion theory generated by $\mathscr{C}$ is hereditary by Prop. 3.3. □

We recall that every hereditary torsion theory may be cogenerated by an injective module (Prop. 3.7). If this injective module is given as the injective envelope of some module M, the corresponding Gabriel topology can be described as follows:

Proposition 5.5. *Let $\mathfrak{F}$ be the Gabriel topology corresponding to the torsion theory cogenerated by $E(M)$. Then $\mathfrak{a} \in \mathfrak{F}$ if and only if $x(\mathfrak{a} : a) \neq 0$ for every $a \in A$ and non-zero $x \in M$.*

Proof. From Lemma 3.8 we obtain that $\mathfrak{a} \in \mathfrak{F}$ if and only if $\mathrm{Hom}(C, M) = 0$ for every cyclic submodule C of $A/\mathfrak{a}$. But the cyclic submodules of $A/\mathfrak{a}$ have the form $A/(\mathfrak{a} : a)$ for $a \in A$, and so the result follows. □

One also easily verifies:

Proposition 5.6. *The Gabriel topology corresponding to the torsion theory cogenerated by $E(M)$ is the strongest Gabriel topology for which M is torsion-free.*

§6. Examples of Gabriel Topologies

1. 1-topologies. A 1-*topology* on A is a Gabriel topology with a basis consisting of principal right ideals. A 1-topology $\mathfrak{F}$ is determined by the set $\Sigma(\mathfrak{F})=\{s\in A \mid sA\in\mathfrak{F}\}$.

Proposition 6.1. *The map $\mathfrak{F}\mapsto\Sigma(\mathfrak{F})$ defines a bijective correspondence between 1-topologies $\mathfrak{F}$ on A and multiplicatively closed subsets S of A satisfying*

$S0$. If $ab\in S$, then $a\in S$.

$S1$. If $s\in S$ and $a\in A$, then there exist $t\in S$ and $b\in A$ such that $sb=at$.

Proof. Suppose $\mathfrak{F}$ is a 1-topology. Then $\Sigma(\mathfrak{F})$ is multiplicatively closed, for if s and t belong to $\Sigma(\mathfrak{F})$, then $(stA:sa)\supset(tA:a)$ for every $a\in A$, and $(tA:a)\in\mathfrak{F}$ by $T3$, so $stA\in\mathfrak{F}$ by $T4$. The property $S0$ for $\Sigma(\mathfrak{F})$ is immediate from $T1$, and also $S1$ is clear, because $T3$ implies that $(sA:a)\supset tA$ for some $tA\in\mathfrak{F}$.

Conversely, let S be a multiplicatively closed set satisfying $S1$, and put $\mathfrak{F}=\{\mathfrak{a}\mid\mathfrak{a}\cap S\neq\emptyset\}$. Then it is easy to verify the axioms $T3$ and 4 for $\mathfrak{F}$, and $\mathfrak{F}$ is thus a 1-topology. $S0$ is a saturation axiom which makes the correspondence $\mathfrak{F}\leftrightarrow S$ bijective. □

For a 1-topology $\mathfrak{F}$, the $\mathfrak{F}$-torsion modules M are characterized by the property that for every $x\in M$ there exists $s\in\Sigma(\mathfrak{F})$ such that $xs=0$. This kind of torsion was studied already in Chap. II (§3), in a somewhat more narrow context.

2. The Goldie torsion theory. The family $\mathfrak{E}$ of essential right ideals of A is a topology, but it does not always satisfy $T4$. The corresponding left exact preradical is usually denoted by Z, and one calls $Z(M)$ the *singular submodule* of M. The Gabriel topology $J(\mathfrak{E})$ is called the *Goldie topology* of A. The corresponding torsion radical G is the smallest radical containing Z. The transfinite process which leads from Z to G (§1) gets stationary at an early stage; in fact we have, in the notation of §1:

Proposition 6.2. $G=Z_2$.

Proof. We first remark that $Z(M)$ is an essential submodule of $G(M)$ for every module M. For if $L\subset G(M)$ and $L\cap Z(M)=0$, then $Z(L)=L\cap Z(M)=0$ and L is thus torsion-free, which implies $L=0$. We now have

$$Z_2(M)/Z(M)=Z(M/Z(M))\supset G(M)/Z(M),$$

so $Z_2(M)=G(M)$. □

Proposition 6.3. *If $\mathfrak{E}$ is the family of essential right ideals, then $J(\mathfrak{E})=\{\mathfrak{a}\mid$ there exists $\mathfrak{b}\in\mathfrak{E}$ such that $\mathfrak{a}\subset\mathfrak{b}$ and $(\mathfrak{a}:b)\in\mathfrak{E}$ for all $b\in\mathfrak{b}\}$.*

Proof. If $\mathfrak{a}$ satisfies the condition, then $\mathfrak{a}\in J(\mathfrak{E})$ by $T4$. If $\mathfrak{a}\in J(\mathfrak{E})$, then $A/\mathfrak{a}$ is a Goldie torsion module, so $A/\mathfrak{a}=Z_2(A/\mathfrak{a})$. We write $Z(A/\mathfrak{a})=\mathfrak{b}/\mathfrak{a}$. The formula defining Z_2 gives $A/\mathfrak{b}=Z(A/\mathfrak{b})$ and hence $\mathfrak{b}\in\mathfrak{E}$. We also have $Z(\mathfrak{b}/\mathfrak{a})=Z^2(A/\mathfrak{a})=Z(A/\mathfrak{a})=\mathfrak{b}/\mathfrak{a}$, so $(\mathfrak{a}:b)\in\mathfrak{E}$ for all $b\in\mathfrak{b}$. □

Another description of $J(\mathfrak{E})$ was obtained in Prop. 5.4. Note that a module is Goldie torsion-free if and only if it is *non-singular*, i.e. its singular submodule is zero.

3. The dense topology. The hereditary torsion theory cogenerated by the injective right module $E(A)$ is of particular importance. The right ideals belonging to the corresponding Gabriel topology $\mathfrak{D}$ are called *dense*. According to Prop. 5.5 they can be described as follows:

Proposition 6.4. *A right ideal $\mathfrak{a}$ is dense if and only if $(\mathfrak{a}:a)$ has no non-zero left annihilators for any $a \in A$.*

Corollary 6.5. *Every dense right ideal is essential in A.*

Corollary 6.6. *A two-sided ideal $\mathfrak{a}$ is dense as a right ideal if and only if $\mathfrak{a}$ has no non-zero left annihilators.*

Proof. Since $(\mathfrak{a}:a) \supset \mathfrak{a}$ for every $a \in A$, this is clear. □

$\mathfrak{D}$ is the strongest Gabriel topology for which A is torsion-free (Prop. 5.6). It is in general weaker than the Goldie topology, but often the two topologies coincide:

Proposition 6.7. *The ring A is right non-singular if and only if every essential right ideal is dense.*

Proof. If $\mathfrak{a}$ is an essential right ideal, then also $(\mathfrak{a}:a)$ is essential for every $a \in A$. Therefore non-singularity implies that $(\mathfrak{a}:a)$ can have no left annihilators. The converse implication is obvious. □

Corollary 6.8. *When A is right non-singular, the Goldie topology and the dense topology coincide.*

Note that when A is a ring satisfying the right Ore condition, every regular element generates a dense right ideal.

4. The topologies $\mathfrak{F}_M^n$. Let M be an arbitrary A-module. Consider the minimal injective resolution of M:

$$0 \longrightarrow M \xrightarrow[\mu_0]{} E_0 \xrightarrow[\mu_1]{} E_1 \xrightarrow[\mu_2]{} E_2 \longrightarrow \cdots$$

where $E_0 = E(M)$ and each E_{i+1} is the injective envelope of Coker μ_i. Let $\mathfrak{F}_M^n$ denote the Gabriel topology corresponding to the hereditary torsion theory cogenerated by the injective module $E_0 \oplus \cdots \oplus E_n$. Note e.g. that $\mathfrak{F}_A^0$ is the dense topology introduced in the previous example.

Proposition 6.9. *A module L is an $\mathfrak{F}_M^n$-torsion module if and only if $\mathrm{Ext}_A^i(L', M) = 0$ for all $i \leq n$ and all [cyclic] submodules L' of L.*

Proof. Put $M_i = \mathrm{Im}\,\mu_i$. A module L is an $\mathfrak{F}_M^n$-torsion module if and only if $\mathrm{Hom}(L, E_i) = 0$ for $i \leq n$, which by Lemma 3.8 is equivalent to the requirement that $\mathrm{Hom}(L', M_i) = 0$ for all [cyclic] $L' \subset L$ and all $i \leq n$. We now proceed by induction on n. For $n = 0$ the assertion is clear. From the exact sequence

$$0 \to M_{i-1} \to E_{i-1} \to M_i \to 0 \tag{1}$$

for $i = n$ we obtain an exact sequence

$$\mathrm{Hom}(L', E_{n-1}) \to \mathrm{Hom}(L', M_n) \to \mathrm{Ext}^1(L', M_{n-1}) \to 0.$$

By an iterated use of sequences (1), we find that $\mathrm{Ext}^1(L', M_{n-1}) \cong \mathrm{Ext}^n(L', M)$. Using the induction hypothesis, we see that an $\mathfrak{F}_M^{n-1}$-torsion module M is $\mathfrak{F}_M^n$-torsion if and only if $\mathrm{Ext}^n(L', M)=0$ for all [cyclic] $L' \subset L$. □

5. Bounded topologies. A topology is *bounded* if it has a basis consisting of two-sided ideals.

Proposition 6.10. *Assume B is a set of two-sided ideals which are finitely generated as right ideals. The set of finite products of ideals belonging to* $\mathfrak{B}$ *is a basis for a bounded Gabriel topology.*

Proof. Let $\mathfrak{B}'$ be the set of all finite products of ideals in $\mathfrak{B}$. Suppose $\mathfrak{a}$ is a right ideal such that $\mathfrak{a} \supset \mathfrak{b}$ for some $\mathfrak{b} \in \mathfrak{B}'$. For each $a \in A$ we have $a\,\mathfrak{b} \subset \mathfrak{b} \subset \mathfrak{a}$, so $(\mathfrak{a} : a) \supset \mathfrak{b}$, and consequently $T3$ is satisfied by the family $\mathfrak{F}$ of right ideals containing ideals in $\mathfrak{B}'$.

Next we verify that $\mathfrak{F}$ satisfies $T4$. Suppose $\mathfrak{a}$ is a right ideal and there exists $\mathfrak{b} \in \mathfrak{F}$ such that $(\mathfrak{a} : b) \in \mathfrak{F}$ for all $b \in \mathfrak{b}$. We may equally well assume $\mathfrak{b} \in \mathfrak{B}'$. Let $b_1, \ldots, b_n$ be generators for $\mathfrak{b}$ as a right ideal. Then there exist $\mathfrak{b}_1, \ldots, \mathfrak{b}_n$ in $\mathfrak{B}'$ such that $b_i\,\mathfrak{b}_i \subset \mathfrak{a}$ for $i=1, \ldots, n$. It follows that $\mathfrak{b} \cdot \mathfrak{b}_1 \cdot \cdots \cdot \mathfrak{b}_n \subset \mathfrak{b}(\mathfrak{b}_1 \cap \cdots \cap \mathfrak{b}_n) \subset \mathfrak{a}$, so $\mathfrak{a} \in \mathfrak{F}$. □

This result may in particular be applied to the case when $\mathfrak{B}$ consists of just one two-sided ideal $\mathfrak{b}$, and then we obtain the $\mathfrak{b}$-adic topology (Example 4.2). If $\mathfrak{b}$ is idempotent, i.e. $\mathfrak{b}^2 = \mathfrak{b}$, one has no need for the hypothesis that $\mathfrak{b}$ should be finitely generated in the above proof. Hence:

Proposition 6.11. *If* $\mathfrak{b}$ *is an idempotent two-sided ideal, then the set of right ideals containing* $\mathfrak{b}$ *is a Gabriel topology.*

This kind of bounded topologies has alternative characterizations of certain interest:

Proposition 6.12. *The following properties of a Gabriel topology* $\mathfrak{F}$ *are equivalent:*

(a) *The class of* $\mathfrak{F}$*-torsion modules is closed under direct products.*

(b) *There is a two-sided ideal* $\mathfrak{b}$ *such that* M *is an* $\mathfrak{F}$*-torsion module if and only if* $M\mathfrak{b}=0$.

(c) $\mathfrak{F}$ *has a basis consisting of a right ideal* $\mathfrak{b}$.

(d) $\mathfrak{F}$ *has a basis consisting of an idempotent two-sided ideal* $\mathfrak{b}$.

Proof. (a) ⇒ (c): Consider the canonical homomorphism $\alpha: A \to \prod A/\mathfrak{a}$ with the product taken over all $\mathfrak{a} \in \mathfrak{F}$. The image of α is by hypothesis an $\mathfrak{F}$-torsion module, so $\mathrm{Ker}\,\alpha = \bigcap \mathfrak{a}$ belongs to $\mathfrak{F}$. This means that $\mathfrak{F}$ has a smallest member $\mathfrak{b} = \bigcap \mathfrak{a}$.

(c) ⇒ (d): The smallest right ideal $\mathfrak{b}$ belonging to $\mathfrak{F}$ must be a two-sided ideal, because $T3$ gives $(\mathfrak{b} : a) \supset \mathfrak{b}$ for all $a \in A$. The idempotency of $\mathfrak{b}$ is a consequence of Lemma 5.3.

The implications (d) ⇒ (b) ⇒ (a) are both obvious. □

We will return to the study of these topologies in §8.

6. Commutative rings. Let A be a commutative ring. Denote by $\mathrm{Spec}(A)$ the set of all prime ideals of A. For each ideal $\mathfrak{a}$, put $V(\mathfrak{a}) = \{\mathfrak{p} \in \mathrm{Spec}(A) \mid \mathfrak{a} \subset \mathfrak{p}\}$.

If $\mathfrak{p}$ is a prime ideal, its complement as a subset of A is a multiplicatively closed set S. For a module M one denotes the module of fractions $M[S^{-1}]$ by $M_{\mathfrak{p}}$. There is a corresponding Gabriel topology

$$\mathfrak{F}_{\mathfrak{p}} = \{\mathfrak{a} \mid \mathfrak{p} \notin V(\mathfrak{a})\}.$$

The $\mathfrak{F}_{\mathfrak{p}}$-torsion modules M are characterized by the property that $M_{\mathfrak{p}} = 0$.

More generally, let $\mathscr{P}$ be a subset of $\operatorname{Spec}(A)$. To $\mathscr{P}$ we associate a Gabriel topology

$$\mathfrak{F}_{\mathscr{P}} = \bigcap_{\mathfrak{p} \in \mathscr{P}} \mathfrak{F}_{\mathfrak{p}} = \{\mathfrak{a} \mid V(\mathfrak{a}) \cap \mathscr{P} = \emptyset\}.$$

The corresponding torsion class consists of all modules M with $M_{\mathfrak{p}} = 0$ for all $\mathfrak{p} \in \mathscr{P}$.

Conversely, to each Gabriel topology $\mathfrak{F}$ on A we associate

$$D(\mathfrak{F}) = \{\mathfrak{p} \in \operatorname{Spec}(A) \mid \mathfrak{p} \notin \mathfrak{F}\} \subset \operatorname{Spec}(A).$$

Then $\mathfrak{F}_{D(\mathfrak{F})} = \{\mathfrak{a} \mid V(\mathfrak{a}) \cap D(\mathfrak{F}) = \emptyset\} = \{\mathfrak{a} \mid V(\mathfrak{a}) \subset \mathfrak{F}\}$. From this one easily obtains:

Proposition 6.13. *The following properties of a Gabriel topology $\mathfrak{F}$ on a commutative ring A are equivalent:*

(a) *$\mathfrak{F}$ is equal to $\mathfrak{F}_{\mathscr{P}}$ for some $\mathscr{P} \subset \operatorname{Spec}(A)$.*

(b) *$\mathfrak{F} = \mathfrak{F}_{D(\mathfrak{F})}$.*

(c) *For every right ideal $\mathfrak{a}$ with $a \notin \mathfrak{F}$, there exists $\mathfrak{p} \in V(\mathfrak{a})$ such that $\mathfrak{p} \notin \mathfrak{F}$.*

Lemma 6.14. *Let $\mathfrak{F}$ be a Gabriel topology on A. Then:*

(i) *If $\mathfrak{a}$ is an ideal which is maximal with respect to $\mathfrak{a} \notin \mathfrak{F}$, then $\mathfrak{a}$ is a prime ideal.*

(ii) *If $\mathfrak{F}$ has a basis consisting of finitely generated ideals and $\mathfrak{a} \notin \mathfrak{F}$, then there exists $\mathfrak{p} \notin V(\mathfrak{a})$ such that $\mathfrak{p} \notin \mathfrak{F}$.*

Proof. (i): Suppose a and b are elements of A not belonging to $\mathfrak{a}$. Then $\mathfrak{a} + Aa$ and $\mathfrak{a} + Ab$ must be members of $\mathfrak{F}$, and also $(\mathfrak{a} + Aa)(\mathfrak{a} + Ab) \in \mathfrak{F}$ by Lemma 5.3. But $(\mathfrak{a} + Aa)(\mathfrak{a} + Ab) \subset \mathfrak{a} + Aab$, and therefore $ab \notin \mathfrak{a}$.

(ii): Since $\mathfrak{a} \notin \mathfrak{F}$, one can use Zorn's lemma to find an ideal $\mathfrak{b} \supset \mathfrak{a}$ which is maximal with respect to $\mathfrak{b} \notin \mathfrak{F}$ (this necessitates the hypothesis concerning finitely generated ideals in $\mathfrak{F}$). $\mathfrak{b}$ is a prime ideal by (i). □

Combining 6.13(c) and 6.14(ii) we obtain:

Corollary 6.15. *If $\mathfrak{F}$ is a Gabriel topology having a basis of finitely generated ideals, then $\mathfrak{F} = \mathfrak{F}_{\mathscr{P}}$ for a certain $\mathscr{P} \subset \operatorname{Spec}(A)$.*

So for a commutative noetherian ring A, all Gabriel topologies are of the form $\mathfrak{F}_{\mathscr{P}}$ with $\mathscr{P} \subset \operatorname{Spec}(A)$.

§7. Stable Torsion

Let A be an arbitrary ring and let $\mathfrak{F}$ be a topology of right ideals on A. Recall that if M is a module, then a fundamental system of neighborhoods of 0 for the $\mathfrak{F}$-topology on M is given by

$$\begin{aligned}\mathfrak{F}(M) &= \{L \subset M \mid (L:x) \in \mathfrak{F} \text{ for every } x \in M\} \\ &= \{L \subset M \mid M/L \text{ is an } \mathfrak{F}\text{-pretorsion module}\}.\end{aligned}$$

If M' is a submodule of M, then the $\mathfrak{F}$-topology on M induces a subspace topology on M', which is easily seen to be weaker than the $\mathfrak{F}$-topology on M'. In this section we want to find out when the $\mathfrak{F}$-topology on M induces the $\mathfrak{F}$-topology on M'. This is the kind of question that is dealt with in the classical Artin-Rees theorem for $\mathfrak{a}$-adic topologies on a commutative noetherian ring (cf. Zariski and Samuel [1], Vol. 2).

Proposition 7.1 (Gabriel [2]). *The following properties of a topology $\mathfrak{F}$ are equivalent:*

(a) *For every module M and submodule M', the $\mathfrak{F}$-topology on M' coincides with the subspace topology induced from the $\mathfrak{F}$-topology on M.*

(b) *The class of $\mathfrak{F}$-pretorsion modules is closed under injective envelopes.*

(c) *For every module M, the $\mathfrak{F}$-pretorsion submodule of M is essentially closed in M.*

(d) *Every injective module is split by its $\mathfrak{F}$-pretorsion submodule.*

Proof. (a) $\Rightarrow$ (b): Let M be an $\mathfrak{F}$-pretorsion module. Since the $\mathfrak{F}$-topology of M is discrete and is induced from the $\mathfrak{F}$-topology of $E(M)$, there exists a submodule L of $E(M)$ such that $L \in \mathfrak{F}(E(M))$ and $L \cap M = 0$. But then $L = 0$, and hence $E(M)$ is an $\mathfrak{F}$-pretorsion module.

(b) $\Rightarrow$ (c): Let $t(M)$ be the $\mathfrak{F}$-pretorsion submodule of M. Considering $E(t(M))$ as imbedded in $E(M)$, we have $E(t(M)) \cap M = t(M)$ since $t(M)$ is the maximal $\mathfrak{F}$-pretorsion module in M. But this means that $t(M)$ is essentially closed in M (cf. Exercise V.5).

(c) $\Rightarrow$ (d) is obvious.

(d) $\Rightarrow$ (b): If M is a pretorsion module, then of course $M \subset t(E(M))$. If $t(E(M))$ splits $E(M)$, we must then have $t(E(M)) = E(M)$ since M is essential in $E(M)$.

(b) $\Rightarrow$ (a): Let M' be a submodule of M. The $\mathfrak{F}$-topology on M' is in any case stronger than the subspace topology induced from the $\mathfrak{F}$-topology on M, as we have already remarked. Suppose on the other hand that $L \in \mathfrak{F}(M')$. We must find $N \in \mathfrak{F}(M)$ such that $N \cap M' = L$. Choose N to be a maximal submodule of M such that $N \cap M' = L$ Then $M'/L = M'/(N \cap M') \cong (M'+N)/N$. But $(M'+N)/N$ is an essential submodule of M/N, for if $M'' \supset N$ and $(M''/N) \cap (M'+N)/N = 0$, then $M'' \cap M' \subset N$ and hence $M'' \cap M' = N \cap M'$, and the maximality of N implies that $M'' = N$. Thus M/N is an essential extension of the $\mathfrak{F}$-pretorsion module $M'/L \cong (M'+N)/N$, and the hypothesis (b) then implies that M/N is $\mathfrak{F}$-pretorsion, as desired. □

A hereditary torsion theory is called *stable* if the class of torsion modules is closed under injective envelopes, i.e. the equivalent conditions of Prop. 7.1 are satisfied.

If r is a left exact preradical, then we know that $\bar{r}(M)$ is an essential extension of $r(M)$ for every module M (Cor. 3.5). In the stable case we can be more precise:

Proposition 7.2. *Let r be a left exact preradical and assume that the torsion theory associated to $\bar{r}$ is stable. Then $\bar{r}(M)$ is the largest essential extension of $r(M)$ in M, for every module M.*

Proof. We want to show that if $r(M)$ is essential in some submodule L of M, then $L \subset \bar{r}(M)$. Choose a pseudo-complement K of $\bar{r}(L)$ in L. Then $r(K)=0$ and hence $r(M) \cap K = 0$. Since $r(M)$ is essential in L, this implies $K=0$. From the maximality of K with respect to $K \cap \bar{r}(L)=0$, it follows that $\bar{r}(L)$ is essential in L. Stability then gives that $L=\bar{r}(L)$ by Prop. 7.1(c), and so $L=\bar{r}(L) \subset \bar{r}(M)$. ▯

Example. Goldie torsion. The customary torsion theory for commutative integral domains is stable; much more generally we have:

Proposition 7.3. *The Goldie torsion theory is stable.*

This is a particular case of the following result:

Proposition 7.4. *Every quotient module M of an injective module E is split by its Goldie torsion submodule.*

Proof. One obtains this by applying the following lemma to the epimorphism $E \to M \to M/t(M)$. ▯

Lemma 7.5. *Let E be an injective module and $\alpha: E \to M$ an epimorphism onto a non-singular module M. Then α splits, and M is thus an injective module.*

Proof. Since M is non-singular, $\operatorname{Ker}\alpha$ must be essentially closed in E. But since E is injective, this implies that $\operatorname{Ker}\alpha$ is a direct summand of E, and hence α splits. ▯

§8. TTF-Classes

A class $\mathcal{T}$ of modules is called a TTF-*class* (TTF stands for "torsion torsion-free") if it is both a torsion class and a torsion-free class. This means that there exist classes $\mathcal{C}$ and $\mathcal{F}$ such that both $(\mathcal{C}, \mathcal{T})$ and $(\mathcal{T}, \mathcal{F})$ are torsion theories. Note that $(\mathcal{T}, \mathcal{F})$ is then hereditary, and that $(\mathcal{C}, \mathcal{T})$ is hereditary if and only if $(\mathcal{T}, \mathcal{F})$ is stable. It will be convenient to call the triple $(\mathcal{C}, \mathcal{T}, \mathcal{F})$ a TTF-*theory*. Clearly:

Proposition 8.1. *A hereditary torsion class is a* TTF-*class if and only if it is closed under direct products.*

Thus by Prop. 6.12, $\mathcal{T}$ is a TTF-class if and only if there is an idempotent two-sided ideal $\mathfrak{c}$ such that $\mathcal{T}$ consists of all modules annihilated by $\mathfrak{c}$. We will examine in more detail the properties of a TTF-theory $(\mathcal{C}, \mathcal{T}, \mathcal{F})$. Let c and t denote the radicals associated to $\mathcal{C}$ and $\mathcal{T}$, respectively. Note that $c(A)$ is precisely the mentioned idempotent two-sided ideal $\mathfrak{c}$, because it is the smallest right ideal $\mathfrak{a}$ of A for which $A/\mathfrak{a} \in \mathcal{T}$. Thus $t(M)=\{x \in M \mid x\,\mathfrak{c}=0\}$, while for c we have

Lemma 8.2. $c(M)=M\,\mathfrak{c}$ *for every module M.*

Proof. $c(M)$ is the smallest submodule L of M for which $M/L \in \mathcal{T}$, i.e. for which $M\mathfrak{c} \subset L$, and this is of course $M\mathfrak{c}$. □

Lemma 8.3. *If* $(\mathcal{C}, \mathcal{T})$ *is hereditary, then* $\mathcal{C} \subset \mathcal{F}$.

Proof. If $M \in \mathcal{C}$, then $t(M) \in \mathcal{C} \cap \mathcal{T} = 0$, so $M \in \mathcal{F}$. □

Lemma 8.4. $\mathcal{F} \subset \mathcal{C}$ *if and only if* $M = c(M) + t(M)$ *for every module* M.

Proof. $\mathcal{F} \subset \mathcal{C}$ holds if and only if $c(M/t(M)) = M/t(M)$ holds for every module M. But

$$c(M/t(M)) = (M/t(M))\,\mathfrak{c} = (M\mathfrak{c} + t(M))/t(M),$$

and the assertion follows. □

We have previously called an arbitrary torsion theory $(\mathcal{T}, \mathcal{F})$ *centrally splitting* if there is a central idempotent e such that $t(M) = Me$ for all modules M (Example 2.2). In that case we have a TTF-theory $(\mathcal{F}, \mathcal{T}, \mathcal{F})$. Conversely:

Proposition 8.5. *The following properties of a* TTF-*theory* $(\mathcal{C}, \mathcal{T}, \mathcal{F})$ *are equivalent:*

(a) $\mathcal{C} = \mathcal{F}$.
(b) $M = c(M) \oplus t(M)$ *for every module* M.
(c) $(\mathcal{C}, \mathcal{T})$ *and* $(\mathcal{T}, \mathcal{F})$ *are splitting.*
(d) $(\mathcal{C}, \mathcal{T})$ *is hereditary and splitting.*
(e) $A = c(A) \oplus t(A)$.
(f) $(\mathcal{C}, \mathcal{T})$ *is hereditary and* $\mathcal{F}$ *is a* TTF-*class.*
(g) $(\mathcal{T}, \mathcal{F})$ *is centrally splitting.*

Proof. (a) ⇒ (b): By Lemma 8.4 we have $M = c(M) + t(M)$. But if $\mathcal{C} = \mathcal{F}$, then $\mathcal{C}$ is hereditary, and $c(M) \cap t(M) \in \mathcal{C} \cap \mathcal{T} = 0$ implies $M = c(M) \oplus t(M)$.

(b) ⇒ (c) is trivial.

(c) ⇒ (d): If $(\mathcal{T}, \mathcal{F})$ is splitting, then $\mathcal{T}$ is closed under injective envelopes by Prop. 7.1, and hence $(\mathcal{C}, \mathcal{T})$ is hereditary by Prop. 3.2.

(d) ⇒ (e): Because of the splitting, there is a right ideal $\mathfrak{a}$ such that $A = c(A) \oplus \mathfrak{a}$. Then $t(A) = t(c(A)) \oplus t(\mathfrak{a}) = t(\mathfrak{a})$ since $t(c(A)) \in \mathcal{C} \cap \mathcal{T} = 0$. But $\mathfrak{a} \cong A/c(A) \in \mathcal{T}$, so $\mathfrak{a} = t(\mathfrak{a}) = t(A)$.

(e) ⇒ (g): We can write $\mathfrak{c} = c(A)$ as $c(A) = eA$ for an idempotent e, which is central because both $c(A)$ and $t(A)$ are two-sided ideals. Then $t(M) = \{x \in M \mid x\mathfrak{c} = 0\} = \{x \in M \mid xe = 0\} = M(1-e)$.

(g) ⇒ (f): It is clear that $\mathcal{C}$ is closed under submodules, and that $\mathcal{F}$ is closed under quotient modules, i.e. $\mathcal{F}$ is a TTF-class.

(f) ⇒ (a): If $\mathcal{F}$ is closed under quotient modules, then $\mathrm{Hom}(F, T) = 0$ for all $F \in \mathcal{F}$, $T \in \mathcal{T}$, so $\mathcal{F} \subset \mathcal{C}$. Also $\mathcal{C} \subset \mathcal{F}$ by Lemma 8.3. □

Examples

1. Commutative noetherian rings. First of all we note:

Lemma 8.6. *If* A *is a commutative ring, then every finitely generated idempotent ideal is generated by a single idempotent element.*

Proof. Suppose the ideal $\mathfrak{a}$ is idempotent and generated by $a_1, \ldots, a_n$. For each a_i one has $a_i = \sum_j b_{ij} a_j$ for certain $b_{ij} \in \mathfrak{a}$. This leads to a system of equations

$$\begin{aligned} (1-b_{11})\, a_1 - b_{12}\, a_2 - \cdots \qquad & - b_{1n}\, a_n = 0 \\ \vdots \qquad\qquad & \\ -b_{n1}\, a_1 - \qquad \cdots \qquad & + (1-b_{nn})\, a_n = 0. \end{aligned}$$

The determinant of this system must annihilate each a_i (see Lang [1], p. 335), and hence it annihilates $\mathfrak{a}$. But the determinant has the form $1-c$ for some $c \in \mathfrak{a}$, and $(1-c)\,\mathfrak{a} = 0$ implies that c is idempotent and generates $\mathfrak{a}$. □

Consequently:

Proposition 8.7. *If A is a commutative noetherian ring, then every* TTF-*class $\mathcal{T}$ defines a centrally splitting torsion theory $(\mathcal{T}, \mathcal{F})$.*

2. Artinian rings. If A is right artinian, then every Gabriel topology has a unique smallest member, and therefore every hereditary torsion class is a TTF-class.

§9. $\mathfrak{F}$-Divisible Modules

Let $\mathfrak{F}$ be a Gabriel topology of right ideals on A. A **left** A-module D is called $\mathfrak{F}$-*divisible* if $D = \mathfrak{a} D$ for every $\mathfrak{a} \in \mathfrak{F}$. (Remark: this generalization of the classical divisibility concept is completely different from the one mentioned in Example 2.3.)

Proposition 9.1. *A left A-module D is $\mathfrak{F}$-divisible if and only if $M \otimes_A D = 0$ for every $\mathfrak{F}$-torsion module M.*

Proof. If $M \otimes D = 0$ for $\mathfrak{F}$-torsion modules M, then in particular for $\mathfrak{a} \in \mathfrak{F}$ we get $0 = A/\mathfrak{a} \otimes D = D/\mathfrak{a} D$ by Example I.8.1, so $D = \mathfrak{a} D$. The converse implication is clear. □

The two concepts "$\mathfrak{F}$-divisible" and "$\mathfrak{F}$-torsion-free" are dual to each other in a sense which we will make precise. For each module ${}_A D$ we put $\hat{D} = \operatorname{Hom}_{\mathbb{Z}}(D, \mathbb{Q}/\mathbb{Z})$, and consider $\hat{D}$ as a right A-module (Chap. I, §9). Since $\mathbb{Q}/\mathbb{Z}$ is an injective cogenerator for **Ab**, we have that $M \otimes D = 0$ if and only if $\operatorname{Hom}_A(M, \hat{D}) = 0$. Hence Prop. 9.1 has the dual formulation:

Proposition 9.2. *A left A-module D is $\mathfrak{F}$-divisible if and only if $\hat{D}$ is $\mathfrak{F}$-torsion-free.*

It is also possible to dualize the construction of the topology $\mathfrak{F}_M^0$, i.e. the strongest Gabriel topology for which M_A is torsion-free.

Proposition 9.3. *Let D be a left A-module. There exists a strongest Gabriel topology $\mathfrak{F}$ for which D is $\mathfrak{F}$-divisible, namely*

$$\{\mathfrak{a} \mid (\mathfrak{a} : a)\, D = D \text{ for every } a \in A\}.$$

Proof. It follows from Prop. 9.2 that the strongest Gabriel topology for which D is divisible, coincides with the strongest Gabriel topology for which $\hat{D}$ is torsion-free. By Prop. 5.5 this topology consists of those right ideals $\mathfrak{a}$ for which $\operatorname{Hom}(A/(\mathfrak{a}:a), \hat{D})=0$ for every $a\in A$. This condition is equivalent to $A/(\mathfrak{a}:a)\otimes D=0$, i.e. $(\mathfrak{a}:a)\,D=D$, for every $a\in A$. ▯

We will use the notation $\mathfrak{F}^D$ for the topology in question. A right A-module M is an $\mathfrak{F}^D$-torsion module if and only if $L\otimes D=0$ holds for every submodule L of M. When D is assumed to be a flat left A-module, the situation is simplified by the fact that $\hat{D}$ is an injective right A-module (Cor. I.10.5), and we obtain:

$$\mathfrak{F}^D=\{\mathfrak{a}\,|\,\mathfrak{a}\,D=D\},$$

M_A is an $\mathfrak{F}^D$-torsion module if and only if $M\otimes D=0$.

We will consider in some more detail the case when D is a projective module. Let P be a projective left A-module, and put

$$P^*=\operatorname{Hom}_A(P,A), \qquad B=\operatorname{Hom}_A(P,P).$$

Define a homomorphism $\alpha\colon P\otimes_B P^*\to A$ of A-A-bimodules as $\alpha(x\otimes\varphi)=\varphi(x)$. Recall from Chap. IV, § 10, that the image of α is the two-sided trace ideal $\tau(P)$ of P in A.

Proposition 9.4. *If P is a projective left A-module, then:*
(i) *$\tau(P)$ is an idempotent ideal, and $\tau(P)\,P=P$.*
(ii) *$\mathfrak{F}^P$ consists of all right ideals containing $\tau(P)$.*

Proof. The fact that $\tau(P)\,P=P$ follows immediately from the existence of "projective coordinates" for P (Prop. I.6.3). Since $\mathfrak{F}^P=\{\mathfrak{a}\,|\,\mathfrak{a}\,P=P\}$, it remains to show that $\mathfrak{a}\,P=P$ implies $\mathfrak{a}\supset\tau(P)$. But $\mathfrak{a}\,P=P$ clearly implies that the image $\tau(P)$ of the A-linear map $\alpha\colon P\otimes_B P^*\to A$ is contained in $\mathfrak{a}$. Finally, $\tau(P)$ is idempotent by Prop. 6.12. ▯

Corollary 9.5. *If P is a projective left A-module, then M is an $\mathfrak{F}^P$-torsion module if and only if $M\,\tau(P)=0$.*

Example. Rings of fractions. Let S be a right denominator set in A. Since $Q=A[S^{-1}]$ is a flat left A-module (Prop. II.3.5), the topology $\mathfrak{F}^Q$ consists of all right ideals $\mathfrak{a}$ such that $\mathfrak{a}\,Q=Q$, i.e. such that $\mathfrak{a}\cap S\neq\emptyset$, and M is an $\mathfrak{F}^Q$-torsion module if and only if $M[S^{-1}]=0$. We thus obtain the same torsion concept as was used in Chap. II.

Exercises

§1

We consider preradicals defined in a complete and cocomplete abelian category **C**.

1. Let r_1 and r_2 be preradicals of **C**. Show that:
 (i) $r_1:r_2$ is actually a preradical.
 (ii) If r_1 and r_2 are idempotent, then $r_1:r_2$ is idempotent.
 (iii) If r_1 and r_2 are left exact, then $r_1:r_2$ is left exact.
 (iv) If r_1 and r_2 are radicals, then $r_1\,r_2$ is a radical.

2. Let r be a preradical of **C**. Show that:
(i) r preserves coproducts.
(ii) If r is left exact and **C** satisfies $AB5$, then r preserves direct unions of subobjects.
3. (i) Let $t \le r$ be preradicals with t idempotent. Show that $tr = rt = t$.
(ii) Deduce that $\hat{r}$ is the largest idempotent preradical smaller than r.
4. Show that:
(i) If r is an idempotent preradical, then $\mathscr{F}_r$ is closed under extensions.
(ii) If r is a radical, then $\mathscr{T}_r$ is closed under extensions.
5. Let r be a preradical of **Mod**-A. Show that the following conditions are equivalent:
(a) r preserves epimorphisms.
(b) $r(M) = M\,r(A)$ for every module M.
(c) r is a radical and $\mathscr{F}_r$ is closed under quotient modules.
6. Let r be a preradical of **C**. Show that the following conditions are equivalent:
(a) r^{-1} is a left exact radical of $\mathbf{C}^{\mathrm{op}}$.
(b) r is idempotent and preserves epimorphisms.
7. Show that the preradicals of **C** form a complete modular lattice, and that this lattice is upper continuous when **C** satisfies $AB5$.

§2

8. Show that a couple $(\mathscr{T}, \mathscr{F})$ of classes of objects of **C** is a torsion theory if and only if it satisfies:
(1) $\mathscr{T} \cap \mathscr{F} = \{0\}$.
(2) $\mathscr{T}$ is closed under quotient objects.
(3) $\mathscr{F}$ is closed under subobjects.
(4) For every object C there is an exact sequence $0 \to T \to C \to F \to 0$ with $T \in \mathscr{T}$ and $F \in \mathscr{F}$.
9. Show that the following properties of a ring A are equivalent:
(a) Every $E(A)$-divisible module is injective.
(b) The injective modules form a torsion class.
(c) A is right hereditary and right noetherian.
10. Let $\mathscr{C}$ be a class of objects, and let $T(\mathscr{C})$ be the class of objects T such that each non-zero quotient object of T has a non-zero subobject in $\mathscr{C}$. Show that:
(i) $T(\mathscr{C})$ is a torsion class.
(ii) If $\mathscr{T}$ is a torsion class containing $\mathscr{C}$, then $\mathscr{T} \supset T(\mathscr{C})$ (but note that $\mathscr{C} \not\subset T(\mathscr{C})$ in general).

§3

11. Let $\mathscr{C}$ be a class of modules closed under submodules, finite products and injective envelopes. Show that $\mathscr{C}$ is closed under extensions.
12. Let r be a preradical of **Mod**-A. For each module M, put $\tilde{r}(M) = M \cap r(E(M))$. Show that:
(i) $\tilde{r}(M)$ is independent of the choice of injective envelope of M, and $\tilde{r}$ is a preradical.
(ii) $\tilde{r}$ is the smallest left exact preradical larger than r.
(iii) $\bar{\tilde{r}}$ is the smallest left exact radical larger than r.
13. Let $\varphi\colon A \to B$ be a monomorphism of rings such that B is a flat left A-module. Put

$$\mathscr{T} = \{M \mid M \otimes_A B = 0\},$$

$$\mathscr{F} = \{M \mid M \to M \otimes_A B \text{ is a monomorphism}\}.$$

Show that $(\mathscr{T}, \mathscr{F})$ is a hereditary torsion theory.
14. Let V be a module, and let $\mathscr{L}$ be the class of modules cogenerated by V (in the sense of Example IV.6.2). Show that:
(i) $\mathscr{L}$ is closed under submodules and direct products.
(ii) $\mathscr{L}$ and $\{V\}$ cogenerate the same torsion theory $(\mathscr{T}, \mathscr{F})$.
(iii) $\mathscr{T} = \{C \mid \mathrm{Hom}(C, V) = 0\}$,
$\mathscr{F} = \{C \mid \mathrm{Hom}(C', V) \neq 0$ for every non-zero submodule C' of $C\}$.
(iv) If one puts $r_V(C) = \bigcap \operatorname{Ker} \alpha$, α ranging over $\mathrm{Hom}(C, V)$, then r_V is a radical.
(v) $r_V(C) = 0$ if and only if $C \in \mathscr{L}$.
(vi) r_V is the idempotent radical corresponding to the torsion theory $(\mathscr{T}, \mathscr{F})$.
(vii) r is idempotent if and only if $\mathscr{L}$ is closed under extensions.
(viii) $(\mathscr{T}, \mathscr{F})$ is hereditary if and only if $r_V = r_{E(V)}$.

(ix) The following assertions are equivalent:
(a) $E(V)$ is cogenerated by V.
(b) $\mathscr{L}$ is closed under injective envelopes.
(c) $(\mathscr{T}, \mathscr{L})$ is a hereditary torsion theory.
(d) $\mathscr{T}$ is closed under submodules and $\mathscr{L}$ is closed under extensions.

§4

15. Show that there is a bijective correspondence between:
(1) Non-empty families of right ideals of A satisfying $T3$.
(2) Non-empty classes $\mathscr{C}$ of A-modules such that
(a) $M \in \mathscr{C}$ if and only if every cyclic submodule of M is in $\mathscr{C}$;
(b) $\mathscr{C}$ is closed under isomorphisms.

16. Let $\mathfrak{F}$ be a linear topology on the ring A. Show that:

(i) if M and N are A-modules and α: $M \to N$ is A-linear, then α is continuous in the $\mathfrak{F}$-topologies of M and N.

(ii) There is a functor L from **Mod**-A to the category **C** of linearly topological right A-modules, such that $L(M)$ is equal to M with its $\mathfrak{F}$-topology.

(iii) The functor L is a left adjoint to the forgetful functor $\mathbf{C} \to$ **Mod**-A.

§5

17. Suppose $\mathfrak{F}$ is a Gabriel topology and let $L \subset M \subset N$ be modules. Show that $L \in \mathfrak{F}(N)$ if and only if $L \in \mathfrak{F}(M)$ and $M \in \mathfrak{F}(N)$.

§6

18. Let S be a multiplicatively closed subset of A, and put $\mathfrak{F}_S = \{\mathfrak{a} \mid (\mathfrak{a} : a) \cap S \neq \emptyset$ for all $a \in A\}$. Show that:

(i) $\mathfrak{F}_S$ is a Gabriel topology.

(ii) $\Sigma(\mathfrak{F}_S) = \{t \in A \mid$ for every $a \in A$ there exist $s \in S$ and $b \in A$ such that $as = tb\}$.

(iii) $\Sigma(\mathfrak{F}_S)$ satisfies axiom $S1$.

19. (i) Show that A_A is a Goldie torsion module if and only if the right singular ideal is an essential right ideal.

(ii) Show that the ring $\mathbb{Z}/p^2\mathbb{Z}$ (p a prime) is a Goldie torsion module.

20. Let $\mathfrak{F}$ be a Gabriel topology with a basis consisting of an idempotent ideal $\mathfrak{b}$. Show that for every module M there is a smallest member of $\mathfrak{F}(M)$, namely $M\mathfrak{b}$.

21. Assume the Goldie topology $\mathfrak{G}$ of A has a basis consisting of an idempotent ideal $\mathfrak{b}$, and let M be a module. Show that:

(i) $M\mathfrak{b}$ is a non-singular module. (Hint: let K be a pseudocomplement in M of the Goldie torsion submodule of $M\mathfrak{b}$, and show that $K \in \mathfrak{G}(M)$.)

(ii) $M\mathfrak{b}$ is contained in the socle of M. (Hint: $M\mathfrak{b}$ is split by all its submodules.)

(iii) Every non-singular module $\neq 0$ has a non-zero socle.

22. Let A be a commutative ring. Show that:

(i) $\mathscr{P} \mapsto \mathfrak{F}_{\mathscr{P}}$ and $\mathfrak{F} \mapsto D(\mathfrak{F})$ define a Galois connection between the lattice of subsets of $\operatorname{Spec}(A)$ and the lattice of Gabriel topologies on A.

(ii) There is induced a lattice anti-isomorphism between the generically stable subsets $\mathscr{P}$ of $\operatorname{Spec}(A)$ and the Gabriel topologies of the form $\mathfrak{F}_{\mathscr{P}}$. ($A$ subset $\mathscr{P}$ of $\operatorname{Spec}(A)$ is called *generically stable* if $\mathfrak{q} \subset \mathfrak{p}$ and $\mathfrak{p} \in \mathscr{P}$ imply $\mathfrak{q} \in \mathscr{P}$.)

23. Show that the following two properties of a commutative ring A are equivalent:
(a) If $\mathfrak{a}$ is an ideal and $(\mathfrak{p}_i)_I$ is a family of prime ideals such that $\mathfrak{a} \subset \bigcup_I \mathfrak{p}_i$, then $\mathfrak{a} \subset \mathfrak{p}_i$ for some $i \in I$.

(b) For every $\mathscr{P} \subset \operatorname{Spec}(A)$ there exists a multiplicatively closed subset S of A such that $\mathfrak{F}_{\mathscr{P}} = \{\mathfrak{a} \mid \mathfrak{a} \cap S \neq \emptyset\}$.

24. Let A be a commutative ring and $\mathscr{P} \subset \operatorname{Spec}(A)$. Show that the $\mathfrak{F}_{\mathscr{P}}$-torsion theory is cogenerated by the injective module $\prod_{\mathscr{P}} E(A/\mathfrak{p})$.

§7

25. Show that every topology $\mathfrak{F}$ satisfying the conditions of Prop. 7.1 is a Gabriel topology.

26. A module M is *quasi-injective* if $\alpha(M) \subset M$ holds for every endomorphism α of $E(M)$. Show that:

(i) If M is quasi-injective and L is an essentially closed submodule of M, then L is a direct summand of M.

(ii) For a stable hereditary torsion theory, every quasi-injective module is split by its torsion submodule.

§8

27. Show that if $(\mathscr{C}, \mathscr{T}, \mathscr{F})$ is a TTF-theory and $(\mathscr{T}, \mathscr{F})$ is splitting, then $\mathscr{C} \subset \mathscr{F}$.

28. Let $(\mathscr{C}, \mathscr{T}, \mathscr{F})$ be a TTF-theory, corresponding to the idempotent ideal $\mathfrak{c}$. Show that $(\mathscr{T}, \mathscr{F})$ is stable if and only if $a \in a\mathfrak{c}$ for every $a \in \mathfrak{c}$. (Cf. Prop. XI.3.13.)

29. Let $\mathscr{T}$ and $\mathscr{T}'$ be TTF-classes with corresponding idempotent ideals $\mathfrak{a}$ and $\mathfrak{b}$. Show that $(\mathscr{T}, \mathscr{T}')$ is a torsion theory if and only if $\mathfrak{a} + \mathfrak{b} = A$ and $\mathfrak{b}\mathfrak{a} = 0$.

§9

30. Let $\mathfrak{F}$ be a Gabriel topology on A. Show that every left A-module M contains a largest $\mathfrak{F}$-divisible submodule $d(M)$, and that d becomes an idempotent radical of **Mod**-A.

31. A left A-module M is called $\mathfrak{F}$-*reduced* for a Gabriel topology $\mathfrak{F}$ if M has no non-zero $\mathfrak{F}$-divisible submodules. Show that if L is an $\mathfrak{F}$-torsion module, then $\hat{L}$ is $\mathfrak{F}$-reduced.

Chapter VII. Hereditary Torsion Theories for Noetherian Rings

For a commutative noetherian ring we have proved that every Gabriel topology $\mathfrak{F}$ arises from a set $\mathscr{P}$ of prime ideals as

$$\mathfrak{F} = \{\mathfrak{a} \mid \mathfrak{a} \not\subset \mathfrak{p}, \text{ all } \mathfrak{p} \in \mathscr{P}\}$$

(Cor. VI.6.15). In the case of a non-commutative right noetherian ring the situation is not quite so simple. The main result of this chapter (Theorem 3.4) shows, however, that there is a large class of right noetherian rings for which the Gabriel topologies are uniquely determined by the prime ideals they contain. An important tool for obtaining this result is the non-commutative theory of associated prime ideals.

Throughout this chapter we assume that A is a right noetherian ring.

§ 1. Associated Prime Ideals

A *prime ideal* in A is a two-sided ideal $\mathfrak{p} \neq A$ with the property that if $\mathfrak{b}$ and $\mathfrak{c}$ are two-sided ideals such that $\mathfrak{b}\,\mathfrak{c} \subset \mathfrak{p}$, then either $\mathfrak{b}$ or $\mathfrak{c}$ is contained in $\mathfrak{p}$. In other words, $\mathfrak{p}$ is a prime ideal if and only if $A/\mathfrak{p}$ is a prime ring (Chap. II, §2). If M is a module, we say that a two-sided ideal $\mathfrak{a}$ is *associated* to M if there exists a non-zero submodule L of M such that $\mathfrak{a} = \mathrm{Ann}(L')$ for all non-zero submodules L' of L (the submodule L may then equally well be chosen to be cyclic). The ideal $\mathfrak{a}$ must then be prime, for if $\mathfrak{b}\,\mathfrak{c} \subset \mathfrak{a}$ but $\mathfrak{b} \not\subset \mathfrak{a}$, then $L\mathfrak{b} \neq 0$ and $\mathfrak{c} \subset \mathrm{Ann}(L\mathfrak{b}) = \mathrm{Ann}(L) = \mathfrak{a}$. We denote the set of all prime ideals associated to M by $\mathrm{Ass}(M)$. It is easy to see that every maximal member of the family of ideals of the form $\mathrm{Ann}(L)$, for submodules $L \neq 0$ of M, is associated to M. Up to this point, our considerations are applicable to an arbitrary ring A, but now we use the assumption that A is right noetherian to conclude that maximal annihilator ideals exist, and hence:

Proposition 1.1. *If M is a non-zero module, then* $\mathrm{Ass}(M) \neq \emptyset$.

We note some useful properties of Ass:

Lemma 1.2. *If $0 \to L \to M \to N \to 0$ is an exact sequence of modules, then*

$$\mathrm{Ass}(L) \subset \mathrm{Ass}(M) \subset \mathrm{Ass}(L) \cup \mathrm{Ass}(N).$$

Proof. Obviously $\mathrm{Ass}(L) \subset \mathrm{Ass}(M)$. Let $\mathfrak{p} = \mathrm{Ann}(K) \in \mathrm{Ass}(M)$, so that $\mathfrak{p} = \mathrm{Ann}(K')$ for all non-zero $K' \subset K$. If $K \cap L = 0$, then K is isomorphic to a submodule of N, and in that case $\mathfrak{p} \in \mathrm{Ass}(N)$. On the other hand, if $K \cap L \neq 0$, then $\mathfrak{p} = \mathrm{Ann}(K')$ for all $0 \neq K' \subset K \cap L$, and $\mathfrak{p} \in \mathrm{Ass}(L)$. □

Lemma 1.3. $\operatorname{Ass}(\bigoplus_I M_i)=\bigcup_I \operatorname{Ass}(M_i)$.

Proof. Since $\mathfrak{p}\in\operatorname{Ass}(M)$ if and only if $\mathfrak{p}\in\operatorname{Ass}(L)$ for some cyclic submodule L of M, we may assume I finite. By induction we are reduced to the case of a direct sum $M_1\oplus M_2$, and then we can apply the previous lemma. □

Lemma 1.4. $\operatorname{Ass}(M)=\operatorname{Ass}(E(M))$.

Proof. This is clear, since every non-zero submodule of $E(M)$ has non-zero intersection with M. □

Proposition 1.5. *If M is a non-zero coirreducible module, then* $\operatorname{Ass}(M)$ *has only one member.*

Proof. If K and L are non-zero submodules of M, then the coirreducibility implies that $K\cap L\neq 0$. Since $\operatorname{Ann}(K)+\operatorname{Ann}(L)\subset\operatorname{Ann}(K\cap L)$, we conclude that there is a unique maximal member of the family of ideals $\operatorname{Ann}(L)$, $L\neq 0$. □

Since the injective envelope of a module of finite rank is a finite direct sum of indecomposable injective modules, we obtain from 1.3, 1.4 and 1.5:

Corollary 1.6. *If M is a module of finite rank, then* $\operatorname{Ass}(M)$ *is finite.*

A module M is called *cotertiary* if $\operatorname{Ass}(M)$ consists of precisely one member, which will then be denoted by $\operatorname{ass}(M)$. A submodule L of M is *tertiary* in M if M/L is a cotertiary module. Prop. 1.5 states that every coirreducible module is cotertiary, and thus irreducible submodules are tertiary. If $\mathfrak{p}$ is a prime ideal, we call a submodule L of M $\mathfrak{p}$*-tertiary* in M if $\operatorname{Ass}(M/L)=\{\mathfrak{p}\}$.

Lemma 1.7. *Each prime ideal $\mathfrak{p}$ is a $\mathfrak{p}$-tertiary right ideal in A.*

Proof. Suppose $\mathfrak{a}$ is a right ideal strictly containing $\mathfrak{p}$. If $\mathfrak{b}=\operatorname{Ann}(\mathfrak{a}/\mathfrak{p})$, then $\mathfrak{a}\,\mathfrak{b}\subset\mathfrak{p}$ and hence $\mathfrak{b}=\mathfrak{p}$. Thus $\operatorname{Ass}(A/\mathfrak{p})=\{\mathfrak{p}\}$. □

Lemma 1.8. *Every intersection of $\mathfrak{p}$-tertiary submodules of M is $\mathfrak{p}$-tertiary in M.*

Proof. If L_i $(i\in I)$ are $\mathfrak{p}$-tertiary submodules of M, the monomorphism $M/\bigcap L_i\to \bigoplus M/L_i$ shows that $\operatorname{Ass}(M/\bigcap L_i)\subset\bigcup\operatorname{Ass}(M/L_i)=\{\mathfrak{p}\}$. Hence $\operatorname{Ass}(M/\bigcap L_i)=\{\mathfrak{p}\}$. □

Let us consider in more detail the module $E(A/\mathfrak{p})$ for a prime ideal $\mathfrak{p}$. This is a cotertiary module, but it is in general not indecomposable, because $\mathfrak{p}$ is not necessarily irreducible. However, there is a decomposition $E(A/\mathfrak{p})=\bigoplus E_i$ into indecomposable modules E_i, for which we have:

Proposition 1.9. *If $\mathfrak{p}$ is a prime ideal and $E(A/\mathfrak{p})=\bigoplus_i E_i$, where E_i are indecomposable injective modules, then all E_i are isomorphic to each other.*

Proof. From Chap. II we know that the noetherian prime ring $A/\mathfrak{p}$ has a simple classical right ring of quotients Q. Since $A/\mathfrak{p}$ is essential in Q as a right $A/\mathfrak{p}$-module, it is essential in Q also as a right A-module. It follows that the inclusion $A/\mathfrak{p}\to E(A/\mathfrak{p})$ may be extended to an A-linear monomorphism $Q\to E(A/\mathfrak{p})$, which also is essential. The simple ring Q may be decomposed as a direct sum of minimal right ideals B_j, which are all isomorphic to each other. We obtain

$$\bigoplus_i E_i = E(A/\mathfrak{p}) = \bigoplus_j E(B_j).$$

The Azumaya theorem (Cor. V.5.5) implies that $E_i \cong E(B_j)$ for all i, and hence all E_i are isomorphic. □

From this proof it is also seen that $E(A/\mathfrak{p})$ is indecomposable if and only if Q is a skew-field. Hence:

Corollary 1.10. *A prime ideal* $\mathfrak{p}$ *in a right noetherian ring is irreducible as a right ideal if and only if* $a\,b \in \mathfrak{p}$ *implies* $a \in \mathfrak{p}$ *or* $b \in \mathfrak{p}$.

Several applications of the theory of associated prime ideals are based upon the following result.

Proposition 1.11. *Let M be a finitely generated module. There exists a chain* $(0) = M_0 \subset \cdots \subset M_n = M$ *such that*

(1) *M_i is a tertiary submodule of M_{i+1};*

(2) *M_{i+1}/M_i is annihilated by its associated prime ideal.*

Proof. Suppose inductively that $M_0, \ldots, M_{i-1}$ have been obtained. If $M_i \neq M$, then $\mathrm{Ass}(M/M_i) \neq \emptyset$, and hence there exists $M_{i+1} \supset M_i$ such that every non-zero submodule of M_{i+1}/M_i is annihilated by a prime ideal $\mathfrak{p}_{i+1}$. Conditions (1) and (2) are then satisfied. □

Note in addition that the module M is annihilated by the ideal $\mathfrak{p}_n \cdot \cdots \cdot \mathfrak{p}_1$, and that $\mathrm{Ass}(M) \subset \{\mathfrak{p}_1, \ldots, \mathfrak{p}_n\}$ by Lemma 1.2. The set $\{\mathfrak{p}_1, \ldots, \mathfrak{p}_n\}$ is usually not uniquely determined by the module M.

We conclude this section with a technical result to be used in § 5:

Proposition 1.12. *For each module M and subset $\mathscr{P}$ of* $\mathrm{Ass}(M)$ *there exists a submodule L of M such that* $\mathrm{Ass}(L) = \mathscr{P}$ *and* $\mathrm{Ass}(M/L) = \mathrm{Ass}(M) - \mathscr{P}$.

Proof. With the help of Zorn's lemma it is easy to see that there is a submodule L of M which is maximal with the property that $\mathrm{Ass}(L) \subset \mathscr{P}$. To make the proof complete, it suffices by Lemma 1.2 to show that $\mathrm{Ass}(M/L) \cap \mathscr{P} = \emptyset$. Let $\mathfrak{p} \in \mathrm{Ass}(M/L)$. Then M/L contains a submodule N/L which is isomorphic to $A/\mathfrak{p}$. One has

$$\mathrm{Ass}(N) \subset \mathrm{Ass}(L) \cup \mathrm{Ass}(N/L) \subset \mathscr{P} \cup \{\mathfrak{p}\}.$$

The maximality of L implies that $\mathfrak{p} \notin \mathscr{P}$. □

Examples

1. Tertiary decomposition (Lesieur and Croisot [1, 4]). Let M be a finitely generated module. A *tertiary decomposition* of a submodule L of M is obtained by writing L as a finite intersection $L = M_1 \cap \cdots \cap M_n$ of tertiary submodules M_i of M, so that

(i) the decomposition is irredundant,

(ii) $\mathrm{ass}(M/M_i) \neq \mathrm{ass}(M/M_j)$ for $i \neq j$.

Proposition 1.13. *Every submodule L of a finitely generated module M has a tertiary decomposition. If* $L = M_1 \cap \cdots \cap M_m = N_1 \cap \cdots \cap N_n$ *are two tertiary decompositions of L in M, then $m = n$ and* $\{\mathrm{ass}(M/M_i)\} = \{\mathrm{ass}(M/N_j)\}$.

Proof. We already know that L can be written as an irredundant intersection $L=M_1\cap\cdots\cap M_k$ of irreducible submodules M_i of M. Each M_i is then tertiary in M. For each prime ideal $\mathfrak{p}\in\{\operatorname{ass}(M/M_i)\}$, let $M(\mathfrak{p})$ denote the intersection of all $\mathfrak{p}$-tertiary M_i. Then $M(\mathfrak{p})$ is $\mathfrak{p}$-tertiary in M by Lemma 1.8, and $L=\bigcap_{\mathfrak{p}} M(\mathfrak{p})$ is the desired tertiary decomposition.

We turn to the question of unicity. Let $L=M_1\cap\cdots\cap M_m$ be any tertiary decomposition of L in M. Consider the monomorphism $M/L\to M/M_1\oplus\cdots\oplus M/M_m$. The irredundancy implies that M/L has non-zero intersection K_i with each M/M_i. Thus there are monomorphisms

$$K_1\oplus\cdots\oplus K_m\to M/L\to M/M_1\oplus\cdots\oplus M/M_m.$$

With the use of Lemmas 1.2 and 1.3 we get

$$\bigcup_i \operatorname{Ass}(K_i)\subset\operatorname{Ass}(M/L)\subset\bigcup_i \operatorname{Ass}(M/M_i).$$

But since each M/M_i is cotertiary and $\operatorname{Ass}(K_i)\neq\emptyset$, we must have $\operatorname{Ass}(K_i)=\operatorname{Ass}(M/M_i)$. It follows that $\{\operatorname{ass}(M/M_i)\}=\operatorname{Ass}(M/L)$, and from this the unicity assertions follow. □

For a commutative noetherian ring the tertiary decomposition theory coincides with the classical primary theory (cf. §4). One can show that the tertiary decomposition is the only reasonable generalization of primary decomposition to non-commutative noetherian rings (Riley [1, 2]).

2. Primary decomposition. We mention one other way of generalizing primary decomposition to non-commutative rings, although it does not give quite as satisfactory results as the tertiary theory. Let M be a finitely generated module. A submodule L of M is *primary* in M if it is tertiary in M and $M\mathfrak{p}^n\subset L$, where $\mathfrak{p}=\operatorname{ass}(M/L)$, for some n. *Primary decompositions* are defined analogously to tertiary decompositions. A sufficient condition for the existence of a primary decomposition of M is that every irreducible submodule is primary.

In particular, a right ideal is primary if and only if it is tertiary and contains a power of the associated prime ideal. This condition can be reformulated in a way which ressembles the traditional definition of primary ideals in a commutative ring. For each right ideal $\mathfrak{a}$ we let $\tilde{\mathfrak{a}}$ be the largest two-sided ideal contained in $\mathfrak{a}$, i.e. $\tilde{\mathfrak{a}}=\{b\in A\,|\,A b\subset\mathfrak{a}\}$. Let $\sqrt{\mathfrak{a}}$ denote the intersection of all prime ideals containing $\tilde{\mathfrak{a}}$. One may show that $\sqrt{\mathfrak{a}}$ is the largest two-sided ideal of A such that some power of it is contained in $\mathfrak{a}$ (a proof of this fact can be found in Chap. XV).

Proposition 1.14. *A right ideal $\mathfrak{a}$ is primary in A if and only if it satisfies: if $bAa\subset\mathfrak{a}$ and $b\notin\mathfrak{a}$, then $a\in\sqrt{\mathfrak{a}}$.*

Proof. Suppose $\mathfrak{a}$ is primary and $bAa\subset\mathfrak{a}$, but $b\notin\mathfrak{a}$. Then a annihilates a non-zero submodule of $A/\mathfrak{a}$, so $a\in\operatorname{ass}(A/\mathfrak{a})$. But $\mathfrak{a}$ primary clearly implies that $\operatorname{ass}(A/\mathfrak{a})\subset\sqrt{\mathfrak{a}}$.

Suppose on the other hand that $\mathfrak{a}$ has the property that $bAa\subset\mathfrak{a}$ and $b\notin\mathfrak{a}$ together imply $a\in\sqrt{\mathfrak{a}}$. Let $\mathfrak{p}\in\operatorname{Ass}(A/\mathfrak{a})$. We then have $\mathfrak{p}=\{a\,|\,bAa\subset\mathfrak{a}\}$ for some $b\notin\mathfrak{a}$. This implies by hypothesis that $\mathfrak{p}\subset\sqrt{\mathfrak{a}}$, but it clearly also implies that $\mathfrak{p}\supset\tilde{\mathfrak{a}}$

and therefore $\mathfrak{p} \supset \sqrt{\mathfrak{a}}$. Hence $\mathfrak{p} = \sqrt{\mathfrak{a}}$. It follows that $\mathfrak{a}$ is tertiary in A, and in fact primary since some power of $\sqrt{\mathfrak{a}}$ is contained in $\mathfrak{a}$. □

Corollary 1.15. *If* $\mathfrak{a}$ *is a primary right ideal, then* $\mathrm{Ass}(A/\mathfrak{a}) = \{\sqrt{\mathfrak{a}}\}$.

3. Isotypic modules. A module M is called *isotypic* if $E(M)$ is a direct sum of isomorphic indecomposable modules (by the Azumaya theorem, this is independent of the choice of direct sum decomposition). Every isotypic module is cotertiary. E.g. the right module $A/\mathfrak{p}$ is isotypic when $\mathfrak{p}$ is a prime ideal (Prop. 1.9).

§2. Fully Bounded Noetherian Rings

We know that for a commutative noetherian ring there is a bijective correspondence between indecomposable injective modules and prime ideals (Prop. V.4.6). For arbitrary noetherian rings this is no longer true; in general there are more indecomposable injectives than there are prime ideals. Denote by $\mathscr{E}(A)$ the set of isomorphism classes of non-zero indecomposable injective right A-modules, and denote by $\mathrm{Spec}(A)$ the set of prime ideals of A. We define a map

$$\Phi\colon\ \mathscr{E}(A) \to \mathrm{Spec}(A)$$

by $\Phi(E) = \mathrm{ass}(E)$, bearing in mind that each indecomposable injective module is cotertiary. Φ is a surjective map, because each prime ideal $\mathfrak{p}$ is associated to the indecomposable injective direct summands of $E(A/\mathfrak{p})$.

Theorem 2.1 (Krause [1]). *The following properties of a right noetherian ring A are equivalent:*

(a) $\Phi\colon \mathscr{E}(A) \to \mathrm{Spec}(A)$ *is bijective.*

(b) *For every prime ideal* $\mathfrak{p}$, *the ring* $A/\mathfrak{p}$ *has the property that every essential right ideal contains a non-zero two-sided ideal.*

(c) *Every cotertiary module is isotypic.*

Proof. We first settle the trivial part of the theorem, namely the equivalence of (a) and (c).

(a) ⇒ (c): Let M be a cotertiary module, and write $E(M) = \bigoplus E_i$ with indecomposable injective modules E_i. Since we must have $\mathrm{Ass}(E_i) = \mathrm{Ass}(M)$ for each E_i, (a) implies that all E_i are isomorphic.

(c) ⇒ (a): If E and E' are two indecomposable injective modules with the same associated prime ideal $\mathfrak{p}$, then $\mathrm{Ass}(E \oplus E') = \{\mathfrak{p}\}$, and thus $E \oplus E'$ is cotertiary. The condition (c) implies that E and E' are isomorphic.

For the proof of (a) ⇒ (b) we need:

Lemma 2.2. *If A satisfies* (a), *then so does every epimorphic image of A.*

Proof. Let $\mathfrak{a}$ be a two-sided ideal of A and put $B = A/\mathfrak{a}$. Let E_1 and E_2 be two non-isomorphic indecomposable injective B-modules. Considered as A-modules, E_1 and E_2 are still coirreducible. Suppose their injective envelopes $E_A(E_1)$ and $E_A(E_2)$ as A-modules are isomorphic. $E_A(E_1)$ would then contain a copy of E_2 and hence a non-zero submodule $E_1 \cap E_2$. But this would imply $E_1 \cong E_B(E_1 \cap E_2)$

$\cong E_2$. Hence we must conclude that $E_A(E_1)$ and $E_A(E_2)$ are non-isomorphic. Condition (a) for A then implies $\operatorname{ass}(E_A(E_1)) \neq \operatorname{ass}(E_A(E_2))$. This means that $\operatorname{ass}_A(E_1) \neq \operatorname{ass}_A(E_2)$. But it is easy to see that $\operatorname{ass}_B(E_i) = (\operatorname{ass}_A(E_i))/\mathfrak{a}$, so we have $\operatorname{ass}_B(E_1) \neq \operatorname{ass}_B(E_2)$. This establishes condition (a) for the ring B. □

We can now prove (a) ⇒ (b): As a consequence of the lemma, it suffices to consider a prime ring A and show that every essential right ideal contains a non-zero two-sided ideal. Suppose there exists an essential right ideal $\mathfrak{a}$ which does not contain any two-sided ideal $\neq 0$. We may suppose $\mathfrak{a}$ is maximal with this property. We first note that $\mathfrak{a}$ must then be irreducible. For suppose $\mathfrak{a} = \mathfrak{b}_1 \cap \mathfrak{b}_2$, where each $\mathfrak{b}_i$ is a right ideal strictly containing $\mathfrak{a}$. Then $\mathfrak{b}_i$ contains a non-zero two-sided ideal $\mathfrak{c}_i$, and $\mathfrak{a} = \mathfrak{b}_1 \cap \mathfrak{b}_2 \supset \mathfrak{c}_1 \cap \mathfrak{c}_2 \supset \mathfrak{c}_1 \mathfrak{c}_2$, with $\mathfrak{c}_1 \mathfrak{c}_2 \neq 0$ since A is a prime ring. We have thereby arrived at a contradiction, so $\mathfrak{a}$ must in fact be irreducible. It follows that $\operatorname{Ass}(A/\mathfrak{a}) = \{\mathfrak{p}\}$, where $\mathfrak{p} = \operatorname{Ann}(\mathfrak{b}/\mathfrak{a})$ for some right ideal $\mathfrak{b}$ strictly containing $\mathfrak{a}$. Then $\mathfrak{b}$ contains a two-sided ideal $\mathfrak{c} \neq 0$, and $\mathfrak{c}\,\mathfrak{p} \subset \mathfrak{b}\,\mathfrak{p} \subset \mathfrak{a}$. But $\mathfrak{a}$ does not contain any non-zero ideals, so $\mathfrak{c}\,\mathfrak{p} = 0$, which entails $\mathfrak{p} = 0$ since A is a prime ring. We have thus come to the conclusion that $\operatorname{Ass}(E(A/\mathfrak{a})) = \{0\}$, and it now just remains to find another indecomposable injective module which also has (0) as its associated prime ideal. For this purpose we use the injective envelope of A_A. Write $E(A) = \bigoplus E_i$ with E_i indecomposable. Then $\operatorname{ass}(E_i) = \operatorname{ass}(A) = (0)$ since A is a prime ring. To arrive at a contradiction we must show that we cannot have $E_i \cong E(A/\mathfrak{a})$. But this is quite easy, because the noetherian prime ring A is non-singular by Theorem II.2.2(c), and then also $E(A)$ is non-singular, whereas $E(A/\mathfrak{a})$ contains an element annihilated by the essential right ideal $\mathfrak{a}$.

(b) ⇒ (a): We want to show that each indecomposable injective module E is uniquely determined by the prime ideal $\operatorname{ass}(E) = \mathfrak{p}$. Now to do this, it suffices to produce a monomorphism $E \to E(A/\mathfrak{p})$. In fact, if we make a decomposition $E(A/\mathfrak{p}) = \bigoplus E_i$, with indecomposable E_i which are all isomorphic to each other by Prop. 1.9, then the Azumaya theorem (Cor. V.5.5) implies that E is isomorphic to E_i.

We have $\mathfrak{p} = \operatorname{Ann}(xA)$ for some $x \in E$, and we put $\mathfrak{a} = \operatorname{Ann}(x)$. So $\mathfrak{a}$ is an irreducible right ideal containing $\mathfrak{p}$. We assert that $\mathfrak{a}/\mathfrak{p}$ is not an essential right ideal in the ring $A/\mathfrak{p}$. For otherwise $\mathfrak{a}/\mathfrak{p}$ would by hypothesis contain a two-sided ideal $\mathfrak{b}/\mathfrak{p}$ with $\mathfrak{b} \supsetneqq \mathfrak{p}$, and $xA\mathfrak{b} = x\mathfrak{b} \subset x\mathfrak{a} = 0$ would contradict the fact that $\operatorname{Ann}(xA) = \mathfrak{p}$. Since $\mathfrak{a}/\mathfrak{p}$ is not essential, there is a right ideal $\mathfrak{c}$ strictly containing $\mathfrak{p}$ such that $\mathfrak{a}/\mathfrak{p} \cap \mathfrak{c}/\mathfrak{p} = 0$, and hence $\mathfrak{p} = \mathfrak{a} \cap \mathfrak{c}$. One may write $\mathfrak{c}$ as an intersection of irreducible right ideals, and this leads to an irredundant irreducible decomposition of $\mathfrak{p}$ as $\mathfrak{p} = \mathfrak{a} \cap \mathfrak{c}_1 \cap \dots \cap \mathfrak{c}_n$. By Cor. V.5.10 it gives

$$E(A/\mathfrak{p}) \cong E(A/\mathfrak{a}) \oplus E(A/\mathfrak{c}_1) \oplus \dots \oplus E(A/\mathfrak{c}_n),$$

and in particular

$$E = E(xA) \cong E(A/\mathfrak{a}) \subset E(A/\mathfrak{p}),$$

as desired. □

Definition. A right noetherian ring A is *fully right bounded* if it satisfies the equivalent conditions of the theorem.

In Cor. 1.10 we obtained a condition for prime ideals to be irreducible. Rings which are fully bounded and have all their prime ideals irreducible may be characterized as follows:

Proposition 2.3. *The following properties of a right noetherian ring A are equivalent:*

(a) *A is fully right bounded and every prime ideal is an irreducible right ideal.*

(b) *For every indecomposable injective module E there exists a prime ideal $\mathfrak{p}$ such that $E \cong E(A/\mathfrak{p})$.*

(c) *For every prime ideal $\mathfrak{p}$ and right ideal $\mathfrak{a} \supsetneqq \mathfrak{p}$, there exists a two-sided ideal $\mathfrak{b}$ such that $\mathfrak{a} \supset \mathfrak{b} \supsetneqq \mathfrak{p}$.*

Proof. (a) $\Rightarrow$ (c): If the prime ideal $\mathfrak{p}$ is irreducible, then $A/\mathfrak{p}$ is a coirreducible module, so every non-zero submodule of $A/\mathfrak{p}$ is essential. This is the same as saying that every non-zero right ideal of the ring $A/\mathfrak{p}$ is essential, and when A is fully bounded, it gives (c).

(c) $\Rightarrow$ (b): Let E be indecomposable injective with $\operatorname{ass}(E) = \operatorname{Ann}(xA) = \mathfrak{p}$ for some $x \in E$. To see that $E \cong E(A/\mathfrak{p})$, it is enough to prove that $\mathfrak{p} = \operatorname{Ann}(x)$. Suppose $\operatorname{Ann}(x) \supsetneqq \mathfrak{p}$. Then by hypothesis there exists a two-sided ideal $\mathfrak{b}$ such that $\operatorname{Ann}(x) \supset \mathfrak{b} \supsetneqq \mathfrak{p}$. But that is impossible since it would imply $\mathfrak{b} \subset \operatorname{Ann}(xA) = \mathfrak{p}$.

(b) $\Rightarrow$ (a): For every indecomposable injective module E we have by hypothesis that $E \cong E(A/\mathfrak{p})$, where necessarily $\mathfrak{p} = \operatorname{ass}(E)$. Hence E is uniquely determined by its associated prime ideal, so A is fully right bounded.

Let $\mathfrak{p}$ be an arbitrary prime ideal. Then $E(A/\mathfrak{p}) = \bigoplus E_i$ with indecomposable injective modules E_i. Now we have $E_i \cong E(A/\mathfrak{q})$ for some prime ideal $\mathfrak{q}$, and $\operatorname{Ass}(A/\mathfrak{p}) = \operatorname{Ass}(E_i)$ implies $\mathfrak{q} = \mathfrak{p}$. Hence $E(A/\mathfrak{p}) \cong E_i$ is indecomposable, and $\mathfrak{p}$ is irreducible. ▯

Examples

1. Commutative rings. Every commutative noetherian ring is fully bounded. More generally, a right noetherian ring for which all right ideals are two-sided is fully right bounded, and its prime ideals are right irreducible.

2. Rings without two-sided ideals. Let A be a right noetherian ring without two-sided ideals $\neq 0$, A. The zero ideal is the only prime ideal. The ring A is fully right bounded if and only if it contains no essential right ideals $\neq A$, i.e. if and only if A is a simple ring. In the non-artinian case this gives examples of rings which are "as far as possible" from being fully bounded. E.g. one can take the ring of formal differential operators $A = K\left[X, \dfrac{d}{dX}\right]$, where K is a field of characteristic zero (Rinehart [1]).

3. The condition (*H*). A useful sufficient condition for A to be fully right bounded has been given by Gabriel [2]:

(H) For every right ideal $\mathfrak{a}$, there exist $b_1, \ldots, b_n \in A$ such that

$$\tilde{\mathfrak{a}} = (\mathfrak{a} : b_1) \cap \cdots \cap (\mathfrak{a} : b_n).$$

Here $\tilde{\mathfrak{a}}$ denotes the largest two-sided ideal contained in $\mathfrak{a}$ (cf. Example 1.2). (H) is clearly equivalent to the stronger condition

(H′) For every finitely generated module M, there exist $x_1, \ldots, x_n \in M$ such that

$$\operatorname{Ann}(M) = \operatorname{Ann}(x_1) \cap \cdots \cap \operatorname{Ann}(x_n).$$

Proposition 2.4. *If the right noetherian ring A satisfies the condition* (H), *then A is fully right bounded.*

Proof. Let E be an indecomposable injective module. The associated prime ideal $\mathfrak{p}$ of E is the annihilator of some cyclic submodule M of E. Then

$$\mathfrak{p}=\operatorname{Ann}(M)=\operatorname{Ann}(x_1)\cap\cdots\cap\operatorname{Ann}(x_n)$$

for some $x_1,\ldots,x_n\in M$. It follows that $A/\mathfrak{p}\subset x_1 A\oplus\cdots\oplus x_n A$, and so $E(A/\mathfrak{p})\subset E^n$. Now it is easy to apply the Azumaya theorem (Cor. V.5.5) to reach the conclusion that E is uniquely determined by its associated prime ideal, and hence that A is fully right bounded. □

4. Finite central extensions. Assume A is a ring such that its center $\operatorname{cen}(A)$ is a noetherian ring and A is finitely generated as a module over $\operatorname{cen}(A)$. Then A is a left and right noetherian ring (Exercise I.12). We assert that A is also fully left and right bounded.

Consider a finitely generated module M. Then M is finitely generated also over $\operatorname{cen}(A)$, say with generators $x_1,\ldots,x_n$. Then $\operatorname{Ann}(M)=\operatorname{Ann}(x_1)\cap\cdots\cap\operatorname{Ann}(x_n)$, for if $a\in\bigcap\operatorname{Ann}(x_i)$ and $x\in M$, then $x=x_1 c_1+\cdots+x_n c_n$ with $c_i\in\operatorname{cen}(A)$ and we obtain $x a=x_1 c_1 a+\cdots+x_n c_n a=x_1 a c_1+\cdots+x_n a c_n=0$. This implies the condition (H).

§3. Topologies for a Fully Bounded Noetherian Ring

Let $\mathfrak{F}$ be a Gabriel topology on the right noetherian ring A. We are going to show that when A is fully right bounded, the family $\mathfrak{F}$ is determined by the prime ideals it contains.

Lemma 3.1. *A prime ideal $\mathfrak{p}$ belongs to $\mathfrak{F}$ if and only if $A/\mathfrak{p}$ is not $\mathfrak{F}$-torsion-free.*

Proof. If $\mathfrak{p}\in\mathfrak{F}$, then $A/\mathfrak{p}$ is a torsion module. Suppose on the other hand that $A/\mathfrak{p}$ is not torsion-free, i.e. $\mathfrak{a}/\mathfrak{p}$ is a torsion module for some right ideal $\mathfrak{a}$ strictly containing $\mathfrak{p}$. Since $A/\mathfrak{p}$ is a right noetherian prime ring, it is a right order in a simple ring (Prop. II.2.6). From this it follows that $A/\mathfrak{p}$ satisfies DCC on right annihilators (cf. the proof of (a) ⇒ (b) in Theorem II.2.2). We now make the following useful observation (for which A need not be right noetherian):

Lemma 3.2. *If A satisfies* DCC *on right annihilators, then every right annihilator is of the form $r(S)$ for a finite subset S of A.*

Proof. Consider an arbitrary right annihilator $r(T)$ with $T\subset A$. By hypothesis there is a minimal right annihilator of the form $r(S)$ for finite subsets S of T. For each other finite subset S' of T we have $r(S\cup S')=r(S)\cap r(S')\subset r(S)$, and hence $r(S\cup S')=r(S)$. Consequently $r(T)=r(S)$. □

The proof of Lemma 3.1 may now be continued. Since $\mathfrak{p}$ is a prime ideal, the right annihilator in $A/\mathfrak{p}$ of $\mathfrak{a}/\mathfrak{p}$ is zero. By Lemma 3.2 there exist $x_1,\ldots,x_n$ in $\mathfrak{a}/\mathfrak{p}$ such that $\bigcap r(x_i)=r(\mathfrak{a}/\mathfrak{p})=0$, where r denotes the right annihilator taken in the ring $A/\mathfrak{p}$. This gives rise to an imbedding

$$A/\mathfrak{p}\subset(x_1 A/\mathfrak{p})\oplus\cdots\oplus(x_n A/\mathfrak{p})\subset\mathfrak{a}/\mathfrak{p}\oplus\cdots\oplus\mathfrak{a}/\mathfrak{p}.$$

$A/\mathfrak{p}$ is therefore a torsion module, and $\mathfrak{p}\in\mathfrak{F}$. □

Proposition 3.3. *Assume A is fully right bounded. If M is an $\mathfrak{F}$-torsion module, then* $\mathrm{Ass}(M)\subset\mathfrak{F}$.

Proof. Write $E(M)=\bigoplus E_i$ with E_i indecomposable injective modules. For each $\mathfrak{p}\in\mathrm{Ass}(M)$, $E(A/\mathfrak{p})$ is a direct sum of indecomposable injective modules which must be isomorphic to one of the modules E_i, because of the bijective correspondence between indecomposable injective modules and prime ideals. It follows that $A/\mathfrak{p}$ contains a non-zero submodule isomorphic to a submodule of M. By Lemma 3.1 we may conclude that $\mathfrak{p}\in\mathfrak{F}$. □

For every set $\mathfrak{E}$ of prime ideals of A one can define a Gabriel topology $\mathfrak{F}$ as

$$\mathfrak{F}=\{\mathfrak{a}\mid\mathfrak{a}\supset\mathfrak{p}_0\cdot\cdots\cdot\mathfrak{p}_n \text{ with } \mathfrak{p}_i\in\mathfrak{E}\},$$

using Prop. VI.6.10. When A is fully right bounded, every Gabriel topology is obtained in this way:

Theorem 3.4 (Gabriel [2]). *Let A be a right noetherian and fully right bounded ring. Every Gabriel topology $\mathfrak{F}$ on A has the form*

$$\mathfrak{F}=\{\mathfrak{a}\mid\mathfrak{a}\supset\mathfrak{p}_0\cdot\cdots\cdot\mathfrak{p}_n \textit{ where } \mathfrak{p}_i\in\mathfrak{F}\cap\mathrm{Spec}(A)\}.$$

Proof. If $\mathfrak{a}$ contains a product of right ideals in $\mathfrak{F}\cap\mathrm{Spec}(A)$, then $\mathfrak{a}\in\mathfrak{F}$ by Lemma VI.5.3. Suppose on the other hand that $\mathfrak{a}\in\mathfrak{F}$. We apply Prop. 1.11 to the torsion module $A/\mathfrak{a}$, and obtain right ideals $\mathfrak{a}=\mathfrak{a}_0\subset\mathfrak{a}_1\subset\cdots\subset\mathfrak{a}_n=A$ such that $\mathfrak{a}_i/\mathfrak{a}_{i-1}$ is annihilated by a prime ideal $\mathfrak{p}_i$. Then $\mathfrak{a}\subset\mathfrak{p}_n\cdot\cdots\cdot\mathfrak{p}_1$, and each $\mathfrak{p}_i\in\mathfrak{F}$ by Prop. 3.3 because $\mathfrak{a}_i/\mathfrak{a}_{i-1}$ is an $\mathfrak{F}$-torsion module. □

The theorem implies that every Gabriel topology on a right noetherian and fully bounded ring is bounded. The proof uses the hypothesis that A is fully right bounded only in the reference to Prop. 3.3, and it therefore works also in the following somewhat more general context. Let A be an arbitrary right noetherian ring and assume $\mathfrak{F}$ is a bounded Gabriel topology on A. Prop. 3.3 is satisfied also in this case. For let M be a torsion module and $\mathfrak{p}\in\mathrm{Ass}(M)$, say $\mathfrak{p}=\mathrm{Ann}(xA)$; since $\mathrm{Ann}(x)\in\mathfrak{F}$, there is a twosided ideal $\mathfrak{b}\in\mathfrak{F}$ contained in $\mathrm{Ann}(x)$, and then $xA\mathfrak{b}=0$ implies $\mathfrak{b}\subset\mathfrak{p}$ and $\mathfrak{p}\in\mathfrak{F}$. The proof of Theorem 3.4 can now be applied to $\mathfrak{F}$, and we obtain:

Proposition 3.5. *Let $\mathfrak{F}$ be a bounded Gabriel topology on a right noetherian ring A. Then*

$$\mathfrak{F}=\{\mathfrak{a}\mid\mathfrak{a}\supset\mathfrak{p}_0\cdot\cdots\cdot\mathfrak{p}_n \textit{ where } \mathfrak{p}_i\in\mathfrak{F}\cap\mathrm{Spec}(A)\}.$$

Example. Commutative noetherian rings. When A is a commutative noetherian ring, we have now available two descriptions of a Gabriel topology $\mathfrak{F}$ in terms of prime ideals, namely

$$\mathfrak{F}=\{\mathfrak{a}\mid\mathfrak{a}\supset\mathfrak{p}_0\cdot\cdots\cdot\mathfrak{p}_n \text{ where } \mathfrak{p}_i\in\mathfrak{F}\cap\mathrm{Spec}(A)\}\quad\text{by Theorem 3.4,}$$

and

$$\mathfrak{F}=\{\mathfrak{a}\mid V(\mathfrak{a})\subset\mathfrak{F}\}\quad\text{by Example VI.6.6.}$$

The $\mathfrak{F}$-torsion modules M are characterized by the property that $M_{\mathfrak{p}}=0$ for all prime ideals $\mathfrak{p}\notin\mathfrak{F}$.

§4. Artin-Rees Modules and Stable Topologies

A linear topology $\mathfrak{F}$ is called *stable* if the $\mathfrak{F}$-topology on a module M always induces the $\mathfrak{F}$-topology on submodules of M (cf. Chap. VI, §7). In this section we study the stability of bounded topologies on a right noetherian ring A.

If $\mathfrak{a}$ is a two-sided ideal of A, the $\mathfrak{a}$-*adic topology* is the bounded Gabriel topology consisting of all right ideals containing some power of $\mathfrak{a}$. When M is a finitely generated module, a fundamental system of neighborhoods of 0 for the $\mathfrak{a}$-adic topology on M is given by the submodules $M\mathfrak{a}^n$. Hence by the definition of stability:

Proposition 4.1. *Assume the* $\mathfrak{a}$*-adic topology is stable and* M *is a finitely generated module. For every submodule* L *of* M *and integer* $n>0$, *there exists an integer* $h(n)$ *such that* $M\mathfrak{a}^{h(n)}\cap L\subset L\mathfrak{a}^n$.

A finitely generated module M is said to be an *Artin-Rees module* if for every two-sided ideal $\mathfrak{a}$ of A, every submodule L of M and $n>0$, there exists $h(n)>0$ such that $M\mathfrak{a}^{h(n)}\cap L\subset L\mathfrak{a}^n$. We need:

Lemma 4.2. *Let* M *be a finitely generated module with submodules* $K\subset L\subset M$. *There exists a submodule* N *of* M *such that* $K=L\cap N$ *and* $\mathrm{Ass}(L/K)=\mathrm{Ass}(M/N)$.

Proof. If $K=L$, take $N=M$. Suppose $K\neq L$. Let $K=M_1\cap\cdots\cap M_n$ be a tertiary decomposition of K in M. Each $M_i\cap L$ is tertiary in L, provided that $M_i\cap L\neq L$, because $L/(M_i\cap L)\cong(L+M_i)/M_i\subset M/M_i$. Therefore we obtain a tertiary decomposition of K in L as $K=(M_1\cap L)\cap\cdots\cap(M_r\cap L)$, after omitting redundant $M_i\cap L$. We now put $N=M_1\cap\cdots\cap M_r$. Then $K=N\cap L$ and

$$\mathrm{Ass}(M/N)=\bigcup_1^r \mathrm{Ass}(M/M_i)=\bigcup_1^r \mathrm{Ass}(L/M_i\cap L)=\mathrm{Ass}(L/K). \quad \square$$

The following fundamental result by Lesieur and Croisot [3] connects the Artin-Rees property with the existence of primary decompositions:

Proposition 4.3. *A finitely generated module* M *is an Artin-Rees module if and only if every tertiary submodule of* M *is primary.*

Proof. Suppose M is an Artin-Rees module and L is a tertiary submodule of M. Let $\mathfrak{p}$ be the associated prime ideal of M/L. Then $\mathfrak{p}=\mathrm{Ann}(M'/L)$ for some $M'\supset L$, and we choose M' maximal, i.e. $M'=\{x\in M\mid x\,\mathfrak{p}\subset L\}$. Since $\mathrm{Ass}(N/L)=\{\mathfrak{p}\}$ for every non-zero submodule N/L of M/L, the maximality of M' implies that M'/L is an essential submodule of M/L. By the Artin-Rees property we have

$$M\mathfrak{p}^n\cap M'\subset M'\mathfrak{p}\subset L$$

for a certain integer n. But M'/L essential in M/L then implies $M\mathfrak{p}^n\subset L$, so L is primary in M.

For the converse, suppose that tertiary submodules of M are primary. Let $\mathfrak{a}$ be a two-sided ideal, L a submodule of M, and n an integer >0. Put $K=L\mathfrak{a}^n$. Then $\mathfrak{a}^n$ annihilates L/K, so for every $\mathfrak{p}\in\mathrm{Ass}(L/K)$ we have $\mathfrak{a}^n\subset\mathfrak{p}$ and hence $\mathfrak{a}\subset\mathfrak{p}$. By Lemma 4.2 there exists a submodule N of M such that $K=L\cap N$ and $\mathrm{Ass}(L/K)=\mathrm{Ass}(M/N)$. Let $N=M_1\cap\cdots\cap M_r$ be a tertiary decomposition of N

in M, with $\operatorname{ass}(M/M_i)=\mathfrak{p}_i$. Then $\mathfrak{p}_i\in \operatorname{Ass}(M/N)=\operatorname{Ass}(L/K)$, and therefore $\mathfrak{a}\subset\mathfrak{p}_i$ as was noted above. Each M_i is primary in M by hypothesis, which means that $M\mathfrak{p}_i^{m(i)}\subset M_i$ for some $m(i)$. Consequently there exists an integer m such that

$$M\mathfrak{a}^m\subset\bigcap_i M\mathfrak{p}_i^m\subset\bigcap_i M_i=N,$$

and then

$$M\mathfrak{a}^m\cap L\subset N\cap L=K=L\mathfrak{a}^n.$$

This establishes the Artin-Rees property for M. □

Theorem 4.4. *The following properties of a right noetherian ring A are equivalent:*
(a) *Every irreducible right ideal is primary.*
(b) *Every tertiary submodule of a finitely generated module M is primary in M.*
(c) *A is an Artin-Rees module (as a right A-module).*
(d) *All finitely generated modules are Artin-Rees modules.*
(e) *All bounded Gabriel topologies on A are stable.*
(f) *For every indecomposable injective module E, with associated prime ideal $\mathfrak{p}$, and every $x\in E$, there exists an integer $n>0$ such that $x\mathfrak{p}^n=0$.*

Proof. Prop. 4.3 entails (b) ⇔ (d) and (c) ⇒ (a). The implication (e) ⇒ (d) follows from Prop. 4.1, while (d) ⇒ (c) is trivial. To complete the proof, it remains to show (a) ⇒ (f) ⇒ (e).

(a) ⇒ (f): If E is an indecomposable injective module and $x\in E$, then $\operatorname{Ann}(x)$ is an irreducible right ideal, so (a) implies that $\operatorname{Ann}(x)\supset\mathfrak{p}^n$ for some n, where $\mathfrak{p}=\operatorname{ass}(E)$. Hence $x\mathfrak{p}^n=0$.

(f) ⇒ (e): Let $\mathfrak{F}$ be a bounded Gabriel topology. By Prop. VI.7.1 we should prove that if M is an $\mathfrak{F}$-torsion module, then also $E(M)$ is $\mathfrak{F}$-torsion. Clearly we may assume $E(M)$ indecomposable. By the remark preceding Prop. 3.5 we have $\operatorname{Ass}(M)\subset\mathfrak{F}$. Since $\operatorname{Ass}(E(M))=\operatorname{Ass}(M)$, condition (f) then implies that $E(M)$ is a torsion module. □

A right noetherian ring A is called *right classical* if it satisfies the equivalent conditions of the Theorem.

Examples

1. Commutative noetherian rings. Let A be a commutative noetherian ring. Since every irreducible ideal is primary (Zariski and Samuel [1], vol. 1, p. 209), A is a classical ring. Thus:

Proposition 4.5. *If A is a commutative noetherian ring, then*
(i) *all Gabriel topologies on A are stable;*
(ii) *all finitely generated modules are Artin-Rees modules.*

2. Separable algebras. Let R be a commutative noetherian ring and let A be a central, separable R-algebra (see Auslander and Goldman [2] for definitions). We assert that A is a classical ring. Let $\mathfrak{a}$ be a two-sided ideal of A, $\mathfrak{b}$ a right ideal of A, and n an integer >0. It is known that $\mathfrak{a}$ has the form $\mathfrak{a}=\mathfrak{c}A$ for some ideal $\mathfrak{c}$ in R (see Auslander and Goldman [2]). Since R is a classical ring and A is a finitely generated R-module, there exists $h(n)>0$ such that $A\mathfrak{c}^{h(n)}\cap\mathfrak{b}\subset\mathfrak{b}\mathfrak{c}^n$. This may be rewritten as $\mathfrak{a}^{h(n)}\cap\mathfrak{b}\subset\mathfrak{b}\mathfrak{a}^n$, so A is an Artin-Rees module. A is thus right classical.

A is also fully right bounded (§2, Example 4), so every Gabriel topology on A is bounded and therefore stable.

3. The Krull intersection theorem. One of the classical consequences of the Artin-Rees lemma is the Krull intersection theorem. Its non-commutative generalization is:

Proposition 4.6. *Let A be a right classical ring. If $\mathfrak{a}$ is a two-sided ideal of A and M is a finitely generated module, then $\bigcap_1^\infty M\mathfrak{a}^n$ consists of those $x \in M$ for which there exists $a \in \mathfrak{a}$ such that $x(1-a)=0$.*

Proof. If $x \in M$ and $x = x\,a$ for some $a \in \mathfrak{a}$, then $x = x\,a^n$ for all n, and hence $x \in \bigcap_1^\infty M\mathfrak{a}^n$. Conversely, if $x \in M' = \bigcap_1^\infty M\mathfrak{a}^n$, then xA is contained in the intersection of all neighborhoods of 0 in the $\mathfrak{a}$-adic topology of M. But this topology induces the $\mathfrak{a}$-adic topology of xA, and since $x\mathfrak{a}$ is a neighborhood of 0 in the latter topology, it follows that $x\mathfrak{a} = xA$. Hence $x = xa$ for some $a \in \mathfrak{a}$. □

§5. Auxiliary Results on Commutative Localization

Consider a commutative noetherian ring A. We will prove a few facts about prime ideals in A which will be needed in the next section. Let S be a multiplicatively closed subset of A, and denote by Φ the set of prime ideals not intersecting S.

Proposition 5.1. *The map $\mathfrak{p} \mapsto \mathfrak{p}A[S^{-1}]$ is a bijection*

$$\Phi \to \operatorname{Spec}(A[S^{-1}]).$$

Proof. If $\mathfrak{p} \in \Phi$, then it is easy to see that $\mathfrak{p}A[S^{-1}]$ is a prime ideal of $A[S^{-1}]$. Conversely, if $\mathfrak{q}$ is a prime ideal of $A[S^{-1}]$, then of course $\varphi^{-1}(\mathfrak{q})$ is a prime ideal of A not intersecting S, where $\varphi\colon A \to A[S^{-1}]$ is the canonical map. Moreover, one clearly has $\varphi^{-1}(\mathfrak{q})A[S^{-1}] = \mathfrak{q}$, and if $\mathfrak{p} \in \Phi$, then

$$\varphi^{-1}(\mathfrak{p}A[S^{-1}]) = \left\{ a \in A \,\middle|\, \frac{a}{1} = \frac{b}{s} \text{ for some } b \in \mathfrak{p},\ s \in S \right\}$$
$$= \{a \in A \mid s\,a \in \mathfrak{p} \text{ for some } s \in S\} = \mathfrak{p}$$

since $S \cap \mathfrak{p} = \emptyset$. Thus the correspondence $\mathfrak{p} \mapsto \mathfrak{p}A[S^{-1}]$ is bijective. □

For this result we did not need the noetherian hypothesis, but when then correspondence is restricted to associated prime ideals, the noetherian condition is crucial for bijectivity:

Proposition 5.2. *For every module M there is a bijection*

$$\operatorname{Ass}_A(M) \cap \Phi \to \operatorname{Ass}_{A[S^{-1}]}(M[S^{-1}]).$$

Proof. If $\mathfrak{p} \in \Phi$ and $\mathfrak{p} = \operatorname{Ann}(x)$ for some $x \in M$, then one easily verifies that $\mathfrak{p}A[S^{-1}] = \operatorname{Ann}\left(\frac{x}{1}\right) \in \operatorname{Ass}_{A[S^{-1}]}(M[S^{-1}])$. Suppose conversely that $\mathfrak{p} \in \Phi$ and

$\mathfrak{p}A[S^{-1}]$ is associated to $M[S^{-1}]$, say $\mathfrak{p}A[S^{-1}]=\operatorname{Ann}\left(\frac{x}{t}\right)$ for some $x\in M$ and $t\in S$. Let $a_1,\ldots,a_n$ be generators of the ideal $\mathfrak{p}$. Then $\frac{a_i}{1}\frac{x}{t}=0$ implies that $s_i\,a_i\,x=0$ for some $s_i\in S$ (we now write all A-modules as left modules). Put $s=s_1\cdot\ldots\cdot s_n\in S$. For each $a\in\mathfrak{p}$ one has $s\,a\,x=0$, so $\mathfrak{p}\subset\operatorname{Ann}(s\,x)$. On the other hand, if $a\in\operatorname{Ann}(s\,x)$, then $\frac{a}{1}\in\operatorname{Ann}\left(\frac{x}{t}\right)=\mathfrak{p}A[S^{-1}]$ and therefore $a\in\mathfrak{p}$. Thus $\mathfrak{p}=\operatorname{Ann}(s\,x)\in\operatorname{Ass}(M)$. □

The *support* of a module M is the set of prime ideals $\mathfrak{p}$ for which $M_{\mathfrak{p}}\neq 0$. We denote it by $\operatorname{Supp}(M)$. It is related to $\operatorname{Ass}(M)$ in the following way:

Proposition 5.3. *A prime ideal belongs to* $\operatorname{Supp}(M)$ *if and only if it contains a prime ideal in* $\operatorname{Ass}(M)$.

Proof. A prime ideal $\mathfrak{p}$ belongs to $\operatorname{Supp}(M)$ if and only if $\operatorname{Ass}_{A_{\mathfrak{p}}}(M_{\mathfrak{p}})\neq\emptyset$ by Prop. 1.1, and by Prop. 5.2 this is equivalent to the condition that there exists $\mathfrak{q}\in\operatorname{Ass}_A(M)$ with $\mathfrak{q}\subset\mathfrak{p}$. □

Corollary 5.4. $\operatorname{Ass}(M)\subset\operatorname{Supp}(M)$, *and these two families have the same minimal members.*

It follows in particular that every minimal prime ideal of A is a member of $\operatorname{Ass}(A)$, and hence there are only a finite number of minimal prime ideals.
For every ideal $\mathfrak{a}$ one has $\operatorname{Supp}(A/\mathfrak{a})=V(\mathfrak{a})$. More generally:

Lemma 5.5. *If* $\mathfrak{a}$ *is an ideal and* M *is a finitely generated module, then*

$$\operatorname{Supp}(M/\mathfrak{a}M)=\operatorname{Supp}(M)\cap V(\mathfrak{a}).$$

Proof. A prime ideal $\mathfrak{p}$ belongs to $\operatorname{Supp}(M/\mathfrak{a}M)$ if and only if $M_{\mathfrak{p}}\neq\mathfrak{a}M_{\mathfrak{p}}$, which occurs if and only if $M_{\mathfrak{p}}\neq 0$ and $\mathfrak{a}\subset\mathfrak{p}$ (e.g. by Prop. 4.6 or, better, by the Nakayama lemma (Prop. VIII.1.3)). □

If M is a module, then an element $a\in A$ is called *M-regular* if $a\,x\neq 0$ for all $x\neq 0$ in M.

Lemma 5.6. *Let M be a module. An element of A is M-regular if and only if it is not contained in any prime ideal associated to M.*

Proof. If $a\in A$ is M-regular, then of course a cannot belong to any $\mathfrak{p}\in\operatorname{Ass}(M)$. On the other hand, if a is not M-regular, then $a\,x=0$ for some $x\neq 0$ in M. Let $\mathfrak{p}$ be a prime ideal associated to the submodule $A\,x$. Then $\mathfrak{p}=\operatorname{Ann}(b\,x)$ for some $b\in A$, and $a\,b\,x=0$ implies $a\in\mathfrak{p}$. Thus a belongs to an associated prime ideal of M. □

The following fact will be needed in the next section.

Lemma 5.7. *Let M be a finitely generated module and $a\in A$ an M-regular element. Then*

$$V(\mathfrak{p}+A\,a)\cap\operatorname{Ass}(M/a\,M)\neq\emptyset \quad \textit{for every } \mathfrak{p}\in\operatorname{Ass}(M).$$

Proof. By Prop. 1.12 there exists a submodule L of M such that $\mathrm{Ass}(L)=\{\mathfrak{p}\}$ and $\mathrm{Ass}(M/L)=\mathrm{Ass}(M)-\{\mathfrak{p}\}$. The M-regular element a is also (M/L)-regular by Lemma 5.6. It follows that $aM\cap L=aL$, and therefore the induced homomorphism $L/aL\to M/aM$ is a monomorphism. Hence $\mathrm{Ass}(L/aL)\subset\mathrm{Ass}(M/aM)$. But by Prop. 5.3 and Lemma 5.5 we have

$$V(\mathfrak{p}+Aa)=V(\mathfrak{p})\cap V(Aa)=\mathrm{Supp}(L)\cap V(Aa)=\mathrm{Supp}(L/aL).$$

Therefore by Prop. 5.3 again, every minimal member of $V(\mathfrak{p}+Aa)$ belongs to $\mathrm{Ass}(L/aL)\subset\mathrm{Ass}(M/aM)$. □

§6. The Topologies $\mathfrak{F}_M^n$ for a Commutative Noetherian Ring

For a given ring A and a module M, we introduced in Chap. VI (§6, Example 4) a sequence of topologies $\mathfrak{F}_M^n$, which could be described as

$$\mathfrak{F}_M^n=\{\mathfrak{a}\,|\,\mathrm{Ext}^i(L,M)=0 \text{ for all } i\le n \text{ and all } L\subset A/\mathfrak{a}\}.$$

We now assume A to be commutative and noetherian, and we set out to describe explicitly the set of prime ideals which determine $\mathfrak{F}_M^n$, according to Theorem 3.4. We begin by considering the case $n=0$, and prepare for this with the following remark:

Lemma 6.1. *If $\mathfrak{a}$ is contained in the union of a finite number of prime ideals, then $\mathfrak{a}$ is contained in one of them.*

Proof. Let $\mathfrak{p}_1,\dots,\mathfrak{p}_n$ be prime ideals and $\mathfrak{a}\subset\bigcup_i\mathfrak{p}_i$. We may as well assume $\mathfrak{p}_i\not\subset\mathfrak{p}_j$ for $i\neq j$. Suppose now that $\mathfrak{a}\not\subset\mathfrak{p}_i$ for every i. From the definition of prime ideals follows quite easily that $\mathfrak{a}\cap\bigcap_{j\neq i}\mathfrak{p}_j\not\subset\mathfrak{p}_i$. If a_i is an element of $\mathfrak{a}\cap\bigcap_{j\neq i}\mathfrak{p}_j$ such that $a_i\notin\mathfrak{p}_i$, then $a_1+\dots+a_n$ belongs to $\mathfrak{a}$ withouth belonging to any $\mathfrak{p}_i$, which is a contradiction. □

Proposition 6.2. *Let M be a finitely generated module. The following properties of an ideal $\mathfrak{a}$ are equivalent:*

(a) $\mathfrak{a}\in\mathfrak{F}_M^0$.
(b) $\mathrm{Hom}(A/\mathfrak{a},M)=0$.
(c) $\mathrm{Ass}(M)\cap V(\mathfrak{a})=\emptyset$.
(d) $\mathfrak{a}$ *contains an M-regular element.*

Proof. (a) ⇒ (b) is trivial.

(b) ⇒ (c): If $\mathfrak{p}\in\mathrm{Ass}(M)$ and $\mathfrak{p}\supset\mathfrak{a}$, then $\mathfrak{p}x=0$ for some non-zero $x\in M$, which implies the existence of a non-zero homomorphism $A/\mathfrak{a}\to A/\mathfrak{p}\to M$, and this contradicts (b).

(c) ⇒ (d): If $\mathfrak{a}$ is not contained in any of the associated prime ideals of M, then since $\mathrm{Ass}(M)$ is finite, Lemma 6.1 implies that $\mathfrak{a}$ is not contained in the union of these prime ideals. It follows from Lemma 5.6 that $\mathfrak{a}$ contains an M-regular element.

(d) $\Rightarrow$ (a): Let L be a submodule of $A/\mathfrak{a}$, and suppose there is a non-zero homomorphism $\alpha: L \to M$. Then there exists $x \in L$ such that $\alpha(x) \neq 0$. But $\mathfrak{a}\, x = 0$ implies $\mathfrak{a}\, \alpha(x) = 0$, and this is impossible when $\mathfrak{a}$ contains M-regular elements. □

For the generalization of condition (d) to $\mathfrak{F}_M^n$ with $n > 0$, we need an extension of the concept of M-regularity. Let M be a finitely generated module. A sequence of elements $a_1, \ldots, a_n$ of A is *M-regular* (of *length* n) if a_1 is M-regular and a_k is $\left(M / \sum_{i=1}^{k-1} a_i M\right)$-regular for $k = 2, \ldots, n$.

Proposition 6.3 (Grothendieck [2]). *Let M be a finitely generated module. The following properties of an ideal $\mathfrak{a}$ are equivalent for each integer $n \geq 0$:*

(a) $\mathfrak{a} \in \mathfrak{F}_M^n$.

(b) $\operatorname{Ext}^i(A/\mathfrak{a}, M) = 0$ for all $i \leq n$.

(c) $\mathfrak{a}$ *contains an M-regular sequence of length $n+1$.*

Proof. (a) $\Rightarrow$ (b) is trivial.

(b) $\Rightarrow$ (c): We proceed by induction on n. The case $n = 0$ is taken care of by Prop. 6.2. Consider an arbitrary $n > 0$. In particular we then assume that $\operatorname{Hom}(A/\mathfrak{a}, M) = 0$, which by Lemma 5.3 implies that $\mathfrak{a}$ contains an M-regular element a. Multiplication by a yields an exact sequence

$$0 \to M \xrightarrow{\cdot a} M \to M/aM \to 0,$$

from which we obtain the exact sequences

$$\operatorname{Ext}^i(A/\mathfrak{a}, M) \to \operatorname{Ext}^i(A/\mathfrak{a}, M/aM) \to \operatorname{Ext}^{i+1}(A/\mathfrak{a}, M).$$

Using condition (b), we conclude that $\operatorname{Ext}^i(A/\mathfrak{a}, M/a\,M) = 0$ for $i \leq n-1$. Now by the induction hypothesis there exists an $M/a\,M$-regular sequence $a_2, \ldots, a_{n+1}$ in $\mathfrak{a}$. The sequence $a, a_2, \ldots, a_{n+1}$ is then M-regular.

(c) $\Rightarrow$ (a): Again we make induction on n. Let $a_1, \ldots, a_{n+1}$ be an M-regular sequence in $\mathfrak{a}$. Then $a_2, \ldots, a_{n+1}$ is an M/a_1M-regular sequence, so by the induction hypothesis we have $\operatorname{Ext}^i(L, M/a_1 M) = 0$ for $i \leq n-1$ and every $L \subset A/\mathfrak{a}$. The exact sequence

$$0 \to M \xrightarrow{\cdot a_1} M \to M/a_1 M \to 0$$

induces a monomorphism

$$0 \to \operatorname{Ext}^i(L, M) \xrightarrow{\cdot a_1} \operatorname{Ext}^i(L, M)$$

for $i \leq n$. But $L \subset A/\mathfrak{a}$ implies that $a_1 L = 0$, so the monomorphism must be zero. Hence $\operatorname{Ext}^i(L, M) = 0$ for $i \leq n$ and $L \subset A/\mathfrak{a}$. □

If the equivalent conditions of Prop. 6.3 are satisfied for a certain integer n, and if $b_1, \ldots, b_r$ is an M-regular sequence in $\mathfrak{a}$ for some $r < n$, then the module $M / \sum_{i=1}^{r} b_i M$ satisfies the same condition for $n - r$. This is easily proved by induction, using the same kind of argument as in the proof of (b) $\Rightarrow$ (c). It follows that all maximal M-regular sequences of elements in $\mathfrak{a}$ have the same length, which we call the $\mathfrak{a}$-*depth* of M and denote by $\operatorname{prof}_{\mathfrak{a}}(M)$ ("prof" stands for the French "profondeur"). We deduce in particular:

Lemma 6.4. *Let $\mathfrak{a}$ be an ideal and M a finitely generated module. If $b \in \mathfrak{a}$ is M-regular, then*

$$\operatorname{prof}_{\mathfrak{a}}(M/bM) = \operatorname{prof}_{\mathfrak{a}}(M) - 1.$$

If $\mathfrak{p}$ is a prime ideal of A, then we write $\operatorname{prof}(M_{\mathfrak{p}})$ for the $\mathfrak{p}A_{\mathfrak{p}}$-depth of the $A_{\mathfrak{p}}$-module $M_{\mathfrak{p}}$.

Lemma 6.5. *Let $\mathfrak{a}$ be an ideal and M a finitely generated module. Then*

$$\operatorname{prof}_{\mathfrak{a}}(M) = \inf_{\mathfrak{p} \in V(\mathfrak{a})} \operatorname{prof}(M_{\mathfrak{p}}).$$

Proof. If M has $\mathfrak{a}$-depth $= n$, then there is an M-regular sequence $a_1, \ldots, a_n$ in $\mathfrak{a}$. For each $\mathfrak{p} \in V(\mathfrak{a})$, the canonical images of the a_i in $A_{\mathfrak{p}}$ form an $M_{\mathfrak{p}}$-regular sequence in $\mathfrak{p}A_{\mathfrak{p}}$, as one easily verifies. Hence $M_{\mathfrak{p}}$ has depth $\geq n$.

To obtain the converse inequality, we prove that if $n \leq \operatorname{prof}(M_{\mathfrak{p}})$ holds for all $\mathfrak{p} \in V(\mathfrak{a})$, then $n \leq \operatorname{prof}_{\mathfrak{a}}(M)$. We do this by induction. The case $n=0$ is trivial, so suppose $n \geq 1$. Then $\mathfrak{p}A_{\mathfrak{p}} \notin \operatorname{Ass}(M_{\mathfrak{p}})$ for every $\mathfrak{p} \in V(\mathfrak{a})$, and it easily follows that $\mathfrak{p} \notin \operatorname{Ass}(M)$. By Lemma 5.6 this entails the existence of an M-regular element $a \in \mathfrak{a}$. Clearly $a/1 \in A_{\mathfrak{p}}$ is $M_{\mathfrak{p}}$-regular, and from Lemma 6.4 it follows that the $A_{\mathfrak{p}}$-module $(M/aM)_{\mathfrak{p}} \cong M_{\mathfrak{p}}/aM_{\mathfrak{p}}$ has depth $\geq n-1$ for all $\mathfrak{p} \in V(\mathfrak{a})$. The induction hypothesis gives that M/aM has $\mathfrak{a}$-depth $\geq n-1$, and hence M has $\mathfrak{a}$-depth $\geq n$. $\square$

Applying Lemma 6.6 to Prop. 6.3, we arrive at the following characterization of the topology $\mathfrak{F}_M^n$:

Proposition 6.6. *Let M be a finitely generated module. Then*

$$\mathfrak{F}_M^n = \{\mathfrak{a} \mid \operatorname{prof}(M_{\mathfrak{p}}) \geq n+1 \text{ for all } \mathfrak{p} \in V(\mathfrak{a})\}.$$

In particular, a prime ideal $\mathfrak{p}$ belongs to $\mathfrak{F}_M^n$ if and only if $M_{\mathfrak{q}}$ has depth $\geq n+1$ for all prime ideals $\mathfrak{q} \supset \mathfrak{p}$. It does not suffice here to require that just $M_{\mathfrak{p}}$ has depth $\geq n+1$, because it may happen for prime ideals $\mathfrak{p} \subset \mathfrak{q}$ that $\operatorname{prof}(M_{\mathfrak{q}}) < \operatorname{prof}(M_{\mathfrak{p}})$ (for an example of this phenomenon we refer to Grothendieck and Dieudonné [1], Chap. 0, Sec. 16.4.6.4). In order to avoid that kind of complications, and obtain Prop. 6.6 in a more pleasant form, we have to impose some additional conditions on A and M. For the sake of simplicity we confine our attention to the case $M = A$ (the general case may be done in about the same way).

But first we make a digression on the relation between depth and dimension. The *Krull dimension* of the ring A is defined to be the maximal length n of chains of prime ideals

$$\mathfrak{p}_0 \subsetneqq \mathfrak{p}_1 \subsetneqq \cdots \subsetneqq \mathfrak{p}_n$$

in A. We denote it by $\dim(A)$. Thus $0 \leq \dim(A) \leq \infty$. If $\mathfrak{p}$ is a prime ideal, the Krull dimension of the local ring $A_{\mathfrak{p}}$ is called the *height* of $\mathfrak{p}$, and is denoted by $\operatorname{ht}\mathfrak{p}$; by Prop. 5.1 it is equal to the maximal length of chains of prime ideals contained in $\mathfrak{p}$. It may be proved that every prime ideal has finite height (Zariski and Samuel [1], Chap. 4, Th. 30), or in other words, that every local ring has finite Krull dimension.

Lemma 6.7. *Let A be a local ring. Then*

$$\operatorname{prof}(A) \leq \inf_{\mathfrak{p} \in \operatorname{Ass}(A)} \dim(A/\mathfrak{p}).$$

Proof. We use induction on $n=\operatorname{prof}(A)$. The case $n=0$ is trivial. Suppose $\operatorname{prof}(A)=n>0$. Let a be an A-regular element in the maximal ideal of A. Then $\operatorname{prof}(A/aA)=n-1$ by Lemma 6.4. For each $\mathfrak{p}\in\operatorname{Ass}(A)$ we may by Lemma 5.7 find a prime ideal $\mathfrak{q}\in\operatorname{Ass}(A/aA)$ such that $\mathfrak{q}\supset\mathfrak{p}$ and $a\in\mathfrak{q}$. Since a is A-regular, we have $a\notin\mathfrak{p}$, so $\mathfrak{p}\neq\mathfrak{q}$. The induction hypothesis therefore gives $\operatorname{prof}(A/aA)\leq\dim(A/\mathfrak{q})$. Then $\mathfrak{p}\neq\mathfrak{q}$ implies

$$\dim(A/\mathfrak{p})>\dim(A/\mathfrak{q})\geq\operatorname{prof}(A/aA)=\operatorname{prof}(A)-1,$$

so $\dim(A/\mathfrak{p})\geq\operatorname{prof}(A)$. □

Definition (Serre). The ring A satisfies condition S_n, for an integer $n\geq 0$, if

$$\operatorname{prof}(A_\mathfrak{p})\geq\inf(n,\operatorname{ht}\mathfrak{p})$$

holds for every prime ideal $\mathfrak{p}$.

Note that if S_n holds, then S_k holds for all $k\leq n$. As a consequence of Lemma 6.7, S_n can be formulated as the combination of the two conditions

$$\operatorname{prof}(A_\mathfrak{p})=\operatorname{ht}\mathfrak{p}\quad\text{when}\quad\operatorname{ht}\mathfrak{p}<n,$$

$$\operatorname{prof}(A_\mathfrak{p})\geq n\quad\text{when}\quad\operatorname{ht}\mathfrak{p}\geq n.$$

Proposition 6.8. *Suppose A satisfies S_{n+1}. Then:*

(i) *$\mathfrak{F}_A^n\cap\operatorname{Spec}(A)$ consists of all prime ideals of height $\geq n+1$.*

(ii) *A module L is an $\mathfrak{F}_A^n$-torsion module if and only if $L_\mathfrak{p}=0$ for all prime ideals of height $\leq n$.*

Proof. (i): By Prop. 6.6, a prime ideal $\mathfrak{p}$ belongs to $\mathfrak{F}_A^n$ if and only if $\operatorname{prof}(A_\mathfrak{q})\geq n+1$ for all prime ideals $\mathfrak{q}\supset\mathfrak{p}$. If $\operatorname{ht}\mathfrak{p}\geq n+1$, then also every $\mathfrak{q}\supset\mathfrak{p}$ has $\operatorname{ht}\mathfrak{q}\geq n+1$, and $\operatorname{prof}(A_\mathfrak{q})\geq n+1$ by S_{n+1}. Hence $\mathfrak{p}\in\mathfrak{F}_A^n$ in this case. When $\operatorname{ht}\mathfrak{p}\leq n$, also $\operatorname{prof}(A_\mathfrak{p})\geq n$ by S_{n+1} and thus $\mathfrak{p}\notin\mathfrak{F}_A^n$.

(ii) now follows by the general results of Chap. VI (§6, Example 6). □

Examples

1. Dense ideals. Since $\mathfrak{F}_A^0$ consists of the dense ideals of A, we have as a particular case of Prop. 6.2:

Proposition 6.9. *An ideal in a commutative noetherian ring is dense if and only if it contains a regular element.*

2. Integrally closed domains. Let A be a (noetherian, as always) integrally closed domain (see Lang [1], Chap. IX). We assert that A satisfies condition S_2. If $\mathfrak{p}$ is a prime ideal of height 0 or 1, then $A_\mathfrak{p}$ obviously has depth 0 respectively 1. To settle the case of $\operatorname{ht}\mathfrak{p}\geq 2$, it suffices to show that every integrally closed local domain A of dimension ≥ 2 has depth ≥ 2 (because every localization of an integrally closed domain is integrally closed). Choose any non-zero element a in the maximal ideal $\mathfrak{m}$ of A. Then every $\mathfrak{p}_i\in\operatorname{Ass}(A/Aa)$ has height 1 (for a proof of this non-trivial fact we refer to Zariski and Samuel [1], Chap. 5, Theorem 14) and is therefore strictly contained in $\mathfrak{m}$. Hence $\mathfrak{m}$ is not contained in $\bigcup\mathfrak{p}_i$ (Lemma 6.1), so there exists an element $b\in\mathfrak{m}$ which does not belong to any $\mathfrak{p}_i\in\operatorname{Ass}(A/Aa)$. The sequence a,b is A-regular by Lemma 5.6, so $\operatorname{prof}(A)\geq 2$.

As a special case of Prop. 6.8 we thus have:

Proposition 6.10. *Let A be a noetherian integrally closed domain. L is an $\mathfrak{F}_A^1$-torsion module if and only if $L_\mathfrak{p}=0$ for every prime ideal of height ≤ 1.*

3. Cohen-Macaulay rings. The ring A is a *Cohen-Macaulay ring* if $\text{prof}(A_\mathfrak{p})=\dim(A_\mathfrak{p})$ for every prime ideal $\mathfrak{p}$. It then satisfies condition S_n for every $n \geq 0$. By the previous Example, every integrally closed domain of dimension ≤ 2 is a Cohen-Macauly ring. For a module M over a Cohen-Macaulay ring A there is a filtration by submodules $M^0 \supset M^1 \supset M^2 \supset \cdots$, where M^n is the $\mathfrak{F}_A^n$-torsion submodule of M and also is the maximal submodule L of M such that $L_\mathfrak{p}=0$ for all prime ideals of height $\leq n$.

Exercises

A is a right noetherian ring in all of the following exercises.

§1

1. Show that every maximal two-sided ideal (in an arbitrary ring) is a prime ideal.

2. Let L be a submodule of a module M. Show that if $\mathfrak{p} \in \text{Ass}(M/L)$ and $\mathfrak{p}$ does not contain $\text{Ann}(L)$, then $\mathfrak{p} \in \text{Ass}(M)$.

3. For every module M, put $T(M)=\{a \in A \mid a \in \text{Ann}(L)$ for some essential submodule L of $M\}$. Show that:
   (i) $T(M)$ is a two-sided ideal of A, and $L \subset M$ implies $T(L) \supset T(M)$.
   (ii) The following properties of M are equivalent:
   (a) M is cotertiary.
   (b) $\text{Ann}(L) \subset T(M)$ for every non-zero submodule L of M.
   (c) $T(L)=T(M)$ for every non-zero submodule L of M.

§2

4. If A is fully right bounded and $\mathfrak{m}$ is a maximal two-sided ideal of A, show that $A/\mathfrak{m}$ is a simple ring.

5. Let A be fully right bounded, and let M be a finitely generated, cotertiary module such that $\text{Ann}(M)=\mathfrak{p}$ is a prime ideal. Show that $\text{Ann}(M)=T(M)$. (Hint: first consider the case of a cyclic module $M=A/\mathfrak{a}$; note that $\mathfrak{a}/\mathfrak{p}$ is not essential in $A/\mathfrak{p}$, and use Exercise 3(i) to conclude that $\mathfrak{p}=T(A/\mathfrak{p})=T(A/\mathfrak{a})$.)

6. Show that every right noetherian ring satisfies condition (H).

§3

7. Show that the following properties of a ring A (which need not be right noetherian) are equivalent:
   (a) A satisfies DCC on right annihilators.
   (b) A satisfies ACC on left annihilators.
   (c) Every right annihilator is of the form $r(S)$ for a finite subset S of A.

8. Let $\mathfrak{F}$ be a bounded Gabriel topology on A, and let M be a finitely generated $\mathfrak{F}$-torsion module. Show that there exist submodules $0=M_0 \subset M_1 \subset \cdots \subset M_n=M$ such that each M_i/M_{i-1} is annihilated by a prime ideal in $\mathfrak{F}$.

§4

9. Let $\mathfrak{a}$ be an ideal and let M be a finitely generated module. Show that if L is a submodule of M, then the closure of L in the $\mathfrak{a}$-adic topology of M is $\bigcap_n (L+M\mathfrak{a}^n)$.

10. Show that if A is right classical and M is a module, then M is an $\mathfrak{F}$-torsion module for the Gabriel topology $\mathfrak{F}$ of right ideals containing finite products of ideals in $\text{Ass}(M)$.

§5

Let A be a commutative noetherian ring in the following exercises.

11. For an arbitrary module M, show that the canonical homomorphism $\varphi\colon M \to \prod_{\mathfrak{p}\in \mathrm{Ass}(M)} M_{\mathfrak{p}}$ is a monomorphism.

12. Let $\mathfrak{p}$ be a prime ideal of A. Put $E=E(A/\mathfrak{p})$ and $E_n=\{x\in E \mid \mathfrak{p}^n x=0\}$ for each integer $n\geq 0$. Show that:

(i) $E=\bigcup_n E_n$.

(ii) $\sqrt{\mathrm{Ann}(x)}$ is the associated prime ideal of Ax, for every $x\in E$.

(iii) $E_{n+1}/E_n=\{x\in E/E_n \mid \mathrm{Ann}(x)=\mathfrak{p} \text{ or } x=0\}$.

(iv) E_{n+1}/E_n is a vector space over the field of fractions of $A/\mathfrak{p}$. (Hint: use Exercise V.19.)

13. Show that if M is a finitely generated module, then $\mathrm{Supp}(M)=V(\mathrm{Ann}(M))$.

14. Let $\mathfrak{a}$ be an ideal and let M be a finitely generated module with a submodule L. Show that L is closed in the $\mathfrak{a}$-adic topology of M if and only if $\mathfrak{p}+\mathfrak{a}\neq A$ holds for all $\mathfrak{p}\in\mathrm{Ass}(M/L)$. (Hint: reduce to the case $L=0$, then use Prop. 4.6 and Lemma 5.6.)

§6

15. Let L and M be finitely generated modules.

(i) Show that the following conditions are equivalent:

(a) L is an $\mathfrak{F}_M^0$-torsion module.

(b) $\mathrm{Ass}(M)\cap\mathrm{Supp}(L)=\emptyset$.

(c) $\mathrm{Ann}(L)$ contains an M-regular element.

(ii) Show that the following conditions are equivalent for an integer $n\geq 0$:

(a) L is an $\mathfrak{F}_M^n$-torsion module.

(b) $\mathrm{Ann}(L)$ contains an M-regular sequence of length $n+1$.

Chapter VIII. Simple Torsion Theories

A hereditary torsion theory is *simple* if it is generated by a class of simple modules. There is a fairly large class of rings, including e.g. the artinian rings, for which all hereditary torsion theories are simple. Therefore it is of a certain interest to study simple torsion theories in some detail. The methods to be used for that purpose are the basic ones of the theory of artinian rings, such as the use of the Jacobson radical and the lifting of idempotents. This chapter is devoted to the development of these methods and to their applications to torsion theory, as well as to an account of the basic theory of rings with various minimum conditions.

§ 1. The Jacobson Radical and Artinian Rings

Let A be a ring. The *(Jacobson) radical* of a right A-module M is defined to be the intersection of all maximal proper submodules of M, and is denoted by $\mathrm{Rad}(M)$. It may happen that there are no maximal proper submodules of M, in which case $\mathrm{Rad}(M)=M$, but for finitely generated modules they exist (Lemma I.6.8). Hence:

Lemma 1.1. *If* $M\neq 0$ *is finitely generated, then* $\mathrm{Rad}(M)\neq M$.

It is clear that the radical also can be described as

$$\mathrm{Rad}(M)=\{x\in M\,|\,\varphi(x)=0 \text{ for every } \varphi\colon M\to S \text{ with } S \text{ simple}\}. \tag{1}$$

From this observation it is easy to deduce that $M\mapsto \mathrm{Rad}(M)$ is a preradical, and that in fact it is a radical in the sense of Chap. VI.

The Jacobson radical of A as a right A-module will be written $J(A)$ or simply J. It is a two-sided ideal of A, and by (1) it can be described as the intersection of all annihilators of simple A-modules.

Lemma 1.2. $MJ(A)\subset \mathrm{Rad}(M)$ *for every module* M.

Proof. Since $J(A)$ annihilates every simple module S, all homomorphisms $M\to S$ are zero on $MJ(A)$, and it follows from (1) that $MJ(A)\subset \mathrm{Rad}(M)$. □

The following result is known as the "Nakayama Lemma".

Proposition 1.3. *Let* M *be a finitely generated module. If* L *is a submodule of* M *such that* $L+MJ(A)=M$, *then* $L=M$.

Proof. $L+MJ(A)=M$ implies $(M/L)\,J(A)=M/L$, and hence $\mathrm{Rad}(M/L)=M/L$ by Lemma 1.2. But then Lemma 1.1 gives $M/L=0$. □

The Jacobson radical of A can be described in a more tangible way:

Proposition 1.4. *$J(A)=\{a\in A \mid 1-ab$ has a right inverse for every $b\in A\}$.*

Proof. $a\in J(A)$ if and only if $\mathfrak{m}+aA\neq A$ for every maximal right ideal $\mathfrak{m}$, and this is equivalent to $1-ab\notin\mathfrak{m}$ for every $b\in A$ and maximal $\mathfrak{m}$, i.e. $1-ab$ has a right inverse. $\square$

Proposition 1.5. *$J(A)$ is the largest two-sided ideal of A consisting of elements a such that $1-a$ is invertible.*

Proof. If $\mathfrak{a}$ is a two-sided ideal such that $1-a$ is invertible for every $a\in\mathfrak{a}$, then $\mathfrak{a}\subset J(A)$ by Prop. 1.4. It remains to see that $a\in J(A)$ implies $1-a$ invertible. We know that there exists c such that $(1-a)c=1$. Then $1-c=-ac\in J(A)$, so there exists c' such that $(1-(1-c))c'=1$, i.e. $cc'=1$. Hence c is invertible, and so is then also $1-a$. $\square$

The characterization of $J(A)$ obtained in this proposition is left-right symmetric, and accordingly we may conclude:

Corollary 1.6. *$J(A)$ is the intersection of all maximal left ideals of A.*

A semi-simple module certainly has zero radical. Conversely we have:

Proposition 1.7. *A module M is semi-simple of finite length if and only if it is artinian and* $\mathrm{Rad}(M)=0$.

Proof. Suppose M is artinian and $\mathrm{Rad}(M)=0$. Let L be a minimal member of the set of finite intersections of maximal submodules of M. For every maximal submodule M' of M one then has $M'\cap L=L$, so $L\subset M'$. But since the intersection of all maximal submodules M' is zero, this implies $L=0$. Hence there exist maximal submodules $M_1,\dots,M_n$ of M such that $M_1\cap\dots\cap M_n=0$. One can then imbed M as a submodule of the semi-simple module $M/M_1\oplus\dots\oplus M/M_n$, and it follows from Cor. I.7.4 that M is semi-simple. Since M is artinian, it is of course of finite length. $\square$

Corollary 1.8. *If M is an artinian module, then $M/\mathrm{Rad}(M)$ is semi-simple of finite length.*

From Prop. 1.7 we also obtain yet another characterization of semi-simple rings (cf. Prop. I.7.7):

Proposition 1.9. *The ring A is semi-simple if and only if A is right artinian and* $J(A)=0$.

Corollary 1.10. *The ring A is simple if and only if A is right artinian and has no two-sided ideals other than* (0) *and A.*

Proof. Use Prop. I.7.8, combined with the fact that $J(A)$ is a two-sided ideal. $\square$

Definition. A is a *semi-primary* ring if $J(A)$ is a nilpotent ideal and $A/J(A)$ is a semi-simple ring.

Proposition 1.11. *Let A be a semi-primary ring. An A-module M is noetherian if and only if it is artinian.*

Proof. Let $J^n=(0)$. Let M be a noetherian or an artinian module. The modules M/MJ, $MJ/MJ^2, \ldots, MJ^{n-1}/MJ^n=MJ^{n-1}$ may be considered as modules over the semi-simple ring A/J, and as such they are semi-simple of finite length. They are then of finite length also as A-modules, and it follows by repeated use of Prop. III.3.8 that M is of finite length. □

The basic properties of artinian rings are summarized in the following statements.

Proposition 1.12. *The ring A is right artinian if and only if it is semi-primary and right noetherian.*

Proof. If A is semi-primary and right artinian, then it is right noetherian by Prop. 1.11. Assume on the other hand that A is right artinian. Then $A/J(A)$ is a semi-simple ring by Cor. 1.8 (or by Prop. 1.9). In order to show that $J(A)$ is nilpotent, consider the descending chain $J\supset J^2\supset\cdots$. Since A is right artinian, we have $J^n=J^{n+1}$ for some n. Assume $J^n\neq(0)$. Let $\mathfrak{a}$ be a minimal right ideal $\neq(0)$ such that $\mathfrak{a}J^n\neq(0)$. Since $(\mathfrak{a}J)J^n=\mathfrak{a}J^{n+1}=\mathfrak{a}J^n\neq(0)$ and $\mathfrak{a}J\subset\mathfrak{a}$, we must have $\mathfrak{a}J=\mathfrak{a}$. But since $\mathfrak{a}$ is a principal right ideal, this contradicts the Nakayama Lemma (Prop. 1.3), and hence $J^n=(0)$. Thus A is semi-primary, and it finally follows from Prop. 1.11 that A is right noetherian. □

Corollary 1.13. *If A is right artinian, then every finitely generated module is of finite length.*

Proposition 1.14. *The following properties of a right noetherian ring A are equivalent:*

(a) *A is right artinian.*

(b) *A is fully right bounded and all prime ideals are maximal (as two-sided ideals).*

Proof. (a) ⇒ (b): If $\mathfrak{p}$ is a prime ideal of a right artinian ring A, then $A/\mathfrak{p}$ is a right artinian prime ring and hence a simple ring. The condition (b) of Theorem VII.2.1 is therefore trivially satisfied, and it also follows that $\mathfrak{p}$ is a maximal two-sided ideal.

(b) ⇒ (a): By Prop. VII.1.11 there exist right ideals $(0)=\mathfrak{a}_0\subset\mathfrak{a}_1\subset\cdots\subset\mathfrak{a}_n=A$ such that each $\mathfrak{a}_i/\mathfrak{a}_{i-1}$ is annihilated by a prime ideal $\mathfrak{p}_i$. One may consider $\mathfrak{a}_i/\mathfrak{a}_{i-1}$ as a module over the ring $A/\mathfrak{p}_i$. But since $\mathfrak{p}_i$ is maximal, $A/\mathfrak{p}_i$ is a simple ring by Theorem VII.2.1 (and Exercise VII.4). Therefore $\mathfrak{a}_i/\mathfrak{a}_{i-1}$ has finite length as an $A/\mathfrak{p}_i$-module, and of course also as an A-module. Hence A has finite length, and it is thus a right artinian ring. □

Proposition 1.15. *The ring A is right artinian if and only if every injective module is a direct sum of injective envelopes of simple modules.*

Proof. If A is right artinian, then it is right noetherian (Prop. 1.12), so every injective module is a direct sum indecomposable injective modules (Prop. V.4.5). But each of these modules has a non-zero socle, and is therefore the injective envelope of a simple module.

Assume conversely that every injective module is a direct sum of injective envelopes of simple modules. Suppose there is an infinite strictly descending chain

$\mathfrak{a}_1 \supset \mathfrak{a}_2 \supset \cdots$ of right ideals of A, and put $\mathfrak{a} = \bigcap \mathfrak{a}_i$. Then

$$E(A/\mathfrak{a}) = E(S_1) \oplus \cdots \oplus E(S_n)$$

for certain simple modules $S_1, \ldots, S_n$. Each $\mathfrak{a}_i/\mathfrak{a}$ has non-zero intersection with $S = S_1 \oplus \cdots \oplus S_n$. Since S is semi-simple of finite length, there must exist i_0 such that $(\mathfrak{a}_i/\mathfrak{a}) \cap S = (\mathfrak{a}_{i_0}/\mathfrak{a}) \cap S$ for all $i \geq i_0$. This contradicts the fact that $\bigcap (\mathfrak{a}_i/\mathfrak{a}) = 0$. We conclude that A is right artinian. □

Examples

1. Nil ideals. If the right ideal $\mathfrak{a}$ is nil (i.e. every element of $\mathfrak{a}$ is nilpotent), then $\mathfrak{a} \subset J(A)$. For if $a \in \mathfrak{a}$ and b is an arbitrary element of A, then $ab \in \mathfrak{a}$ implies that ab is nilpotent, and consequently $1 - ab$ is invertible, and this means that $a \in J(A)$ by Prop. 1.4.

But $J(A)$ does not necessarily contain all nilpotent elements of A (consider e.g. matrix rings $M_n(K)$ over a field K).

2. Regular rings. If A is a regular ring, i.e. for every $a \in A$ there exists $x \in A$ such that $a = axa$, then it is clear from Prop. 1.4 that $J(A) = 0$.

3. Morita invariance of the radical. Let P be a finitely generated projective generator for **Mod**-A, and put $B = \mathrm{Hom}_A(P, P)$. Morita theory states that the functor $F = \mathrm{Hom}_A(P, \cdot)$ is an equivalence between **Mod**-A and **Mod**-B. It clearly follows that for every right A-module M one has $F(\mathrm{Rad}(M)) = \mathrm{Rad}(F(M))$. For $M = A^n$ one obtains in particular that $J(M_n(A)) = M_n(J(A))$, i.e. the radical of $M_n(A)$ consists of all matrices with entries in $J(A)$.

4. Semi-local rings. The ring A is local in the sense of Chap. V, § 5, if and only if $A/J(A)$ is a skew-field. More generally, A is said to be a *semi-local* ring if $A/J(A)$ is a semi-simple ring. Note that a semi-local ring has only finitely many maximal right ideals. A commutative ring is semi-local if and only if the number of maximal ideals is finite.

§ 2. Semi-Artinian Modules and Rings

A module is called *semi-artinian* if it belongs to the hereditary torsion class generated by all simple modules. It follows from Prop. VI.2.5 that M is semi-artinian if and only if every non-zero quotient module of M has non-zero socle. This implies in particular that a semi-artinian module is an essential extension of its socle. Every artinian module is of course semi-artinian.

Let s denote the left exact preradical which to each module assigns its socle. Recall that the corresponding radical $\bar{s}$ is obtainable in the following manner:

$$s_0(M) = 0,$$
$$s_{\alpha+1}(M)/s_\alpha(M) = s(M/s_\alpha(M)),$$
$$s_\alpha(M) = \sum_{\beta<\alpha} s_\beta(M) \quad \text{when } \alpha \text{ is a limit ordinal,}$$
$$\bar{s}(M) = s_\alpha(M) \quad \text{where } \alpha \text{ is the first ordinal for which } s_\alpha(M) = s_{\alpha+1}(M).$$

M is thus semi-artinian if and only if $M = s_\alpha(M)$ for some α. The simple modules appearing in the various quotients $s_{\alpha+1}(M)/s_\alpha(M)$ are called the *composition factors* of the semi-artinian module M.

If M is a noetherian module, then $\bar{s}(M) = s_n(M)$ for some integer n, and each $s_{i+1}(M)/s_i(M)$ is semi-simple of finite length. Thus:

Proposition 2.1. *A noetherian and semi-artinian module has finite length.*

Corollary 2.2. *A semi-artinian module over a right noetherian ring is the sum of its submodules of finite length.*

Corollary 2.3. *Let A be a right noetherian ring. The Gabriel topology corresponding to the class of semi-artinian modules consists of the right ideals $\mathfrak{a}$ for which $A/\mathfrak{a}$ has finite length.*

If $\mathcal{T}$ is a hereditary torsion class consisting entirely of semi-artinian modules, then $\mathcal{T}$ is *simple*, i.e. generated by a class of simple modules, as a consequence of the following easily established corollary of Prop. VI.2.5:

Lemma 2.4. *Let $\mathcal{T}$ be a torsion class generated by a class $\mathcal{C}$ closed under quotient modules. If $\mathcal{T}'$ is a hereditary torsion class contained in $\mathcal{T}$, then $\mathcal{T}'$ is generated by $\mathcal{C} \cap \mathcal{T}'$.*

The ring A is said to be *right semi-artinian* if it is semi-artinian as a right A-module, i.e. every non-zero cyclic module has non-zero socle. There are a number of alternative characterizations of semi-artinian rings:

Proposition 2.5. *The following properties of a ring A are equivalent:*
(a) *A is right semi-artinian.*
(b) *Every module is semi-artinian.*
(c) *Every non-zero module has non-zero socle.*
(d) *Every module is an essential extension of its socle.*
(e) *Every hereditary torsion theory is simple.*

Proof. (b) $\Rightarrow$ (e) follows from the remark preceding Lemma 2.4, while (e) $\Rightarrow$ (b) follows by considering the torsion class of all A-modules. The equivalence of (b) with the remaining conditions is quite obvious. ▯

Every right noetherian and right semi-artinian ring is right artinian by Prop. 2.1. The Jacobson radical of a semi-artinian ring is in general not nilpotent, but it nearly is so. More generally, if M is a module over an arbitrary ring A, we say that the Jacobson radical is *T-nilpotent on M* if for every $x \in M$ and every sequence $a_1, a_2, \ldots$ in J there exists an integer n such that $x a_1 a_2 \cdot \cdots \cdot a_n = 0$. In particular, J is *left T-nilpotent* if it is T-nilpotent on the right A-module A, i.e. for every sequence $a_1, a_2, \ldots$ in J one has $a_1 \cdot \cdots \cdot a_n = 0$ for some n. In that case J is a nil ideal.

Proposition 2.6. *J is T-nilpotent on every semi-artinian module.*

Proof. Let M be a semi-artinian module. For each $x \in M$ we can define $o(x)$ to be the smallest ordinal β for which $x \in s_\beta(M)$. Then $o(x)$ cannot be a limit ordinal, because $x \in \sum_{\alpha<\beta} s_\alpha(M)$ implies $x \in s_\alpha(M)$ for some $\alpha < \beta$. Hence we can write $o(x) = \alpha + 1$ for some α (unless $x = 0$). But $s_{\alpha+1}(M)/s_\alpha(M)$ is a semi-simple module and is therefore annihilated by J, so $s_{\alpha+1}(M) \cdot J \subset s_\alpha(M)$. It follows that $o(xa) < o(x)$ for every non-zero $x \in M$ and $a \in J$. If now $a_1, a_2, \ldots$ is a sequence of elements in J such that $x a_1 a_2 \cdot \cdots \cdot a_n \neq 0$ for every n, then $o(x) > o(x a_1) > o(x a_1 a_2) > \cdots$. But since every strictly descending chain of ordinals is finite, this is impossible. □

Corollary 2.7. *If A is a right semi-artinian ring, then J is a left T-nilpotent ideal.*

It is possible to prove a converse to this corollary:

Proposition 2.8. *The following properties of the ring A are equivalent:*
(a) *A is a right semi-artinian ring.*
(b) *J is left T-nilpotent and A/J is a right semi-artinian ring.*

Proof. (a) ⇒ (b): We have already seen that J is left T-nilpotent, and it is easy to show that every epimorphic image of a right semi-artinian ring is right semi-artinian.

(b) ⇒ (a): We first note:

Lemma 2.9. *If J is left T-nilpotent on a non-zero module M, then there exists $x \neq 0$ in M such that $xJ = 0$.*

Proof. Suppose $yJ \neq 0$ for all $y \neq 0$ in M. Choose arbitrarily $x \neq 0$ in M. There exists $a_1 \in J$ such that $x a_1 \neq 0$. In this way we can go on and obtain $a_2, a_3, \ldots$ in J such that $x a_1 a_2 \ldots \neq 0$ for all k. This contradicts the T-nilpotence of J on M. □

We apply this lemma to an arbitrary non-zero module M, and obtain a non-zero submodule L of M such that $LJ = 0$. Then L is a module over the right semi-artinian ring A/J, and hence contains a simple A/J-module, which evidently is simple also as an A-module. Thus M has non-zero socle, and it follows that A is right semi-artinian. □

Examples

1. Semi-primary rings. Every semi-primary ring is right and left semi-artinian, as a consequence of Prop. 2.8. In particular, every right artinian ring is both right and left semi-artinian.

2. Group rings. Let A be a commutative semi-artinian ring and let G be an arbitrary group. The group ring $A[G]$ is then right and left semi-artinian. For let M be a non-zero right $A[G]$-module. M contains a simple A-module S, and it is easy to verify that $S[G] = \{\sum x_i g_i \mid x_i \in S, g_i \in G\}$ is a simple $A[G]$-module contained in M.

3. Abelian groups. The semi-artinian $\mathbb{Z}$-modules are precisely the torsion groups.

§3. Simple Torsion Theories

We wish to study the influence of each isomorphism class of simple modules on the class of all semi-artinian modules. A hereditary torsion class is *simple* if it is generated by a class of simple modules. It then consists entirely of semi-artinian modules. The converse also holds, as we have already remarked, i.e. every hereditary torsion class consisting of semi-artinian modules is simple.

Let Ω denote the set of isomorphism classes of simple right A-modules. For each subset Ω_1 of Ω we let $\mathscr{T}(\Omega_1)$ be the torsion class generated by simple modules with isomorphism class belonging to Ω_1. (For brevity we will often abuse language by saying that a simple module S belongs to Ω_1 when the isomorphism class of S belongs to Ω_1.) The Ω_1-*socle* of a module M is the sum of all simple submodules of M belonging to Ω_1, and it is denoted by $s_{\Omega_1}(M)$. From s_{Ω_1} we obtain a radical $\bar{s}_{\Omega_1}$ in the way described in §2 for $s = s_{\Omega}$. Prop. VI.2.5 gives:

Proposition 3.1. *$M \in \mathscr{T}(\Omega_1)$ if and only if $s_{\Omega_1}(M/L) \neq 0$ for every proper submodule L of M.*

Proposition 3.2. *$\mathscr{T}(\Omega_1)$ consists of those semi-artinian modules for which each composition factor belongs to Ω_1.*

Proof. $\mathscr{T}(\Omega_1)$ is closed under quotient modules and submodules, so $M \in \mathscr{T}(\Omega_1)$ implies that the socle of $M/s_\alpha(M)$ consists of simple modules in Ω_1. Thus every composition factor of $M \in \mathscr{T}(\Omega_1)$ belongs to Ω_1. Conversely, if M is semi-artinian and all its composition factors are in Ω_1, then it is evident that $s_\alpha(M) = s_{\Omega_1,\alpha}(M)$ for each ordinal α, and therefore $\bar{s}_{\Omega_1}(M) = \bar{s}(M) = M$. Hence M belongs to $\mathscr{T}(\Omega_1)$. □

As an application of this result we can describe simple torsion classes defined in terms of a cogenerating injective module:

Proposition 3.3. *Let $\mathscr{T}$ be a simple torsion class cogenerated by an injective module E. Then $\mathscr{T} = \mathscr{T}(\Omega_1)$ with $\Omega_1 = \{\omega \in \Omega \mid s_\omega(E) = 0\}$.*

Proof. Suppose $M \in \mathscr{T}(\Omega_1)$. If $\varphi\colon M \to E$ is a non-zero homomorphism, let α be the first ordinal for which there exists $x \in s_\alpha(M)$ with $\varphi(x) \neq 0$. Then α cannot be a limit ordinal. Since φ is zero on $s_{\alpha-1}(M)$, φ induces a non-zero homomorphism $s_\alpha(M)/s_{\alpha-1}(M) \to E$. This contradicts the fact that $s_{\Omega_1}(E) = 0$, and we must have $\operatorname{Hom}(M, E) = 0$. Hence $M \in \mathscr{T}$.

Suppose on the other hand that $M \in \mathscr{T}$. By Prop. 3.2 we ought to show that every composition factor S of M belongs to Ω_1. So suppose S is a simple submodule of $M/s_\alpha(M)$. If S were isomorphic to a submodule of E, then the monomorphism $S \to E$ could be extended to a homomorphism $M/s_\alpha(M) \to E$, which would induce a non-zero homomorphism $M \to E$, and this is impossible when $M \in \mathscr{T}$. □

To the left exact preradical s_{Ω_1} there corresponds by Prop. VI.4.2 a topology, consisting of all finite intersections of maximal right ideals $\mathfrak{m}$ such that $A/\mathfrak{m} \in \omega$ for some $\omega \in \Omega_1$. This topology generates the Gabriel topology $\mathfrak{F}(\Omega_1)$ associated to $\mathscr{T}(\Omega_1)$.

The following result gives a necessary condition for a simple torsion class to be closed under direct products (Dlab [4]):

Proposition 3.4. *If every simple torsion class is closed under direct products, then A is a semi-local ring.*

Proof. For each $\Omega_1 \subset \Omega$ we let $J(\Omega_1)$ denote the intersection of all maximal right ideals in $\mathfrak{F}(\Omega_1)$. We have $J(\Omega_1) \in \mathfrak{F}(\Omega_1)$ by Prop. VI.6.12. In particular, the Jacobson radical $J(\Omega)$ is a member of $\mathfrak{F}(\Omega)$. It follows, using Cor. VI.3.5, that the socle of $A/J(A)$ is an essential submodule of $A/J(A)$. To prove that $A/J(A)$ is semi-simple, it suffices to show that it coincides with its socle, which we can write as $S/J(A)$ where S is a two-sided ideal of A containing $J(A)$. Suppose $S \neq A$. Then there exists a maximal right ideal $\mathfrak{m}$ of A such that $S \subset \mathfrak{m}$. Let $A/\mathfrak{m} \in \omega_0 \in \Omega$. Each maximal right ideal $\mathfrak{m}'$ in $\mathfrak{F}(\omega_0)$ has the form $(\mathfrak{m} : a)$ with $a \in A$, because $A/\mathfrak{m}' \in \omega_0$. The two-sidedness of S therefore implies that $S \subset J(\omega_0)$. Put $\Omega_1 = \{\omega \in \Omega \mid \omega \neq \omega_0\}$. Then

$$J(A) = J(\Omega) = J(\Omega_1) \cap J(\omega_0) \supset J(\Omega_1) \cap S \supset J(A),$$

and hence $J(A) = J(\Omega_1) \cap S$. Since $S/J(A)$ is essential in $A/J(A)$, this implies $J(A) = J(\Omega_1)$. But $J(\Omega_1) \in \mathfrak{F}(\Omega_1)$, and since $\mathfrak{m} \supset J(A)$, we get that $\mathfrak{m} \in \mathfrak{F}(\Omega_1)$. We must then have $A/\mathfrak{m} \in \Omega_1$, which is a contradiction. $\square$

The converse of this proposition does not hold, unless some further conditions are imposed on $J(A)$ (see Dlab [4], and cf. §6).

Example. Primary decomposition. It has been proved by Dickson [1] that if all simple torsion theories for A are stable, then one has

$$\bar{s}(M) = \bigoplus_{\omega \in \Omega} \bar{s}_\omega(M)$$

for every module M. This generalizes the classical primary decomposition theory for abelian torsion groups.

§4. Semi-Perfect Rings

If the ring A is decomposed as $A = \mathfrak{a}_1 \oplus \cdots \oplus \mathfrak{a}_n$ into a direct sum of right ideals $\mathfrak{a}_i$, then one can write $1 = e_1 + \cdots + e_n$ with $e_i \in \mathfrak{a}_i$. It is easily seen that $e_1, \ldots, e_n$ is a set of orthogonal idempotents, i.e. $e_i^2 = e_i$ and $e_i e_j = 0$ when $i \neq j$, and that $\mathfrak{a}_i = e_i A$. Conversely, if $e_1, \ldots, e_n$ is a set of orthogonal idempotents of A and $e = e_1 + \cdots + e_n$, then e is an idempotent and $eA = e_1 A \oplus \cdots \oplus e_n A$. A non-zero idempotent e is *primitive* if it cannot be written as a sum of two non-zero orthogonal idempotents, i.e. the module eA (or Ae) is indecomposable.

Suppose A is a right artinian ring. Then A/J is semi-simple, so $\bar{1} \in A/J$ can be written as a sum $\bar{1} = f_1 + \cdots + f_n$ of primitive orthogonal idempotents f_i in A/J. It is a fundamental property of artinian rings that such a decomposition can be lifted to A.

More generally, let $\mathfrak{a}$ be a two-sided ideal of an arbitrary ring A. We say that *idempotents can be lifted modulo* $\mathfrak{a}$ if for every idempotent $f \in A/\mathfrak{a}$ there exists an idempotent $e \in A$ such that $\bar{e} = f$. It turns out that it is then possible to lift entire families of orthogonal idempotents:

Proposition 4.1. *Assume idempotents can be lifted modulo an ideal* $\mathfrak{a} \subset J(A)$. *If* $f_1, \ldots, f_n$ *is a set of orthogonal idempotents of* $A/J(A)$, *then there exists a set* $e_1, \ldots, e_n$ *of orthogonal idempotents of* A *such that* $\bar{e}_i = f_i$.

Proof. We proceed by induction. Suppose we have already obtained $e_1, \ldots, e_{k-1}$. Then $e = e_1 + \cdots + e_{k-1}$ is an idempotent, and $\bar{e}$ is orthogonal to f_k. Let g be an idempotent of A such that $\bar{g}=f_k$. Then $ge\in\mathfrak{a}\subset J(A)$, so $1-ge$ is invertible. We now define e_k as

$$e_k=(1-e)(1-ge)^{-1}g(1-e).$$

Then $\bar{e}_k=\bar{g}=f_k$, and $e_k e_i=e_i e_k=0$ for $i<k$. We also obtain $e_k^2=e_k$ by noting that $g(1-e)=g(1-ge)$. So e_k has the required properties. □

Note that the induction argument shows more generally that any countable set of orthogonal idempotents can be lifted. The possibility of using idempotent lifting in the classical theory of artinian rings depends on the following general result:

Proposition 4.2. *If $\mathfrak{a}$ is a nil ideal, then idempotents can be lifted modulo $\mathfrak{a}$.*

Proof. Suppose f is an idempotent of $A/\mathfrak{a}$. Choose $u\in A$ such that $\bar{u}=f$. Then $u^2-u\in\mathfrak{a}$, and hence $(u^2-u)^r=0$ for some r. From this we obtain

$$0=u^r(1-u)^r=u^r-u^{r+1}g(u), \tag{1}$$

where $g=g(u)$ is a polynomial expression in u. Now put $e=u^r g^r$. By the use of (1) we get

$$\begin{aligned} e^2=u^{2r}g^{2r}&=u^{r-1}\cdot u^{r+1}g\cdot g^{2r-1}=u^{r-1}\cdot u^r\cdot g^{2r-1}=u^{2r-1}g^{2r-1}\\ &=\cdots=u^r g^r=e. \end{aligned}$$

We also have $\bar{e}=f$, because (1) gives $f=f\bar{g}=\cdots=f\bar{g}^r$. □

Definition (Bass [1]). The ring A is *semi-perfect* if A is semi-local and idempotents can be lifted modulo $J(A)$.

Proposition 4.3. *Let A be a semi-perfect ring. Then there exist primitive orthogonal idempotents $e_1,\dots,e_n$ such that $1=e_1+\cdots+e_n$. If also $e'_1,\dots,e'_m$ are primitive orthogonal idempotents such that $1=e'_1+\cdots+e'_m$, then $m=n$ and there exists a permutation π of $\{1,\dots,n\}$ such that $e_iA\cong e'_{\pi(i)}A$.*

Proof. Since A/J is a semi-simple ring, one can write $1\in A/J$ as $1=f_1+\cdots+f_n$, where $f_1,\dots,f_n$ are primitive orthogonal idempotents of A/J. They can be lifted to orthogonal idempotents $e_1,\dots,e_n$ of A. If we put $e=e_1+\cdots+e_n$, then $\bar{1}=\bar{e}$ implies $1-e\in J$, and hence $e=1$. (Note that J can never contain non-zero idempotents.) We also assert that each e_i is a primitive idempotent. Suppose $e_i=g+h$, where g and h are orthogonal idempotents. Then $f_i=\bar{e}_i=\bar{g}+\bar{h}$, and the primitivity of f_i implies that either $g\in J$ or $h\in J$, i.e. either $g=0$ or $h=0$.

To prove the unicity assertion, we shall use the Azumaya theorem (Prop. V.5.4), and for this it suffices to show that $\operatorname{Hom}_A(eA,eA)$ is a local ring for every primitive idempotent e. We do this in a number of steps.

Lemma 4.4. *If e is an idempotent of A, then there is a ring isomorphism*

$$\operatorname{Hom}_A(eA,eA)\cong eAe.$$

Proof. Every $\varphi\colon eA\to eA$ is uniquely determined by the element $\varphi(e)$, for which one can choose any $a\in eA$ such that $ex=0$ implies $ax=0$, which means that one can choose any $a\in eAe$. It is a routine matter to verify that $\varphi\mapsto\varphi(e)$ defines an isomorphism between the rings $\operatorname{Hom}_A(eA,eA)$ and eAe. □

Lemma 4.5. *If e is an idempotent of A, then*

$$J(eAe)=eAe\cap J(A)=eJ(A)e.$$

Proof. It is trivial that $eJ(A)e\subset eAe\cap J(A)$, and if $a\in eAe\cap J(A)$, then $a=eae\in eJ(A)e$. Thus $eAe\cap J(A)=eJ(A)e$. Suppose $a\in J(eAe)$, and let b be an arbitrary element of A. The element $e-abe$ has a right inverse c in eAe by Prop. 1.4, which implies that $e=(e-abe)c=(1-ab)c$. Then

$$\begin{aligned}(1-ab)(1-e+c)(1+ab(1-e))&=(1-ab-e+abe+e)(1+ab(1-e))\\ &=(1-ab(1-e))(1+ab(1-e))=1-ab(1-e)ab(1-e)=1\end{aligned}$$

since $(1-e)a=0$. So $1-ab$ has a right inverse in A, and $a\in J(A)$ by Prop. 1.4. Hence $J(eAe)\subset eAe\cap J(A)$. To show the reverse inclusion it suffices to show that $e-a$ is right invertible in eAe for each $a\in eAe\cap J(A)$. But there exists $b\in A$ such that $(1-a)b=1$, which immediately gives $(e-a)ebe=e$. ▯

Lemma 4.6. *Let A be a semi-local ring and e an idempotent of A. Then:*

(i) *eAe is a semi-local ring.*

(ii) *If A is semi-perfect and e is primitive, then eAe is a local ring.*

Proof. (i) If $\mathfrak{a}$ is a right ideal of eAe such that $\mathfrak{a}\supset J(eAe)$, then it follows that $(\mathfrak{a}A+J(A))\cap eAe=\mathfrak{a}$, for if $\sum a_i b_i+c=a\in eAe$ with $a_i\in\mathfrak{a}$, $b_i\in A$ and $c\in J(A)$, then

$$a=eae=\sum a_i b_i e+ece\in\mathfrak{a}+eJ(A)e\subset\mathfrak{a}$$

by Lemma 4.4. Since A satisfies DCC on right ideals containing $J(A)$, it follows from this formula that eAe satisfies DCC on right ideals containing $J(eAe)$, and hence $eAe/J(eAe)$ is semi-simple (Prop. 1.9).

(ii) From Lemma 4.5 we see that there is a monomorphism $eAe/J(eAe)\to A/J$ of abelian groups. It follows from the semi-perfectness of A that $\bar{e}$ must be a primitive idempotent of A/J, and it is then clear that $\bar{e}$ is primitive also as an idempotent of $eAe/J(eAe)$. But if the identity element of a semi-simple ring is a primitive idempotent, then the ring must be a skew-field. Hence $eAe/J(eAe)$ is a skew-field, and thus eAe is local. ▯

This also completes the proof of Prop. 4.3. The unicity assertion of Prop. 4.3 can be strengthened somewhat, because one can show that in the context of Prop 4.3 there exists an inner automorphism of A mapping $\{e_1,\ldots,e_n\}$ onto $\{e'_1,\ldots,e'_n\}$ (see e.g. Lambek [4], §3.7).

Examples

1. Semi-artinian rings. When A is a right semi-artinian ring, $J(A)$ is a nil ideal by Cor. 2.7, so idempotents may be lifted modulo $J(A)$.

2. Semi-primary rings. Every semi-primary ring, and thus in particular every right or left artinian ring, is semi-perfect.

3. Local rings. Every local ring is semi-perfect. In fact, a ring is local if and only if it is semi-perfect and 1 is a primitive idempotent.

4. Commutative rings. Since a commutative semi-simple ring is a finite product of fields, every commutative semi-perfect ring is a finite product of local rings.

§5. Perfect Rings

We have noticed that a semi-primary ring is both semi-perfect and semi-artinian. The class of semi-perfect and right semi-artinian rings thus strictly contains the class of right artinian rings, but its members share several of the crucial properties of artinian rings, and therefore it should be of interest to study this class of rings.

Proposition 5.1. *The following properties of a ring A are equivalent:*
(a) *A is right semi-artinian and semi-local.*
(b) *A is right semi-artinian and semi-perfect.*
(c) *J is left T-nilpotent and A is semi-local.*
(d) *A is right semi-artinian and contains no infinite set of orthogonal idempotents.*

Proof. (a) $\Rightarrow$ (b) was remarked in Example 4.1.
(b) $\Rightarrow$ (c) follows directly from Cor. 2.7.
(c) $\Rightarrow$ (d): A is right semi-artinian by Prop. 2.8. It is clear that A cannot contain any infinite set of orthogonal idempotents when A/J is semi-simple.
(d) $\Rightarrow$ (a): Since A is right semi-artinian, countable sets of orthogonal idempotents may be lifted modulo J, and it follows that also A/J cannot contain any infinite set of orthogonal idempotents. It remains to show that A/J is semi-simple. Since the Jacobson radical of A/J is zero, every minimal right ideal of A/J is the complement of a maximal right ideal and is therefore generated by an idempotent. Since there are only finite sets of orthogonal idempotents in A/J, it follows that the socle of A/J is the direct sum of finitely many minimal right ideals, and so the socle itself is a direct summand of A/J. But since A/J is semi-artinian, A/J is equal to its socle and thus is a semi-simple ring. □

Definition (Bass [1]). The ring A is *left perfect* if it satisfies the conditions of Prop. 5.1.

The reason for using the prefix "left" is that these rings may also be characterized by the property that every left module has a projective cover (see Example V.2.2 for definition), but this is a property which will not be needed in the sequel.

A right noetherian ring which is either left or right perfect, must be right artinian. In fact, a more general result holds:

Proposition 5.2. *If A is right noetherian and right or left semi-artinian, then A is right artinian.*

Proof. We have noticed already in §2 that a right noetherian and right semi-artinian ring must be right artinian. If A is right noetherian and left semi-artinian, then A is semilocal by Prop. 5.1, and the following Proposition implies that A is semi-primary and hence right artinian. □

Proposition 5.3 (Faith [2]). *If A satisfies* ACC *on right annihilators and is left semi-artinian, then J is nilpotent.*

Proof. Consider the ascending chain of right annihilators $r(J)\subset r(J^2)\subset\cdots$. By hypothesis we have $r(J^n)=r(J^{n+1})$ for some n. If $r(J^n)\neq A$, the left A-module $A/r(J^n)$ has non-zero socle, which is of the form $\mathfrak{a}/r(J^n)$ for some left ideal $\mathfrak{a}\supset r(J^n)$.

But J annihilates the semi-simple module $\mathfrak{a}/r(J^n)$, so $J\,\mathfrak{a} \subset r(J^n)$ and hence $\mathfrak{a} \subset r(J^{n+1}) = r(J^n)$. But then $\mathfrak{a} = r(J^n)$, which is a contradiction, and we must conclude that $r(J^n) = A$ and hence $J^n = 0$. ▯

Corollary 5.4. *If A is a right perfect ring and satisfies* ACC *on right annihilators, then A is semi-primary.*

Perfect rings may also be characterized in terms of minimum conditions on certain ideals. This is due to Bass [1], who proved (a) $\Leftrightarrow$ (b) of the following proposition, and Björk [1], who established (b) $\Rightarrow$ (c).

Proposition 5.5. *The following properties of a ring A are equivalent:*
(a) *A is left perfect.*
(b) *A satisfies* DCC *on principal right ideals.*
(c) *A satisfies* DCC *on finitely generated right ideals.*

Proof. (c) $\Rightarrow$ (b) is trivial, and also (b) $\Rightarrow$ (a) is easy: an infinite set of orthogonal idempotents in A would give rise to an infinite descending chain of principal right ideals. It therefore only remains to show that every non-zero module M has a simple submodule. We may assume M cyclic, say $M = A/\mathfrak{a}$ for a right ideal $\mathfrak{a}$. Choose $a \in A$ so that aA is minimal with $aA \not\subset \mathfrak{a}$. Then $(aA + \mathfrak{a})/\mathfrak{a}$ is a simple submodule of $A/\mathfrak{a}$.

(a) $\Rightarrow$ (c): We prove the following more general assertion:

Proposition 5.6. *If A is a left perfect ring, then every right A-module satisfies* DCC *on finitely generated submodules.*

Proof. Let M be an arbitrary right A-module. With the use of Zorn's lemma we can find a submodule L of M which is maximal with respect to the property that DCC holds for finitely generated submodules of L. We assume $L \neq M$ and seek to deduce a contradiction from this. By Lemma 2.9 there exists $x \in M$ such that $x \notin L$ and $xJ \subset L$. Since A is semi-perfect, we can write $1 = e_1 + \cdots + e_r$, where e_i are orthogonal primitive idempotents. We must have $x e_i \notin L$ for at least one e_i, say $x e_1 \notin L$. Replacing x by $x e_1$, we may assume $x e_1 = x$. We shall now obtain a contradiction by proving that $L + xA$ satisfies DCC on finitely generated submodules. Let $M_1 \supset M_2 \supset \cdots$ be a strictly decreasing sequence of finitely generated submodules of $L + xA$. We must then of course have $M_n \not\subset L$ for each n.

By induction we will construct finitely generated submodules $L_1 \supset L_2 \supset \cdots$ contained in L, and a sequence of elements $x_1, x_2, \ldots$, such that $x_n \in L_{n-1}$ and

$$M_n = (x + x_1 + \cdots + x_n)\, e_1 A + L_n.$$

We use $L_0 = L$ to start off the induction. Since $M_n \not\subset L$, there exists $y_n \in M_n$ such that $y_n \notin L$. Since $y_n \in M_{n-1}$, we have by the induction hypothesis that

$$y_n = (x + x_1 + \cdots + x_n)\, e_1\, a + x'$$

for some $a \in A$ and $x' \in L_{n-1}$. Since $xJ \subset L$ and $y_n \notin L$, we must have $e_1 a \notin J$. We now use:

Lemma 5.7. *If A is a left perfect ring and $\mathfrak{a}$ is a right ideal which is not contained in J, then $\mathfrak{a}$ contains a non-zero idempotent.*

Proof. The image of $\mathfrak{a}$ in the semi-simple ring A/J contains a non-zero idempotent $\bar{u}$, where $u \in \mathfrak{a}$. A glance at the proof of Prop. 4.2 reveals that the idempotent e of A which lifts $\bar{u}$ can be constructed as a polynomial expression in u with integer coefficients, and therefore $e \in \mathfrak{a}$. □

We apply the Lemma to the right ideal $e_1 a A$, which thus contains a non-zero idempotent f. Then e_1 is the sum of two orthogonal idempotents $f e_1$ and $e_1 - f e_1$, and since e_1 is primitive, we must have $e_1 = f e_1$. This implies that $e_1 = e_1 a b$ for some $b \in A$. M_n contains the element

$$y_n b = (x + x_1 + \cdots + x_{n-1}) e_1 + x' b e_1 = (x + x_1 + \cdots + x_n) e_1$$

with $x_n = x' b \in L_{n-1}$. Since $M_n \subset M_{n-1}$, an arbitrary element $u \in M_n$ can be expressed as

$$u = (x + x_1 + \cdots + x_{n-1}) e_1 a' + u'$$

with $a' \in A$ and $u' \in L_{n-1}$, and this we rewrite as

$$u = (x + x_1 + \cdots + x_{n-1} + x_n) e_1 a' + u' - x_n e_1 a'$$

where the first term is in M_n and $u' - x_n e_1 a' \in L_{n-1}$. We define L_n to be the submodule of M generated by the elements $u' - x_n e_1 a'$ corresponding to a finite family of generators u for M_n. Then $L_n \subset L_{n-1}$ and $M_n = (x + x_1 + \cdots + x_n) e_1 A + L_n$.

Since L was assumed to satisfy DCC on finitely generated submodules, we have $L_n = L_{n+1}$ for some n. It clearly follows that also $M_n = M_{n+1}$, and we have arrived at the desired contradiction. □

Examples

1. Commutative rings. A commutative ring is perfect if and only if it is a finite product of local rings whose maximal ideals are T-nilpotent.

2. Group rings. Let A be a commutative perfect ring and let G be a finite group. The group ring $A[G]$ is left and right perfect. In fact, every finitely generated right ideal of $A[G]$ is a finitely generated submodule of the right A-module $A[G]$, so we can apply Prop. 5.6.

§6. Hereditary Torsion Theories for a Perfect Ring

Let A be a semi-perfect ring. We can write 1 as a sum of orthogonal primitive idempotents

$$1 = e_{11} + \cdots + e_{1k_1} + \cdots + e_{n1} + \cdots + e_{nk_n}$$

where $e_{ij} A \cong e_{rs} A$ if and only if $i = r$. Put

$$e_i = e_{i1} + \cdots + e_{ik_i}$$

for each $i = 1, \ldots, n$. Then $\bar{e}_i$ is a central idempotent in the semi-simple ring A/J, and it generates one of the simple components of that ring. It follows that for each simple A-module S there is a unique e_i such that $S e_i \neq 0$, and we say that this e_i is *associated* to S.

Let Ω_1 be a subset of the (finite) set of isomorphism classes of simple modules. Let e_{Ω_1} be the sum of all idempotents e_i associated to simple modules of isomorphism class not belonging to Ω_1. We then have the following description of the torsion class $\mathcal{T}(\Omega_1)$ (Alin and Armendariz [1]):

Proposition 6.1. *Let A be a semi-perfect ring, and Ω_1 a set of isomorphism classes of simple modules. A semi-artinian module M belongs to $\mathcal{T}(\Omega_1)$ if and only if $M e_{\Omega_1}=0$.*

Proof. Suppose $M e_{\Omega_1}=0$. Every composition factor S of M then clearly also satisfies $S e_{\Omega_1}=0$, and by definition of e_{Ω_1} it follows that the isomorphism class of S belongs to Ω_1. Consequently $M\in\mathcal{T}(\Omega_1)$ by Prop. 3.2.

Suppose on the other hand that $M\in\mathcal{T}(\Omega_1)$. Using transfinite induction, it clearly suffices to show that $s_\alpha(M)e_{\Omega_1}=0$ implies $s_{\alpha+1}(M)e_{\Omega_1}=0$. Since $s_{\alpha+1}(M)/s_\alpha(M)$ consists of simple modules annihilated by e_{Ω_1}, we have $s_{\alpha+1}(M)\,e_{\Omega_1}\subset s_\alpha(M)$. But e_{Ω_1} is an idempotent, so it follows that $s_{\alpha+1}(M)\,e_{\Omega_1}=s_\alpha(M)\,e_{\Omega_1}=0$. □

When A is both semi-perfect and right semi-artinian, i.e. A is left perfect, we obtain:

Proposition 6.2. *Suppose A is a left perfect ring and $\Omega_1\subset\Omega$. Then:*
(i) *A module M belongs to $\mathcal{T}(\Omega_1)$ if and only if $M e_{\Omega_1}=0$.*
(ii) *$\mathfrak{F}(\Omega_1)$ consists of all right ideals containing $A e_{\Omega_1} A$.*

Corollary 6.3. *The ring A is left perfect if and only if every hereditary torsion class is simple and closed under direct products.*

Proof. If A is left perfect, then A is right semi-artinian, so every hereditary torsion class is simple, and is closed under direct products by the proposition. The converse statement follows by the use of Props. 3.4 and 5.1. □

From Prop. 6.2(ii) we also obtain the following result of Michler [2]:

Corollary 6.4. *Suppose A is a left perfect ring. Then:*
(i) *Every idempotent two-sided ideal of A has the form AeA, where e is an idempotent of A which is central modulo J.*
(ii) *There is a bijective correspondence between idempotent two-sided ideals of A and two-sided ideals of A/J.*

Here an idempotent e in A is called "central modulo J" if its image in A/J is a central idempotent.

Examples

1. Stability under direct products. It may happen that the class of semi-artinian modules is closed under direct products, although not every simple torsion class has this property. In fact, this occurs for every right semi-artinian ring which is not left perfect.

2. Dense right ideals of a perfect ring. Let A be a left perfect ring. Then there exists a smallest dense right ideal of A, which by Prop. 3.3 has the form $A e_{\Omega_1} A$, where $\Omega_1=\{\omega\in\Omega\,|\,s_\omega(A)=0\}$.

Exercises

§1

1. Let M be a module. Show that:

(i) If L is a finitely generated submodule of M and $L \subset \mathrm{Rad}(M)$, then L is small in M (Example V.2.2).

(ii) $\mathrm{Rad}(M)$ is the sum of all small submodules of M.

2. Show that if M is a module, then $\mathrm{Rad}(M)=0$ if and only if M is a submodule of a direct product of simple modules.

3. Show that if M is a module of finite length n, then $MJ^n=0$.

4. (i) Show that if $a \in A$ and $r(a) \subset \mathfrak{a}$ for a right ideal $\mathfrak{a}$, then $E(A/\mathfrak{a})\,a \neq 0$.

(ii) Show that if $E(S)$ has finite length for every simple module S, then J is a nil ideal and $\bigcap_n J^n=0$, and that J is nilpotent if furthermore there are only finitely many isomorphism classes of simple modules. (Hint: if $a \in J$ and $\bigcup r(a^i)$ is contained in a maximal right ideal $\mathfrak{m}$, then $E(A/\mathfrak{m})\,a^i \neq 0$ for all i, contradicting Exercise 3.)

5. Let A be a right artinian ring. Show that a module M is faithful (Exercise I.3) if and only if A is a submodule of a finite direct sum of copies of M.

6. Show that if A is right artinian and there exists a faithful simple A-module, then A is a simple ring.

7. Let K be a field, and let F be an extension field of K such that $[F:K]=\infty$. Show that the ring $\begin{pmatrix} K & F \\ 0 & F \end{pmatrix}$ is right artinian with $J^2=0$, but that it is not left artinian. Also show that it is right and left hereditary.

8. Let M be a non-singular, artinian and coirreducible module. Show that $\mathrm{End}_A(M)$ is a skew-field.

§2

9. Show that a semi-artinian module is non-singular if and only if its socle is a projective module.

10. For an arbitrary ring A and an A-module M we define the set $\mathrm{Ass}(M)$ of associated prime ideals in the same way as was done over a noetherian ring in Chap. VII. Show that:

(i) If S is a simple module, then $\mathrm{Ass}(S)=\{\mathrm{Ann}(S)\}$ and $\mathrm{Ann}(S)$ is the intersection of all maximal right ideals $\mathfrak{m}$ such that $A/\mathfrak{m} \cong S$.

(ii) If M is a non-zero semi-artinian module, then $\mathrm{Ass}(M) \neq \emptyset$.

(iii) If M is a non-zero artinian module, then $\mathrm{Ass}(M)$ is finite.

11. Let A be a right semi-artinian ring. Show that:

(i) If A has no zero-divisors, then A is a skew field.

(ii) If $\mathfrak{a}$ is a two-sided ideal of A, then $A/\mathfrak{a}$ is a right semi-artinian ring.

12. Let $A_i=\mathbb{Z}/p\mathbb{Z}$ for $i=1,2,\ldots$, and let A be the ring $A=\prod_{i=1}^{\infty} A_i$. Consider the subring B of A generated by $\bigoplus_{i=1}^{\infty} A_i$ together with $1 \in A$. Show that B is a semi-artinian ring with $J(B)=0$.

13. Show that if A is a right semi-artinian ring, then $Z_r(A)=l(S)$, where $Z_r(A)$ is the right singular ideal and S is the right socle of A.

§3

14. Show that $\bigoplus_{\omega \in \Omega} \bar{s}_\omega(M)$ is an essential submodule of $\bar{s}_\Omega(M)$.

§4

15. Show that a non-zero idempotent in A is primitive if and only if the ring eAe has no other idempotents but 0 and e.

16. Show that if e is a primitive idempotent in a regular ring A, then eAe is a skew-field.

17. Let e and f be idempotents in A. Show that:

(i) The A-modules eA and fA are isomorphic if and only if there exist $x, y \in A$ such that $xy=e$ and $yx=f$ (e and f are then called *equivalent* idempotents).

(ii) If two central idempotents are equivalent, then they are equal.

18. Let A be a semi-perfect ring. Show that:

(i) If M is a finitely generated module such that $MJ=0$, then $M \cong P/PJ$ for a projective module P.

(ii) Every finitely generated module M has a projective cover (Example V.2.2). (Hint: apply (i) to the module M/MJ.)

(iii) If $(\mathscr{T}, \mathscr{F})$ is a hereditary torsion theory for **Mod**-A, then $\mathscr{T}$ is closed under projective covers if and only if $\mathscr{F}$ is closed under quotient modules.

§5

19. Let K be the field consisting of two elements and let $K\{X_0, X_1, \ldots\}$ denote the polynomial ring over K in a countably infinite number of *non-commuting* variables. Let $\mathfrak{a}$ be the two-sided ideal generated by the elements of the form $X_n X_{i_1} \cdot \cdots \cdot X_{i_n}$, where $n = 1, 2, \ldots$ and i_k are arbitrary, together with the element X_0^2. Show that:

(i) $A = K\{X_0, X_1, \ldots\}/\mathfrak{a}$ satisfies DCC on principal right ideals (i.e. A is a left perfect ring).

(ii) $\overline{X}_0 \in A$ belongs to $J(A)$, but the right ideal $\overline{X}_0 A$ is not nilpotent.

20. Let F be a field, and let $M_\omega(F)$ consist of all matrices with a countably infinite number of rows and columns, but where each row has only finitely many non-zero entries. Let A be the subring of $M_\omega(F)$ generated by the set N of all strictly lower triangular matrices together with the identity matrix. Show that:

(i) $N = J(A)$ and $A/J(A) \cong F$.

(ii) $J(A)$ is left T-nilpotent but not right T-nilpotent (thus A is left perfect but not right perfect).

§6

21. Let A be a left perfect ring with minimal dense right ideal $\mathfrak{d}$. Show that:

(i) $\mathfrak{d}$ is the smallest two-sided ideal with zero left annihilator.

(ii) If S is the right socle of A, then $S \subset \mathfrak{d}$, and $S = \mathfrak{d}$ holds if and only if A is right non-singular. (Cf. Exercise 13.)

Chapter IX. Rings and Modules of Quotients

The development of a general theory of rings of quotients may be said to have started with the construction by Johnson [1] of the maximal ring of quotients $Q_{\max}$ of a non-singular ring A. This was done before the theory of injective envelopes had become available, but it was later proved that $Q_{\max}$ could be used as an injective envelope of the ring A. The maximal ring of quotients of an arbitrary ring A was defined by Utumi [1] and studied by Findlay and Lambek [1]. In particular, Lambek [2] proved that it could be interpreted as the bicommutator of the injective envelope of A.

Rings and modules of quotients with respect to an additive topology $\mathfrak{F}$ were introduced by Gabriel [2] (also in a series of exercises in Bourbaki [2]). Our presentation of the theory in § 1 will closely follow Gabriel's approach in Bourbaki. This construction of modules of quotients is a special instance of a more general method of constructing associated sheaves of presheaves for an additive Grothendieck topology (cf. Exposé II in Artin-Grothendieck-Verdier [1] for the corresponding non-additive theory). The results of § 1 will then be interpreted in terms of injective envelopes in §§ 2 and 3.

§ 1. Construction of Modules of Quotients

Let $\mathfrak{F}$ be a Gabriel topology of right ideals on the ring A. For each right module M we will define its module of quotients with respect to $\mathfrak{F}$. This we do in two steps. The first step we take is to define

$$M_{(\mathfrak{F})} = \varinjlim \operatorname{Hom}_A(\mathfrak{a}, M), \qquad \mathfrak{a} \in \mathfrak{F},$$

where the direct limit is taken over the downwards directed family $\mathfrak{F}$ of right ideals. Every element in $M_{(\mathfrak{F})}$ is thus represented by a homomorphism $\xi\colon \mathfrak{a} \to M$ for some $\mathfrak{a} \in \mathfrak{F}$, with the understanding that ξ represents the same element in $M_{(\mathfrak{F})}$ as does $\eta\colon \mathfrak{b} \to M$ if and only if ξ and η coincide on some $\mathfrak{c} \in \mathfrak{F}$ such that $\mathfrak{c} \subset \mathfrak{a} \cap \mathfrak{b}$. We want to give $A_{(\mathfrak{F})}$ the structure of a ring and $M_{(\mathfrak{F})}$ that of a right $A_{(\mathfrak{F})}$-module. For this we need:

Lemma 1.1. *If* $\mathfrak{a}, \mathfrak{b} \in \mathfrak{F}$ *and* $\alpha\colon \mathfrak{a} \to A$ *is a homomorphism, then* $\alpha^{-1}(\mathfrak{b}) \in \mathfrak{F}$.

Proof. For each $a \in \mathfrak{a}$ we have $(\alpha^{-1}(\mathfrak{b}) : a) = \{b \mid \alpha(a\,b) \in \mathfrak{b}\} = (\mathfrak{b} : \alpha(a)) \in \mathfrak{F}$ by $T3$, so $\alpha^{-1}(\mathfrak{b}) \in \mathfrak{F}$ by $T4$. ▯

We define a pairing $M_{(\mathfrak{F})} \times A_{(\mathfrak{F})} \to M_{(\mathfrak{F})}$ as follows: suppose $x \in M_{(\mathfrak{F})}$, $a \in A_{(\mathfrak{F})}$ are represented by $\xi\colon \mathfrak{b} \to M$ and $\alpha\colon \mathfrak{a} \to A$; we then define $x\,a \in M_{(\mathfrak{F})}$ to be represented by the composed map

$$\alpha^{-1}(\mathfrak{b}) \to \mathfrak{b} \to M,$$

using Lemma 1.1. It is easy to see that $x\,a$ is well-defined, i.e. is independent of the choice of the representing homomorphisms ξ and α. One also easily verifies that the pairing $M_{(\mathfrak{F})} \times A_{(\mathfrak{F})} \to M_{(\mathfrak{F})}$ is biadditive. When $M = A$, this makes $A_{(\mathfrak{F})}$ into a ring, and in the general case it makes $M_{(\mathfrak{F})}$ into an $A_{(\mathfrak{F})}$-module.

There is a canonical homomorphism

$$\varphi_M\colon\ M \simeq \operatorname{Hom}_A(A, M) \to \varinjlim \operatorname{Hom}_A(\mathfrak{a}, M) = M_{(\mathfrak{F})}.$$

In particular, φ_A is a ring homomorphism. By pullback along φ_A we may consider each $M_{(\mathfrak{F})}$ as an A-module. φ_M is then A-linear. The assignment $M \mapsto M_{(\mathfrak{F})}$ is a functor **Mod**-$A \to$ **Mod**-$A_{(\mathfrak{F})}$ which is left exact, because Hom is left exact and direct limits are exact in **Mod**-A. We denote by L: **Mod**-$A \to$ **Mod**-A this functor followed by the forgetful functor **Mod**-$A_{(\mathfrak{F})} \to$ **Mod**-A.

Let t denote the torsion radical associated with the Gabriel topology $\mathfrak{F}$.

Lemma 1.2. Ker $\varphi_M = t(M)$.

Proof. If $\varphi_M(x) = 0$, then x goes to zero already in some $\operatorname{Hom}(\mathfrak{a}, M)$, $\mathfrak{a} \in \mathfrak{F}$. This means that the map $a \mapsto x\,a$ is zero on $\mathfrak{a}$, i.e. $x\,\mathfrak{a} = 0$ and $x \in t(M)$. This argument is reversible. $\square$

Lemma 1.3. *M is an $\mathfrak{F}$-torsion module if and only if $M_{(\mathfrak{F})} = 0$.*

Proof. If $M_{(\mathfrak{F})} = 0$, then $t(M) = \operatorname{Ker} \varphi_M = M$ by Lemma 1.2. Suppose on the other hand that M is a torsion module. Let $x \in M_{(\mathfrak{F})}$ be represented by $\xi\colon \mathfrak{b} \to M$ with $\mathfrak{b} \in \mathfrak{F}$. If we can show that Ker $\xi \in \mathfrak{F}$, it will follow that $x = 0$. For each $b \in \mathfrak{b}$ there exists $\mathfrak{a}_b \in \mathfrak{F}$ such that $\xi(b)\,\mathfrak{a}_b = 0$. Put $\mathfrak{c} = \sum_{b \in \mathfrak{b}} b\,\mathfrak{a}_b$. Then $\mathfrak{c} \subset \operatorname{Ker} \xi$, and for each $b \in \mathfrak{b}$ we have $(\mathfrak{c} : b) \supset \mathfrak{a}_b$, so $(\mathfrak{c} : b) \in \mathfrak{F}$ and it follows that $\mathfrak{c} \in \mathfrak{F}$ by $T4$. Hence Ker $\xi \in \mathfrak{F}$. $\square$

Lemma 1.4. *If $x \in M_{(\mathfrak{F})}$ is represented by $\xi\colon \mathfrak{a} \to M$ with $\mathfrak{a} \in \mathfrak{F}$, then the diagram*

$$\begin{array}{ccc} \mathfrak{a} & \longrightarrow & A \\ {\scriptstyle \xi}\downarrow & & \downarrow{\scriptstyle \beta} \\ M & \xrightarrow[\varphi_M]{} & M_{(\mathfrak{F})} \end{array}$$

commutes, where $\beta(a) = x\,a$.

Proof. If $a \in \mathfrak{a}$, then $x\,a$ is represented by the composed morphism $A = (\mathfrak{a} : a) \to \mathfrak{a} \to M$ given by $b \mapsto \xi(a\,b) = \xi(a)\,b$. Hence $x\,a = \varphi_M\,\xi(a)$. $\square$

An immediate consequence of this is:

Lemma 1.5. Coker φ_M *is an $\mathfrak{F}$-torsion module.*

The second step in our construction of modules of quotients is to apply the functor L once more to $M_{(\mathfrak{F})}$. This results in a ring $A_{\mathfrak{F}}$ (called the *ring of quotients* of A with respect to $\mathfrak{F}$) and an $A_{\mathfrak{F}}$-module $M_{\mathfrak{F}}$ for each A-module M. The description of all this is considerably simplified by:

Lemma 1.6. *L carries the monomorphism $M/t(M) \to M_{(\mathfrak{F})}$ into an isomorphism $(M/t(M))_{(\mathfrak{F})} \cong M_{\mathfrak{F}}$.*

Proof. Apply the left exact functor L to the exact sequence

$$0 \to M/t(M) \to M_{(\mathfrak{F})} \to \operatorname{Coker} \varphi_M \to 0$$

and use Lemmas 1.5 and 1.3. □

We have thus obtained the formula

$$M_{\mathfrak{F}} = \varinjlim \operatorname{Hom}_A(\mathfrak{a}, M/t(M)), \qquad \mathfrak{a} \in \mathfrak{F},$$

for the module of quotients. One verifies that the ring structure of $A_{\mathfrak{F}}$ and the module structure of $M_{\mathfrak{F}}$ are given by the following pairing $M_{\mathfrak{F}} \times A_{\mathfrak{F}} \to M_{\mathfrak{F}}$:

let $x \in M_{\mathfrak{F}}$ be represented by $\xi\colon \mathfrak{b} \to M/t(M)$,

$a \in A_{\mathfrak{F}}$ be represented by $\alpha\colon \mathfrak{a} \to A/t(A)$;

ξ induces $\mathfrak{b}/t(\mathfrak{b}) \to M/t(M)$, and $\mathfrak{b}/t(\mathfrak{b})$ is a submodule of $A/t(A)$ by the left exactness of t; $x\,a \in M_{\mathfrak{F}}$ is represented by

$$\alpha^{-1}(\mathfrak{b}/t(\mathfrak{b})) \to \mathfrak{b}/t(\mathfrak{b}) \to M/t(M).$$

For each homomorphism $f\colon M \to N$ in **Mod**-A one gets in a natural way an induced homomorphism $f_{\mathfrak{F}}\colon M_{\mathfrak{F}} \to N_{\mathfrak{F}}$ in **Mod**-$A_{\mathfrak{F}}$. As a result we have a functor

$$q\colon \textbf{Mod-}A \to \textbf{Mod-}A_{\mathfrak{F}}$$

where $q(M) = M_{\mathfrak{F}}$ and $q(f) = f_{\mathfrak{F}}$. There are canonical homomorphisms $\psi_M\colon M \to M_{\mathfrak{F}}$ of A-modules; in particular $\psi_A\colon A \to A_{\mathfrak{F}}$ is a ring homomorphism. For each $f \in \operatorname{Hom}_A(M, N)$ there is a commutative diagram

$$\begin{array}{ccc} M & \xrightarrow{f} & N \\ {\scriptstyle \psi_M}\downarrow & & \downarrow{\scriptstyle \psi_N} \\ M_{\mathfrak{F}} & \xrightarrow[f_{\mathfrak{F}}]{} & N_{\mathfrak{F}} \end{array}$$

which means that ψ may be considered as a natural transformation of functors. Also note that $\operatorname{Ker} \psi_M = t(M)$ and that $\operatorname{Coker} \psi_M$ likewise is a torsion module.

When the torsion theory is stable (Chap. VI, §7), the formula for $M_{\mathfrak{F}}$ simplifies somewhat:

Proposition 1.7. *When $\mathfrak{F}$-torsion is stable, one has $M_{\mathfrak{F}} = \varinjlim \operatorname{Hom}_A(\mathfrak{a}, M)$ for every module M.*

Proof. Since the direct limit functor is exact, the sequence

$$0 \to t(M) \to M \to M/t(M) \to 0$$

induces an exact sequence

$$0 \to \varinjlim \operatorname{Hom}(\mathfrak{a}, t(M)) \to \varinjlim \operatorname{Hom}(\mathfrak{a}, M) \to \varinjlim \operatorname{Hom}(\mathfrak{a}, M/t(M)) \to \varinjlim \operatorname{Ext}^1(\mathfrak{a}, t(M)).$$

The first term is zero by Lemma 1.3. If E is an injective envelope of $t(M)$, then E

is a torsion module by hypothesis. The sequence

$$0 \to t(M) \to E \to E/t(M) \to 0$$

induces the exact sequence

$$\varinjlim \operatorname{Hom}(\mathfrak{a}, E/t(M)) \to \varinjlim \operatorname{Ext}^1(\mathfrak{a}, t(M)) \to 0$$

where the first term is zero, again by Lemma 1.3. Hence the last term of the long exact sequence is zero, and the middle arrow is an isomorphism. □

We wish to give a characterization of those A-modules which may be considered as modules of quotients with respect to $\mathfrak{F}$ (similar to the characterization of modules of fractions in Prop. II.3.7). For this purpose we introduce:

Definition. An A-module M is *$\mathfrak{F}$-closed* (resp. *$\mathfrak{F}$-injective*) if the canonical homomorphisms

$$M \cong \operatorname{Hom}_A(A, M) \to \operatorname{Hom}_A(\mathfrak{a}, M)$$

are isomorphisms (resp. epimorphisms) for all $\mathfrak{a} \in \mathfrak{F}$.

Thus M is $\mathfrak{F}$-closed if and only if M is both $\mathfrak{F}$-torsion-free and $\mathfrak{F}$-injective, and in that case there is an isomorphism $\psi_M: M \to M_{\mathfrak{F}}$. Conversely, every module of quotients $M_{\mathfrak{F}}$ is $\mathfrak{F}$-closed (and is therefore its own module of quotients):

Proposition 1.8. *$M_{\mathfrak{F}}$ is an $\mathfrak{F}$-closed A-module for every A-module M.*

Proof. To show that $M_{\mathfrak{F}}$ is torsion-free, it suffices to show that if M is torsion-free, then so is also $M_{(\mathfrak{F})}$. Suppose $x \in M_{(\mathfrak{F})}$ and $x\,\mathfrak{b} = 0$ for some $\mathfrak{b} \in \mathfrak{F}$. Let x be represented by $\xi: \mathfrak{a} \to M$. By Lemma 1.4 there is a commutative diagram

$$\begin{array}{ccc} \mathfrak{a} & \longrightarrow & A \\ {\scriptstyle \xi}\downarrow & & \downarrow{\scriptstyle \beta} \\ M & \xrightarrow[\varphi_M]{} & M_{(\mathfrak{F})} \end{array} \qquad \beta(a) = x\,a$$

so $\varphi_M\,\xi$ is zero when restricted to $\mathfrak{a} \cap \mathfrak{b} \in \mathfrak{F}$. But φ_M is a monomorphism, so $\xi \mid \mathfrak{a} \cap \mathfrak{b} = 0$ and $x = 0$.

Next we show that $M_{\mathfrak{F}}$ is $\mathfrak{F}$-injective. Suppose we are given $f: \mathfrak{a} \to M_{\mathfrak{F}}$ with $\mathfrak{a} \in \mathfrak{F}$. Taking the pullback of $M/t(M) \to M_{\mathfrak{F}}$ with respect to f, we obtain a commutative diagram

$$\begin{array}{ccccccccc} 0 & \longrightarrow & \mathfrak{b} & \longrightarrow & \mathfrak{a} & \longrightarrow & \mathfrak{a}/\mathfrak{b} & \longrightarrow & 0 \\ & & {\scriptstyle g}\downarrow & & \downarrow{\scriptstyle f} & & \downarrow{\scriptstyle \cong} & & \\ 0 & \longrightarrow & M/t(M) & \longrightarrow & M_{\mathfrak{F}} & \longrightarrow & \operatorname{Coker}\psi_M & \longrightarrow & 0 \end{array}$$

with exact rows. Since $\operatorname{Coker}\psi_M$ is a torsion module by Lemma 1.5, we conclude that $\mathfrak{a}/\mathfrak{b}$ is a torsion module and hence $\mathfrak{b} \in \mathfrak{F}$ by $T4$. Lemma 1.4 now tells us that we can extend $g: \mathfrak{b} \to M/t(M)$ to a homomorphism $h: A \to M_{\mathfrak{F}}$. h then extends also $f: \mathfrak{a} \to M_{\mathfrak{F}}$, by the following lemma:

Lemma 1.9. *Let $\mathfrak{b} \subset \mathfrak{a}$ be right ideals in $\mathfrak{F}$ and M a torsion-free module. Suppose f and g are homomorphisms $\mathfrak{a} \to M$ such that $f \mid \mathfrak{b} = g \mid \mathfrak{b}$. Then $f = g$.*

Proof. Since f and g are equal on $\mathfrak{b}$, their difference $f-g$ factors over the torsion module $\mathfrak{a}/\mathfrak{b}$. But M torsion-free then implies that $f-g=0$. □

Corollary 1.10. *The full subcategory of* **Mod**-$A_{\mathfrak{F}}$ *consisting of modules of the form* $M_{\mathfrak{F}}$ *is equivalent to the full subcategory of* **Mod**-A *consisting of* $\mathfrak{F}$-*closed modules.*

We let **Mod**-$(A, \mathfrak{F})$ denote the full subcategory of **Mod**-A consisting of $\mathfrak{F}$-closed modules, and we call it the *quotient category* of **Mod**-A with respect to $\mathfrak{F}$. In practice we will usually not make any distinction between this category and the category of $A_{\mathfrak{F}}$-modules of the form $M_{\mathfrak{F}}$. It is important to notice that with this convention, every A-linear map between $\mathfrak{F}$-closed modules becomes automatically $A_{\mathfrak{F}}$-linear. There are a number of interesting functors:

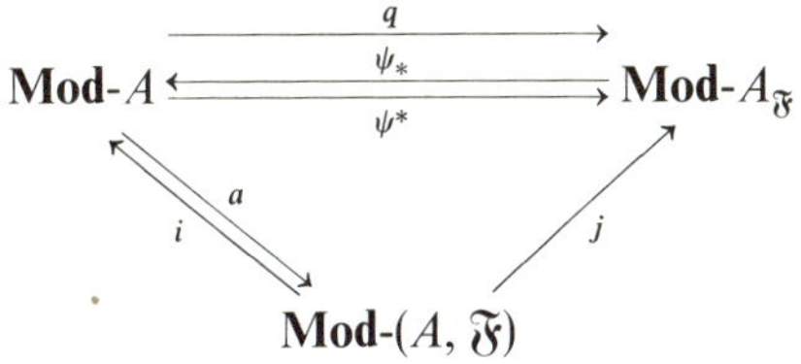

ψ_* is the forgetful functor,
ψ^* is the functor $M \mapsto M \otimes_A A_{\mathfrak{F}}$,
q is the functor $M \mapsto M_{\mathfrak{F}}$,
i is the inclusion functor,
a is the functor $M \mapsto M_{\mathfrak{F}}$,
j considers each $\mathfrak{F}$-closed module as an $A_{\mathfrak{F}}$-module, and is full and faithful.

We have $j\,a=q$ and $i=\psi_*\,j$. ψ induces a natural equivalence between $a\,i$ and the identity functor on **Mod**-$(A, \mathfrak{F})$. On the other hand, $i\,a=LL=\psi_*\,q$. The functor q is left exact, for we have noticed that L is left exact. (Conditions for q to be exact are given in Exercise 4.) Also note that there is a natural transformation $\Theta\colon \psi^* \to q$ with $\Theta_M\colon M \otimes_A A_{\mathfrak{F}} \to M_{\mathfrak{F}}$ given by $\Theta_M(x \otimes q) = \psi_M(x)\,q$. In many important cases it turns out that Θ is a natural equivalence, and then the right exactness of ψ^* forces q to be exact (this will be discussed in Chap. XI). In any case, there is always a commutative diagram

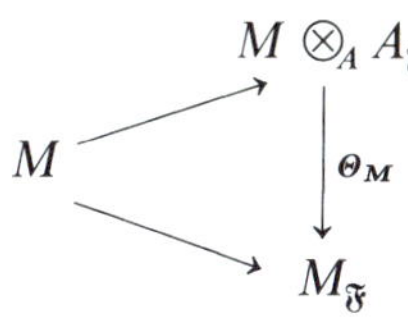

Proposition 1.11. *a is a left adjoint of i.*

Proof. We must show that the canonical map

$$\operatorname{Hom}_A(M_{\mathfrak{F}}, N) \to \operatorname{Hom}_A(M, N)$$

is an isomorphism when N is $\mathfrak{F}$-closed. It clearly is an epimorphism, so it remains to show that it is a monomorphism. The exact sequence

$$0 \to M/t(M) \to M_{\mathfrak{F}} \to \operatorname{Coker} \psi_M \to 0$$

induces

$$\operatorname{Hom}(\operatorname{Coker} \psi_M, N) \to \operatorname{Hom}(M_{\mathfrak{F}}, N) \to \operatorname{Hom}(M/t(M), N),$$

where the first term is zero since Coker ψ_M is a torsion module. The desired conclusion now follows from the observation that $\operatorname{Hom}(M/t(M), N) = \operatorname{Hom}(M, N)$. □

It will be shown in the following chapter that the quotient category **Mod**-$(A, \mathfrak{F})$ is very well-behaved, in fact it is a Grothendieck category, although the inclusion functor i is not exact in general.

Examples

1. The maximal ring of quotients. Let $\mathfrak{D}$ be the family of dense right ideals. The ring $A_{\mathfrak{D}}$ is called the *maximal* (or *complete*) *right ring of quotients* of A, and it will be denoted by $Q^r_{\max}(A)$ or simply $Q_{\max}$. Since A is $\mathfrak{D}$-torsion-free, we have $Q_{\max} = \varinjlim \operatorname{Hom}(\mathfrak{a}, A)$, $\mathfrak{a} \in \mathfrak{D}$, and A is a subring of $Q_{\max}$.

If $\mathfrak{F}$ is a Gabriel topology on A such that A is $\mathfrak{F}$-torsion-free, then $\mathfrak{F} \subset \mathfrak{D}$ and $A_{\mathfrak{F}}$ may be imbedded as a subring of $Q_{\max}$.

2. Rings of fractions. Let S be a multiplicatively closed subset of A consisting of non-zero-divisors. Assume S satisfies the generalized Ore condition $S\,1$:

given $s \in S$ and $a \in A$, there exist $t \in S$ and $b \in A$ such that $s\,b = a\,t$.

The ring of fractions $A[S^{-1}]$ then exists (Chap. II, Prop. 1.4), and the family $\mathfrak{F}$ of right ideals intersecting S is a Gabriel topology (Example VI.6.1). Let M be a module. We will show that $M_{\mathfrak{F}} \cong M[S^{-1}]$.

By definition we have

$$M_{\mathfrak{F}} = \varinjlim_{s \in S} \operatorname{Hom}(s\,A, M/t(M))$$

where $t(M) = \{x \in M \mid x\,s = 0 \text{ for some } s \in S\}$. An element in $M_{\mathfrak{F}}$ is represented by a homomorphism $\xi\colon s\,A \to M/t(M)$, i.e. by the pair (s, x) where $s \in S$, $x \in M$ and $\xi(s) = \bar{x}$. Two pairs (s, x) and (t, y) represent the same element in $M_{\mathfrak{F}}$ if and only if the two homomorphisms $\xi\colon s\,A \to M/t(M)$ and $\eta\colon t\,A \to M/t(M)$, given by $\xi(s) = \bar{x}$ and $\eta(t) = \bar{y}$, coincide on some $u\,A \subset s\,A \cap t\,A$ with $u \in S$. This happens if and only if there exist $a, b \in A$ such that $u = s\,a = t\,b$ and $x\,a - y\,b \in t(M)$, i.e. $(x\,a - y\,b)\,v = 0$ for some $v \in S$. Putting $c = a\,v$ and $d = b\,v$, we find that this is precisely the definition of $M[S^{-1}]$ according to Prop. 3.2 of Chap. II.

It is worth noting that the natural transformation Θ, where $\Theta_M\colon M \otimes_A A_{\mathfrak{F}} \to M_{\mathfrak{F}}$, is a natural equivalence in this case. We will return to this example in Chap. XI, where we will also treat the case when S may contain zero-divisors.

3. Topologies bounded by an idempotent ideal. Let $\mathfrak{a}$ be an idempotent two-sided ideal of A, and let $\mathfrak{F}$ be the Gabriel topology of right ideals containing $\mathfrak{a}$ (Prop. VI.6.12). The category **Mod**-$(A, \mathfrak{F})$ will be denoted as **Mod**-$(A, \mathfrak{a})$, and consists of all modules M for which the canonical map $M \to \operatorname{Hom}_A(\mathfrak{a}, M)$ is an isomorphism. Recall that a module M is an $\mathfrak{F}$-torsion module if and only if $M\,\mathfrak{a} = 0$. The functor L has the simple form $L(M) = \operatorname{Hom}_A(\mathfrak{a}, M)$, and therefore $a = LL$: **Mod**-$A \to$ **Mod**-$(A, \mathfrak{a})$ is given by

$$a(M) = \operatorname{Hom}_A(\mathfrak{a}, \operatorname{Hom}_A(\mathfrak{a}, M)) \cong \operatorname{Hom}_A(\mathfrak{a} \otimes_A \mathfrak{a}, M).$$

Note that a in this case not only has a right adjoint i, but also has the left adjoint $C \mapsto C \otimes_A \mathfrak{a} \otimes_A \mathfrak{a}$.

§2. $\mathfrak{F}$-Injective Envelopes

In Example V.2.3 we saw that the module of fractions of a torsion-free module M can be canonically imbedded as a submodule of the injective envelope $E(M)$ of M, namely $M[S^{-1}] \cong \{x \in E(M) \mid (M:x) \cap S \neq \emptyset\}$. We will now see that the same thing holds for modules of quotients with respect to a Gabriel topology $\mathfrak{F}$, and that $M_{\mathfrak{F}}$ for an $\mathfrak{F}$-torsion-free module M can be interpreted as an "$\mathfrak{F}$-injective envelope" of M.

Recall that $\mathfrak{F}(M)$ denotes the filter of open submodules of M in the $\mathfrak{F}$-topology, i.e. the submodules L of M for which M/L is an $\mathfrak{F}$-torsion module (Chap. VI, §4).

Proposition 2.1. *The following properties of a module E are equivalent:*

(a) *E is $\mathfrak{F}$-injective.*

(b) *If M is a module and $L \in \mathfrak{F}(M)$, then every homomorphism $L \to E$ may be extended to a homomorphism $M \to E$.*

Proof. (a) $\Rightarrow$ (b): Let $L \in \mathfrak{F}(M)$ and $\beta\colon L \to E$ be given. In the usual way (cf. the proof of Prop. I.6.5) we may assume that there is a maximal extension $\beta'\colon L' \to E$ of β, where $L \subset L' \subset M$. Then also $L' \in \mathfrak{F}(M)$ by TM 1. Suppose there exists $x \in M$ such that $x \notin L'$. Put $\mathfrak{a} = (L' : x) \in \mathfrak{F}$, and let $\alpha\colon \mathfrak{a} \to E$ be the homomorphism $\alpha(a) = \beta'(xa)$. By (a) we can extend α to A, i.e. there exists $y \in E$ such that $\alpha(a) = ya$. We may then define $\gamma\colon L' + xA \to E$ as $\gamma(z + xa) = \beta'(z) + ya$. Then γ extends β', which is a contradiction.

(b) $\Rightarrow$ (a) is obvious. □

Definition. An *$\mathfrak{F}$-injective envelope* of M is an essential monomorphism $M \to E$ such that E is $\mathfrak{F}$-injective and $M \in \mathfrak{F}(E)$.

An $\mathfrak{F}$-injective envelope of M is unique up to isomorphism. In fact, the proof of the corresponding result for $E(M)$ (Prop. V.2.3) can be generalized to the present case without difficulties. We will denote the $\mathfrak{F}$-injective envelope of M by $E_{\mathfrak{F}}(M)$. It can of course be imbedded in $E(M)$, and as a submodule of $E(M)$ it can be described in the following manner (which also establishes the existence of $E_{\mathfrak{F}}(M)$):

Proposition 2.2. $E_{\mathfrak{F}}(M) = \{x \in E(M) \mid (M:x) \in \mathfrak{F}\}$.

Proof. Clearly the subset $E' = \{x \in E(M) \mid (M:x) \in \mathfrak{F}\}$ is a submodule of $E(M)$, and $M \in \mathfrak{F}(E')$. It therefore only remains to show that E' is $\mathfrak{F}$-injective. Suppose we are given $\alpha\colon \mathfrak{a} \to E'$ with $\mathfrak{a} \in \mathfrak{F}$. α may in any case be extended to a homomorphism $\beta\colon A \to E(M)$, so we get a commutative diagram

$$\begin{array}{ccccccccc}
0 & \longrightarrow & \mathfrak{a} & \longrightarrow & A & \longrightarrow & A/\mathfrak{a} & \longrightarrow & 0 \\
 & & \downarrow{\scriptstyle\alpha} & & \downarrow{\scriptstyle\beta} & & \downarrow{\scriptstyle\gamma} & & \\
0 & \longrightarrow & E' & \longrightarrow & E(M) & \longrightarrow & E(M)/E' & \longrightarrow & 0 \\
 & & \downarrow & & \downarrow & & \| & & \\
0 & \longrightarrow & E'/M & \longrightarrow & E(M)/M & \longrightarrow & E(M)/E' & \longrightarrow & 0
\end{array}$$

If $\gamma \neq 0$, then $E(M)/E'$ contains a non-zero torsion submodule, and from the lowest row we see that this contradicts the fact that E' is the maximal submodule of $E(M)$ containing M as an $\mathfrak{F}$-open submodule. Hence $\gamma = 0$, and β actually maps A into E'. ☐

We will give another useful description of the $\mathfrak{F}$-injective envelope. Recall that every hereditary torsion theory is cogenerated by an injective module (Prop. VI.2.10).

Proposition 2.3. *If C is an injective module cogenerating the torsion theory associated to $\mathfrak{F}$, then*

$$E_{\mathfrak{F}}(M) = \{x \in E(M) \mid \alpha(x) = 0 \text{ for all } \alpha\colon E(M) \to C \text{ with } \alpha(M) = 0\}$$

for every module M.

Proof. Suppose $x \in E_{\mathfrak{F}}(M)$ and consider any $\alpha\colon E(M) \to C$ with $\alpha(M) = 0$. There exists $\mathfrak{a} \in \mathfrak{F}$ such that $x\,\mathfrak{a} \subset M$. Then $\alpha(x)\,\mathfrak{a} = \alpha(x\,\mathfrak{a}) = 0$, and C torsion-free implies that $\alpha(x) = 0$. Conversely, assume $x \in E(M)$ has the property that $\alpha(x) = 0$ for every $\alpha\colon E(M) \to C$ with $\alpha(M) = 0$. We want to show that the right ideal $\mathfrak{a} = (M : x)$ is in $\mathfrak{F}$, i.e. to show that $\operatorname{Hom}(A/\mathfrak{a}, C) = 0$. If $\beta\colon A/\mathfrak{a} \to C$ is given, we may extend it over the monomorphism $\iota\colon A/\mathfrak{a} \to E(M)/M$, defined as $\iota(\bar{a}) = \overline{xa}$, to a homomorphism $\gamma\colon E(M)/M \to C$. If $\alpha'\colon E(M) \to E(M)/M$ is the canonical map, then $\gamma\alpha'(M) = 0$ implies $\gamma\alpha'(x) = 0$ by hypothesis, and hence $\beta = 0$. ☐

We now connect the concept of $\mathfrak{F}$-injective envelope to the theory of modules of quotients.

Proposition 2.4. *If M is a torsion-free module, then $M_{\mathfrak{F}} \cong E_{\mathfrak{F}}(M)$ as $A_{\mathfrak{F}}$-modules.*

Proof. $M_{\mathfrak{F}}$ is an $\mathfrak{F}$-injective module by Prop. 1.8, and $M_{\mathfrak{F}}/M$ is a torsion module by Lemma 1.5. Hence $M_{\mathfrak{F}}$ is an $\mathfrak{F}$-injective envelope of M. Both $M_{\mathfrak{F}}$ and $E_{\mathfrak{F}}(M)$ are $\mathfrak{F}$-closed modules, and they are therefore $A_{\mathfrak{F}}$-modules. The isomorphism between them is $A_{\mathfrak{F}}$-linear by Cor. 1.10. ☐

For an arbitrary module M one of course gets $M_{\mathfrak{F}} \cong E_{\mathfrak{F}}(M/t(M))$. Let us describe "explicitly" the $A_{\mathfrak{F}}$-module structure of the $\mathfrak{F}$-closed module $E(M)$ for a torsion-free module M. Suppose $x \in E(M)$, and $q \in A_{\mathfrak{F}}$ is represented by $\alpha\colon \mathfrak{a} \to A/t(A)$. Since $E(M)$ is $\mathfrak{F}$-closed, there exists a unique $y \in E(M)$ such that $x\alpha(a) = ya$ for all $a \in \mathfrak{a}$ (note that $E(M)$ is a module over $A/t(A)$), and then $xq = y$.

Proposition 2.5. *If M is a torsion-free module, then $E(M)$ is an injective envelope of $M_{\mathfrak{F}}$ in* **Mod**-$A_{\mathfrak{F}}$.

Proof. The inclusion map $M_{\mathfrak{F}} \cong E_{\mathfrak{F}}(M) \to E(M)$ is $A_{\mathfrak{F}}$-linear by Cor. 1.10. Since it is an essential monomorphism in **Mod**-A, it is obviously essential also in **Mod**-$A_{\mathfrak{F}}$. It only remains to show that $E(M)$ is injective as an $A_{\mathfrak{F}}$-module. This follows from:

Lemma 2.6. *Every torsion-free injective A-module is injective over $A_{\mathfrak{F}}$.*

Proof. Let E be torsion-free injective over A. Suppose $N' \to N$ is any monomorphism in **Mod**-$A_{\mathfrak{F}}$ and $\alpha\colon N' \to E$ is $A_{\mathfrak{F}}$-linear. α extends to a homomorphism $\beta\colon N \to E$ in **Mod**-A. For each $x \in N$, consider the two maps $A_{\mathfrak{F}} \to E$ given by

$\beta'(q)=\beta(xq)$ and $\beta''(q)=\beta(x)q$. They are both A-linear and coincide on A. Hence $\beta'-\beta''$ factors over a homomorphism $\operatorname{Coker}\psi_A \to E$. But $\operatorname{Coker}\psi_A$ is a torsion module (Lemma 1.5) and E is torsion-free, so $\beta'=\beta''$, and β is $A_{\mathfrak{F}}$-linear. □

One can give a rather simple condition on M for $M_{\mathfrak{F}}$ to become $A_{\mathfrak{F}}$-injective:

Proposition 2.7. *The following properties of a torsion-free A-module M are equivalent:*

(a) *$M_{\mathfrak{F}}$ is an injective module over $A_{\mathfrak{F}}$.*

(b) *$M_{\mathfrak{F}}$ is an injective module over A.*

(c) *For every right ideal $\mathfrak{a}$ of A and homomorphism $\alpha\colon \mathfrak{a}\to M$, there exist $\mathfrak{b}\in\mathfrak{F}$ and a homomorphism $\beta\colon \mathfrak{b}\to M$, such that $\mathfrak{a}\subset\mathfrak{b}$ and $\beta|\mathfrak{a}=\alpha$.*

Proof. (a) ⇒ (b): Let $K\to L$ be a monomorphism of A-modules, and assume given a homomorphism $\alpha\colon K\to M_{\mathfrak{F}}$. Then α induces an $A_{\mathfrak{F}}$-linear map $K_{\mathfrak{F}}\to M_{\mathfrak{F}}$, which can be extended to a map $L_{\mathfrak{F}}\to M_{\mathfrak{F}}$, since $M_{\mathfrak{F}}$ is injective over $A_{\mathfrak{F}}$. The composed map $L\to L_{\mathfrak{F}}\to M_{\mathfrak{F}}$ extends α.

(b) ⇒ (a) is precisely Lemma 2.6.

(b) ⇒ (c): Assume given a right ideal $\mathfrak{a}$ of A and a homomorphism $\alpha\colon \mathfrak{a}\to M$. By hypothesis one can extend α to $\gamma\colon A\to M_{\mathfrak{F}}$ so that the diagram

$$\begin{array}{ccc} \mathfrak{a} & \longrightarrow & A \\ {\scriptstyle\alpha}\downarrow & & \downarrow{\scriptstyle\gamma} \\ M & \longrightarrow & M_{\mathfrak{F}} \end{array}$$

commutes. Put $\mathfrak{b}=\gamma^{-1}(M)$ and $\beta=\gamma|\mathfrak{b}$. Then $\mathfrak{b}\supset\mathfrak{a}$, and γ induces a monomorphism $A/\mathfrak{b}\to M_{\mathfrak{F}}/M$, from which it follows that $\mathfrak{b}\in\mathfrak{F}$.

(c) ⇒ (b): Let $\mathfrak{a}$ be a right ideal of A and $\alpha\colon \mathfrak{a}\to M_{\mathfrak{F}}$ a homomorphism. Put $\alpha^{-1}(M)=\mathfrak{a}'$. By hypothesis we can extend $\alpha|\mathfrak{a}'$ to $\beta\colon \mathfrak{b}\to M$ for some $\mathfrak{b}\in\mathfrak{F}$ such that $\mathfrak{a}'\subset\mathfrak{b}$. Finally, since $M_{\mathfrak{F}}$ is $\mathfrak{F}$-injective, β induces $\gamma\colon A\to M_{\mathfrak{F}}$. It remains to show that γ extends α. Now $\gamma|\mathfrak{a}$ and α coincide on $\mathfrak{a}'$, and their difference therefore induces a map $\mathfrak{a}/\mathfrak{a}'\to M_{\mathfrak{F}}$. But this map must be zero, because $M_{\mathfrak{F}}$ is torsion-free, while $\mathfrak{a}/\mathfrak{a}'$ is a torsion module since α induces a monomorphism $\mathfrak{a}/\mathfrak{a}'\to M_{\mathfrak{F}}/M$. □

The following is a useful observation:

Proposition 2.8. *Let I and M be right A-submodules of $A_{\mathfrak{F}}$, and assume that $I\in\mathfrak{F}(A_{\mathfrak{F}})$. Then*

$$\operatorname{Hom}_A(I,M)\cong\{q\in A_{\mathfrak{F}}\mid qI\subset M\}.$$

Proof. First, it is clear that left multiplication by $q\in A_{\mathfrak{F}}$ is a homomorphism $I\to M$ whenever $qI\subset M$. On the other hand, consider a homomorphism $\varphi\colon I\to M$. Since $I\in\mathfrak{F}(A_{\mathfrak{F}})$ and $A_{\mathfrak{F}}$ is $\mathfrak{F}$-closed, there exists a unique A-homomorphism $\alpha\colon A_{\mathfrak{F}}\to A_{\mathfrak{F}}$ which makes the square

$$\begin{array}{ccc} I & \longrightarrow & A_{\mathfrak{F}} \\ {\scriptstyle\varphi}\downarrow & & \downarrow{\scriptstyle\alpha} \\ M & \longrightarrow & A_{\mathfrak{F}} \end{array}$$

commutative. Since α automatically is $A_{\mathfrak{F}}$-linear, it has the form of left multiplication by a $q\in A_{\mathfrak{F}}$. Hence $\varphi(a)=qa$ for all $a\in I$. □

Corollary 2.9. *If $\mathfrak{F}$ is bounded by a two-sided idempotent dense ideal $\mathfrak{a}$, then*

$$A_{\mathfrak{F}}\cong \operatorname{Hom}_A(\mathfrak{a},\mathfrak{a}).$$

Proof. For every $q\in A_{\mathfrak{F}}$ we have $(\mathfrak{a}:q)=\{b\in A\,|\,qb\in\mathfrak{a}\}\in\mathfrak{F}$, and hence $(\mathfrak{a}:q)\supset\mathfrak{a}$. Therefore $\operatorname{Hom}_A(\mathfrak{a},\mathfrak{a})=\{q\in A_{\mathfrak{F}}\,|\,q\mathfrak{a}\subset\mathfrak{a}\}=A_{\mathfrak{F}}$. □

Examples

1. S-injective modules. Let S be a right denominator set in A, and let $\mathfrak{F}$ be the topology of right ideals intersecting S. The $\mathfrak{F}$-injective modules are precisely those modules which were called S-injective in §3 of Chap. II.

2. The Goldie topology. Let $\mathfrak{G}$ be the Goldie topology, i.e. the Gabriel topology generated by the topology of essential right ideals of A (Example VI.6.2).

Lemma 2.10. *Every $\mathfrak{G}$-injective module is injective.*

Proof. Suppose E is $\mathfrak{G}$-injective, and consider a homomorphism $\alpha\colon \mathfrak{a}\to E$ from a right ideal $\mathfrak{a}$ of A. Choose a pseudo-complement $\mathfrak{b}$ of $\mathfrak{a}$ in A. Then α can trivially be extended to a homomorphism $\alpha'\colon \mathfrak{a}\oplus\mathfrak{b}\to E$. But since $\mathfrak{a}\oplus\mathfrak{b}$ is essential in A (Prop. III.6.4), α' can be extended to all of A. □

Combining this result with Lemma 2.6, we obtain:

Proposition 2.11. *$M_{\mathfrak{G}}$ is an injective $A_{\mathfrak{G}}$-module, for every module M.*

In particular we may conclude that $A_{\mathfrak{G}}$ is a right self-injective ring (i.e. injective as a right module over itself). When A is right non-singular (i.e. $\mathfrak{G}$-torsion-free), then

$$Q^r_{\max}(A)=A_{\mathfrak{G}}\cong E(A).$$

Proposition 2.12. *The functor q:* **Mod**-$A\to$ **Mod**-$A_{\mathfrak{G}}$ *is exact.*

Proof. Let $0\to L\to M\to N\to 0$ be an exact sequence of A-modules. There is a commutative diagram

$$\begin{array}{ccccccc}
0\longrightarrow & L & \xrightarrow{\ \alpha\ } & M & \xrightarrow{\ \beta\ } & N & \longrightarrow 0\\
 & \downarrow{\scriptstyle \psi_L} & & \downarrow{\scriptstyle \psi_M} & & \downarrow{\scriptstyle \psi_N} & \\
0\longrightarrow & L_{\mathfrak{G}} & \xrightarrow[\alpha_{\mathfrak{G}}]{} & M_{\mathfrak{G}} & \xrightarrow[\beta_{\mathfrak{G}}]{} & N_{\mathfrak{G}} &
\end{array}$$

with exact rows. Since $L_{\mathfrak{G}}$ and $M_{\mathfrak{G}}$ are injective modules (by Prop. 2.11), the sequence $0\to L_{\mathfrak{G}}\to M_{\mathfrak{G}}\to \operatorname{Im}\beta_{\mathfrak{G}}\to 0$ splits and also $\operatorname{Im}\beta_{\mathfrak{G}}$ is an injective module. In order to show that $\beta_{\mathfrak{G}}$ is an epimorphism, it therefore suffices to show that $\operatorname{Im}\beta_{\mathfrak{G}}$ is essential in $N_{\mathfrak{G}}$. But for each non-zero $x\in N_{\mathfrak{G}}$, there exists $\mathfrak{a}\in\mathfrak{G}$ such that $x\mathfrak{a}\subset\operatorname{Im}\psi_N=\operatorname{Im}(\beta_{\mathfrak{G}}\psi_M)\subset\operatorname{Im}\beta_{\mathfrak{G}}$, and $x\mathfrak{a}\neq 0$. □

3. Rational completion (Findlay and Lambek [1]). For a given module M we let $\mathfrak{F}^0_M$ denote the strongest Gabriel topology for which M is torsion-free (Example VI.6.4). It is the Gabriel topology corresponding to the hereditary torsion theory cogenerated by $E(M)$.

Definitions. (i) A monomorphism $L \to M$ is *rational* if $L \in \mathfrak{F}_M^0(M)$.
(ii) M is *rationally complete* if it is $\mathfrak{F}_M^0$-injective.
(iii) A *rational completion* of M is a rational monomorphism $M \to N$ such that N is rationally complete.

Lemma 2.13. *A monomorphism $L \to M$ is rational if and only if it is essential and $L \in \mathfrak{F}_L^0(M)$.*

Proof. Since $\mathfrak{F}_L^0 = \mathfrak{F}_M^0$ whenever $L \to M$ is an essential monomorphism, it suffices to show that every rational monomorphism $L \to M$ is essential. Suppose K is a submodule of M such that $K \cap L = 0$. K may then be considered as a submodule of the $\mathfrak{F}_M^0$-torsion module M/L. Since M is $\mathfrak{F}_M^0$-torsion-free, this implies $K = 0$. □

From this we can conclude:

Proposition 2.14. (i) $M \to E_{\mathfrak{F}_M^0}(M)$ *is a rational completion of M.*
(ii) *The rational completion of M is unique up to isomorphism.*
(iii) *M is rationally complete if and only if it has no proper rational extensions.*

Note that from Prop. 2.3 we have

$$E_{\mathfrak{F}_M^0}(M) = \{x \in E(M) \mid \alpha(x) = 0 \text{ for all } \alpha\colon E(M) \to E(M) \text{ with } \alpha(M) = 0\}.$$

Thus M is rationally complete if and only if it consists of all elements of $E(M)$ annihilated by the endomorphisms of $E(M)$.

In particular, the rational completion of A is precisely the maximal right ring of quotients of A, and so A is rationally complete if and only if $A = Q_{\max}^r(A)$. R. Courter [1] has determined those rings for which every module is rationally complete.

4. The canonical topology. For a given module M we let $\mathfrak{F}_M^1$ denote the Gabriel topology corresponding to the hereditary torsion theory cogenerated by $E(M) \oplus E(E(M)/M)$ (cf. Example VI.6.4).

Proposition 2.15. *$\mathfrak{F}_M^1$ is the strongest Gabriel topology $\mathfrak{F}$ for which M is $\mathfrak{F}$-closed.*

Proof. It follows from Prop. 2.2 that M is $\mathfrak{F}$-injective if and only if $E(M)/M$ is an $\mathfrak{F}$-torsion module. Hence the strongest Gabriel topology for which M is $\mathfrak{F}$-closed is $\mathfrak{F}_{M \oplus E(M)/M}^0 = \mathfrak{F}_M^1$. □

In particular, the topology $\mathfrak{F}_A^1$ is called the *canonical topology* on A, and it is thus the strongest Gabriel topology on A for which A is its own ring of quotients. (Remember that we have already determined the canonical topology for a commutative noetherian ring satisfying Serre's condition S 2 (Prop. VII.6.8).)

§3. The Ring of Quotients is a Bicommutator

In this section it will be proved that each ring of quotients is the bicommutator of a suitably chosen injective module. Let E be an injective right A-module, and let H be its endomorphism ring. E may then be considered as a left H-module, and we call ${}_H E$ the *countermodule* of E. The bicommutator $\operatorname{Bic}(E)$ of E is the endomorphism ring of the countermodule of E (Example I.1.4). There is a canonical ring homomorphism $\eta\colon A \to \operatorname{Bic}(E)$ given by $\eta(a)\colon x \mapsto xa$ for $x \in E$.

Lemma 3.1. *The countermodule of E is cyclic if and only if there exists $x \in E$ such that $xa=0$ implies $Ea=0$ for $a \in A$.*

Proof. If $E=Hx$ for some $x \in E$, then obviously $xa=0$ implies $Ea=0$. Conversely, assume $x \in E$ has the property that $xa=0$ implies $Ea=0$. For every $y \in E$ there is then a well-defined homomorphism $xA \to E$ taking x to y, and which can be extended to an endomorphism h of E. Then $y=h(x)$, and hence $E=Hx$. □

The injective module E will be called *countercyclic* when its countermodule is cyclic.

Lemma 3.2. *If E is an injective module, then there is a direct product of copies of E which is countercyclic.*

Proof. Let $\tilde{E}$ denote the direct product of copies of E over the index set E, and choose $u \in \tilde{E}$ to be the element whose x-th coordinate is precisely x. Then

$$\operatorname{Ann}(u) = \bigcap_{x \in E} \operatorname{Ann}(x) = \operatorname{Ann}(E) = \operatorname{Ann}(\tilde{E}),$$

so $\tilde{E}$ is a countercyclic module by Lemma 3.1. □

Consider a hereditary torsion theory cogenerated by an injective module E. By Lemma 3.2 we may as well assume E to be countercyclic, because a direct product of copies of E cogenerates the same torsion theory as does E. Let $\mathfrak{F}$ denote the Gabriel topology corresponding to this torsion theory. Since E is $\mathfrak{F}$-closed, the endomorphism ring H of E_A is equal to the endomorphism ring of E as a right $A_{\mathfrak{F}}$-module. Thus there is a commutative diagram of ring homomorphisms

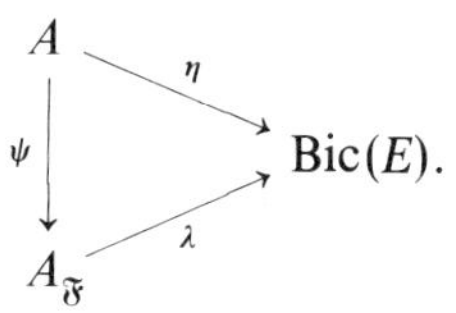

Theorem 3.3. *If E is a countercyclic injective module, then λ is a ring isomorphism*

$$A_{\mathfrak{F}} \cong \operatorname{Bic}(E).$$

Proof. Choose $x \in E$ such that $xa=0$ implies $Ea=0$, and define $\alpha\colon A \to E$ as $\alpha(a)=xa$. Let $\mathfrak{a}=\operatorname{Ker}\psi$ be the $\mathfrak{F}$-torsion ideal of A. Since E is $\mathfrak{F}$-torsion-free, we have $\mathfrak{a} \subset \operatorname{Ker}\alpha = \operatorname{Ann}(x)$. We want to show that $\mathfrak{a}=\operatorname{Ann}(x)$, and to accomplish this it suffices to show that $\operatorname{Ann}(x)$ is a torsion module, i.e. $\operatorname{Hom}(\operatorname{Ann}(x), E)=0$. But every $\beta\colon \operatorname{Ann}(x) \to E$ is of the form $\beta(a)=ya$ for some $y \in E$, and $\operatorname{Ann}(x)=\operatorname{Ann}(E)$ immediately implies that $\beta=0$. Hence $\operatorname{Ann}(x)=\mathfrak{a}=\operatorname{Ker}\psi$. We may thus consider $A/\mathfrak{a}$ as a submodule of E, with $\bar{1} \in A/\mathfrak{a}$ and $x \in E$ identified. There results a splitting of E as $E=E(A/\mathfrak{a}) \oplus F$. Let $p\colon E \to E(A/\mathfrak{a})$ denote the canonical projection.

We will now exhibit an inverse μ of λ. Suppose $\beta \in \operatorname{Bic}(E)=\operatorname{Hom}_H(E, E)$. We assert that $p\beta(\bar{1}) \in A_{\mathfrak{F}}$, where $A_{\mathfrak{F}}$ is considered as a submodule of $E(A/\mathfrak{a})$

according to Prop. 2.3, i.e.

$$A_{\mathfrak{F}} = \{x \in E(A/\mathfrak{a}) \mid f(x) = 0 \text{ for all } f\colon E(A/\mathfrak{a}) \to E \text{ with } f(\bar{1}) = 0\}.$$

So suppose $f\colon E(A/\mathfrak{a}) \to E$ has the property that $f(\bar{1}) = 0$. f may be extended to an endomorphism $f' \in H$ by putting $f' = 0$ on the complementary summand F of E. Then

$$f(p\,\beta(\bar{1})) = f'(\beta(\bar{1})) = \beta(f'(\bar{1})) = 0,$$

using the H-linearity of β. Hence $p\,\beta(\bar{1}) \in A_{\mathfrak{F}}$ as asserted, and we can define μ as $\mu(\beta) = p\,\beta(\bar{1})$. μ is an additive map, and clearly $\mu\lambda$ is the identity map of $A_{\mathfrak{F}}$. Then $\mu\lambda\mu = \mu$, and it only remains to show that μ is a monomorphism. Thus we must show that if $\beta(\bar{1}) \in F$, then $\beta = 0$. But this is easy, because

$$\beta(E) = \beta(Hx) = H\,\beta(x) = H\,\beta(\bar{1}). \quad \square$$

Examples

1. The maximal ring of quotients. The injective module $E(A)$ is countercyclic by Lemma 3.1, so the maximal right ring of quotients of A is equal to the bicommutator of $E(A)$. This fact was first discovered by Lambek [2].

2. Artinian rings. If A is right artinian and E is an injective module, then already a finite product of copies of E is countercyclic. For $\operatorname{Ann}(E) = \bigcap_{x \in E} \operatorname{Ann}(x)$, and the minimum condition implies that actually $\operatorname{Ann}(E) = \operatorname{Ann}(x_1) \cap \cdots \cap \operatorname{Ann}(x_n)$ for a finite number of elements $x_1, \ldots, x_n$ of E. Then $(x_1, \ldots, x_n) \in E^n$ has the same annihilator as E^n.

§4. The Lattice of Saturated Submodules

Let $\mathfrak{F}$ be a Gabriel topology on the ring A. For each module M and submodule L of M we define a new submodule L^c of M by

$$L^c = \{x \in M \mid (L:x) \in \mathfrak{F}\}.$$

Then $L^c \supset L$ and L^c/L is the $\mathfrak{F}$-torsion submodule of M/L. The assignment $L \mapsto L^c$ is a closure operator on the lattice $\mathsf{L}(M)$ of submodules of M. The submodules L for which $L^c = M$ are just the $\mathfrak{F}$-open submodules of M, constituting the filter we have denoted by $\mathfrak{F}(M)$. On the other hand, we say that a submodule L of M is $\mathfrak{F}$-*saturated* in M if $L^c = L$, and we let $\mathsf{Sat}_{\mathfrak{F}}(M)$ be the collection of all $\mathfrak{F}$-saturated submodules of M. Note that

$$\mathsf{Sat}_{\mathfrak{F}}(M) = \{L \subset M \mid M/L \text{ is } \mathfrak{F}\text{-torsion-free}\}.$$

By Prop. 1.8 and 1.9 of Chap. III we obtain:

Proposition 4.1. $\mathsf{Sat}_{\mathfrak{F}}(M)$ *is a complete modular lattice.*

Proposition 4.2. *If M is an $\mathfrak{F}$-closed module, then $\mathsf{Sat}_{\mathfrak{F}}(M)$ consists of the $\mathfrak{F}$-closed submodules of M.*

Proof. For every $\mathfrak{a}\in\mathfrak{F}$ and submodule L of M we have a commutative diagram

$$\begin{array}{ccccccc} 0\longrightarrow & \mathrm{Hom}(A,L) & \longrightarrow & \mathrm{Hom}(A,M) & \longrightarrow & \mathrm{Hom}(A,M/L) & \longrightarrow 0 \\ & \downarrow{\scriptstyle\lambda} & & \downarrow{\scriptstyle\mu} & & \downarrow{\scriptstyle\eta} & \\ 0\longrightarrow & \mathrm{Hom}(\mathfrak{a},L) & \longrightarrow & \mathrm{Hom}(\mathfrak{a},M) & \longrightarrow & \mathrm{Hom}(\mathfrak{a},M/L) & \end{array}$$

with exact rows and μ an isomorphism. It is then easy to verify that λ is an epimorphism if and only if η is a monomorphism, i.e. L is $\mathfrak{F}$-closed if and only if M/L is $\mathfrak{F}$-torsion-free. (If one knows homological algebra, the conclusion can also be drawn from the exact sequence

$$\mathrm{Hom}(A/\mathfrak{a},M)\to\mathrm{Hom}(A/\mathfrak{a},M/L)\to\mathrm{Ext}^1(A/\mathfrak{a},L)\to\mathrm{Ext}^1(A/\mathfrak{a},M),$$

using Exercise 6.) □

Proposition 4.3. *For every module M there is a lattice isomorphism*

$$\mathsf{Sat}_{\mathfrak{F}}(M)\to\mathsf{Sat}_{\mathfrak{F}}(M_{\mathfrak{F}})$$

given by $L\mapsto L_{\mathfrak{F}}$.

Proof. If $L\subset M$, then $L_{\mathfrak{F}}$ belongs to $\mathsf{Sat}_{\mathfrak{F}}(M_{\mathfrak{F}})$ by the preceding Proposition. Let $\psi\colon M\to M_{\mathfrak{F}}$ denote the canonical homomorphism. If $N\in\mathsf{Sat}_{\mathfrak{F}}(M_{\mathfrak{F}})$, then $\psi^{-1}(N)$ is saturated in M because $M/\psi^{-1}(N)$ may be considered as a submodule of $M_{\mathfrak{F}}/N$, and is therefore $\mathfrak{F}$-torsion-free. We assert that $N\mapsto\psi^{-1}(N)$ is the inverse of the map $L\mapsto L_{\mathfrak{F}}$. For each $L\subset M$ we have $\psi^{-1}(L_{\mathfrak{F}})=L^c$, so $L=\psi^{-1}(L_{\mathfrak{F}})$ when L is $\mathfrak{F}$-saturated. On the other hand, for every $\mathfrak{F}$-closed submodule N of $M_{\mathfrak{F}}$ we have $(\psi^{-1}(N))_{\mathfrak{F}}=N$. $L\mapsto L_{\mathfrak{F}}$ thus defines an isomorphism of complete lattices. □

From Prop. 4.2 and 4.3 we can draw the following conclusion, which explains why the lattice $\mathsf{Sat}_{\mathfrak{F}}(M)$ is of particular interest:

Corollary 4.4. *$\mathsf{Sat}_{\mathfrak{F}}(M)$ is isomorphic to the complete lattice of subobjects of $M_{\mathfrak{F}}$ in the category* **Mod**-$(A,\mathfrak{F})$ *of $\mathfrak{F}$-closed modules.*

Since the category **Mod**-$(A,\mathfrak{F})$ actually is abelian (as we will see in the next chapter), this corollary also "explains" once more why the lattice $\mathsf{Sat}_{\mathfrak{F}}(M)$ is modular.

The lattice $\mathsf{Sat}_{\mathfrak{F}}(A)$ deserves some special attention. It consists of all right ideals $\mathfrak{a}$ for which $A/\mathfrak{a}$ is $\mathfrak{F}$-torsion-free. Since a module is torsion-free if and only if every cyclic submodule is torsion-free, we have:

Proposition 4.5. *A module M is $\mathfrak{F}$-torsion-free if and only if $\mathrm{Ann}(x)\in\mathsf{Sat}_{\mathfrak{F}}(A)$ for every $x\in M$.*

$\mathsf{Sat}_{\mathfrak{F}}(A)$ has the following neat description in terms of a cogenerating injective:

Proposition 4.6. $\mathsf{Sat}_{\mathfrak{F}}(A)=\{\mathrm{Ann}(S)\mid S\subset E\}$, *where E is an injective module cogenerating the torsion theory corresponding to $\mathfrak{F}$.*

Proof. If $\mathfrak{a}$ is a right ideal of A, then $A/\mathfrak{a}$ is torsion-free if and only if $A/\mathfrak{a}$ can be imbedded as a submodule of a direct product E^I of copies of E (Prop. VI.2.12). Hence $\mathfrak{a}\in\mathsf{Sat}_{\mathfrak{F}}(A)$ if and only if $\mathfrak{a}=\bigcap_{i\in I}\mathrm{Ann}(x_i)$, $x_i\in E$. □

Example. Modules of fractions. Let S be a right denominator set in A, and let $\mathfrak{F}$ be the Gabriel topology of right ideals intersecting S. The $\mathfrak{F}$-saturated submodules were considered in Exercise 11 of Chap. II under the name of "S-saturated submodules".

§5. $\mathfrak{F}$-Invertible Ideals

In this section we will continue the theory of invertible fractional ideals that was initiated for right orders in Chap. II. Let $\mathfrak{F}$ be a Gabriel topology on A, and assume that A is $\mathfrak{F}$-torsion-free (so every right ideal in $\mathfrak{F}$ is dense). Let I be a right A-submodule of $A_{\mathfrak{F}}$. Then $I \in \mathfrak{F}(A_{\mathfrak{F}})$ holds if and only if $I \cap A \in \mathfrak{F}$, as one sees from the exact sequence

$$0 \to A/(I \cap A) \to A_{\mathfrak{F}}/I \to A_{\mathfrak{F}}/(A+I) \to 0.$$

Thus $\mathfrak{F}(A_{\mathfrak{F}})$ consists of those right A-submodules of $A_{\mathfrak{F}}$ which contain a right ideal in $\mathfrak{F}$. The "conductor" of I into A is $I^* = \{q \in A_{\mathfrak{F}} \mid q I \subset A\}$. By Prop. 2.8 we have:

Lemma 5.1. *If* $I \in \mathfrak{F}(A_{\mathfrak{F}})$, *then* $I^* \cong \mathrm{Hom}_A(I, A)$.

Definition. A right A-submodule I of $A_{\mathfrak{F}}$ is called *$\mathfrak{F}$-invertible* if there exist $a_1, \ldots, a_n \in I$ and $q_1, \ldots, q_n \in I^*$ such that $1 = \sum a_i q_i$.

Similarly to Prop. II.4.3 we have:

Proposition 5.2. *A right A-submodule I of $A_{\mathfrak{F}}$ is $\mathfrak{F}$-invertible if and only if $I \in \mathfrak{F}(A_{\mathfrak{F}})$ and I is a finitely generated, projective A-module.*

Proof. Suppose I is invertible, i.e. $1 = \sum a_i q_i$ with $a_i \in I$ and $q_i I \subset A$ $(i = 1, \ldots, n)$. Define homomorphisms φ_i: $I \to A$ by $\varphi_i(a) = q_i a$. The family (φ_i, a_i) clearly satisfies the requirements of Prop. I.6.3, and therefore makes I into a finitely generated, projective module. We have

$$I \cap A = \{a \in A \mid q_i a \in A \text{ for } i = 1, \ldots, n\} = \bigcap_1^n (A : q_i) \in \mathfrak{F},$$

because if $q_i a \in A$ for all i, then $a = \sum a_i(q_i a) \in I$. Hence $I \in \mathfrak{F}(A_{\mathfrak{F}})$.

Assume conversely that $I \in \mathfrak{F}(A_{\mathfrak{F}})$ is finitely generated projective. Let (φ_i, a_i) be finite projective coordinates for I, as given by Prop. I.6.3. From Lemma 5.1 we know that each φ_i: $I \to A$ has the form $\varphi_i(a) = q_i a$ for some $q_i \in I^*$. For each $a \in I$ we have $a = \sum a_i \varphi_i(a) = \sum a_i q_i a$, so $(1 - \sum a_i q_i) I = 0$. But $A_{\mathfrak{F}}$ is $\mathfrak{F}$-torsion-free, so this implies $1 = \sum a_i q_i$, as desired. □

If $\mathfrak{a}$ is an $\mathfrak{F}$-invertible right ideal of A and M is an arbitrary right module, then the canonical map

$$\Theta: M \otimes_A \mathrm{Hom}_A(\mathfrak{a}, A) \to \mathrm{Hom}_A(\mathfrak{a}, M),$$

given by $\Theta(x \otimes \alpha)$: $a \mapsto x\,\alpha(a)$ for $a \in \mathfrak{a}$, is an isomorphism. For if (φ_i, a_i) are finite projective coordinates for $\mathfrak{a}$, then Θ^{-1} is the map $\xi \mapsto \sum \xi(a_i) \otimes \varphi_i$.

We now assume that $\mathfrak{F}$ has a basis consisting of $\mathfrak{F}$-invertible right ideals, i.e. every member of $\mathfrak{F}$ contains an $\mathfrak{F}$-invertible right ideal. Then it is clear that

$\mathfrak{a} A_{\mathfrak{F}} = A_{\mathfrak{F}}$ holds for every $\mathfrak{a} \in \mathfrak{F}$, which means that $A_{\mathfrak{F}}$ is an $\mathfrak{F}$-divisible left A-module (Chap. VI, §9). If M is a right A-module, then $M_{\mathfrak{F}}$ is gotten by taking the direct limit of the groups $\operatorname{Hom}_A(\mathfrak{a}, M)$ with $\mathfrak{a}$ running through the cofinal family of $\mathfrak{F}$-invertible right ideals. For a torsion-free module M one thus obtains by means of the isomorphism Θ that $M_{\mathfrak{F}} \cong M \otimes_A A_{\mathfrak{F}}$. The same result is obtained for arbitrary M, using the right exactness of $\otimes$ and the fact that $t(M) \otimes_A A_{\mathfrak{F}} = 0$ (Prop. VI.9.1). Thus

Proposition 5.3. *If $\mathfrak{F}$ has a basis of $\mathfrak{F}$-invertible right ideals, then for every module M one has*

$$M_{\mathfrak{F}} \cong M \otimes_A A_{\mathfrak{F}}.$$

Examples

1. S-invertible modules. Let S be a right denominator set in A, and assume that S consists of regular elements. A right A-submodule of $A[S^{-1}]$ is *right S-invertible* if it is $\mathfrak{F}$-invertible for the topology $\mathfrak{F}$ of right ideals intersecting S. The case when S consists of all regular elements was studied in Chap. II.

2. Hereditary noetherian rings. When A is right noetherian and right hereditary, then every Gabriel topology $\mathfrak{F}$ consists of $\mathfrak{F}$-invertible right ideals.

Exercises

In the following exercises, $\mathfrak{F}$ denotes a Gabriel topology on A.

§1

1. Show that if $\mathfrak{E} \subset \mathfrak{F}$ are Gabriel topologies on A, then there is a ring homomorphism $A_{\mathfrak{E}} \to A_{\mathfrak{F}}$.

2. Show that if A is a commutative ring, then also every $A_{\mathfrak{F}}$ is commutative. (Hint: one can reduce the problem to showing that if $\mathfrak{a} \in \mathfrak{F}$ and α, β: $\mathfrak{a} \to A$, then $\alpha\beta(x) = \beta\alpha(x)$ for $x \in \mathfrak{a}^2$.)

3. Let $A = K[X, Y]$ where K is a field, and let $\mathfrak{m}$ be the maximal ideal $\mathfrak{m} = (X, Y)$. Let $\mathfrak{F}$ be the $\mathfrak{m}$-adic topology of A. Show that:

(i) $A_{\mathfrak{F}} = A$.

(ii) If $M = A/(X)$, then $M_{\mathfrak{F}} = K[\bar{Y}, 1/\bar{Y}]$, where $\bar{Y}$ is the class of Y in M. (So if f: $A \to M$ is the canonical epimorphism, then $f_{\mathfrak{F}}$ is not an epimorphism, and the functor q is not exact.)

4. Show that the following properties of a Gabriel topology $\mathfrak{F}$ are equivalent:

(a) The functor ia: $M \mapsto M_{\mathfrak{F}}$ is exact.

(b) If the sequence $0 \to L \to M \to N \to 0$ is exact with L and M $\mathfrak{F}$-closed, then N is $\mathfrak{F}$-closed.

(c) If M is $\mathfrak{F}$-closed, then $E(M)/M$ is $\mathfrak{F}$-closed.

5. Show that:

(i) If $\mathfrak{a}$ is a dense right ideal of A, then $\mathfrak{a} Q_{\max}$ is a dense right ideal of $Q_{\max}$.

(ii) If $\mathfrak{q}$ is a dense right ideal of $Q_{\max}$, then $\mathfrak{q} \cap A$ is a dense right ideal of A.

6. Suppose A is a subring of a ring B, and that for each $b \in B$ there exists a dense right ideal $\mathfrak{a}$ of A such that $b\mathfrak{a} \subset A$. Show that:

(i) There is a unique homomorphism $B \to Q^r_{\max}(A)$ of right A-modules such that the diagram

$$\begin{array}{ccc} A & \longrightarrow & B \\ & \searrow \quad \swarrow & \\ & Q_{\max} & \end{array}$$

commutes, and this homomorphism is a ring monomorphism.

(ii) B is a maximal right ring of quotients of A if and only if for every dense right ideal $\mathfrak{a}$ of A and homomorphism $\alpha: \mathfrak{a} \to A$, there exists $b \in B$ such that $\alpha(a) = b\,a$ for all $a \in \mathfrak{a}$.

§2

7. Show that for any ring A and module M one has

$$E(M)/Z(M) = \varinjlim \operatorname{Hom}_A(\mathfrak{a}, M)$$

where $\mathfrak{a}$ runs through the downwards directed family of essential right ideals of A.

8. Show that a module M is $\mathfrak{F}$-injective if and only if $\operatorname{Ext}_A^1(L, M) = 0$ for every $\mathfrak{F}$-torsion module L.

9. Let E be an injective module. The *E-dominant dimension* of a module M is said to be $\geq n$ (notation: E-dom.dim $M \geq n$) if there exists an exact sequence

$$0 \to M \to E_1 \to \cdots \to E_n$$

where each E_i is a direct product of copies of E. Show that with respect to the hereditary torsion theory cogenerated by E one has:

(i) M is $\mathfrak{F}$-torsion-free if and only if E-dom.dim $M \geq 1$.

(ii) M is $\mathfrak{F}$-closed if and only if E-dom.dim $M \geq 2$.

10. Let $R^k t$ denote the k-th right derived functor of the $\mathfrak{F}$-torsion radical. Let M be a module.

(i) Show that E-dom.dim $M \geq n$ if and only if $R^k t(M) = 0$ for all $k < n$, where E is an injective module cogenerating the $\mathfrak{F}$-torsion theory (cf. Exercise 9).

Assume the $\mathfrak{F}$-torsion theory is stable. Show that:

(ii) If M is $\mathfrak{F}$-torsion, then $R^k t(M) = 0$ for $k > 0$.

(iii) Coker $\psi_M = R^1 t(M)$, where $\psi_M: M \to M_{\mathfrak{F}}$ is the canonical map.

11. Show that if $M_{\mathfrak{F}}$ is an injective module for every A-module M, then the functor q: **Mod**-$A \to$ **Mod**-$A_{\mathfrak{F}}$ is exact. (Hint: use Exercise 4.)

12. Suppose M is a submodule of an $\mathfrak{F}$-closed module N. Show that $M_{\mathfrak{F}} \cong \{x \in N \mid (M:x) \in \mathfrak{F}\}$.

13. Suppose the Gabriel topology $\mathfrak{F}$ consists of dense right ideals. Show that

$$A_{\mathfrak{F}} \cong \{q \in Q_{\max} \mid q\,\mathfrak{a} \subset A \text{ for some } \mathfrak{a} \in \mathfrak{F}\}.$$

14. Let $\mathfrak{a}$ be a two-sided ideal of A. Show that:

(i) The torsion theory generated by the module $A/\mathfrak{a}$ is hereditary.

(ii) The corresponding quotient category consists of those modules M for which the canonical map $M \to \operatorname{Hom}_A(\mathfrak{a}, M)$ is an isomorphism. (We denote this category by **Mod**-$(A, \mathfrak{a})$.)

15. Assume M and N are $A_{\mathfrak{F}}$-modules, with N $\mathfrak{F}$-torsion-free as an A-module. Show that:

(i) The canonical map $\operatorname{Hom}_{A_{\mathfrak{F}}}(M, N) \to \operatorname{Hom}_A(M, N)$ is an isomorphism.

(ii) The canonical map $\operatorname{Hom}_{A_{\mathfrak{F}}}(\mathfrak{a} A_{\mathfrak{F}}, N) \to \operatorname{Hom}_A(\mathfrak{a}, N)$ is an isomorphism for every right ideal $\mathfrak{a}$ of A.

16. Let $\mathfrak{F}_i$ $(i \in I)$ be a family of Gabriel topologies on A, and consider the Gabriel topology $\mathfrak{F} = \bigcap_I \mathfrak{F}_i$. Show that:

(i) M is an $\mathfrak{F}$-torsion module if and only if M is an $\mathfrak{F}_i$-torsion module for each $i \in I$.

(ii) If the $\mathfrak{F}_i$-torsion theory is cogenerated by the injective module E_i, then the $\mathfrak{F}$-torsion theory is cogenerated by $\prod_I E_i$.

(iii) If M is $\mathfrak{F}$-torsion-free, then $M_{\mathfrak{F}} \cong \{x \in \prod_I M_{\mathfrak{F}_i} \mid \text{for each } i \in I \text{ there exists } \mathfrak{a}_i \in \mathfrak{F}_i \text{ such that } x\,\mathfrak{a}_i \subset M\}$.

(Hint: show that $\prod_I M_{\mathfrak{F}_i}$ is $\mathfrak{F}$-closed, and use Exercise 10.)

17. Let A be a right non-singular ring. Show that:

(i) Every essential monomorphism is rational.

(ii) If every module is rationally complete, then A is semi-simple.

§4

18. Show that if $K \subset L \subset M$ and $K \in \mathfrak{F}(L)$, then $K^c = L^c$ in M.

19. Denote the lattice operations in $\operatorname{Sat}_{\mathfrak{F}}(M)$ by $\vee$ and $\wedge$. Show that:

(i) $K^c \vee L^c = (K+L)^c$.

(ii) $K^c \wedge L^c = (K \cap L)^c$.

20. Show that if L is an essentially closed submodule of M, then $L \in \mathsf{Sat}_{\mathfrak{F}_M^0}(M)$.

21. Show that the following properties of a proper right ideal $\mathfrak{a}$ are equivalent:

(a) There is a Gabriel topology $\mathfrak{F}$ such that $\mathfrak{a}$ is maximal among the proper $\mathfrak{F}$-saturated right ideals of A.

(b) There is a Gabriel topology $\mathfrak{F}$ such that $\mathfrak{a}$ is $\mathfrak{F}$-saturated in A and $\mathfrak{b} \in \mathfrak{F}$ for every right ideal $\mathfrak{b} \supsetneqq \mathfrak{a}$.

(c) $\operatorname{Hom}_A(A/\mathfrak{b}, E(A/\mathfrak{a})) = 0$ for every right ideal $\mathfrak{b} \supsetneqq \mathfrak{a}$.

(d) There is an indecomposable injective module E such that $\mathfrak{a}$ is a maximal member of

$$\{\operatorname{Ann}(x) \mid 0 \neq x \in E\}.$$

A right ideal $\mathfrak{a}$ with these properties is called *critical.*

22. Let A be a commutative ring. Show that every critical ideal is a prime ideal.

23. An exact sequence $0 \to L \to M \to N \to 0$ of modules is called $\mathfrak{F}$-*pure* if for every x belonging to the $\mathfrak{F}$-torsion submodule of N there exists $y \in M$ mapping onto x such that $\operatorname{Ann}(x) = \operatorname{Ann}(y)$. L is then an $\mathfrak{F}$-*pure* submodule of M. Show that:

(i) L is $\mathfrak{F}$-pure in M if and only if for every $x \in M$ such that $(L:x) \in \mathfrak{F}$ there exists $y \in L$ such that $y\,a = x\,a$ for all $a \in (L:x)$.

(ii) The following properties of a module L are equivalent:

(a) L is $\mathfrak{F}$-injective.

(b) Every exact sequence $0 \to L \to M \to N \to 0$ is $\mathfrak{F}$-pure.

(c) L is an $\mathfrak{F}$-pure submodule of $E(L)$.

(iii) The $\mathfrak{F}$-torsion submodule of M is $\mathfrak{F}$-pure in M.

(iv) The following properties of a submodule L of M are equivalent:

(a) $L \in \mathsf{Sat}_{\mathfrak{F}}(M)$.

(b) $(L:x) \in \mathsf{Sat}_{\mathfrak{F}}(A)$ for every $x \in M$.

(c) L contains the $\mathfrak{F}$-torsion submodule of M and is $\mathfrak{F}$-pure in M.

24. Show that the following properties of an exact sequence $0 \to L \to M \to N \to 0$ of left A-modules are equivalent:

(a) $\mathfrak{a} L = \mathfrak{a} M \cap L$ for all $\mathfrak{a} \in \mathfrak{F}$.

(b) $0 \to \hat{N} \to \hat{M} \to \hat{L} \to 0$ is an $\mathfrak{F}$-pure sequence (Exercise 23), where $(\hat{\cdot}) = \operatorname{Hom}_{\mathbb{Z}}(\cdot, \mathbb{Q}/\mathbb{Z})$.

A sequence with these properties is called $\mathfrak{F}$-*copure.*

25. Show that the following properties of a left A-module F are equivalent:

(a) $\mathfrak{a} \otimes_A F \to F$ is a monomorphism when $\mathfrak{a} \in \mathfrak{F}$.

(b) $L \otimes_A F \to M \otimes_A F$ is a monomorphism when $L \in \mathfrak{F}(M)$.

(c) All exact sequences $0 \to H \to K \to F \to 0$ of left A-modules are $\mathfrak{F}$-copure (Exercise 24).

(d) $\hat{F} = \operatorname{Hom}_{\mathbb{Z}}(F, \mathbb{Q}/\mathbb{Z})$ is $\mathfrak{F}$-injective.

A module F with these properties is called $\mathfrak{F}$-*flat.*

§5

26. Show that the following properties of $\mathfrak{F}$ are equivalent:

(a) Every right ideal belonging to $\mathfrak{F}$ is a projective module.

(b) If P is a projective module and $L \in \mathfrak{F}(P)$, then L is projective.

(c) Every quotient module of an $\mathfrak{F}$-injective module is $\mathfrak{F}$-injective.

27. Assume that $\mathfrak{F}$ is a Gabriel topology on A such that every finitely generated right ideal belonging to $\mathfrak{F}$ is projective. Show that if M is a free module and L is a finitely generated submodule of M such that $L \in \mathfrak{F}(M)$, then L is isomorphic to a direct sum of finitely generated right ideals in $\mathfrak{F}$. (Hint: cf. Prop. I.6.9.)

Chapter X. The Category of Modules of Quotients

In this chapter we will prove that every quotient category **Mod**-$(A, \mathfrak{F})$ is a Grothendieck category. The proof uses arguments of a rather general nature, and it can also be used to show that the category of abelian sheaves on a topological space is a Grothendieck category. The second main result of this chapter is the Popescu-Gabriel theorem, which states that every Grothendieck category actually is a quotient category of a module category.

§ 1. Giraud Subcategories

Let $\mathbf{A}$ be a complete Grothendieck category.

Definition. A full subcategory $\mathbf{C}$ of $\mathbf{A}$ is *reflective* if the inclusion functor $i\colon \mathbf{C} \to \mathbf{A}$ has a left adjoint a.

(Note: some authors use the term "coreflective" instead of "reflective".) We now assume that $\mathbf{C}$ is reflective in $\mathbf{A}$, which means that there are natural isomorphisms

$$\eta_{B,C}\colon \operatorname{Hom}(B, C) \to \operatorname{Hom}(a(B), C)$$

for $B \in \mathbf{A}$ and $C \in \mathbf{C}$, with corresponding natural transformations $\xi\colon a\,i \to 1_{\mathbf{C}}$ and $\zeta\colon 1_{\mathbf{A}} \to i\,a$. The couple $(a(B), \zeta_B)$ solves a universal problem: for every morphism $\alpha\colon B \to C$, where $C \in \mathbf{C}$, there exists a unique $\bar{\alpha}\colon a(B) \to C$ such that $\bar{\alpha}\,\zeta_B = \alpha$.

$$\begin{array}{ccc} B & \xrightarrow{\zeta_B} & a(B) \\ & {\scriptstyle\alpha}\searrow & \downarrow{\scriptstyle\bar{\alpha}} \\ & & C \end{array}$$

From this it is clear that $\xi\colon a\,i \to 1_{\mathbf{C}}$ is a natural equivalence, i.e. $a(C) \cong C$ for every $C \in \mathbf{C}$.

Lemma 1.1. *If there exists* $\alpha\colon a(B) \to B$ *such that* $\alpha\,\zeta_B = 1_B$, *then* ζ_B *is an isomorphism.*

Proof. $\zeta_B\colon B \to a(B)$ has the two factorizations $\zeta_B = 1 \cdot \zeta_B = (\zeta_B\,\alpha)\,\zeta_B$ over $a(B)$, so the unicity of factorizations implies $\zeta_B\,\alpha = 1$. Hence ζ_B is an isomorphism. □

Proposition 1.2. *The reflective subcategory* $\mathbf{C}$ *is complete and cocomplete.*

Proof. First note that **C** is preadditive, since it is a full subcategory of **A**. Let **I** be a small category and $G\colon \mathbf{I}\to\mathbf{C}$ a functor. Then $\varprojlim iG$ exists and we denote it by B, with canonical morphisms $\pi_i\colon B\to G(i)$. Since $G(i)\in\mathbf{C}$, there exist $\bar\pi_i\colon a(B)\to G(i)$ such that $\bar\pi_i\zeta_B=\pi_i$. The family $(\bar\pi_i)_{i\in\mathbf{I}}$ is compatible with the morphisms in **I**, for if $\lambda\colon i\to j$ in **I**, then

$$G(\lambda)\,\bar\pi_i\,\zeta_B=G(\lambda)\,\pi_i=\pi_j=\bar\pi_j\,\zeta_B$$

and hence $G(\lambda)\,\bar\pi_i=\bar\pi_j$. Consequently there is induced a morphism $\beta\colon a(B)\to \varprojlim G=B$ such that $\pi_i\,\beta=\bar\pi_i$ for all i. Then $\pi_i\,\beta\,\zeta_B=\bar\pi_i\,\zeta_B=\pi_i$ for all i, so $\beta\,\zeta_B=1_B$. It follows from Lemma 1.1 that ζ_B is an isomorphism, and it is then clear that $a(B)$ is a limit for G in **C**. We have by the way obtained the formula

$$i(\varprojlim G)=\varprojlim iG$$

which also follows from the fact that the right adjoint functor i commutes with limits (when these exist) (Prop. IV.9.4).

To prove that $\varinjlim G$ exists in **C** is easier, because the left adjoint a preserves colimits and we therefore have $a(\varinjlim iG)=\varinjlim a\,i\,G=\varinjlim G$, since $a\,i\cong 1$. ▯

In other words, limits in **C** may be computed in **A**, while colimits in **C** are taken in **A** and then reflected into **C**.

Proposition 1.3. *If* **C** *is reflective in* **A** *and* a: $\mathbf{A}\to\mathbf{C}$ *preserves kernels, then* **C** *is an abelian category with exact direct limits and a generator.*

Proof. We have proved that **C** is preadditive and has limits and colimits. To prove that **C** is abelian, it only remains to show that if α is a morphism in **C**, then the canonical morphism

$$\bar\alpha\colon\ \mathrm{Coker}(\ker\alpha)\to\mathrm{Ker}(\mathrm{coker}\,\alpha)$$

is an isomorphism. If we denote kernels and cokernels taken in **A** by underlining them, we have $\mathrm{Coker}(\ker\alpha)=a(\underline{\mathrm{Coker}}(\underline{\ker}\,\alpha))$ and $\mathrm{Ker}(\mathrm{coker}\,\alpha)=\underline{\mathrm{Ker}}(a(\underline{\mathrm{coker}}\,\alpha))$ $=a(\underline{\mathrm{Ker}}(\underline{\mathrm{coker}}\,\alpha))$ since a preserves kernels. $\bar\alpha$ is therefore an isomorphism.

Next we show that direct limits are exact. Let **I** be a small directed category and $G, G'\colon\mathbf{I}\to\mathbf{C}$ two functors with a monomorphism $G\to G'$. The induced morphism $\varinjlim iG\to\varinjlim iG'$ is a monomorphism in **A**, and since a preserves monomorphisms, it follows that $\varinjlim G\to\varinjlim G'$ is a monomorphism in **C**. Finally, it is easy to see that if U is a generator for **A**, then $a(U)$ is a generator for **C**. ▯

Definition. A reflective subcategory of **A** is called a *Giraud subcategory* if the left adjoint of the inclusion functor preserves kernels.

Thus if **C** is a Giraud subcategory of **A**, then **C** is a Grothendieck category and the left adjoint a of $i\colon\mathbf{C}\to\mathbf{A}$ is an exact functor. It is important to notice that the inclusion functor i is in general not exact.

Proposition 1.4. *Let* **C** *be a Giraud subcategory of* **A**. *An object* C *of* **C** *is injective in* **C** *if and only if* $i(C)$ *is injective in* **A**.

Proof. Since i preserves monomorphisms, it is clear that if C is injective in **A**, then C is injective also in **C**. Suppose conversely that C is injective in **C**. Let there be given a monomorphism $\beta\colon B\to B'$ in **A** and a morphism $\varphi\colon B\to C$. Then φ

induces a morphism $a(B) \to C$, which may be extended in $\mathbf{C}$ to a morphism $a(B') \to C$. The composed morphism $B' \to a(B') \to C$ then extends φ. $\square$

If $\mathbf{C}$ is a Giraud subcategory of $\mathbf{A}$, and $a: \mathbf{C} \to \mathbf{A}$ is the left adjoint of the inclusion functor, then we consider the class $\mathscr{T}$ of objects B of $\mathbf{A}$ for which $a(B)=0$, and also the class $\mathscr{F}$ of objects B of $\mathbf{A}$ for which the reflection $B \to i\,a(B)$ is a monomorphism. Then:

Proposition 1.5. *$(\mathscr{T}, \mathscr{F})$ is a hereditary torsion theory for* $\mathbf{A}$.

Proof. From the exactness of a follows immediately that $\mathscr{T}$ is closed under subobjects, quotient objects and extensions. Since a has a right adjoint, a preserves coproducts, and hence $\mathscr{T}$ is closed under coproducts. $\mathscr{T}$ is thus a hereditary torsion class.

Clearly $\mathrm{Hom}(T, C)=0$ for every $T \in \mathscr{T}$ and $C \in \mathbf{C}$, and it follows that $\mathrm{Hom}(T, F)=0$ for $T \in \mathscr{T}$ and $F \in \mathscr{F}$. Conversely, if B is an object such that $\mathrm{Hom}(T, B)=0$ for all $T \in \mathscr{T}$, then $B \in \mathscr{F}$ because the kernel of $B \to i\,a(B)$ belongs to $\mathscr{T}$. $\square$

Let us now specialize to the case when $\mathbf{A}$ is the category of right modules over a ring A. Let $\mathfrak{F}$ be a Gabriel topology on A.

Theorem 1.6. $\mathbf{Mod}$-$(A, \mathfrak{F})$ *is a Giraud subcategory of* $\mathbf{Mod}$-A.

Proof. It was proved in Prop. IX.1.11 that $\mathbf{Mod}$-$(A, \mathfrak{F})$ is a reflective subcategory of $\mathbf{Mod}$-A. The reflector a is determined by $i\,a = LL$, where $L = \varinjlim \mathrm{Hom}(\mathfrak{a}, \cdot)$. The functors L and i preserve kernels, and since i is full and faithful, it follows that a preserves kernels. $\square$

$\mathbf{Mod}$-$(A, \mathfrak{F})$ is thus a Grothendieck category, and the functor a: $\mathbf{Mod}$-$A \to \mathbf{Mod}$-$(A, \mathfrak{F})$ is exact. By Prop. 1.4 we have:

Proposition 1.7. *An $\mathfrak{F}$-closed module M is an injective object in* $\mathbf{Mod}$-$(A, \mathfrak{F})$ *if and only if M is injective in* $\mathbf{Mod}$-A.

Since the injective envelope of an $\mathfrak{F}$-torsion-free module is $\mathfrak{F}$-torsion-free, it follows that $\mathbf{Mod}$-$(A, \mathfrak{F})$ has enough injectives:

Corollary 1.8. *Every object in* $\mathbf{Mod}$-$(A, \mathfrak{F})$ *is a subobject of an injective object.*

Proposition 1.9. *Let the injective module E cogenerate the torsion theory associated to $\mathfrak{F}$. E is then an injective cogenerator for the category* $\mathbf{Mod}$-$(A, \mathfrak{F})$.

Proof. E is an injective object in $\mathbf{Mod}$-$(A, \mathfrak{F})$, and $\mathrm{Hom}_{\mathbf{Mod}\text{-}(A,\mathfrak{F})}(M, E) = \mathrm{Hom}_A(M, E) \neq 0$ for every $\mathfrak{F}$-closed module M. $\square$

Examples

1. The Goldie topology. If $\mathfrak{G}$ denotes the Goldie topology on A, then every $\mathfrak{G}$-injective module is injective, so it follows from Prop. 1.7 that $\mathbf{Mod}$-$(A, \mathfrak{G})$ is a spectral category.

2. Abelian sheaves. Let X be a topological space and let $\mathbf{P}$ be the category of presheaves of abelian groups on X. This is a Grothendieck category (Example V.1.3). A presheaf F is a *sheaf* if it satisfies the following condition:

let $(U_i)_I$ be a family of open subsets of X, and U the union of the U_i; if for each $i \in I$ there is given an element $s_i \in F(U_i)$, such that s_i and s_j have a common restriction to $U_i \cap U_j$ for each couple (i, j), then there exists a unique $s \in F(U)$ with restriction s_i to U_i for each $i \in I$.

E.g. the presheaf of continuous real-valued functions (Example IV.7.2) is a sheaf on X.

We will show that the sheaves form a Giraud subcategory of $\mathbf{P}$. For each presheaf F and each point $x \in X$ we define the *stalk* of F at x as $F_x = \varinjlim F(U)$, where the direct limit is taken over the downwards directed family of open neighborhoods U of x. For each open set U and $x \in U$ there is a canonical homomorphism $\rho_{U,x}$: $F(U) \to F_x$. Construct a presheaf $\tilde{F}$ by putting

$$\tilde{F}(U) = \{(s_x)_{x \in U} \mid s_x \in F_x, \text{ and there is a covering of } U \text{ by open subsets } U_i \text{ with elements } s_i \in F(U_i) \text{ such that } \rho_{U_i,x}(s_i) = s_x \text{ for each } x \in U_i\}.$$

It may be verified that $\tilde{F}$ is a sheaf and that the assignment $F \mapsto \tilde{F}$ is a functor a: $\mathbf{P} \to \mathbf{S}$, where $\mathbf{S}$ denotes the full subcategory of sheaves. Furthermore, it turns out that a is a left adjoint of the inclusion functor i: $\mathbf{S} \to \mathbf{P}$, and that a preserves kernels. $\mathbf{S}$ is thus a Giraud subcategory of $\mathbf{P}$.

§ 2. Gabriel Topologies and Giraud Subcategories

To each Gabriel topology $\mathfrak{F}$ on the ring A we have associated the Giraud subcategory **Mod**-$(A, \mathfrak{F})$ of **Mod**-A. We will now show that every Giraud subcategory of **Mod**-A arises in this way. The following result is essentially due to Gabriel [2]:

Theorem 2.1. *There is a bijective correspondence between Gabriel topologies on A and strict Giraud subcategories of* **Mod**-A.

Proof. Here we call a subcategory $\mathbf{C}$ of a category $\mathbf{D}$ *strict* if the class of objects of $\mathbf{C}$ is closed under isomorphisms in $\mathbf{D}$. If $\mathfrak{F}$ is a Gabriel topology on A, then **Mod**-$(A, \mathfrak{F})$ is a strict Giraud subcategory of **Mod**-A. Conversely, if $\mathbf{C}$ is a strict Giraud subcategory of **Mod**-A, and a: $\mathbf{C} \to$ **Mod**-A is the left adjoint of the inclusion functor, then we have defined in the previous section a hereditary torsion class $\mathscr{T}$, consisting of the modules M for which $a(M) = 0$. In this way we have assigned a Gabriel topology $\{\mathfrak{a} \mid a(A/\mathfrak{a}) = 0\}$ to $\mathbf{C}$.

We now have to verify that these two maps

$$\{\text{Gabriel topologies on } A\} \underset{\Phi}{\overset{\Gamma}{\rightleftarrows}} \{\text{Giraud subcategories of } \mathbf{Mod}\text{-}A\}$$

are the inverses of each other. To show that $\Phi\Gamma = \mathrm{id}$ is easy, because if $\mathfrak{F}$ is a Gabriel topology on A and M is a module such that $M_{\mathfrak{F}} = a(M) = 0$, then M certainly is an $\mathfrak{F}$-torsion module.

It remains to show that $\Gamma\Phi = \mathrm{id}$. Let $\mathbf{D}$ be a Giraud subcategory of **Mod**-A with the inclusion i': $\mathbf{D} \to$ **Mod**-A and its left adjoint a'. Let $\mathfrak{F}$ be the Gabriel topology of right ideals $\mathfrak{a}$ for which $a'(A/\mathfrak{a}) = 0$. We must show that the category

Mod-$(A, \mathfrak{F})$ is equivalent to **D**. We have a diagram

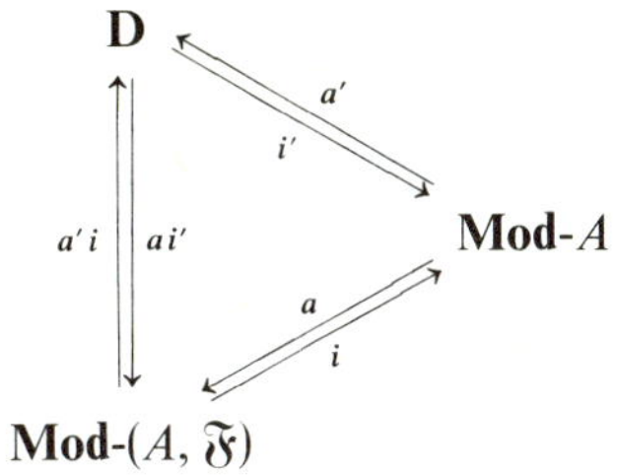

We first note that $a' i a \cong a'$, for if $M \in$ **Mod**-A then the exact sequence

$$0 \to \operatorname{Ker} \psi_M \to M \xrightarrow{\psi_M} M_{\mathfrak{F}} \to \operatorname{Coker} \psi_M \to 0$$

gives $a'(M) \cong a'(M_{\mathfrak{F}})$ since a' is zero on $\mathfrak{F}$-torsion modules. Similarly we have $a i' a' \cong a$, because the adjointness transformation ϕ: $\mathrm{id} \to i' a'$ gives the exact sequence

$$0 \to \operatorname{Ker} \phi_M \to M \xrightarrow{\phi_M} i' a'(M) \to \operatorname{Coker} \phi_M \to 0$$

and $a'(\phi_M)$ is an isomorphism, so $\operatorname{Ker} \phi_M$ and $\operatorname{Coker} \phi_M$ are $\mathfrak{F}$-torsion modules, and it follows that $a(\phi_M)$ is an isomorphism $a(M) \cong a i' a'(M)$.

From these two natural equivalences we obtain $a' i \cdot a i' \cong a' i' \cong \mathrm{id}$ and $a i' \cdot a' i \cong a i \cong \mathrm{id}$, and hence **D** and **Mod**-$(A, \mathfrak{F})$ are equivalent categories. □

Let $\mathfrak{F}$ be a Gabriel topology on A, and recall the diagram of functors (Chap. IX, §1)

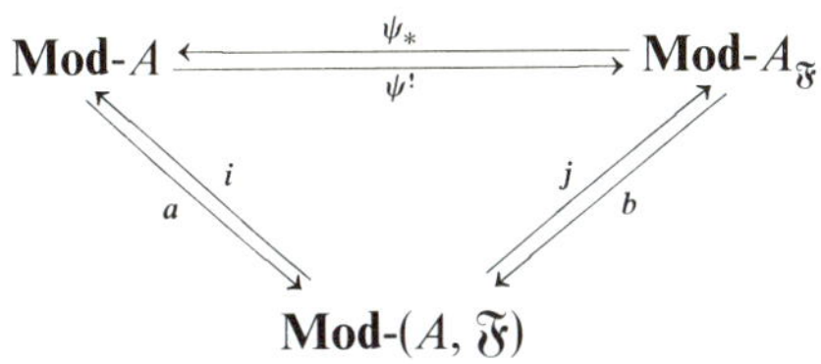

where we have added the two new functors $b = a\psi_*$ and $\psi^! = \operatorname{Hom}_A(A_{\mathfrak{F}}, \cdot)$. We know that $\psi^!$ is a right adjoint of ψ_* (Example IV.9.4), and $\psi^! i = j$ since M $\mathfrak{F}$-closed implies $\operatorname{Hom}_A(A_{\mathfrak{F}}, M) \cong \operatorname{Hom}_A(A, M) \cong M$. The functor j then has a left adjoint which is the composition of the left adjoints of i and $\psi^!$, i.e. $b = a\psi_*$ is a left adjoint of j. Since a and ψ_* are exact functors, also b is exact. The category **Mod**-$(A, \mathfrak{F})$ is thus a Giraud subcategory also of **Mod**-$A_{\mathfrak{F}}$. Let us determine the Gabriel topology $\mathfrak{F}^e$ on $A_{\mathfrak{F}}$ which corresponds to this Giraud subcategory. ψ_A: $A \to A_{\mathfrak{F}}$ denotes as usual the canonical ring homomorphism.

Proposition 2.2. *$\mathfrak{F}^e$ consists of those right ideals $\mathfrak{u}$ of $A_{\mathfrak{F}}$ for which $\psi_A^{-1}(\mathfrak{u}) \in \mathfrak{F}$.*

Proof. An $A_{\mathfrak{F}}$-module M is an $\mathfrak{F}^e$-torsion module if and only if $b(M) = 0$, and since $b = a\psi_*$, this happens if and only if $\psi_*(M)$ is an $\mathfrak{F}$-torsion module. In particular, a right ideal $\mathfrak{u}$ of $A_{\mathfrak{F}}$ belongs to $\mathfrak{F}^e$ if and only if $A_{\mathfrak{F}}/\mathfrak{u}$ is an $\mathfrak{F}$-torsion module, i.e. $\mathfrak{u} \in \mathfrak{F}(A_{\mathfrak{F}})$. The result then follows from:

Lemma 2.3. *Let M be an A-module and let L be an A-submodule of $M_{\mathfrak{F}}$. Then $L \in \mathfrak{F}(M_{\mathfrak{F}})$ if and only if $\psi_M^{-1}(L) \in \mathfrak{F}(M)$.*

Proof. This is immediately seen from the commutative diagram

$$\begin{array}{ccccccc}
 & & M & \xrightarrow{\psi_M} & M_{\mathfrak{F}} & \longrightarrow & \operatorname{Coker}\psi_M \longrightarrow 0 \\
 & & \downarrow & & \downarrow & & \| \\
0 & \longrightarrow & M/\psi_M^{-1}(L) & \longrightarrow & M_{\mathfrak{F}}/L & \longrightarrow & \operatorname{Coker}\psi_M \longrightarrow 0
\end{array}$$

where the rows are exact. □

For an $A_{\mathfrak{F}}$-module M, the module of quotients with respect to $\mathfrak{F}^e$ will be $M_{\mathfrak{F}^e}=b(M)=a\psi_*(M)=(\psi_*(M))_{\mathfrak{F}}$. Thus $M_{\mathfrak{F}^e}=M$ if and only if M is $\mathfrak{F}$-closed as an A-module. In particular, $A_{\mathfrak{F}}$ is its own ring of quotients with respect to $\mathfrak{F}^e$, and hence $\mathfrak{F}^e$ is weaker than the canonical topology on $A_{\mathfrak{F}}$. Let E be an injective A-module cogenerating the torsion theory corresponding to $\mathfrak{F}$. E is then an $A_{\mathfrak{F}}$-module, and it is easily seen that E cogenerates the torsion theory for **Mod**-$A_{\mathfrak{F}}$ corresponding to $\mathfrak{F}^e$.

Example. The maximal ring of quotients. $E(A)$ cogenerates the torsion theory corresponding to the topology $\mathfrak{D}$ of dense right ideals. Since $E(A)$ also cogenerates the torsion theory on $Q^r_{\max}(A)$ corresponding to $\mathfrak{D}^e$, it follows that $\mathfrak{D}^e$ consists of the dense right ideals of $Q^r_{\max}(A)$. In particular, a right ideal $\mathfrak{q}$ of $Q^r_{\max}(A)$ is dense if and only if $\mathfrak{q}\cap A$ is a dense right ideal of A.

§ 3. Rings of Quotients of Morita Equivalent Rings

Let A and B be two Morita equivalent rings, with the equivalence given by functors

$$T=\cdot\otimes_A P\colon \mathbf{Mod}\text{-}A\to\mathbf{Mod}\text{-}B,$$
$$S=\cdot\otimes_B Q\colon \mathbf{Mod}\text{-}B\to\mathbf{Mod}\text{-}A,$$

for bimodules ${}_AP_B$ and ${}_BQ_A$. Assume there is given a hereditary torsion theory $(\mathcal{T},\mathcal{F})$ for **Mod**-A, with the corresponding Gabriel topology $\mathfrak{F}$. It is easily seen that $(T(\mathcal{T}), T(\mathcal{F}))$ is a hereditary torsion theory for **Mod**-B, and the corresponding Gabriel topology will be denoted by $T(\mathfrak{F})$. Clearly we obtain in this way a bijective correspondence between hereditary torsion theories for **Mod**-A and for **Mod**-B.

Proposition 3.1. *Let A and B be Morita equivalent rings, with the equivalence T:* **Mod**-$A\to$ **Mod**-B. *If $\mathfrak{F}$ is a Gabriel topology on A, then:*

(i) *The rings $A_{\mathfrak{F}}$ and $B_{T(\mathfrak{F})}$ are Morita equivalent.*

(ii) *The categories* **Mod**-$(A,\mathfrak{F})$ *and* **Mod**-$(B,T(\mathfrak{F}))$ *are equivalent.*

Proof. (i) We use Theorem IX.3.3 to describe the two rings of quotients. Let E be an injective A-module cogenerating the $\mathfrak{F}$-torsion theory. Then $T(E)$ is an injective B-module, and it cogenerates the $T(\mathfrak{F})$-torsion theory because

$$\operatorname{Hom}_B(N,T(E))=0\Leftrightarrow\operatorname{Hom}_A(S(N),E)=0\Leftrightarrow S(N)$$
$$\text{is }\mathfrak{F}\text{-torsion}\Leftrightarrow N\cong TS(N)\text{ is }T(\mathfrak{F})\text{-torsion}.$$

We may assume that both E and $T(E)$ are countercyclic modules, for otherwise we can replace E by a suitable product E^I, since T preserves direct products.

The functor T induces an isomorphism

$$\mathrm{Hom}_B(T(E), T(E)) \cong \mathrm{Hom}_A(E, E) = H.$$

Now $T(E) = E \otimes_A P$, where the left A-module P is a finitely generated projective generator, i.e. $P \sim A$ in the notation of Chap. IV, §10. It follows that $E \otimes_A P \sim E$ as left H-modules. But then the rings

$$B_{T(\mathfrak{F})} = \mathrm{Bic}_B(E \otimes_A P) = \mathrm{Hom}_H(E \otimes_A P, E \otimes_A P)$$

and

$$A_{\mathfrak{F}} = \mathrm{Bic}_A(E) = \mathrm{Hom}_H(E, E)$$

are Morita equivalent by Prop. IV.10.5.

(ii) There is a diagram of functors (with obvious notations):

$$\begin{array}{ccc} \mathbf{Mod}\text{-}A & \underset{i}{\overset{a}{\rightleftarrows}} & \mathbf{Mod}\text{-}(A, \mathfrak{F}) \\ S \downarrow \uparrow T & & a'Si \downarrow \uparrow aTi' \\ \mathbf{Mod}\text{-}B & \underset{i'}{\overset{a'}{\rightleftarrows}} & \mathbf{Mod}\text{-}(B, T(\mathfrak{F})) \end{array}$$

Mod-$(B, T(\mathfrak{F}))$ may be considered as (equivalent to) a Giraud subcategory of **Mod**-A, by means of the functors Ti' and $i'S$. The proof of Theorem 2.1 shows that the functors $a'Si$ and aTi' define an equivalence between **Mod**-$(A, \mathfrak{F})$ and **Mod**-$(B, T(\mathfrak{F}))$. □

Examples

1. Matrix rings. Let $B = M_n(A)$, so we may choose ${}_AP$ to be the free module $P = A^n$. If $\mathfrak{F}$ is a Gabriel topology on A, then with the notation used in the proof of Prop. 3.1 (i) we have

$$M_n(A)_{T(\mathfrak{F})} \cong \mathrm{Hom}_H(E \otimes_A A^n, E \otimes_A A^n) \cong \mathrm{Hom}_H(E^n, E^n) \cong M_n(A_{\mathfrak{F}}).$$

Thus, the two processes of formation of rings of quotients and of rings of matrices "commute".

2. Maximal ring of quotients. The maximal ring of quotients is Morita invariant:

Proposition 3.2. *If A and B are Morita equivalent rings, then $Q_{\max}(A)$ and $Q_{\max}(B)$ are Morita equivalent.*

Proof. If A and B are Morita equivalent, then there are bimodules ${}_AP_B$ and ${}_BQ_A$ such that $P \otimes_B Q \cong A$ and $Q \otimes_A P \cong B$. In the proof of Prop. 3.1 (i) we now choose $E = E(A)$. Since $A \sim Q$, we have $E(A) \sim E(Q)$ and consequently

$$E(A) \sim E(A) \otimes_A P \sim E(Q) \otimes_A P.$$

It is easily seen that equivalences preserve essential monomorphisms, so it follows that $E(Q) \otimes_A P \cong E(Q \otimes_A P) \cong E(B)$. Hence $\mathrm{Bic}_B(E(B)) = Q_{\max}(B)$ and $\mathrm{Bic}_A(E(A)) = Q_{\max}(A)$ are Morita equivalent. □

Combining this result with the previous example, we get (Utumi [1]):

Corollary 3.3. $Q_{\max}(M_n(A)) \cong M_n(Q_{\max}(A))$.

§4. Representation of Grothendieck Categories

The theorem of Popescu and Gabriel [1] reduces (at least in principle) the theory of Grothendieck categories to a study of categories of modules of quotients. A reduction of the theory of Grothendieck categories to ring theory in an entirely different direction (and deeper in nature) is indicated by the Gabriel-Roos-Oberst theory, which establishes a duality between Grothendieck categories and certain categories of topological modules over linearly topological rings (Roos [5], Oberst [1]).

Theorem 4.1 (Popescu and Gabriel [1]). *Let* $\mathbf{C}$ *be a Grothendieck category with a generator* U. *Put* $A=\mathrm{Hom}_{\mathbf{C}}(U,U)$ *and let* $T\colon \mathbf{C}\to\mathbf{Mod}\text{-}A$ *be the functor* $T(C)=\mathrm{Hom}_{\mathbf{C}}(U,C)$. *Then:*

(i) T *is full and faithful.*

(ii) T *induces an equivalence between* $\mathbf{C}$ *and the category* $\mathbf{Mod}\text{-}(A,\mathfrak{F})$, *where* $\mathfrak{F}$ *is the strongest Gabriel topology on* A *for which all modules* $T(C)$ *are* $\mathfrak{F}$*-closed.*

Proof. (i): Since U is a generator, the functor $T=\mathrm{Hom}(U,\cdot)$ is faithful. To see that it is full, we must show that if C and D are objects in $\mathbf{C}$ and $\Phi\colon \mathrm{Hom}_{\mathbf{C}}(U,C)\to\mathrm{Hom}_{\mathbf{C}}(U,D)$ is A-linear, then Φ is of the form $\Phi(f)=\varphi f$ for some $\varphi\colon C\to D$. Let I be the set of all morphisms $U\to C$. For each $\alpha\in I$ we let $i_\alpha\colon U\to U^{(I)}$ denote the corresponding injection. There is a unique morphism $\lambda\colon U^{(I)}\to C$ such that $\lambda\, i_\alpha=\alpha$ for each $\alpha\in I$, and λ is an epimorphism since U is a generator. Similarly there is a unique morphism $\mu\colon U^{(I)}\to D$ such that $\mu\, i_\alpha=\Phi(\alpha)$ for each $\alpha\in I$. Let $\kappa\colon K\to U^{(I)}$ be the kernel of λ. If we can show that $\mu\,\kappa=0$, then μ factors as $\mu=\varphi\,\lambda$ for some $\varphi\colon C\to D$, and for each $\alpha\colon U\to C$ we get $\Phi(\alpha)=\mu\, i_\alpha=\varphi\,\lambda\, i_\alpha=\varphi\,\alpha$, and our assertion would be proved.

So we set about proving that $\mu\,\kappa=0$. For each finite subset J of I and each $\alpha\in J$ there are canonical morphisms $\pi'_\alpha\colon U^{(J)}\to U$, $i'_\alpha\colon U\to U^{(J)}$ and $i_J\colon U^{(J)}\to U^{(I)}$. Let $\kappa_J\colon K_J\to U^{(J)}$ be the kernel of the composed morphism $\lambda\, i_J\colon U^{(J)}\to C$. Since K is the direct limit of the kernels K_J for all finite subsets J of I, it suffices to show that $\mu\, i_J\,\kappa_J=0$. Now for each $\beta\colon U\to K_J$ we have, using the A-linearity of Φ, that

$$\begin{aligned}\mu\, i_J\,\kappa_J\,\beta&=\mu\, i_J\Big(\sum_{\alpha\in J} i'_\alpha\,\pi'_\alpha\Big)\kappa_J\,\beta=\sum_{\alpha\in J}\mu\, i_\alpha\,\pi'_\alpha\,\kappa_J\,\beta=\sum_{\alpha\in J}\Phi(\alpha)\,\pi'_\alpha\,\kappa_J\,\beta\\&=\sum_{\alpha\in J}\Phi(\alpha\,\pi'_\alpha\,\kappa_J\,\beta)=\Phi\Big(\sum_{\alpha\in J}\lambda\, i_J\, i'_\alpha\,\pi'_\alpha\,\kappa_J\,\beta\Big)=\Phi(\lambda\, i_J\,\kappa_J\,\beta)=0\end{aligned}$$

since $\lambda\, i_J\,\kappa_J=0$. Since this holds for arbitrary $\beta\colon U\to K_J$, it follows that $\mu\, i_J\,\kappa_J=0$.

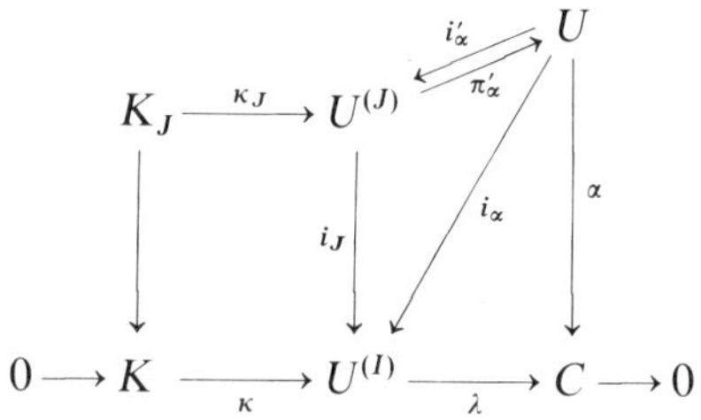

(ii): Let $\mathfrak{F}$ be the strongest Gabriel topology on A for which all modules of the form $T(C)$, with C in $\mathbf{C}$, are $\mathfrak{F}$-closed. This topology exists by the results of Example IX.2.4; the corresponding torsion theory is cogenerated by the class of injective modules of the form $E(T(C)\oplus E(T(C))/T(C))$. Since every module $T(C)$ is $\mathfrak{F}$-closed, there results a diagram of functors

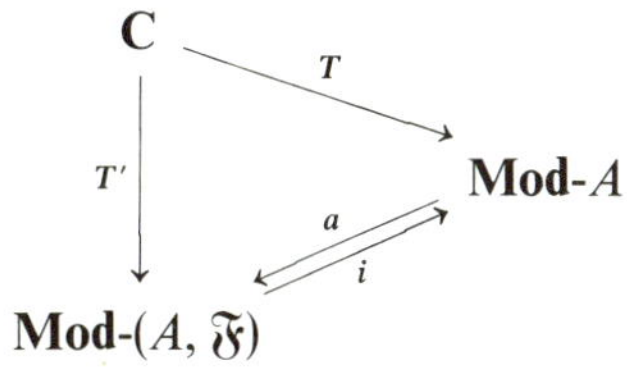

with $T=iT'$. We want to show that T' is an equivalence. Since $iT'=T$ is full and faithful by (i), also T' is full and faithful, and we only have to show that every $\mathfrak{F}$-closed module is isomorphic to a module of the form $T(C)$.

For each A-module M we choose an exact sequence

$$A^{(I)} \xrightarrow{\alpha=(\alpha_{ij})} A^{(J)} \longrightarrow M \longrightarrow 0 \tag{1}$$

where the homomorphism α is represented by a row-finite matrix (α_{ij}). Since A is the endomorphism ring of U, α induces in a natural way a morphism $\beta=(\beta_{ij})\colon U^{(I)} \to U^{(J)}$ in $\mathbf{C}$ such that $T(\beta_{ij})=\alpha_{ij}$. We define $\tilde{M}\in\mathbf{C}$ by the exact sequence

$$U^{(I)} \xrightarrow{\beta} U^{(J)} \longrightarrow \tilde{M} \longrightarrow 0. \tag{2}$$

We now apply the functor T' to (2). If we knew that T' preserves colimits, we would then get a commutative diagram in $\mathbf{Mod}\text{-}(A,\mathfrak{F})$

$$\begin{array}{ccccccc} A^{(I)} & \xrightarrow{\alpha} & A^{(J)} & \longrightarrow & a(M) & \longrightarrow & 0 \\ \| & & \| & & & & \\ A^{(I)} & \xrightarrow[T'(\beta)=\alpha]{} & A^{(J)} & \longrightarrow & T'(\tilde{M}) & \longrightarrow & 0 \end{array}$$

with exact rows (exact in $\mathbf{Mod}\text{-}(A,\mathfrak{F})$, not in $\mathbf{Mod}\text{-}A$!), and M $\mathfrak{F}$-closed would imply $M=T'(\tilde{M})$. To conclude the proof it thus remains to show:

Lemma 4.2. *The functor $T'\colon \mathbf{C}\to\mathbf{Mod}\text{-}(A,\mathfrak{F})$ is exact and preserves direct sums.*

Proof. $T'=aT$ is left exact, so to prove exactness it suffices to show that T' preserves epimorphisms. This means that if $\varphi\colon C'\to C''$ is an epimorphism in $\mathbf{C}$, then we should show that $\operatorname{Coker} T(\varphi)$ is an $\mathfrak{F}$-torsion module. By Prop. VI.6.9 it is equivalent to show that for each $x\in T(C'')$ one has

$$\operatorname{Hom}_A(A, T(C)) \cong \operatorname{Hom}_A((\operatorname{Im} T(\varphi):x), T(C))$$

for all $C\in\mathbf{C}$. Define $\xi\colon U\to C''$ such that $T(\xi)\colon A\to T(C'')$ maps 1 into x. From the pullback in $\mathbf{C}$

$$\begin{array}{ccccccccc} 0 & \longrightarrow & K & \xrightarrow{\lambda} & P & \xrightarrow{\mu} & U & \longrightarrow & 0 \\ & & \| & & \downarrow & & \downarrow{\scriptstyle\xi} & & \\ 0 & \longrightarrow & K & \longrightarrow & C' & \xrightarrow[\varphi]{} & C'' & \longrightarrow & 0 \end{array}$$

we obtain a pullback diagram with exact rows in **Mod**-A

$$\begin{array}{ccccccc} 0 & \longrightarrow & T(K) & \xrightarrow{T(\lambda)} & T(P) & \xrightarrow{T(\mu)} & A \\ & & \| & & \downarrow & & \downarrow \\ 0 & \longrightarrow & T(K) & \longrightarrow & T(C') & \xrightarrow[T(\varphi)]{} & T(C'') \end{array}$$

and $(\operatorname{Im} T(\varphi):x)=\operatorname{Im} T(\mu)$. Now suppose we have a homomorphism $\operatorname{Im} T(\mu)\to T(C)$ for some $C\in\mathbf{C}$. Since T is full, the composed homomorphism $T(P)\to\operatorname{Im} T(\mu)\to T(C)$ is of the form $T(\alpha)$ for some $\alpha: P\to C$ in $\mathbf{C}$. Since T also is faithful, $T(\alpha)\,T(\lambda)=0$ implies that $\alpha\lambda=0$. Hence α factors over U, and it follows that the given homomorphism $\operatorname{Im} T(\mu)\to T(C)$ extends to a homomorphism $A=T(U)\to T(C)$, which is precisely what we wanted to show.

It remains to prove that T' preserves direct sums. Actually we prove a little more, namely that aT preserves direct unions. Let (C_α) be a directed family of subobjects of $C\in\mathbf{C}$. We must show that the cokernel of the monomorphism

$$\varphi\colon \sum_\alpha T(C_\alpha)\to T(\sum_\alpha C_\alpha)$$

is a torsion module. By Prop. VI.6.9 this means that for each $x\in T(\sum C_\alpha)$ we shall show that $\operatorname{Hom}(A, T(C'))\cong\operatorname{Hom}((\operatorname{Im}\varphi:x), T(C'))$ for all $C'\in\mathbf{C}$. Define $\xi: U\to\sum C_\alpha$ such that $T(\xi)\colon A\to T(\sum C_\alpha)$ maps 1 to x. From the pullback diagrams

$$\begin{array}{ccccc} 0 & \longrightarrow & P_\alpha & \longrightarrow & U \\ & & \downarrow & & \downarrow{\scriptstyle\xi} \\ 0 & \longrightarrow & C_\alpha & \longrightarrow & \sum C_\alpha \end{array}$$

we obtain a commutative diagram

$$\begin{array}{ccccc} 0 & \longrightarrow & \sum T(P_\alpha) & \xrightarrow{\psi} & A \\ & & \downarrow & & \downarrow{\scriptstyle T(\xi)} \\ 0 & \longrightarrow & \sum T(C_\alpha) & \xrightarrow[\varphi]{} & T(\sum C_\alpha) \end{array}$$

which is a pullback diagram because pullbacks are preserved both by T and when taking direct limits in **Mod**-A. Therefore we have $(\operatorname{Im}\varphi:x)=\operatorname{Im}\psi\cong\sum T(P_\alpha)$. Now

$$\begin{aligned} \operatorname{Hom}_A(\textstyle\sum T(P_\alpha), T(C'))&\cong\varprojlim\operatorname{Hom}_A(T(P_\alpha), T(C'))\cong\varprojlim\operatorname{Hom}_{\mathbf{C}}(P_\alpha, C')\\ &\cong\operatorname{Hom}_{\mathbf{C}}(\textstyle\sum P_\alpha, C')=\operatorname{Hom}_{\mathbf{C}}(U, C')\cong\operatorname{Hom}_A(A, T(C')), \end{aligned}$$

where we have utilized the fact that exactness of $\varinjlim$ implies $\sum P_\alpha=U$.

This concludes the proof of the Popescu-Gabriel theorem. $\square$

As a corollary of the theorem we obtain, with the help of Cor. 1.6, the following important property of Grothendieck categories:

Corollary 4.3. *Every object in a Grothendieck category is a subobject of an injective object.*

We also note that Prop. 1.2 gives:

Corollary 4.4. *Every Grothendieck category is complete.*

Examples

1. Standard representations. In the proof of the Popescu-Gabriel theorem we represented the Grothendieck category **C** as the category of $\mathfrak{F}$-closed modules, where $\mathfrak{F}$ is the strongest Gabriel topology on $A = \mathrm{Hom}_{\mathbf{C}}(U, U)$, U an arbitrary generator for **C**, such that all A-modules $\mathrm{Hom}_{\mathbf{C}}(U, C)$ are $\mathfrak{F}$-closed. This topology is weaker than the canonical topology on A.

In general, if a Grothendieck category **C** is represented by

$$\mathbf{C} \underset{a}{\overset{i}{\rightleftarrows}} \mathbf{Mod}\text{-}A$$

as a Giraud subcategory of module category, we call this a *standard* representation if $i\,a(A) \cong A$.

For example, if $\mathfrak{F}$ is a Gabriel topology on an arbitrary ring A and $\mathbf{C} = \mathbf{Mod}\text{-}(A, \mathfrak{F})$, then

$$\mathbf{Mod}\text{-}(A, \mathfrak{F}) \underset{b}{\overset{j}{\rightleftarrows}} \mathbf{Mod}\text{-}A_{\mathfrak{F}}$$

gives a standard representation of $\mathbf{Mod}\text{-}(A, \mathfrak{F})$, according to §2.

2. Module categories. If the Grothendieck category **C** has a generator U which is a finitely generated, projective object, then the functor T is exact and preserves direct sums, and a slight modification in the proof of Theorem 4.1 (ii) shows that every module M is isomorphic to an object $T(M)$. Hence **C** is equivalent to the module category $\mathbf{Mod}\text{-}A$, where $A = \mathrm{Hom}_{\mathbf{C}}(U, U)$.

3. Objects generated by U. Let U be an arbitrary object of a Grothendieck category, and assume the object C is generated by U, i.e. there is an epimorphism $U^{(I)} \to C$ for some I. Consider $T(C) = \mathrm{Hom}_{\mathbf{C}}(U, C)$ as a right module over $A = \mathrm{Hom}_{\mathbf{C}}(U, U)$. The proof of Theorem 4.1 (i) shows that there is an isomorphism

$$\mathrm{Hom}_{\mathbf{C}}(C, C) \cong \mathrm{Hom}_A\big(T(C), T(C)\big).$$

Exercises

§1

1. Show that for every torsion theory on $\mathbf{Mod}\text{-}A$, the torsion-free modules form a reflective subcategory of $\mathbf{Mod}\text{-}A$.

2. Verify the assertions made at the end of Example 2:
(i) if F is a presheaf on X, then $\tilde{F}$ is a sheaf;
(ii) the assignment $F \mapsto \tilde{F}$ is a functor, which is a left adjoint of the inclusion functor $\mathbf{S} \to \mathbf{P}$, and it preserves kernels.

3. Let $\mathfrak{F}$ be a Gabriel topology on A, and suppose $T\colon \mathbf{Mod}\text{-}A \to \mathbf{B}$ is an exact functor into an abelian category **B**, such that $T(M) = 0$ for all $\mathfrak{F}$-torsion modules M. Show that T has a unique factorization $T = T'a$ over the category $\mathbf{Mod}\text{-}(A, \mathfrak{F})$.

4. Prove that if $\mathfrak{F}$ is a stable Gabriel topology on A and E is an injective A-module, then $E_{\mathfrak{F}}$ is an injective object of $\mathbf{Mod}\text{-}(A, \mathfrak{F})$.

§2

5. Let $\mathfrak{F}$ be a Gabriel topology on A. Show that $\mathfrak{F}^e$ is the strongest Gabriel topology $\mathfrak{F}'$ on $A_{\mathfrak{F}}$ such that all $\mathfrak{F}$-closed modules (considered as $A_{\mathfrak{F}}$-modules) are $\mathfrak{F}'$-closed.

6. Show that the following properties of a Gabriel topology $\mathfrak{F}$ on A are equivalent:

(a) $M_{\mathfrak{F}}$ is an injective $A_{\mathfrak{F}}$-module for every A-module M.

(b) $\mathfrak{F}^e$ consists of the essential right ideals of the ring $A_{\mathfrak{F}}$.

7. Let A be a left perfect ring with minimal dense right ideal $\mathfrak{a}$. Show that $\mathfrak{a}Q_{\max}$ is the minimal dense right ideal of $Q^r_{\max}(A)$.

§3

8. Let A and B be Morita equivalent rings, with the equivalence given by $T = \cdot \otimes_A P$ and $S = \cdot \otimes_B Q$ for bimodules ${}_AP_B$ and ${}_BQ_A$. Let $\mathfrak{F}$ be a Gabriel topology on A. Show that:

(i) $T(\mathfrak{F})$ consists of the right ideals $\mathfrak{b}$ of B for which $Q/\mathfrak{b}Q$ is an $\mathfrak{F}$-torsion module.

(ii) $T(A_{\mathfrak{F}}) \cong P_{T(\mathfrak{F})}$ and $S(B_{T(\mathfrak{F})}) \cong Q_{\mathfrak{F}}$.

§4

9. Let $\mathbf{C}$ be a Grothendieck category with a generator U, and let $\mathbf{Mod}\text{-}(A, \mathfrak{F})$ be the standard representation of $\mathbf{C}$ determined by U.

(i) Show that the following conditions are equivalent:

(a) $E(U)$ is a cogenerator for $\mathbf{C}$.

(b) $\mathfrak{F} = \mathfrak{D}$ (the topology of dense right ideals).

(ii) Show similarly that the following conditions are equivalent:

(a) U is a cogenerator for $\mathbf{C}$.

(b) $\mathfrak{F} = \mathfrak{D}$ and $E(A)$ is an A-torsionless right A-module.

Chapter XI. Perfect Localizations

Let $\mathfrak{F}$ be a Gabriel topology on A, and let q: **Mod**-$A \to$ **Mod**-$A_{\mathfrak{F}}$ be the localization functor $q(M) = M_{\mathfrak{F}}$. There is a natural transformation $\Theta: \cdot \otimes_A A_{\mathfrak{F}} \to q$ (Chap. IX, § 1), which in many important cases actually is a natural equivalence (e.g. for rings of fractions). In these cases, q must be an exact functor and thus $A_{\mathfrak{F}}$ is flat as a left A-module. But there are several other nice properties of rings of fractions which also extend to these cases. One such property, which turns out to be of particular significance, is that every ring homomorphism from $A_{\mathfrak{F}}$ is completely determined by its restriction to A, i.e. $A \to A_{\mathfrak{F}}$ is an epimorphism in the category of rings.

§ 1. Epimorphisms of Rings

We begin with a discussion of the general properties of epimorphisms in the category of rings. Let $\varphi: A \to B$ be a ring homomorphism. φ is an *epimorphism* if for any ring C and homomorphisms $\alpha, \beta: B \to C$, $\alpha\varphi = \beta\varphi$ implies $\alpha = \beta$. More generally we say that $b \in B$ is *dominated* by φ if $\alpha\varphi = \beta\varphi$ always implies $\alpha(b) = \beta(b)$. The set of elements of B dominated by φ is a subring of B, called the *dominion* of φ. φ is an epimorphism if and only if its dominion equals B.

Proposition 1.1. *If $\varphi: A \to B$ is a ring homomorphism, then the following properties of an element $b_0 \in B$ are equivalent:*

(a) *b_0 is dominated by φ.*

(b) *If M is a B-B-bimodule and $x \in M$ has the property that $ax = xa$ for all $a \in A$, then $b_0 x = x b_0$.*

(c) *$b_0 \otimes 1 = 1 \otimes b_0$ in $B \otimes_A B$.*

(d) *If M and N are right B-modules and $\alpha: M \to N$ is A-linear, then $\alpha(x b_0) = \alpha(x) b_0$ for all $x \in M$.*

Proof. (a) $\Rightarrow$ (b): Consider the trivial extension $B \times M$ of B by M (see Exercise I.9 for the details of its definition). Define two maps $\alpha, \beta: B \to B \times M$ as $\alpha(b) = (b, 0)$ and $\beta(b) = (b, bx - xb)$. Both α and β are ring homomorphisms, and $\alpha\varphi = \beta\varphi$ by the hypothesis on x. So if b_0 is dominated by φ, then $\alpha(b_0) = \beta(b_0)$ and hence $b_0 x = x b_0$.

(b) $\Rightarrow$ (c): Apply (b) to the element $x = 1 \otimes 1$ in $B \otimes_A B$.

(c) $\Rightarrow$ (d): Given the A-linear map $\alpha: M \to N$ and $x \in M$, we define $\beta: B \otimes_A B \to N$ as $\beta(b \otimes b') = \alpha(xb)b'$. This is well-defined because of the A-linearity of α. Now $b_0 \otimes 1 = 1 \otimes b_0$ immediately gives $\alpha(x b_0) = \alpha(x) b_0$.

(d) $\Rightarrow$ (a): Let α, β: $B \to C$ be ring homomorphisms such that $\alpha\varphi = \beta\varphi$. If we consider C as a right B-module by means of β, then α becomes an A-linear map. But then $\alpha(b_0) = \alpha(1 \cdot b_0) = \alpha(1)\,\beta(b_0) = \beta(b_0)$. $\square$

We will elaborate on the consequences of Prop. 1.1. Let

$$\mathbf{Mod}\text{-}A \underset{\varphi_*}{\overset{\varphi^*}{\rightleftarrows}} \mathbf{Mod}\text{-}B$$

be the functors $\varphi^*(M) = M \otimes_A B$ and $\varphi_*(N) = N$. φ^* is of course a left adjoint of φ_*. Let Φ: $\varphi^* \varphi_* \to \mathrm{id}$ denote the adjunction transformation. In particular,

$$\Phi_B\colon\ B \otimes_A B \to B$$

is the bimodule homomorphism induced by the multiplication in B.

Proposition 1.2. *The following properties of the ring homomorphism φ: $A \to B$ are equivalent:*

(a) *φ is an epimorphism.*
(b) *Φ_B: $B \otimes_A B \to B$ is an isomorphism.*
(c) *Φ: $\varphi^* \varphi_* \to \mathrm{id}$ is a natural equivalence of functors.*
(d) *The functor φ_*: $\mathbf{Mod}\text{-}B \to \mathbf{Mod}\text{-}A$ is full.*
(e) *$B \otimes_A B/\varphi(A) = 0$.*

Proof. (a) $\Rightarrow$ (d): This is just the implication (a) $\Rightarrow$ (d) of Prop. 1.1.

(d) $\Rightarrow$ (c): We have to show that the homomorphism μ: $M \otimes_A B \to M$ given by $\mu(x \otimes b) = x\,b$ is an isomorphism for every B-module M. The map $M \to M \otimes_A B$ given by $x \mapsto x \otimes 1$ is clearly A-linear, and by hypothesis therefore B-linear. It is the desired inverse of μ.

(c) $\Rightarrow$ (b) is obvious, while (b) $\Rightarrow$ (a) follows from Prop. 1.1.

(b) $\Leftrightarrow$ (e): The composition

$$B \cong B \otimes_A A \xrightarrow{1 \otimes \varphi} B \otimes_A B \xrightarrow{\Phi_B} B$$

is the identity map, so Φ_B is an isomorphism if and only if $1 \otimes \varphi$ is an isomorphism, which is equivalent to $B \otimes_A B/\varphi(A) = \mathrm{Coker}(1 \otimes \varphi) = 0$. $\square$

An immediate consequence of the condition (d) is:

Corollary 1.3. *Let φ: $A \to B$ be an epimorphism. If M is a right B-module such that M_A is injective, then M_B is injective.*

Examples

1. Dominant rings. A ring A is called *dominant* if A is the dominion of every ring monomorphism φ: $A \to B$. A dominant ring cannot have any proper epimorphic extensions.

Proposition 1.4. *Every regular ring is dominant.*

Proof. Suppose A is regular and is a subring of a ring B. Tensoring the monomorphism $A \to B$ with the left A-modules B and B/A, we obtain a commutative diagram

$$\begin{array}{ccc} B & \longrightarrow & B \otimes_A B \\ \downarrow & & \downarrow{\scriptstyle\gamma} \\ B/A & \xrightarrow[\beta]{} & B \otimes_A B/A \end{array}$$

If $b \in B$ is dominated by $A \to B$, then $b \otimes 1 = 1 \otimes b$ in $B \otimes_A B$ implies that in $B \otimes_A B/A$ one has $\beta(\bar{b}) = 1 \otimes \bar{b} = \gamma(1 \otimes b) = \gamma(b \otimes 1) = b \otimes \bar{1} = 0$. Since A is regular, β is a monomorphism, so $\bar{b} = 0$ and hence $b \in A$. □

2. Commutative rings. The results of this section also hold true for epimorphisms in the category of commutative rings. The proof of the analogue of Prop. 1.1 has to be done differently in this case, because the use of the trivial extension $B \times M$ would introduce a non-commutative ring. However, the proof is easy since $B \otimes_A B$ is itself a ring when A is commutative. Epimorphisms of commutative rings have been studied quite extensively (e.g. Séminaire Samuel [1], Storrer [1]).

§2. Flat Epimorphisms of Rings

The following theorem of Popescu and Spircu [1] shows that flat epimorphisms correspond to a kind of generalized calculus of fractions:

Theorem 2.1. *Let $\varphi\colon A \to B$ be a ring homomorphism. The following assertions are equivalent:*

(a) *φ is an epimorphism and makes B into a flat left A-module.*

(b) *The family $\mathfrak{F}$ of right ideals $\mathfrak{a}$ of A such that $\varphi(\mathfrak{a})B = B$ is a Gabriel topology, and there is a ring isomorphism $\sigma\colon B \to A_{\mathfrak{F}}$ such that $\sigma\varphi\colon A \to A_{\mathfrak{F}}$ is the canonical homomorphism.*

(c) *The following two conditions are satisfied:*

(i) *For every $b \in B$ there exist $s_1, \dots, s_n \in A$ and $b_1, \dots, b_n \in B$ such that $b\,\varphi(s_i) \in \varphi(A)$ and $\sum \varphi(s_i)\,b_i = 1$.*

(ii) *If $\varphi(a) = 0$, then there exist $s_1, \dots, s_n \in A$ and $b_1, \dots, b_n \in B$ such that $a s_i = 0$ and $\sum \varphi(s_i)\,b_i = 1$.*

Proof. (a) ⇒ (b): The forgetful functor φ_*: **Mod**-$B \to$ **Mod**-A makes **Mod**-B equivalent to a full subcategory of **Mod**-A by Prop. 1.2. The left adjoint $\varphi^* = \cdot \otimes_A B$ of φ_* is in this case exact, since ${}_AB$ is flat. **Mod**-B may thus be considered as a Giraud subcategory of **Mod**-A. Theorem 2.1 of Chap. X states that the family $\mathfrak{F}$ of right ideals $\mathfrak{a}$ for which $\varphi^*(A/\mathfrak{a}) = 0$, i.e. $\varphi(\mathfrak{a})B = B$, is a Gabriel topology and that there is an equivalence of categories $a\varphi_*$: **Mod**-$B \to$ **Mod**-$(A, \mathfrak{F})$.

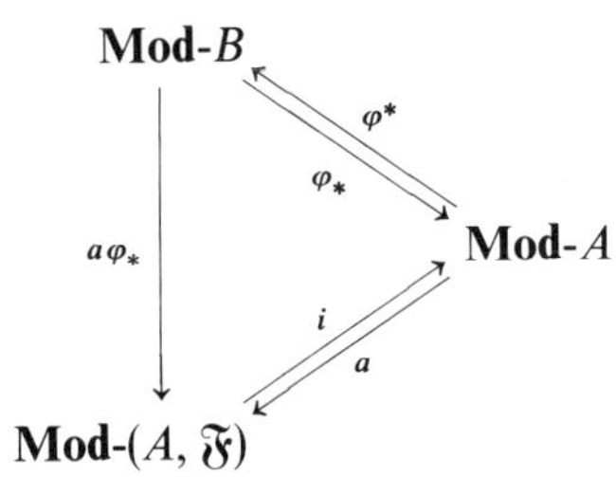

From the proof of the cited theorem we also have that $\mathfrak{a}\,\varphi_*\,\varphi^* \cong \mathfrak{a}$. This gives the desired isomorphism $B \to A_{\mathfrak{F}}$, once we have verified that B is $\mathfrak{F}$-closed. (We leave to the reader to prove that $B \to A_{\mathfrak{F}}$ actually is a ring homomorphism.) It is clear that B is $\mathfrak{F}$-torsion-free when one has $\varphi(\mathfrak{a})\,B = B$ for every $\mathfrak{a} \in \mathfrak{F}$. Let $\alpha\colon \mathfrak{a} \to B$ be a homomorphism with $\mathfrak{a} \in \mathfrak{F}$. Write $1 \in B$ as $1 = \sum \varphi(a_i)\,b_i$ with $a_i \in \mathfrak{a}$. Since $\mathfrak{a} \otimes_A B \cong B$, we have for each $a \in \mathfrak{a}$ that $a \otimes 1 = \sum a_i \otimes b_i\,\varphi(a)$ in $\mathfrak{a} \otimes_A B$. The composed homomorphism

$$\mathfrak{a} \otimes_A B \xrightarrow{\alpha \otimes 1} B \otimes_A B \longrightarrow B$$

then gives $\alpha(a) = \sum \alpha(a_i)\,b_i\,\varphi(a)$. Thus α consists in multiplying to the left by $\sum \alpha(a_i)\,b_i$, so α may be extended to A, and B is $\mathfrak{F}$-injective.

(b) $\to$ (c). If $b \in D \cong A_{\mathfrak{F}}$, then there exists $\mathfrak{a} \in \mathfrak{F}$ such that $b\,\mathfrak{a} \subset \varphi(A)$. By the definition of $\mathfrak{F}$ we can write $1 = \sum \varphi(s_i)\,b_i$ with $s_i \in \mathfrak{a}$, $b_i \in B$. This establishes (i).

If $\varphi(a) = 0$, then a belongs to the $\mathfrak{F}$-torsion submodule of A, so there exists $\mathfrak{a} \in \mathfrak{F}$ with $a\,\mathfrak{a} = 0$. Again we may write $1 = \sum \varphi(s_i)\,b_i$ with $s_i \in \mathfrak{a}$, $b_i \in B$, and also (ii) is satisfied.

For the proof of (c) $\Rightarrow$ (a) we need:

Lemma 2.2. *Let there be given modules L_A and ${}_AM$. The following assertions are equivalent:*

(a) *$L' \otimes_A M = 0$ for every submodule L' of L.*

(b) *$C \otimes_A M = 0$ for every cyclic submodule C of L.*

(c) *For every $y \in M$ and $x_1, \ldots, x_n \in L$ there exist $y_1, \ldots, y_m \in M$ and $a_1, \ldots, a_m \in A$ such that $y = \sum_i a_i\,y_i$ and $x_i\,a_j = 0$ for all i and j.*

Proof. (a) $\Leftrightarrow$ (b) is clear, because every module can be written as a quotient module of the direct sum of its cyclic submodules.

(a) $\Rightarrow$ (c): By induction on n it is easy to show that (a) implies that $K \otimes M = 0$ for every submodule K of a direct sum of n copies of L. Applying Prop. I.8.8 to the element $(x_1, \ldots, x_n) \otimes y = 0$, one obtains (c).

(c) $\Rightarrow$ (b) is immediate with the help of Prop. I.8.8. □

We also note:

Lemma 2.3. *If $\varphi\colon A \to B$ is a ring homomorphism satisfying* (i) *of* 2.1(c), *then $C \otimes_A B = 0$ for every submodule C of $B/\varphi(A)$.*

Proof. For each $b \in B$ we choose $s_1, \ldots, s_n \in A$ and $b_1, \ldots, b_n \in B$ such that $b\,\varphi(s_i) \in \varphi(A)$ and $\sum \varphi(s_i)\,b_i = 1$. If $\bar{b} \in C$, then we get for any $b' \in B$ that

$$\bar{b} \otimes b' = \bar{b} \otimes \sum \varphi(s_i)\,b_i\,b' = \sum \bar{b}\,\varphi(s_i) \otimes b_i\,b' = 0$$

in $C \otimes_A B$. □

So if we assume condition (c), then in particular φ is an epimorphism by Prop. 1.2. To show that ${}_AB$ is flat, we will use Prop. I.10.7. Suppose $s_1, \ldots, s_n \in A$ and $b_1, \ldots, b_n \in B$ have the property that $\sum \varphi(s_i)\,b_i = 0$. We apply condition (c) of Lemma 2.2 to $1 \in B$ and $\bar{b}_1, \ldots, \bar{b}_n \in B/\varphi(A)$, and obtain elements $a_1, \ldots, a_m \in A$ and $b'_1, \ldots, b'_m \in B$ such that

$$1 = \sum_j \varphi(a_j)\,b'_j,$$

$$b_i\,\varphi(a_j) = \varphi(c_{ij}) \quad \text{for some } c_{ij} \in A.$$

Then $\sum_i \varphi(s_i c_{ij}) = \sum_i \varphi(s_i) b_i \varphi(a_j) = 0$, and thus $\sum_i s_i c_{ij} \in \operatorname{Ker} \varphi$ for each j. We now make use of part (ii) of condition (c), and obtain for each j elements $t_{j1}, \ldots, t_{jr} \in A$ and $b''_{j1}, \ldots, b''_{jr} \in B$ such that

$$\sum_i s_i c_{ij} t_{jk} = 0,$$

$$\sum_k \varphi(t_{jk}) b''_{jk} = 1.$$

We then have

$$b_i = \sum_j b_i \varphi(a_j) b'_j = \sum_j \varphi(c_{ij}) b'_j = \sum_{j,k} \varphi(c_{ij} t_{jk}) b''_{jk} b'_j,$$

and since $\sum_i s_i c_{ij} t_{jk} = 0$, this means that the given relation $\sum_i \varphi(s_i) b_i = 0$ emanates from a relation in A, as was to be shown. ▯

Remark 1. Note that condition (c) of the theorem implies in particular that each $b \in B$ may be written as

$$b = \sum b \varphi(s_i) b_i = \sum \varphi(a_i) b_i \quad \text{with} \quad \sum \varphi(s_i) b_i = 1.$$

B is thus obtained by a kind of generalized calculus of fractions.

Remark 2. In the proof of (c) ⇒ (a) we obtained the following variant of condition (i) of (c):

(i′) For every family $b_1, \ldots, b_r \in B$ there exist $s_1, \ldots, s_n \in A$ and $b'_1, \ldots, b'_n \in B$ such that $b_j \varphi(s_i) \in \varphi(A)$ and $\sum_i \varphi(s_i) b'_i = 1$.

Definition. If $\varphi: A \to B$ is an epimorphism of rings which makes B into a flat left A-module, then we will call B a *perfect right localization* of A (B is also called a *flat epimorphic right ring of quotients* of A). We write $\mathfrak{F}^B$ for the Gabriel topology of right ideals $\mathfrak{a}$ of A such that $\varphi(\mathfrak{a}) B = B$ (cf. Chap. VI, §9).

When an overring B of A is contained in a ring of quotients of A, one may be interested in knowing if also B is obtainable as a ring of quotients of A. A particular result in that direction is:

Proposition 2.4. *Assume $A \subset B \subset C$ are rings such that C is a perfect right localization of A, and that B is a flat right A-module. Then B is a perfect left localization of A.*

Proof. It remains to show that $A \to B$ is a ring epimorphism, and for this we use Prop. 1.2. Consider the commutative diagram

$$\begin{array}{ccccc}
B \otimes_A B & & \xrightarrow{\quad\beta\quad} & & B \\
\downarrow{\scriptstyle 1 \otimes \lambda} & & & & \downarrow{\scriptstyle \lambda} \\
B \otimes_A C & \xrightarrow[\lambda \otimes 1]{} & C \otimes_A C & \xrightarrow[\gamma]{} & C
\end{array}$$

induced by ring multiplication and the monomorphism $\lambda: B \to C$. The maps γ, $\lambda \otimes 1$ and $1 \otimes \lambda$ are all monomorphisms by the hypothesis, so β is a monomorphism. ▯

Example. Rings of fractions. If S is a right denominator set in A, then it is easily seen that $A \to A[S^{-1}]$ is a ring epimorphism, and it is known that $A[S^{-1}]$ is a flat left A-module (Prop. II.3.5).

§ 3. Perfect Gabriel Topologies

The flat epimorphisms may be approached from a direction somewhat different from the one employed in the previous section. Starting with a Gabriel topology $\mathfrak{F}$ on A, one may look for conditions which make $\psi\colon A \to A_{\mathfrak{F}}$ into a perfect localization. We first recall the notations for the functors involved (Chap. IX, §1):

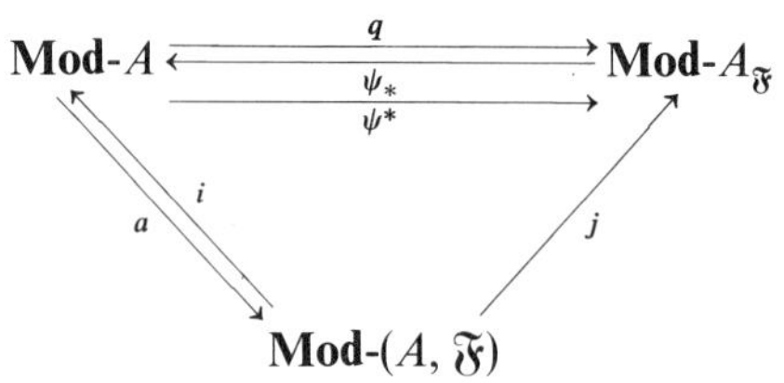

where $j\,a = q$ and $\psi_*\, q = i\,a$. Both i and q are left exact. Exactness of i means that if $\alpha\colon L \to M$ is a homomorphism of $\mathfrak{F}$-closed modules, then also Coker α is $\mathfrak{F}$-closed. It is therefore clear that i is exact if and only if q is exact. There is a natural transformation $\Theta\colon \psi^* \to q$ with $\Theta_M\colon M \otimes_A A_{\mathfrak{F}} \to M_{\mathfrak{F}}$ given by $\Theta_M(x \otimes c) = \psi_M(x)\, c$.

Proposition 3.1. *If the functor q is exact, then $\Theta_M\colon M \otimes_A A_{\mathfrak{F}} \to M_{\mathfrak{F}}$ is an isomorphism for every finitely presented module M.*

Proof. Since M is finitely presented, there is an exact sequence $A^m \to A^n \to M \to 0$. We apply the functors q and $\cdot \otimes_A A_{\mathfrak{F}}$ to this sequence and obtain a commutative diagram

$$\begin{array}{ccccccc}
A_{\mathfrak{F}}^m & \longrightarrow & A_{\mathfrak{F}}^n & \longrightarrow & M \otimes_A A_{\mathfrak{F}} & \longrightarrow & 0 \\
\| & & \| & & \downarrow{\scriptstyle \Theta_M} & & \\
A_{\mathfrak{F}}^m & \longrightarrow & A_{\mathfrak{F}}^n & \longrightarrow & M_{\mathfrak{F}} & \longrightarrow & 0
\end{array}$$

with exact rows. It follows that Θ_M must be an isomorphism. □

Corollary 3.2. *If A is right coherent and q is exact, then $A_{\mathfrak{F}}$ is flat as a left A-module.*

Proof. Every finitely generated right ideal $\mathfrak{a}$ is finitely presented, so it follows from the Proposition and the left exactness of q that the canonical map $\mathfrak{a} \otimes_A A_{\mathfrak{F}} \to A_{\mathfrak{F}}$ is a monomorphism. Hence $A_{\mathfrak{F}}$ is flat by Prop. I.10.6. □

As an example of a case where q is exact, we mention:

Proposition 3.3. *If $\mathfrak{F}$ has a basis consisting of projective right ideals, then the functor q is exact.*

Proof. Consider an epimorphism $\alpha: L \to M$. An element $x \in M_{\mathfrak{F}}$ is represented by a homomorphism $\xi: \mathfrak{a} \to M/t(M)$ for some $\mathfrak{a} \in \mathfrak{F}$, and by hypothesis we may assume $\mathfrak{a}$ is projective. Then ξ can be lifted to a homomorphism $\eta: \mathfrak{a} \to L/t(L)$, which represents an element in $L_{\mathfrak{F}}$ mapping onto x. Hence $q(\alpha): L_{\mathfrak{F}} \to M_{\mathfrak{F}}$ is an epimorphism, and thus q is exact. □

When the functor q preserves both exactness and arbitrary direct sums, the proof of Prop. 3.1 works for all modules M, and Θ is then a natural equivalence. In more detail we have the following result (where the equivalence of conditions (a)–(g) was established by Walker and Walker [1]):

Proposition 3.4. *The following properties of a Gabriel topology $\mathfrak{F}$ are equivalent:*
(a) *The functor j:* **Mod**-$(A, \mathfrak{F}) \to$ **Mod**-$A_{\mathfrak{F}}$ *is an equivalence.*
(b) *The functor i:* **Mod**-$(A, \mathfrak{F}) \to$ **Mod**-A *has a right adjoint.*
(c) *The functor q:* **Mod**-$A \to$ **Mod**-$A_{\mathfrak{F}}$ *is exact and preserves direct sums.*
(d) *$\mathfrak{F}$ has a basis consisting of finitely generated right ideals, and q is exact.*
(e) *$\Theta: \psi^* \to q$ is a natural equivalence of functors.*
(f) $\operatorname{Ker}(M \to M \otimes_A A_{\mathfrak{F}})$ *is the $\mathfrak{F}$-torsion submodule of M, for every module M.*
(g) *The left A-module $A_{\mathfrak{F}}$ is $\mathfrak{F}$-divisible.*
(h) *$A_{\mathfrak{F}}$ is a perfect right localization of A, and* $\mathfrak{F} = \{\mathfrak{a} \mid \psi(\mathfrak{a}) A_{\mathfrak{F}} = A_{\mathfrak{F}}\}$.

Definition. A Gabriel topology $\mathfrak{F}$ is called *perfect* if it satisfies these equivalent conditions.

Proof. (a) ⇒ (b): $i = \psi_* j$, where j has a right adjoint by hypothesis and ψ_* has the right adjoint $\psi^! = \operatorname{Hom}_A(A_{\mathfrak{F}}, \cdot)$, and hence i has a right adjoint.

(b) ⇒ (c): If i has a right adjoint, then i preserves exactness and direct sums, and so does then also q.

(c) ⇒ (e): This was noticed above.

(e) ⇒ (f) is clear, because the $\mathfrak{F}$-torsion submodule of M is the kernel of the canonical map $M \to M_{\mathfrak{F}}$.

(f) ⇒ (g): If $\mathfrak{a} \in \mathfrak{F}$, then $A/\mathfrak{a}$ is an $\mathfrak{F}$-torsion module, and hence the map $A/\mathfrak{a} \to A/\mathfrak{a} \otimes A_{\mathfrak{F}} \cong A_{\mathfrak{F}}/\mathfrak{a} A_{\mathfrak{F}}$ is zero. This implies $A_{\mathfrak{F}} = \mathfrak{a} A_{\mathfrak{F}}$, which by definition (Chap. VI, §9) means that $A_{\mathfrak{F}}$ is $\mathfrak{F}$-divisible.

(g) ⇒ (h): By Theorem 2.1 it suffices to show that if $\mathfrak{a}$ is a right ideal such that $\mathfrak{a} A_{\mathfrak{F}} = A_{\mathfrak{F}}$, then $\mathfrak{a} \in \mathfrak{F}$. Now if $\psi(\mathfrak{a}) A_{\mathfrak{F}} = A_{\mathfrak{F}}$, then $1 = \sum \psi(a_i) q_i$ with $a_i \in \mathfrak{a}$ and $q_i \in A_{\mathfrak{F}}$. There exists $\mathfrak{b} \in \mathfrak{F}$ such that $q_i \mathfrak{b} \subset \psi(A)$ for all i, and then

$$\psi(\mathfrak{b}) = \sum \psi(a_i) q_i \psi(\mathfrak{b}) \subset \psi(\mathfrak{a}).$$

This implies $\mathfrak{a} \in \mathfrak{F}$, as is easily verified.

(h) ⇒ (a): In the proof of Theorem 2.1 we noticed that the functor $a \psi_*$: **Mod**-$A_{\mathfrak{F}} \to$ **Mod**-$(A, \mathfrak{F})$ is an equivalence. Since $a \psi_* j$ is naturally equivalent to the identity functor on **Mod**-$(A, \mathfrak{F})$, it follows that j must be an equivalence.

(h) + (c) ⇒ (d): If $\mathfrak{a} \in \mathfrak{F}$, then $\psi(\mathfrak{a}) A_{\mathfrak{F}} = A_{\mathfrak{F}}$ implies that $1 = \sum_1^n \psi(a_i) q_i$ with $a_i \in \mathfrak{a}$, $q_i \in A_{\mathfrak{F}}$. The elements $a_1, \ldots, a_n$ generate a right ideal $\mathfrak{b}$ such that $\psi(\mathfrak{b}) A_{\mathfrak{F}} = A_{\mathfrak{F}}$, and hence $\mathfrak{b} \in \mathfrak{F}$.

(d) ⇒ (g): It suffices to show that $\psi(\mathfrak{a}) A_{\mathfrak{F}} = A_{\mathfrak{F}}$ holds when $\mathfrak{a}$ is a finitely generated right ideal in $\mathfrak{F}$. But then $A/\mathfrak{a}$ is a finitely presented $\mathfrak{F}$-torsion module, so Prop. 3.1 gives $0 = (A/\mathfrak{a})_{\mathfrak{F}} \cong A/\mathfrak{a} \otimes A_{\mathfrak{F}} \cong A_{\mathfrak{F}}/\mathfrak{a} A_{\mathfrak{F}}$, and hence $A_{\mathfrak{F}} = \mathfrak{a} A_{\mathfrak{F}}$. □

Corollary 3.5. *If $\mathfrak{F}$ has a basis of $\mathfrak{F}$-invertible right ideals, then $\mathfrak{F}$ is perfect.*

Proof. Use condition (d), together with Prop. IX.5.2 and Prop. 3.3. (Note that this corollary is just a restatement of Prop. IX.5.3.) □

Corollary 3.6. *If A is a right noetherian hereditary ring, then every Gabriel topology is perfect.*

Recall that $\mathsf{Sat}_{\mathfrak{F}}(M)$ denotes the lattice of $\mathfrak{F}$-saturated submodules of M (Chap. IX, §4). When $\mathfrak{F}$ is perfect and M is $\mathfrak{F}$-closed, then $\mathsf{Sat}_{\mathfrak{F}}(M)$ consists of all $A_{\mathfrak{F}}$-submodules of M. Therefore Prop. IX.4.3 gives:

Proposition 3.7. *If $\mathfrak{F}$ is perfect, then for every module M the lattice $\mathsf{Sat}_{\mathfrak{F}}(M)$ is isomorphic to the lattice of all $A_{\mathfrak{F}}$-submodules of $M \otimes_A A_{\mathfrak{F}}$.*

In particular we obtain for $M = A$:

Corollary 3.8. *If $\mathfrak{F}$ is a perfect topology, then there is a bijective correspondence between the set of right ideals of $A_{\mathfrak{F}}$ and the set of right ideals $\mathfrak{a}$ of A for which $\mathfrak{a} = \psi^{-1}(\mathfrak{a} A_{\mathfrak{F}})$.*

An important consequence of this corollary is:

Proposition 3.9. *Let B be a perfect right localization of A. If A is right noetherian or right artinian, then so is also B.*

Results concerning the behavior of prime ideals under a flat epimorphic ring extension can be found in Silver [1].

When $A \to B$ is a perfect localization, one can compare the homological dimensions of A and B. (Here we assume some knowledge of homological algebra.) Generally, if $A \to B$ is a ring homomorphism such that ${}_A B$ is flat, then one has the following formulas (Cartan and Eilenberg [1], Chap. 6, §4):

$$\operatorname{Ext}^n_A(M, N) \cong \operatorname{Ext}^n_B(M \otimes_A B, N) \quad \text{for } M_A, N_B; \tag{1}$$

$$\operatorname{Tor}^A_n(M, N) \cong \operatorname{Tor}^A_n(M \otimes_A B, N) \quad \text{for } M_A, {}_B N. \tag{2}$$

Combining these formulas with Prop. 1.2(c) we obtain:

Proposition 3.10. *If $A \to B$ is a perfect right localization, then:*

$$\operatorname{Ext}^n_A(M, N) \cong \operatorname{Ext}^n_B(M, N) \quad \textit{for } M_B, N_B;$$

$$\operatorname{Tor}^A_n(M, N) \cong \operatorname{Tor}^B_n(M, N) \quad \textit{for } M_B, {}_B N;$$

$$\text{r.gl.dim}\, B \leq \text{r.gl.dim}\, A;$$

$$\text{w.gl.dim}\, B \leq \text{w.gl.dim}\, A.$$

Here r.gl.dim is the right global dimension and w.gl.dim. is the weak global dimension. From (1) and (2) we also get (cf. Prop. X.1.4):

Proposition 3.11. *Suppose B is a perfect right localization of A.*
(i) *If N is a right B-module, then N_B is injective if and only if N_A is injective.*
(ii) *If N is a left B-module, then ${}_B N$ is flat if and only if ${}_A N$ is flat.*

Examples

1. The Goldie topology. When A is a right noetherian ring, the Goldie topology is perfect, since q is exact by Prop. IX.2. More generally, if A is right coherent, then the Goldie ring of quotients $A_{\mathfrak{G}}$ is flat as a left A-module (Cor. 3.2).

2. Coherent rings. The class of coherent rings is closed under perfect localizations:

Proposition 3.12. *If B is a perfect right localization of a right coherent ring A, then B is right coherent.*

Proof. Consider a family $(N_i)_I$ of flat left B-modules. Each N_i is flat as an A-module by (2), and then $\prod N_i$ is flat over A. But then $\prod N_i$ is flat also over B by Prop. 3.11, so B is right coherent. □

3. Flat quotient rings. Let $\mathfrak{a}$ be a two-sided ideal of the ring A. There is an epimorphism $A \to A/\mathfrak{a}$, and one may ask when this makes $A/\mathfrak{a}$ into a perfect right localization of A.

Proposition 3.13. *The following conditions are equivalent for a two-sided ideal $\mathfrak{a}$ of A:*

(a) *$A/\mathfrak{a}$ is flat as a left A-module.*
(b) *$\mathfrak{a}$ is pure as a left ideal in A.*
(c) *For every $a \in \mathfrak{a}$ there exists $c \in \mathfrak{a}$ such that $a = a\,c$.*
(d) *Every injective right $A/\mathfrak{a}$-module is injective as an A-module.*
(e) *Every flat left $A/\mathfrak{a}$-module is flat as an A-module.*

Proof. (a) $\Leftrightarrow$ (b) follows from Prop. I.11.1.

(a) $\Rightarrow$ (c): The flatness of $A/\mathfrak{a}$ gives in the usual way (cf. the proof of Prop. I.12.1) that $a A \cap \mathfrak{a} = a\,\mathfrak{a}$ for every $a \in A$. In particular, $a \in \mathfrak{a}$ implies $a \in a\,\mathfrak{a}$.

(c) $\Rightarrow$ (d): Let E be an injective right $A/\mathfrak{a}$-module, and consider an A-homomorphism $\varphi\colon \mathfrak{b} \to E$ from a right ideal $\mathfrak{b}$ of A. The condition (c) implies that φ induces a well-defined homomorphism $\bar{\varphi}\colon (\mathfrak{b}+\mathfrak{a})/\mathfrak{a} \to E$, for if $a \in \mathfrak{b} \cap \mathfrak{a}$ and $a = a\,c$ with $c \in \mathfrak{a}$, then $\varphi(a) = \varphi(a\,c) = \varphi(a)\,c = 0$ since $E\,\mathfrak{a} = 0$. Then $\bar{\varphi}$ extends to a homomorphism $A/\mathfrak{a} \to E$, and the composed map $A \to A/\mathfrak{a} \to E$ extends φ, so it follows that E is injective over A.

(d) $\Rightarrow$ (a) follows by applying Cor. I.10.5 to the module $\mathrm{Hom}_{\mathbb{Z}}(A/\mathfrak{a}, \mathbb{Q}/\mathbb{Z})$.

(a) $\Rightarrow$ (e) follows from Prop. 3.11 (and so does also (a) $\Rightarrow$ (d)).

(e) $\Rightarrow$ (a) is trivial. □

§4. The Maximal Flat Epimorphic Ring of Quotients

The following theorem states that every ring A has a maximal flat epimorphic right ring of quotients, which is unique up to isomorphism. It will be denoted by $Q^r_{\mathrm{tot}}(A)$, or simply Q_{tot}. We use the term *bimorphism* for a morphism which is both a monomorphism and an epimorphism.

Theorem 4.1. *For every ring A there exists a ring $Q^r_{\mathrm{tot}}(A)$ and a ring homomorphism $\varphi\colon A \to Q^r_{\mathrm{tot}}(A)$ such that*

(i) *φ is a bimorphism and $Q^r_{tot}(A)$ is flat as a left A-module.*

(ii) *For every bimorphism $\alpha: A \to B$ of rings such that ${}_A B$ is flat, there is a unique ring homomorphism $\beta: B \to Q^r_{tot}(A)$ such that $\beta\alpha = \varphi$ (β is then necessarily a monomorphism).*

Proof. Every flat epimorphism is obtained as the canonical homomorphism $\psi: A \to A_{\mathfrak{F}}$ for a perfect Gabriel topology $\mathfrak{F}$ (Theorem 2.1). ψ is a monomorphism if and only if A has no $\mathfrak{F}$-torsion, i.e. if and only if $\mathfrak{F}$ is a subset of the set of dense right ideals. So if Q_{tot} exists, it should be a subring of the maximal ring of quotients Q_{max}. This leads us to consider the family $\mathscr{P}$ of all subrings B of Q_{max} which are perfect right localizations of A. Note that inclusion of subrings in $\mathscr{P}$ corresponds to inclusion of the corresponding perfect topologies.

Lemma 4.2. *The family $\mathscr{P}$ is directed under inclusion.*

Proof. We will show that if B and C are members of $\mathscr{P}$, then the smallest subring D of Q_{max} containing both B and C is also a member of $\mathscr{P}$. Every element of D is a sum of elements of the form

$$d = b_1 c_1 \dots b_n c_n \quad \text{with} \quad b_i \in B,\ c_i \in C, \tag{1}$$

and we may assume that each d appearing in a given sum has the same length n. We will verify condition (i) of 2.1(c), i.e.:

given $d_1, \dots, d_r$ of length n, there exist $d'_1, \dots, d'_m$ in D and $s_1, \dots, s_m \in A$ such that $d_i s_k \in A$ for all i, k and $\sum s_k d'_k = 1$.

We do this by induction on the length n. For $n=0$, i.e. $d=1$, the condition is clearly satisfied. Suppose $n>1$ and the condition has been verified for $n-1$. To simplify the notation somewhat, we only consider the case $r=1$; it is easy to see that our argument extends immediately to the case of a family $d_1, \dots, d_r \in D$. So suppose we are given d of the form (1). By the induction hypothesis there exist $d'_1, \dots, d'_m$ in D and $t_1, \dots, t_m$ in A such that

$$x_i = b_2 c_2 \dots b_n c_n t_i \in A \quad \text{for all } i, \quad \text{and} \quad \sum t_i d'_i = 1.$$

By the Remark 2 at the end of §2, applied to C, there exist $c'_1, \dots, c'_p$ in C and $s_1, \dots, s_p$ in A such that

$$x_{ij} = c_1 x_i s_j \in A \quad \text{for all } i, j, \quad \text{and} \quad \sum s_j c'_j = 1.$$

Similarly there exist $b'_1, \dots, b'_q$ in B and $r_1, \dots, r_q$ in A such that

$$b_1 x_{ij} r_k \in A \quad \text{for all } i, j, k, \quad \text{and} \quad \sum r_k b'_k = 1.$$

We have thus got elements $b'_k c'_j d'_i \in D$ and $t_i s_j r_k \in A$ such that

$$d t_i s_j r_k = b_1 c_1 x_i s_j r_k \in A \quad \text{and} \quad \sum_{i,j,k} t_i s_j r_k b'_k c'_j d'_i = 1. \quad \square$$

This finishes the proof of the lemma, and we may continue the proof of the Theorem. Define Q_{tot} to be the union of all rings in $\mathscr{P}$. It is obvious that $\varphi: A \to Q_{tot}$ is an epimorphism, and Q_{tot} is flat as a left A-module because it is a direct limit of flat A-modules.

Suppose $\alpha: A \to B$ is another flat bimorphism. There is a corresponding perfect topology $\mathfrak{F}$, and as was noticed above, $\mathfrak{F}$ is contained in the set of dense right ideals. Hence $B=A_{\mathfrak{F}}$ may be imbedded in $Q_{\max}$; more precisely, there is a ring homomorphism $\beta\colon B\to Q_{\max}$ such that

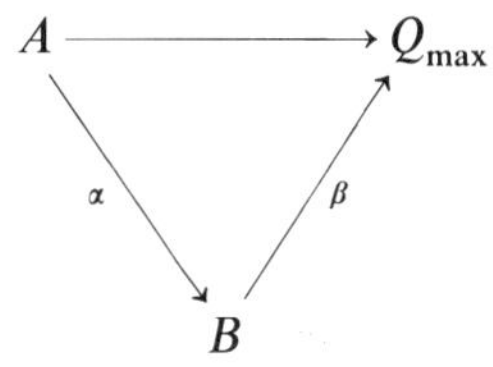

commutes (Exercise IX.6). The image of B lies within Q_{tot}. Since β obviously is unique (α is an epimorphism), the theorem is thereby proved. □

Since $Q^r_{\mathrm{tot}}(A)$ is the solution of a "universal problem", it is uniquely determined by A, up to isomorphism. The existence of Q_{tot} has been proved by a number of authors, using various methods (Findlay [2], Knight [1], Lazard [1], Morita [5], Popescu and Spircu [1]).

Examples

1. Perfectly complete rings. The ring A is called *(right) perfectly complete* if $A=Q^r_{\mathrm{tot}}(A)$. The following rings are perfectly complete:

(i) Dominant rings (Example 1.1).
(ii) Regular rings (Prop. 1.4).
(iii) Rationally complete rings (Example IX.2.3).
(iv) $Q^r_{\mathrm{tot}}(A)$ for arbitrary A.

2. The classical ring of quotients. Suppose A has a classical right ring of quotients Q_{cl}. Then we have $Q_{\mathrm{cl}}\subset Q_{\mathrm{tot}}$. If Q_{cl} is regular, then necessarily $Q_{\mathrm{cl}}=Q_{\mathrm{tot}}$, because the inclusion map $Q_{\mathrm{cl}}\to Q_{\mathrm{tot}}$ is also an epimorphism.

3. Hereditary noetherian rings. When A is right noetherian and right hereditary, Cor. 3.6 shows that $Q_{\mathrm{tot}}=Q_{\max}$. This result will be considerably improved in the next Chapter.

§5. The Maximal Ring of Quotients as a Perfect Localization

A natural question to ask is: when is the maximal right ring of quotients of A a perfect localization? In other words, when is $Q^r_{\mathrm{tot}}(A)=Q^r_{\max}(A)$? The complete answer to this question is not known, and we can only give a sufficient condition on $Q_{\max}$ for this to happen.

We say that a ring Q is a *right Kasch ring* if it has no proper dense right ideals, and in that case it follows trivially that Q is rationally complete, i.e. Q coincides with its maximal right ring of quotients. There are other useful characterizations of Kasch rings:

Lemma 5.1. *The following properties of a ring Q are equivalent:*
(a) *Q is a right Kasch ring.*
(b) *$E(Q)$ is an injective cogenerator for* **Mod-Q**.

(c) *Every simple right Q-module is isomorphic to a minimal right ideal of Q.*
(d) $\mathrm{Hom}(C, Q) \neq 0$ *for every non-zero cyclic right Q-module C.*
(e) $l(\mathfrak{q}) \neq 0$ *for every right ideal* $\mathfrak{q} \neq Q$.

Proof. (a) $\Rightarrow$ (b): A right ideal $\mathfrak{q}$ is dense if and only if $\mathrm{Hom}(Q/\mathfrak{q}, E(Q)) = 0$. So if there are no dense right ideals $\neq Q$, then $\mathrm{Hom}(C, E(Q)) \neq 0$ for every cyclic module $C \neq 0$, and this means that $E(Q)$ is a cogenerator.

(b) $\Rightarrow$ (c): If S is a simple module, then there is a monomorphism $\alpha: S \to E(Q)$. Since Q is essential in $E(Q)$, the image of α must lie within Q.

(c) $\Rightarrow$ (d) is clear, because C has a simple quotient module.

(d) $\Rightarrow$ (e) is obvious.

(e) $\Rightarrow$ (a) follows from Prop. VI.6.4. □

We will now see what happens when $Q_{\max}$ is assumed to be a Kasch ring. Let $\mathfrak{D}$ denote the topology of dense right ideals of A.

Proposition 5.2. *The following properties of A are equivalent:*
(a) $Q^r_{\max}(A)$ *is a right Kasch ring.*
(b) $\mathfrak{D}$ *is a perfect topology.*
(c) $E(A)$ *is an injective cogenerator for* **Mod**-$Q^r_{\max}(A)$.

Proof. (a) $\Leftrightarrow$ (c) follows from Lemma 5.1.

(a) $\Rightarrow$ (b): If $\mathfrak{a} \in \mathfrak{D}$, then $\mathfrak{a} Q_{\max}$ is a dense right ideal of $Q_{\max}$, so $\mathfrak{a} Q_{\max} = Q_{\max}$.

(b) $\Rightarrow$ (a): If $\mathfrak{q}$ is a dense right ideal of $Q_{\max}$, then $\mathfrak{q} \cap A \in \mathfrak{D}$. It follows that $Q_{\max} = (\mathfrak{q} \cap A) Q_{\max} \subset \mathfrak{q}$, and hence $\mathfrak{q} = Q_{\max}$. □

An immediate consequence of this result is the following sufficient condition for $Q_{\max}$ to be a perfect localization of A:

Proposition 5.3. *If $Q^r_{\max}(A)$ is a right Kasch ring, then $Q^r_{\max}(A) = Q^r_{\mathrm{tot}}(A)$.*

As an application we derive a sufficient condition (due to Morita [5] and Masaike [3]) for $Q^r_{\max}(A)$ to coincide with $Q^l_{\max}(A)$.

Proposition 5.4. *If $Q^r_{\max}(A)$ is a two-sided Kasch ring, then the following conditions are equivalent:*
(a) $Q^r_{\max}(A) = Q^l_{\max}(A)$.
(b) *Every finitely generated submodule of the right A-module $Q^r_{\max}(A)$ is isomorphic to a submodule of a free A-module.*
(c) $Q^r_{\max}(A)$ *is flat as a right A-module.*

Proof. (a) $\Rightarrow$ (c) is clear.

(c) $\Rightarrow$ (b): Since $Q = Q^r_{\max}(A)$ is a right Kasch ring, it is an epimorphic extension of A. When Q is flat also as a right A-module, it is a perfect left localization of A, and the left ideals $\mathfrak{a}$ of A for which $Q\mathfrak{a} = Q$ form a perfect Gabriel topology $\mathfrak{F}$. If M is a finitely generated submodule of Q_A, then there exists $\mathfrak{a} \in \mathfrak{F}$ such that $\mathfrak{a} M \subset A$. Let $a_1, \ldots, a_n$ generate $\mathfrak{a}$. Define $\varphi: M \to A^n$ as $\varphi(x) = (a_1 x, \ldots, a_n x)$. Then φ is a monomorphism, because ${}_A Q$ is $\mathfrak{F}$-torsion-free.

(b) $\Rightarrow$ (a): In order to show that $Q^r_{\max}(A) = Q^l_{\max}(A)$, it is enough to show that $Q^r_{\max}(A)$ is a subring of $Q^l_{\max}(A)$, for since $Q^r_{\max}(A)$ is a left Kasch ring, it coincides with its maximal left ring of quotients and must therefore contain $Q^l_{\max}(A)$.

So it suffices to show that for each $q\in Q^r_{\max}(A)$, the left ideal $\mathfrak{a}=\{a\in A\,|\,a\,q\in A\}$ of A is dense (Exercise IX.6). By Prop. VI.6.4, $\mathfrak{a}$ is dense if $(\mathfrak{a}:b)$ has no right annihilators for $b\in A$, i.e. for each couple $b, d\neq 0$ of elements of A there exists $c\in A$ such that $c\,b\,q\in A$ and $c\,d\neq 0$. Consider the right A-submodule L of $Q^r_{\max}(A)$ generated by $b\,q$, d and 1. From condition (b) follows that there exists a homomorphism α: $L\to A$ such that $\alpha(d)\neq 0$. But α is of the form $\alpha(x)=c\,x$ for some $c\in Q$ (Prop. IX.2.8), and $1\in L$ implies that $c\in A$. Then $c\,b\,q\in A$ and $c\,d\neq 0$. □

Examples

1. Semi-simple rings. A semi-simple ring is a two-sided Kasch ring. The case when $Q_{\max}$ is semi-simple will be discussed in the next Chapter (§2).

2. Non-singular rings. If A is right non-singular and a right Kasch ring, then A has no proper essential right ideals, so A is semi-simple.

3. Perfect rings. If A is a left perfect ring, it is known that there is a minimal dense right ideal $\mathfrak{d}$ (Cor. VIII.6.4). For this situation we have:

Proposition 5.5. *Let A be a left perfect ring with minimal dense right ideal $\mathfrak{d}$. The following conditions are equivalent:*

(a) *$Q^r_{\max}(A)$ is a right Kasch ring.*

(b) *$\mathfrak{d}$ is finitely generated and projective as a right A-module.*

Proof. (a) ⇒ (b): From $\mathfrak{d}Q_{\max}=Q_{\max}$ we obtain the existence of elements $d_1,\ldots,d_n\in\mathfrak{d}$ and $q_1,\ldots,q_n\in Q_{\max}$ such that $\sum d_i\,q_i=1$. If we define φ_i: $\mathfrak{d}\to A$ by $\varphi_i(d)=q_i\,d$, then $d=\sum d_i\,\varphi_i(d)$ for every $d\in\mathfrak{d}$, so $\mathfrak{d}$ is finitely generated projective by Prop. I.6.3.

(b) ⇒ (a): If (d_i,φ_i) are finite projective coordinates for $\mathfrak{d}$, then there exist $q_i\in Q$ such that $\varphi_i(d)=q_i\,d$ for all $d\in\mathfrak{d}$. But then $(1-\sum d_i\,q_i)\,d=0$ for all $d\in\mathfrak{d}$, and hence $1=\sum d_i\,q_i$, so $Q_{\max}=\mathfrak{d}\,Q_{\max}$. □

4. Commutative noetherian rings. When A is commutative and noetherian, every dense ideal contains a regular element (Prop. VII.6.9), so $Q_{\mathrm{cl}}(A)=Q_{\mathrm{tot}}(A)=Q_{\max}(A)$.

§6. 1-Topologies and Rings of Fractions

A *1-topology* on A is a Gabriel topology with a basis consisting of principal right ideals. A 1-topology $\mathfrak{F}$ is determined by the set $\Sigma(\mathfrak{F})=\{s\in A\,|\,sA\in\mathfrak{F}\}$.

Proposition 6.1. *The map $\mathfrak{F}\mapsto\Sigma(\mathfrak{F})$ defines a bijective correspondence between 1-topologies $\mathfrak{F}$ on A and multiplicatively closed subsets S of A satisfying*

S0. If $ab\in S$, then $a\in S$.

S1. If $s\in S$ and $a\in A$, then there exist $t\in S$ and $b\in A$ such that $sb=at$.

Proof. Let $\mathfrak{F}$ be a 1-topology. $\Sigma(\mathfrak{F})$ is multiplicatively closed, for if s and t belong to $\Sigma(\mathfrak{F})$, then $(stA:sa)\supset(tA:a)$, and $(tA:a)\in\mathfrak{F}$ by axiom $T3$ for topologies, so $stA\in\mathfrak{F}$ by $T4$. Property $S0$ for S is immediate from $T1$, and $S1$ is clear since $T3$ implies that $(sA:a)\supset tA$ for some $tA\in\mathfrak{F}$.

Conversely, if S is multiplicatively closed and satisfies $S1$, then it is easy to verify that $\mathfrak{F}=\{\mathfrak{a}|\mathfrak{a}\cap S\neq\emptyset\}$ is a 1-topology. $S0$ is a saturation axiom which makes the correspondence $\mathfrak{F}\leftrightarrow S$ bijective. □

For a 1-topology one can describe the modules of quotients quite explicitly;

Proposition 6.2. *If $\mathfrak{F}$ is a 1-topology and M is a right A-module, then*

$$M_{\mathfrak{F}}=\{(x,s)\in M\times S\,|\,s\,a=0 \text{ in } A \text{ implies } x\,a\,t=0 \text{ for some } t\in S\}/\sim$$

where $\sim$ is the equivalence relation

$$(x,s)\sim(y,t) \text{ if there exist } a,b\in A \text{ such that } s\,a=t\,b\in S \quad\text{and}\quad x\,a=y\,b.$$

Proof. Recall that we have $M_{\mathfrak{F}}=\varinjlim_{\mathfrak{F}}\operatorname{Hom}_A(sA, M/t(M))$ with $s\in S$. A homomorphism $\varphi\colon sA\to M/t(M)$ is determined by an element $x\in M$ such that $s\,a=s\,b$ in A implies $x\,a-x\,b\in t(M)$, i.e. $x\,a\,t=x\,b\,t$ for some $t\in S$. In the limit, φ gives the same element of $M_{\mathfrak{F}}$ as $\psi\colon tA\to M/t(M)$, determined by $y\in M$, if and only if φ and ψ coincide on some $uA\subset sA\cap tA$ with $u\in S$, i.e. if and only if there exist $a,b\in A$ such that $u=s\,a=t\,b\in S$ and $x\,a-y\,b\in t(M)$. This clearly corresponds to the relation $\sim$. □

With the description of $M_{\mathfrak{F}}$ given in the Proposition, the module operations take the form:

$$(x,s)+(y,t)=(x\,a+y\,b,u) \quad\text{where}\quad s\,a=t\,b=u\in S;$$

$$(x,s)\cdot(y,t)=(x\,b,t\,v) \quad\text{for some}\quad b\in A,\ v\in S \quad\text{such that}\quad a v=s b.$$

Proposition 6.3. *A 1-topology $\mathfrak{F}$ is perfect if and only if $S=\Sigma(\mathfrak{F})$ satisfies:*

$S2'$. For every $s\in S$ there exists $a\in A$ such that $s\,a\in S$ and such that $s\,a\,b=0$ implies $a\,b\,u=0$ for some $u\in S$.

Proof. Perfectness of $\mathfrak{F}$ means that for every $s\in S$ there exists $q\in A_{\mathfrak{F}}$ such that $s\,q=1$ (Prop. 3.4(g)). Suppose that so is the case. Using 6.2, we write $q=(a,t)$. Then $(s,1)(a,t)=(s\,a,t)\sim(1,1)$, so there exist $a', b'\in A$ such that $t\,a'=1\cdot b'\in S$ and $s\,a\,a'=1\cdot b'$. We may then take $a\,a'$ as the element a in $S2'$, because $s\,a\,a'=b'\in S$ and if $b\in A$ is an element such that $s\,a\,a'\,b=0$, then $t\,a'\,b=b'\,b=s\,a\,a'\,b=0$ and $(a,t)\in A_{\mathfrak{F}}$ implies $a\,a'\,b\,u=0$ for some $u\in S$ (by (6.2)).

Conversely, if for each $s\in S$ there exists $a\in A$ as in $S2'$, then $(a, s\,a)$ represents an element in $A_{\mathfrak{F}}$ and one has $(s,1)(a,s\,a)=1$. □

The axiom $S2'$ is a weakened form of the condition $S2$ that was introduced in Chap. II:

$S2$. If $s\,a=0$ with $s\in S$, then $a\,t=0$ for some $t\in S$.

From Chap. II we recall:

Proposition 6.4. *Let S be a multiplicatively closed subset of A. The ring of fractions $A[S^{-1}]$ exists if and only if S satisfies $S1$ and $S2$. If $A[S^{-1}]$ exists, then $A[S^{-1}]=A_{\mathfrak{F}}$ where the topology $\mathfrak{F}=\{\mathfrak{a}\,|\,\mathfrak{a}\cap S\neq\emptyset\}$ is perfect.*

We also recall from Chap. II (Prop. 3.2) the formula for the module of fractions: $M[S^{-1}]=M\times S/\sim$, where $\sim$ is defined as above in Prop. 6.2.

§7. Finite Localizations

A ring epimorphism $A \to B$ is a *finite right localization* if it makes B into a finitely generated projective left A-module; it is of course then a perfect right localization. As before we let $\mathfrak{F}^B$ denote the corresponding Gabriel topology, and let $\tau(B)$ be the trace ideal in A of the left A-module B.

Proposition 7.1 (Cunningham [1]). *Let $\varphi: A \to B$ be a perfect right localization. The following assertions are equivalent:*

(a) *φ is a finite right localization.*

(b) *$\mathfrak{F}^B$ consists of all right ideals containing $\tau(B)$.*

(c) *The $\mathfrak{F}^B$-torsion modules form a* TTF*-class.*

Proof. (a) ⇒ (b) follows immediately from Prop. VI.9.4, and also (b) ⇒ (c) is clear.

(c) ⇒ (a): The condition (c) implies that the localization functor $\cdot \otimes_A B$ preserves direct products (cf. Example IX.1.3), and consequently that B is finitely presented as a left A-module, by Lemma I.13.2. It then follows from Cor. I.11.5 that B is a projective module. □

Proposition 7.2. *Suppose A is a subring of B such that $A \to B$ is a finite right localization. Then $\tau(B)$ is a finitely generated projective right ideal of A, and $B \cong \mathrm{Hom}_A(\tau(B), \tau(B))$.*

Proof. Since the topology $\mathfrak{F}^B$ is perfect, the right ideal $\tau(B)$ must be finitely generated. Let F be a free module with an epimorphism $\alpha: F \to \tau(B)$. Since A is a subring of B, both F and $\tau(B)$ are $\mathfrak{F}^B$-torsion-free. But on $\mathfrak{F}^B$-torsion-free modules, the localization functor has the form $\mathrm{Hom}_A(\tau(B), \cdot)$, and its exactness implies that α splits. Hence $\tau(B)$ is a projective module. Finally it follows from Cor. IX.2.9 that $B \cong \mathrm{Hom}_A(\tau(B), \tau(B))$. □

When A is a left perfect ring, every hereditary torsion class is a TTF-class (Cor. VIII.6.3), and therefore we obtain (Cunningham [1], Storrer [6]):

Proposition 7.3. *When A is a left perfect ring, every perfect right localization is finite.*

Example. Commutative rings. Suppose $\varphi: A \to B$ is a finite localization of commutative rings. It is a general property of rings of quotients that for every $b \in B$ there exists $\mathfrak{a} \in \mathfrak{F}^B$ such that $b\,\mathfrak{a} \subset \varphi(A)$, and hence $B\,\tau(B) \subset \varphi(A)$. But $\tau(B)\,B = B$ by perfectness, so φ must be a surjective map. Since B is a projective A-module, it follows that φ is just the projection $A \to Ae$ for an idempotent e of A.

§8. The Endomorphism Ring of a Finitely Generated Projective Module

Let P be a finitely generated projective left A-module. Put $B = \mathrm{Hom}_A(P, P)$, and consider P as an A-B-bimodule. In the particular case when P also is a generator for **Mod**-A, we know that the ring B is Morita equivalent to A. In the general case, we know at least (from Prop. IV.6.7) that P is a generator for **Mod**-B and that P is finitely generated projective also as a left $\tilde{A}$-module, where we use $\tilde{A}$ to denote the bicommutator $\mathrm{Bic}_A(P)$ of P.

The purpose of this section is to show that there is an equivalence between **Mod**-B and the category **Mod**-$(A, \tau(P))$ of modules of quotients with respect to the Gabriel topology

$$\mathfrak{F}^P = \{\mathfrak{a} \mid \mathfrak{a} P = P\} = \{\mathfrak{a} \mid \mathfrak{a} \supset \tau(P)\},$$

where $\tau(P)$ is the trace ideal in A of P (cf. Prop. VI.9.4 and Example IX.1.3 for the notation), and to investigate when this is a perfect localization.

A crucial role for this theory will be played by the two functors

$$F = \cdot \otimes_A P\colon \qquad \mathbf{Mod}\text{-}A \to \mathbf{Mod}\text{-}B,$$

$$H = \mathrm{Hom}_B(P, \cdot)\colon \ \mathbf{Mod}\text{-}B \to \mathbf{Mod}\text{-}A.$$

F is an exact left adjoint of H, and there are natural transformations $\xi\colon FH \to 1$ and $\zeta\colon 1 \to HF$.

Proposition 8.1. *$\xi\colon FH \to 1$ is a natural equivalence.*

Proof. We have to show that the homomorphism

$$\xi_M\colon \mathrm{Hom}_B(P, M) \otimes_A P \to M,$$

defined for each right B-module M as $\xi_M(\sum \alpha_i \otimes x_i) = \sum \alpha_i(x_i)$, is an isomorphism. Since P is finitely generated and projective as a left A-module, there exist $x_1, \ldots, x_n \in P$ and $\varphi_1, \ldots, \varphi_n\colon P \to A$ such that $x = \sum \varphi_i(x)\, x_i$ for every $x \in P$. By using these projective coordinates, we can define a map

$$\mu\colon M \to \mathrm{Hom}_B(P, M) \otimes_A P$$

by setting $\mu(y) = \sum_1^n \alpha_i \otimes x_i$, where $\alpha_i\colon P \to M$ is defined as $\alpha_i(z) = y\, \varphi_i(\cdot)\, z$. Here $\varphi_i(\cdot)\, z$ is an endomorphism of ${}_AP$ and thus an element of B, so $y\, \varphi_i(\cdot)\, z$ is in fact an element of M_B. We have $\xi_M\, \mu = 1$, for

$$\xi_M\, \mu(y) = \xi_M(\sum \alpha_i \otimes x_i) = \sum \alpha_i(x_i) = \sum y\, \varphi_i(\cdot)\, x_i = y,$$

since $\sum \varphi_i(\cdot)\, x_i = 1 \in B$. It remains to verify $\mu\, \xi_M = 1$. Now

$$\mu\, \xi_M(\alpha \otimes x) = \mu(\alpha(x)) = \sum \alpha_i \otimes x_i$$

with

$$\alpha_i(z) = \alpha(x)\, \varphi_i(\cdot)\, z = \alpha(\varphi_i(x)\, z)$$

by the B-linearity of α. Hence

$$\mu\, \xi_M(\alpha \otimes x) = \sum \alpha_i \otimes x_i = \sum \alpha(\varphi_i(x) \cdot) \otimes x_i = \sum \alpha(\cdot) \otimes \varphi_i(x)\, x_i = \alpha \otimes x,$$

where the A-module structure of $\mathrm{Hom}_B(P, M)$ has been used to move $\varphi_i(x)$ to the right of the $\otimes$-sign. □

As for the natural transformation $\zeta\colon 1 \to HF$ we have:

Proposition 8.2. *The modules* $\mathrm{Ker}\, \zeta_M$ *and* $\mathrm{Coker}\, \zeta_M$ *are $\mathfrak{F}^P$-torsion modules, for every right A-module M.*

Proof. We must show that $F(\operatorname{Ker}\zeta_M)=F(\operatorname{Coker}\zeta_M)=0$. Since F is an exact functor, it suffices for this to show that $F(\zeta_M)$ is an isomorphism. But that is easy, because $\xi_{F(M)}F(\zeta_M)=1_{F(M)}$ by Prop. IV.9.3, and $\xi_{F(M)}$ is an isomorphism by Prop. 8.1. □

Consider the diagram of functors

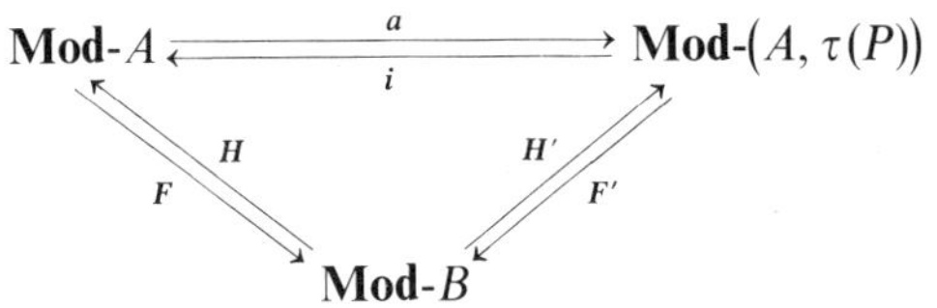

where $a(M)=M_{\mathfrak{F}^P}$, i is the inclusion functor, $F'=Fi$ and $H'=aH$.

Proposition 8.3. *If N is a right B-module, then $H(N)=\operatorname{Hom}_B(P,N)$ is an $\mathfrak{F}^P$-closed right A-module.*

Proof. Let $0\to M_0\to M_1\to M_2\to 0$ be an exact sequence of right A-modules, with M_2 an $\mathfrak{F}^P$-torsion module. Then $F(M_0)\cong F(M_1)$, and since H is a right adjoint of F, it follows that $\operatorname{Hom}_A(M_0,H(N))\cong\operatorname{Hom}_A(M_1,H(N))$. Hence $H(N)$ is $\mathfrak{F}^P$-closed. □

Combining this result with Prop. 8.2 we obtain:

Corollary 8.4. *The functors $i\,a$ and HF are naturally equivalent.*

In particular, we obtain $i\,a(A)\cong HF(A)=\operatorname{Hom}_B(P,P)$, and thus (recall that $\tilde{A}$ denotes the bicommutator of P):

Proposition 8.5. $A_{\mathfrak{F}^P}\cong\tilde{A}$.

Proposition 8.6. *The functors F' and H' define an equivalence between* **Mod**-B *and* **Mod**-$(A,\tau(P))$.

Proof. We have $F'H'=Fi\,aH\cong FHFH\cong 1$ by Cor. 8.3 and Prop. 8.1, and similarly $H'F'=aHFi\cong a\,i\,a\,i\cong 1$. □

A particularly interesting case to investigate is when **Mod**-$(A,\tau(P))$ coincides with **Mod**-$\tilde{A}$.

Proposition 8.7 (Cunningham [1]). *The following properties of a finitely generated projective left A-module P are equivalent:*

(a) *F' and H' induce a Morita equivalence between B and $\tilde{A}$.*
(b) *P_B is finitely generated and projective.*
(c) *P is a generator for $\tilde{A}$-**Mod**.*
(d) *The topology $\mathfrak{F}^P$ is perfect.*
(e) $\tau(P)\,\tilde{A}=\tilde{A}$.

Proof. (a) ⇒ (b): If B and $\operatorname{Hom}_B(P,P)$ are Morita equivalent, with the equivalence induced by the functor $H'=\operatorname{Hom}_B(P,\cdot)$, then P_B is finitely generated projective (Prop. IV.10.2).

(b) ⇒ (c) is a consequence of Prop. IV.6.7.

(c) ⇒ (e): The trace ideal $\tau_{\tilde{A}}(P)$ is the image of the canonical map

$$P\otimes_B\operatorname{Hom}_{\tilde{A}}(P,\tilde{A})\to\tilde{A}.$$

But since $P=\tau(P)P$ by Prop. VI.9.4, it follows that this image is contained in $\tau(P)\tilde{A}$. So if ${}_AP$ is a generator, then $\tau_{\tilde{A}}(P)=\tau(P)\tilde{A}=\tilde{A}$.

(e) $\Rightarrow$ (d) is clear from Prop. 3.1.

(d) $\Rightarrow$ (a) is immediate from Props. 8.5 and 8.6. $\square$

Note that when the conditions of the Proposition are satisfied, then $\mathfrak{F}^{\tilde{A}}=\mathfrak{F}^P$ and consequently $A\to\tilde{A}$ is a finite right localization (by Prop. 7.3; cf. also Exercise 19). However, the converse is not true, i.e. $A\to\tilde{A}$ may be a finite localization without satisfying the conditions of Prop. 8.7 (see Exercise 21).

Examples

1. Morita equivalence. When ${}_AP$ is not only finitely generated and projective, but also a generator for A-**Mod**, then $\mathfrak{F}^P=\{A\}$ and consequently $\tilde{A}=A$. Prop. 8.7 then just asserts the known fact that there is a Morita equivalence between A and $\mathrm{Hom}_A(P,P)$.

2. Noetherian hereditary rings. When A is right noetherian and right hereditary, then $\mathfrak{F}^P$ is perfect by Cor. 3.4, and hence every finitely generated projective left A-module P satisfies the conditions of Prop. 8.7.

3. Idempotents. Suppose e is an idempotent element of A, and consider the projective left A-module Ae. Its trace ideal in A is AeA, so $\mathfrak{F}^{Ae}=\{\mathfrak{a}\mid\mathfrak{a}\supset AeA\}$, and M_A is an $\mathfrak{F}^{Ae}$-torsion module if and only if $Me=0$. The endomorphism ring of Ae is eAe, and Cor. 8.4 gives

$$M_{\mathfrak{F}^{Ae}}\cong\mathrm{Hom}_{eAe}(Ae, M\otimes_A Ae)\cong\mathrm{Hom}_{eAe}(Ae, Me).$$

Exercises

§1

1. Let $\varphi\colon A\to B$ be a ring homomorphism, and suppose $b\in B$ is dominated by φ. Show that if L_B and ${}_BM$ are B-modules, then $xb\otimes y=x\otimes by$ holds in $L\otimes_A M$ for all $x\in L$, $y\in M$.
2. Prove the analogue of Prop. 1.2 for epimorphisms of commutative rings.
3. Show that the canonical map $\mathbb{Z}\to\mathbb{Q}\times\mathbb{Z}/n\mathbb{Z}$ $(n>1)$ is an epimorphism of rings.
4. Show that if A is FP-injective as a right A-module (Exercise I.43), then A is a dominant ring.

§§2 and 3

5. Show that if A is right hereditary, then $A_{\mathfrak{F}}$ is flat as a left A-module, for every Gabriel topology $\mathfrak{F}$.
6. Show that a Gabriel topology $\mathfrak{F}$ is perfect if and only if all $A_{\mathfrak{F}}$-modules are $\mathfrak{F}$-torsion-free as A-modules.
7. Suppose $\mathfrak{F}$ is a perfect Gabriel topology. Show that:

(i) M_A is an $\mathfrak{F}$-torsion module if and only if $M\otimes_A A_{\mathfrak{F}}=0$.

(ii) The $\mathfrak{F}$-torsion submodule of M is isomorphic to $\mathrm{Tor}_1^A(M, A_{\mathfrak{F}}/A)$, assuming that $A_{\mathfrak{F}}$ is $\mathfrak{F}$-torsion-free.

8. Let $A\subset B$ be commutative rings. Show that B is a perfect localization of A if and only if $(A:b)B=B$ holds for every $b\in B$. (Hint: note that if $\mathfrak{a}_i B=B$ holds for a finite family of ideals $\mathfrak{a}_i$, then $(\bigcap\mathfrak{a}_i)B=B$.)
9. Let $A\subset B$ be a flat epimorphism of commutative rings. Show that the induced homomorphism $A[X]\to B[X]$ also is a flat epimorphism.
10. Assume B is a perfect right localization of a right noetherian subring A. Show that if $\mathfrak{p}$ is a prime ideal of A, then either $\mathfrak{p}B=B$ or $\mathfrak{p}B\cap A=\mathfrak{p}$.
11. Show that if B is a perfect right localization of a right semi-artinian ring A, then also B is right semi-artinian.
12. Show that if A is a regular commutative ring, then every simple module is injective.

§4

13. Suppose every finitely generated right ideal of A is principal. Show that if $Q_{cl}^r(A)$ exists, then $Q_{cl}^r(A) = Q_{tot}^r(A)$.

14. Show that $Q_{tot}^r(M_n(A)) \cong M_n(Q_{tot}^r(A))$.

§5

15. Show that if Q is a right Kasch ring, then also $M_n(Q)$ is a right Kasch ring.

16. Let A be a right Kasch ring. Show that:

(i) $J(A) = r(S)$, where S is the right socle of A.

(ii) If A satisfies DCC on right annihilators, then A is semi-local.

§6

17. Let A be a regular ring and put $S = \{a \in A \mid ba = 0 \text{ implies } b = 0\}$. Show that S satisfies $S0$, $S1$ and $S2'$, and that S satisfies $S2$ if and only if all elements of S are invertible. (Illustration: A is the endomorphism ring of an infinite-dimensional vector space.)

18. Let $\mathfrak{a}$ be a two-sided ideal of A such that $A/\mathfrak{a}$ is flat as a left A-module. Show that it follows from condition (c) of Prop. 3.11 that $S = \{1 - a \mid a \in \mathfrak{a}\}$ satisfies $S1$ and $S2$, and that $A/\mathfrak{a} \cong A[S^{-1}]$.

§7

19. Let $\mathfrak{a}$ be an idempotent, two-sided ideal of A, and assume $\mathfrak{a}$ is finitely generated, projective and dense as a right ideal (cf. Prop. 5.5). Let $\mathfrak{F}$ be the Gabriel topology of right ideals containing $\mathfrak{a}$. Show that:

(i) $\mathfrak{F}$ is a perfect topology.

(ii) $A \to A_{\mathfrak{F}}$ is a finite right localization.

(iii) The trace ideal in A of the right A-module $A_{\mathfrak{F}}$ is $\mathfrak{a}$.

20. Assume B is a ring with a left ideal $\mathfrak{a}$ such that $\mathfrak{a}B = B$. Let A be a subring of B containing $\mathfrak{a}$. Show that B is a finite right localization of A, with $\tau(B) = \mathfrak{a}A$.

§8

21. Suppose P is a finitely generated, projective left A-module such that P also is finitely generated and projective as a right B-module, where $B = \operatorname{Hom}_A(P, P)$. Show by a direct argument (using only the results of Chap. IV, §10) that $\operatorname{Hom}_B(P, P)$ is finitely generated and projective as a left A-module.

22. Let K be a field and let F be an extension field of K such that $[F:K] = \infty$. Consider the ring

$$A = \begin{pmatrix} F & F \\ 0 & K \end{pmatrix}$$

with the idempotent $e = \begin{pmatrix} 0 & 0 \\ 0 & 1 \end{pmatrix}$. Put $P = Ae$ and $B = \operatorname{Hom}_A(P, P)$. Show that:

(i) $B \cong eAe \cong K$, and accordingly $P \cong F \oplus K$ (so P is not finitely generated as a right B-module).

(ii) $\tilde{A} = \operatorname{Hom}_B(P, P) \cong \operatorname{Hom}_K(F, F)$ (which is not Morita equivalent to B).

23. Consider the ring

$$A = \begin{pmatrix} \mathbb{Q} & \mathbb{Q} \\ 0 & \mathbb{Z} \end{pmatrix}$$

with the idempotent $e = \begin{pmatrix} 0 & 0 \\ 0 & 1 \end{pmatrix}$. Put $P = Ae$ and $B = \operatorname{Hom}_A(P, P)$ (cf. the preceding exercise). Show that:

(i) P is not projective as a left B-module.

(ii) $\operatorname{Hom}_B(P, P) \cong A$ (so $A \to \tilde{A}$ is trivially a finite localization).

24. Let $\langle A, B, P, Q, \alpha, \beta \rangle$ be a Morita context (Exercise IV.35) with trace ideals $\tau_A = \operatorname{Im} \alpha$ and $\tau_B = \operatorname{Im} \beta$. There are functors $G = \operatorname{Hom}_A(Q, \cdot)$ and $H = \operatorname{Hom}_B(P, \cdot)$ with natural transformations $\Phi: 1 \to HG$ and $\psi: 1 \to GH$ induced by α and β. Show that for every right A-module M:

(i) $\operatorname{Ker} \Phi_M = \{x \in M \mid x\tau_A = 0\}$.

(ii) If $x\tau_A = 0$ for all non-zero $x \in M$, then the canonical map $\operatorname{Hom}_A(\tau_A, M) \to \operatorname{Hom}_A(P \otimes_B Q, M)$ is an isomorphism.

(iii) Φ_M is an isomorphism if and only if M belongs to the quotient category **Mod**-(A, τ_A) (Exercise IX.12).

(iv) Deduce that Φ and ψ induce an equivalence between the categories **Mod**-(A, τ_A) and **Mod**-(B, τ_B).

Chapter XII. The Maximal Ring of Quotients of a Non-Singular Ring

When A is a right non-singular ring, its maximal right ring of quotients Q_{max} is obtained as $A_{\mathfrak{E}}$, where $\mathfrak{E}$ is the family of all essential right ideals of A, and **Mod**-$(A, \mathfrak{E})$ consists of the non-singular injective A-modules. It follows from Prop. X.1.7 that every object in the category **Mod**-$(A, \mathfrak{E})$ is injective. Thus **Mod**-$(A, \mathfrak{E})$ is a spectral category. In view of this observation, it is natural to begin this chapter with a study of the properties of spectral categories in terms of a standard representation as a Giraud subcategory of a module category.

§ 1. Standard Representation of Spectral Categories

In a while we will see that the description of spectral categories involves the canonical topologies of regular, right self-injective rings. Let us therefore first briefly have a look at these rings.

Lemma 1.1. *Every regular ring A is non-singular.*

Proof. Suppose $\mathfrak{a}$ is an essential right ideal and $b\,\mathfrak{a}=0$ for some non-zero $b \in A$. Write $b = b\,x\,b$. Then $x\,b \neq 0$, which implies that $0 \neq x\,b\,c \in \mathfrak{a}$ for some c. This gives $b\,c = b\,x\,b\,c \in b\,\mathfrak{a} = 0$, which contradicts $x\,b\,c \neq 0$. ▯

Now suppose A is a regular and right self-injective ring (a ring is *right self-injective* if it is injective as a right module over itself). The essential right ideals of A form a Gabriel topology $\mathfrak{E}$, coinciding with the canonical topology of A. The category **Mod**-$(A, \mathfrak{E})$ is a spectral category, which consists of all non-singular injective right A-modules. Since A is an injective cogenerator for **Mod**-$(A, \mathfrak{E})$ by Prop. X.1.9, the spectral category **Mod**-$(A, \mathfrak{E})$ consists precisely of all direct summands of powers of A.

Proposition 1.2. *Let* **C** *be a spectral category, C any object of* **C** *and $A =$* $\mathrm{Hom}_{\mathbf{C}}(C, C)$. *Then:*

(i) *A is regular and right self-injective ring.*

(ii) *If D is any other object of* **C**, *then the right A-module* $\mathrm{Hom}_{\mathbf{C}}(C, D)$ *is non-singular and injective.*

Proof. The regularity of A was pointed out in Prop. V.6.1. The assertion that A is right self-injective is a special case of (ii), so we will prove that $\mathrm{Hom}_{\mathbf{C}}(C, D)$ is injective over A. Let $\mathfrak{a}$ be any right ideal of A. $\mathfrak{a}$ is a direct union of finitely generated right ideals, which are of the form eA with $e^2=e$, since A is regular. A homomorphism $\phi: eA \to \mathrm{Hom}_{\mathbf{C}}(C, D)$ is given by $\phi(e)=\varphi$, with the requirement that $\varphi(1-e)=0$. Since $C=\mathrm{Im}\, e \oplus \mathrm{Im}(1-e)$, it follows that $\mathrm{Hom}_A(eA, \mathrm{Hom}_{\mathbf{C}}(C, D)) \cong \mathrm{Hom}_{\mathbf{C}}(\mathrm{Im}\, e, D)$. As a result we obtain

$$\begin{aligned}\mathrm{Hom}_A(\mathfrak{a}, \mathrm{Hom}_{\mathbf{C}}(C, D)) &= \mathrm{Hom}_A(\varinjlim eA, \mathrm{Hom}_{\mathbf{C}}(C, D))\\ &\cong \varprojlim \mathrm{Hom}_A(eA, \mathrm{Hom}_{\mathbf{C}}(C, D)) \cong \varprojlim \mathrm{Hom}_{\mathbf{C}}(\mathrm{Im}\, e, D)\\ &\cong \mathrm{Hom}_{\mathbf{C}}(\varinjlim \mathrm{Im}\, e, D) = \mathrm{Hom}_{\mathbf{C}}(C', D),\end{aligned}$$

where $C' = \varinjlim \mathrm{Im}\, e$. Since C' is a direct summand of C, there is an epimorphic map

$$\mathrm{Hom}_{\mathbf{C}}(C, D) \to \mathrm{Hom}_{\mathbf{C}}(C', D) \cong \mathrm{Hom}_A(\mathfrak{a}, \mathrm{Hom}_{\mathbf{C}}(C, D)),$$

which is induced by the inclusion $\mathfrak{a} \to A$, as one may verify.

It remains to show that $\mathrm{Hom}_{\mathbf{C}}(C, D)$ is non-singular. Suppose $\mathfrak{a}$ is an essential right ideal in A and that $\varphi\,\mathfrak{a}=0$ for some non-zero $\varphi \in \mathrm{Hom}_{\mathbf{C}}(C, D)$. Then $C= C' \oplus \mathrm{Ker}\, \varphi$. Let e: $C \to C$ be the projection of C onto C'. Then e is a non-zero element of A, and hence $eA \cap \mathfrak{a} \neq 0$. For each non-zero $e\alpha \in eA \cap \mathfrak{a}$ we have $\varphi\, e\, \alpha=0$, since $\varphi\,\mathfrak{a}=0$. But this contradicts the fact that $\mathrm{Ker}\, \varphi \cap \mathrm{Im}\, e = \mathrm{Ker}\, \varphi \cap C' =0$. Hence $\mathrm{Hom}_{\mathbf{C}}(C, D)$ is non-singular. ▯

Let U be a generator for the spectral category $\mathbf{C}$. Since every morphism splits, it is easy to see that U will also be an injective cogenerator for $\mathbf{C}$. Therefore every object of $\mathbf{C}$ is a direct summand of a direct product of copies of U. We now apply Prop. 1.2 to the standard representation of $\mathbf{C}$ determined by U:

Theorem 1.3 (Gabriel and Oberst [1]). *Let* $\mathbf{C}$ *be a spectral category,* U *a generator for* $\mathbf{C}$ *and* $A=\mathrm{Hom}_{\mathbf{C}}(U, U)$. *Then:*

(i) *A is a regular and right self-injective ring.*

(ii) $\mathbf{C}$ *is equivalent to the category* **Mod**-$(A, \mathfrak{E})$, *where* $\mathfrak{E}$ *is the set of essential right ideals of* A.

(iii) *A right A-module is $\mathfrak{E}$-closed if and only if it is a direct summand of a power of* A.

Proof. (i) was proved in the preceding Proposition.

(ii): Every module $\mathrm{Hom}_{\mathbf{C}}(U, D)$, with arbitrary $D \in \mathbf{C}$, is $\mathfrak{E}$-closed for the Gabriel topology $\mathfrak{E}$ of essential right ideals of A, by Prop. 1.2. $\mathfrak{E}$ is clearly the strongest Gabriel topology with this property (it is even the strongest topology making A torsion-free). It therefore follows from the Popescu-Gabriel theorem (Theorem X.4.1) that $\mathbf{C}$ is equivalent to **Mod**-$(A, \mathfrak{E})$.

(iii): Every $\mathfrak{E}$-closed A-module is isomorphic to some $\mathrm{Hom}_{\mathbf{C}}(U, C)$. As was noted above, C is a direct summand of some power U^I. $\mathrm{Hom}_{\mathbf{C}}(U, C)$ is then correspondingly a direct summand of A^I. Conversely, since A is $\mathfrak{E}$-closed, so is also every direct summand of a power A^I. ▯

When $\mathbf{C}$ is a discrete spectral category, every object is semi-simple, and the endomorphisms take a more specialized form. The situation is summarized in:

Proposition 1.4 (Roos [2]). *Let* $\mathbf{C}$ *be a spectral category,* U *a generator for* $\mathbf{C}$ *and* $A=\mathrm{Hom}_{\mathbf{C}}(U, U)$. *The following statements are equivalent:*

(a) $\mathbf{C}$ *is a discrete spectral category.*

(b) *A is the direct product of endomorphism rings of vector spaces over skew-fields.*

(c) *Every non-zero right ideal of A contains a minimal right ideal.*

Proof. (a) $\Rightarrow$ (b): If $\Omega = \{\omega\}$ is the set of isomorphism classes of simple objects of $\mathbf{C}$, then U may be written as $U = \bigoplus_{\Omega} U_\omega$, where U_ω is the coproduct of simple objects in ω. Then we have

$$A = \mathrm{Hom}_{\mathbf{C}}(U, U) = \prod_{\Omega} \mathrm{Hom}_{\mathbf{C}}(U_\omega, U_\omega).$$

For each ω we let S_ω be a simple object in ω, and put

$$K_\omega = \mathrm{Hom}_{\mathbf{C}}(S_\omega, S_\omega), \qquad V_\omega = \mathrm{Hom}_{\mathbf{C}}(S_\omega, U_\omega).$$

Then K_ω is a skew-field, and V_ω is a vector space over K. It follows from Example X.4.3 that there is an isomorphism

$$\mathrm{Hom}_{\mathbf{C}}(U_\omega, U_\omega) \cong \mathrm{Hom}_{K_\omega}(V_\omega, V_\omega).$$

Hence $A \cong \prod_{\Omega} \mathrm{Hom}_{K_\omega}(V_\omega, V_\omega)$.

(b) $\Rightarrow$ (c): Suppose $A = \prod_I A_i$, where each A_i is an endomorphism ring of a vector space. If $\mathfrak{a}$ is a right ideal of A, let $\mathfrak{a}_i$ be the image of $\mathfrak{a}$ under the projection onto A_i. Then $\mathfrak{a} \supset \bigoplus_I \mathfrak{a}_i$, and therefore it suffices to consider the case when $A = \mathrm{Hom}_K(V, V)$ for a vector space V over a skew-field K. For every idempotent $e \in \mathrm{Hom}_K(V, V)$ one may take $f: V \to V$ to be the projection onto a one-dimensional subspace of $e(V)$, and then fA is a minimal right ideal contained in eA.

(c) $\Rightarrow$ (a): The condition (c) asserts that the right socle $\mathfrak{a}$ of A is essential in A, and $\mathfrak{a}$ is then the smallest essential right ideal of A. The inclusion $\mathfrak{a} \to A$ becomes an isomorphism in **Mod**-$(A, \mathfrak{a})$. It follows that A is a semi-simple generator of **Mod**-$(A, \mathfrak{a})$, which is equivalent to $\mathbf{C}$. ▯

Corollary 1.5. *The following properties of a ring A are equivalent:*

(a) *A is regular and right self-injective, and is an essential extension of its right socle.*

(b) *A is a direct product of endomorphism rings of vector spaces over skew-fields.*

Example. One-sided self-injective rings. If A is the endomorphism ring of an infinite-dimensional vector space over a skew-field, then A is right self-injective, but it may be proved that A cannot be left self-injective (see Prop. XIV.3.7).

§2. The Maximal Ring of Quotients

The results of the preceding section will now be applied to the maximal right ring of quotients $Q_{\max}$ of a right non-singular ring A. Let $\mathfrak{E}$ denote the Gabriel topology of essential right ideals of A. For each module M we have $M_{\mathfrak{E}} \cong E(M/Z(M))$. The functor q: **Mod**-$A \to$ **Mod**-$Q_{\max}$, given by $q(M) = M_{\mathfrak{E}}$, is exact by Prop. IX.2.12.

The category **Mod**-$(A, \mathfrak{E})$ consists of the non-singular injective A-modules, and it is a spectral category by Prop. X.1.7. Since the endomorphism ring of $Q_{\max}$ as an object of **Mod**-$(A, \mathfrak{E})$ is equal to $Q_{\max}$, it follows from Prop. 1.2 that $Q_{\max}$ is a regular and right self-injective ring. The category **Mod**-$(A, \mathfrak{E})$ is equivalent to the category **Mod**-$(Q_{\max}, \mathfrak{E}^e)$, where $\mathfrak{E}^e$ denotes the family of essential right ideals of $Q_{\max}$ (cf. Example of Chap. X, §2). From the remark preceding Prop. 1.2 we obtain:

Proposition 2.1. *Let A be a right non-singular ring. Then:*

(i) *$Q_{\max}$ is a regular and right self-injective ring.*

(ii) *The full subcategory of* **Mod**-*A consisting of the non-singular injective modules is a spectral category, and it consists of all $Q_{\max}$-modules which are direct summands of powers of $Q_{\max}$.*

For the sake of completeness we note a converse result:

Proposition 2.2. *A is a right non-singular ring if and only if $Q_{\max}$ is a regular ring.*

Proof. Suppose $Q_{\max}$ is regular. Let $\mathfrak{a}$ be an essential right ideal of A and assume $b\mathfrak{a}=0$ for some $b \neq 0$ in A. We can write $b=bqb$ for some $q \in Q_{\max}$. Then $qb \neq 0$ and $0 \neq qbc \in \mathfrak{a}$ for some $c \in A$. But then $bc=bqbc \in b\mathfrak{a}=0$, contradicting $qbc \neq 0$. ☐

The following particular consequence of the argument leading to Prop. 2.1 is worth remarking:

Corollary 2.3. *If A is right non-singular, then*

$$Q_{\max} \cong \operatorname{Hom}_A(E(A), E(A)).$$

The case when the category of modules of quotients becomes a discrete spectral category has been treated by Chase and Faith [1]:

Proposition 2.4. *The following properties of a non-singular ring A are equivalent:*

(a) **Mod**-$(A, \mathfrak{E})$ *is a discrete spectral category.*

(b) *$Q_{\max}$ is a direct product of endomorphism rings of vector spaces over skew-fields.*

(c) *The ring $Q_{\max}$ is an essential extension of its right socle.*

(d) *Every non-zero right ideal of A contains a non-zero coirreducible right ideal.*

Proof. The equivalence of (a), (b) and (c) is immediate from the equivalence of the corresponding conditions in Prop. 1.4.

(c) $\Rightarrow$ (d): If $\mathfrak{a}$ is a non-zero right ideal of A, then $E(\mathfrak{a})$ may be considered as a right ideal of $Q_{\max}$. Let S be a minimal right $Q_{\max}$-ideal contained in $E(\mathfrak{a})$. Then $S \cap \mathfrak{a}$ is a non-zero coirreducible right ideal of A.

(d) $\Rightarrow$ (c): If $\mathfrak{c}$ is a non-zero right ideal of $Q_{\max}$, let $\mathfrak{a}$ be a coirreducible right ideal of A contained in $\mathfrak{c}$. Then $E(\mathfrak{a})$ is indecomposable, and hence simple, as a right ideal of $Q_{\max}$. Clearly $E(\mathfrak{a}) \subset \mathfrak{c}$. ☐

The case when $Q_{\max}$ is a semi-simple ring is of particular importance, and is described by the following result (Johnson [5], Walker and Walker [1] and Sandomierski [1]):

Theorem 2.5. *Let A be a right non-singular ring. The following assertions are equivalent:*

(a) *Q_{max} is a semi-simple ring.*

(b) *A is of finite right rank.*

(c) *Every essential right ideal of A contains a finitely generated essential right ideal.*

(d) *$\mathfrak{a} Q_{max} = Q_{max}$ for every essential right ideal $\mathfrak{a}$ of A.*

Proof. (a) ⇒ (d): If $\mathfrak{a}$ is an essential right ideal of A, then $\mathfrak{a} Q_{max}$ is an essential right ideal of Q_{max}, as one easily verifies. But a semi-simple ring has no proper essential right ideals, so $\mathfrak{a} Q_{max} = Q_{max}$.

(d) ⇒ (c) follows from Prop. XI.3.4.

(c) ⇒ (b): Suppose $(\mathfrak{a}_i)_I$ is an independent family of right ideals of A. We may assume that $\mathfrak{a} = \sum_I \mathfrak{a}_i$ is essential in A, by adding a pseudo-complement of $\mathfrak{a}$ if necessary. Then $\mathfrak{a}$ contains a finitely generated essential right ideal $\mathfrak{b}$. But $\mathfrak{b}$ is contained in the sum of finitely many $\mathfrak{a}_i$, and since $\mathfrak{b}$ is essential, it follows that the family $(\mathfrak{a}_i)_I$ is finite.

(b) ⇒ (a) is clear, since $Q_{max} \cong E(A)$. □

Corollary 2.6. *If the maximal right ring of quotients of A is semi-simple, then it is a perfect right localization of A.*

Corollary 2.7. *If $Q_{max}(A)$ is semi-simple, then:*

(i) *Every right Q_{max}-module is injective as an A-module.*

(ii) *Every left Q_{max}-module is flat as an A-module.*

Proof. (i) follows with the use of Prop. X.1.7.

(ii) follows from the fact that for every left Q_{max}-module M there is a natural equivalence

$$\cdot \otimes_A M \cong (\cdot \otimes_A Q_{max}) \otimes_{Q_{max}} M,$$

where the right hand functor is the composition of two exact functors. □

Corollary 2.8. *If $Q_{max}(A)$ is semi-simple and M is a non-singular A-module, then $M \otimes_A Q_{max}$ is an injective envelope of M.*

Proof. An injective envelope of M is furnished by the module of quotients of M with respect to the topology of essential right ideals, and this module is isomorphic to $M \otimes_A Q_{max}$ by Prop. XI.3.1. □

Examples

1. Q_{max} is a skew-field. If Q_{max} is a skew-field, then A is a ring of finite right rank and without zero-divisors, so it follows from Goldie's Theorem (Theorem II.2.2) that Q_{max} is a classical right ring of quotients of A.

2. Non-singular rings with essential socle. Let A be a right non-singular ring. Then the singular modules form a TTF-class if and only if A contains a smallest essential right ideal, and by Prop. III.6.7 this is equivalent to the condition that the right socle S of A is essential as a right ideal. When this is satisfied, we have $Q_{max} \cong \mathrm{Hom}_A(S, S)$ by Cor. IX.2.9. In particular we obtain from Prop. XI.7.1 and 7.2:

Proposition 2.9. *The following properties of a right non-singular ring A are equivalent:*

(a) *$A \to Q_{\max}$ is a finite right localization.*

(b) *The right socle S of A is a finitely generated essential right ideal of A.*

When these conditions are satisfied, then S is a projective right ideal and $Q_{\max} \cong \operatorname{Hom}_A(S, S)$.

3. Prime rings with socle. Let A be a prime ring with a non-zero right socle S. Since every non-zero two-sided ideal of a prime ring is essential, S must be the smallest essential right ideal of A. Hence $Q_{\max} \cong \operatorname{Hom}_A(S, S)$ by the preceding example. Since A is prime, there is only one isomorphism class of minimal right ideals. So if we put $K = \operatorname{Hom}_A(\mathfrak{a}, \mathfrak{a})$ for some minimal right ideal $\mathfrak{a}$ of A, then K is a skew-field, and $Q_{\max} \cong \operatorname{Hom}_A(S, S)$ is the endomorphism ring of $\operatorname{Hom}_A(\mathfrak{a}, S)$ as a right vector space over K.

4. Coherent rings. When A is a right coherent and non-singular ring, then $Q_{\max}$ is flat as a left A-module by Cor. XI.3.2.

§3. The Maximal Ring of Quotients of a Boolean Ring

Let A be a boolean ring and let $\mathsf{B}(A)$ denote the associated boolean algebra. First of all we recall that $\mathsf{B}(A)$ is isomorphic to the lattice of principal ideals of A (Prop. III.4.7), so A is complete as a regular ring if and only if $\mathsf{B}(A)$ is complete.

Lemma 3.1. *If $\mathfrak{a}$ is an ideal of A and $\alpha\colon \mathfrak{a} \to A$ is a homomorphism, then $\alpha(\mathfrak{a}) \subset \mathfrak{a}$ and $\alpha^2 = \alpha$.*

Proof. If $a \in \mathfrak{a}$, then $\alpha(a) = \alpha(a^2) = \alpha(a)\, a \in A a \subset \mathfrak{a}$. Further,

$$\alpha^2(a) = \alpha(\alpha(a)) = \alpha(\alpha(a)\, a) = \alpha(a)\, \alpha(a) = \alpha(a),$$

so $\alpha^2 = \alpha$. □

Proposition 3.2. *The maximal ring of quotients of a boolean ring is a boolean ring.*

Proof. $Q_{\max}(A)$ is obtained as a direct limit of homomorphisms $\mathfrak{a} \to A$, for certain ideals $\mathfrak{a}$, and Lemma 3.1 therefore implies that all elements of $Q_{\max}$ are idempotents. □

Proposition 3.3. *Every complete boolean ring is self-injective.*

Proof. Let $\alpha\colon \mathfrak{a} \to A$ be a homomorphism from an ideal $\mathfrak{a}$. Put

$$s = \sup\{a \in \mathfrak{a} \mid \alpha(a) = a\}.$$

We assert that $\alpha(a) = a s$ holds for every $a \in \mathfrak{a}$. From Lemma 3.1 we obtain that $\alpha(\alpha(a)) = \alpha^2(a) = \alpha(a)$ for all $a \in \mathfrak{a}$, so $\alpha(a) \le s$, and therefore

$$\alpha(a) = \alpha(a^2) = \alpha(a)\, a \le s a.$$

On the other hand, if $b \in \mathfrak{a}$ has the property that $\alpha(b) = b$, then for $a \in \mathfrak{a}$ we get

$$b a = \alpha(b)\, a = \alpha(b a) = \alpha(a)\, b \le \alpha(a).$$

But since $\mathsf{B}(A)$ is upper continuous (Prop. III.4.5), we have $sa=\sup\{ba\,|\,\alpha(b)=b\}$, and it follows that $sa\leq\alpha(a)$ for all $a\in\mathfrak{a}$. $\square$

We now come to the main result of this section, which is due to Brainerd and Lambek [1].

Theorem 3.4. *Let $\varphi\colon A\to B$ be a monomorphism of boolean rings. Then $\mathsf{B}(B)$ is a completion of $\mathsf{B}(A)$ if and only if B is the maximal ring of quotients of A.*

Proof. Suppose $\mathsf{B}(B)$ is a completion of $\mathsf{B}(A)$. Then B is a self-injective ring by Prop. 3.3. To prove that $B=Q_{\max}(A)$ it then suffices to show that B is an essential extension of the A-module A. But each $b\in B$ can be written as

$$b=\sup\{a\in A\,|\,a\leq b\}=\sup\{a\in A\,|\,ab=a\},$$

so if $b\neq 0$, then there must exist $a\neq 0$ in A such that $ab=a$.

On the other hand, we know that each boolean algebra A has a completion, which also is a boolean algebra (Prop. III.8.4). As we have just proved, this completion coincides with $Q_{\max}(A)$, and we may conclude that $Q_{\max}(A)$ always is a completion of A. $\square$

Corollary 3.5. *A boolean ring A is self-injective if and only if $\mathsf{B}(A)$ is complete.*

§ 4. The Lattice of Essentially Closed Ideals

Throughout this section we let A be a right non-singular ring, and $\mathfrak{E}$ denotes the filter of essential right ideals of A. If M is a non-singular module, then for each submodule L of M we put

$$L^c=\{x\in M\,|\,(L:x)\in\mathfrak{E}\}.$$

Lemma 4.1. *L^c is the unique maximal essential extension of L in M.*

Proof. L^c is an essential extension of L, for if $0\neq x\in L^c$, then there exists $\mathfrak{a}\in\mathfrak{E}$ such that $x\,\mathfrak{a}\subset L$, and $x\,\mathfrak{a}\neq 0$ by the non-singularity of M. On the other hand, if L is an essential submodule of a submodule M' of M, then $(L:x)\in\mathfrak{E}$ for every $x\in M'$, so $M'\subset L^c$. $\square$

Corollary 4.2. *The folllowing properties of a submodule L of a non-singular module M are equivalent:*

(a) *$L^c=L$.*
(b) *M/L is non-singular.*
(c) *L is essentially closed in M.*

It follows from Prop. IX.4.1 that the essentially closed submodules of a non-singular module M form a complete modular lattice, which we here will denote by $\mathsf{Cl}(M)$. If M is a non-singular injective module, then $\mathsf{Cl}(M)$ consists of the direct summands of M (cf. Prop. IX.4.2).

Proposition 4.3. *For every non-singular module M there is a lattice isomorphism $\mathsf{Cl}(M)\simeq\mathsf{Cl}(E(M))$.*

Proof. This follows from Prop. IX.4.3. The isomorphism is explicitly defined as $L \mapsto E(L)$, with the inverse $E \mapsto E \cap M$. □

Corollary 4.4. *If M is a non-singular module, then $\mathsf{Cl}(M)$ is a complemented modular lattice.*

This corollary of course also follows from the fact that the category **Mod**-$(A, \mathfrak{E})$ is a spectral category (by Cor. IX.4.4).

We now turn our attention to the lattice of essentially closed right ideals of A, which we denote by $\mathsf{Cl}(A.)$ (while $\mathsf{Cl}(.A)$ denotes the lattice of essentially closed left ideals). From Prop. IX.4.6 we know that $\mathsf{Cl}(A.)$ consists of the annihilators of subsets of the injective envelope $E(A)$ of A. It follows that every annihilator right ideal $r(S)$, for $S \subset A$, belongs to $\mathsf{Cl}(A.)$. More generally we have:

Proposition 4.5. *If $\mathfrak{a}$ is a right ideal, then $\mathfrak{a}^c \subset r(l(\mathfrak{a}))$.*

Proof. Suppose $a \in \mathfrak{a}^c$, which means that $\mathfrak{b} = (\mathfrak{a} : a)$ is an essential right ideal of A. For each $c \in l(\mathfrak{a})$ we have $c\,a\,\mathfrak{b} \subset c\,\mathfrak{a} = 0$. Since A is non-singular, this implies $c\,a = 0$, and hence $a \in r(l(\mathfrak{a}))$. □

The lattice $\mathsf{Cl}(A.)$ is related to the ideal lattice of $Q_{\max}$ by the following result, which is a consequence of Cor. IX.4.4:

Proposition 4.6. *The lattice $\mathsf{Cl}(A.)$ is isomorphic to the lattice of principal right ideals of $Q_{\max}$.*

When A is regular and right self-injective, the essentially closed right ideals are precisely the principal right ideals. In the general case, the essentially closed right ideals of A are of the form $e\,Q_{\max} \cap A$ for idempotents $e \in Q_{\max}$. The following result by Utumi [4] gives necessary and sufficient conditions for $\mathsf{Cl}(A.)$ to be equal to the lattice $\mathsf{Ann}(A.)$ of annihilator right ideals:

Proposition 4.7. *The following properties of a right non-singular ring A are equivalent:*

(a) *Every essentially closed right ideal is a right annihilator.*

(b) $\mathfrak{a}^c = r(l(\mathfrak{a}))$ *for every right ideal* $\mathfrak{a}$.

(c) $l(\mathfrak{a}) \neq 0$ *for every right ideal* $\mathfrak{a} \in \mathfrak{E}$.

(d) *Every non-zero left ideal of $Q_{\max}$ has non-zero intersection with A.*

Proof. (a) $\Rightarrow$ (b) is clear by Prop. 4.5.

(b) $\Rightarrow$ (c): If $l(\mathfrak{a}) = 0$, then $\mathfrak{a}^c = r(l(\mathfrak{a})) = A$, and thus $\mathfrak{a}$ is essential in A.

(c) $\Rightarrow$ (d): Assume there is a non-zero idempotent $e \in Q = Q_{\max}$ such that $Q\,e \cap A = 0$. Now $(1-e)\,Q \cap A$ is an essentially closed right ideal of A, so $(1-e)\,Q \cap A$ cannot be essential in A, and hence there exists a non-zero element $a \in l((1-e)\,Q \cap A)$. If $r_Q(a)$ denotes the right annihilator of a in $Q_{\max}$, then

$$(1-e)\,Q \cap A \subset r_Q(a) \cap A\,.$$

But the map $\mathfrak{a} \mapsto \mathfrak{a} \cap A$ is a lattice isomorphism between principal right ideals of $Q_{\max}$ and $\mathsf{Cl}(A.)$, so we must have $(1-e)\,Q \subset r_Q(a)$. Then $a(1-e)\,Q = 0$ implies that $a = a\,e \in Q\,e \cap A = 0$, which is a contradiction.

(d) $\Rightarrow$ (a): Suppose there is an essentially closed right ideal $\mathfrak{a}$ such that $\mathfrak{a} \neq r(l(\mathfrak{a}))$. Then $r(l(\mathfrak{a}))$ is not an essential extension of $\mathfrak{a}$, so there exists a right ideal $\mathfrak{c} \neq 0$ such that $\mathfrak{c} \subset r(l(\mathfrak{a}))$ and $\mathfrak{a} \cap \mathfrak{c} = 0$. Let $\mathfrak{d}$ be a pseudo-complement of $\mathfrak{c}$ in A with $\mathfrak{d} \supset \mathfrak{a}$. Then $\mathfrak{d}$ is essentially closed, so $\mathfrak{d} = eQ \cap A$ for some idempotent $e \neq 1$ of $Q = Q_{\max}$. By hypothesis there exists a non-zero element $a \in Q(1-e) \cap A$. Now $a\mathfrak{d} = 0$, which implies $a\mathfrak{a} = 0$ and hence $a \in l(\mathfrak{a})$. But then also $a \cdot r(l(\mathfrak{a})) = 0$, so $a\mathfrak{c} = 0$. This gives $a(\mathfrak{c} + \mathfrak{d}) = 0$, which is a contradiction since $\mathfrak{c} + \mathfrak{d}$ is essential. $\square$

Definition. The ring A is a *right Utumi ring* if it is right non-singular and satisfies the equivalent conditions of Prop. 4.7.

Since right annihilator ideals always are essentially closed, the condition (a) implies:

Corollary 4.8. *If A is a right Utumi ring, then* $\mathsf{Cl}(A.) = \mathsf{Ann}(A.)$.

We mention without proof the following result by Utumi [4]:

Proposition 4.9. *Assume A is both right and left non-singular. The right and left maximal rings of quotients of A coincide if and only if A is a two-sided Utumi ring.*

For a complete regular ring A, one can make a nice reformulation of the conditions of Prop. 4.7. For an arbitrary regular ring A one has

$$\mathsf{Cl}(A.) \supset \mathsf{Ann}(A.) \supset \mathsf{Princ}(A.). \tag{1}$$

A is called *right (left) continuous* if the lattice $\mathsf{Princ}(A.)$ (resp. $\mathsf{Princ}(.A)$) is complete and upper continuous, and A is *continuous* if it is both right and left continuous.

Proposition 4.10 (Utumi [3]). *The following properties of a regular ring A are equivalent:*

(a) *A is right continuous.*
(b) *Every right ideal is an essential submodule of a principal right ideal.*
(c) *Every essentially closed right ideal is principal.*
(d) *A is a complete, right Utumi ring.*

Proof. (a) $\Rightarrow$ (b): If $\mathfrak{a}$ is a right ideal, then let eA be the join in $\mathsf{Princ}(A.)$ of all principal right ideals contained in $\mathfrak{a}$. Then of course $\mathfrak{a} \subset eA$. If a is an element of A such that $\mathfrak{a} \cap eA = 0$, then by the upper continuity we have

$$eA \cap aA = \Big(\bigvee_{x \in \mathfrak{a}} xA\Big) \cap aA = \bigvee_{x \in \mathfrak{a}} (xA \cap aA) = 0.$$

It follows that $\mathfrak{a}$ is essential in eA.

(b) $\Rightarrow$ (c) is obvious.

(c) $\Rightarrow$ (d) follows from (1) since $\mathsf{Cl}(A.)$ is complete.

(d) $\Rightarrow$ (a): We use the criterion stated in Prop. III.5.1. Let $(\mathfrak{a}_i)_I$ be a directed family of principal right ideals, and suppose there exists $c \neq 0$ in A such that $cA \cap \mathfrak{a}_i = 0$ for all $i \in I$. Put $\mathfrak{a} = \sum_I \mathfrak{a}_i$. Then $cA \cap \mathfrak{a} = 0$, so we have $l(\mathfrak{a}) \neq 0$ by condition 4.7(c). Let e be a non-zero idempotent belonging to $l(\mathfrak{a})$. Since $\mathfrak{a} \subset (1-e)A$, we cannot have $\mathfrak{a} = A$. $\square$

Corollary 4.11. *Every right self-injective regular ring is right continuous.*

Proof. This follows e.g. from condition (c). $\square$

Examples

1. Commutative rings. A non-singular commutative ring is always a Utumi ring, because condition 4.7(d) is trivially satisfied.

2. Strongly regular rings. We have already noticed that a complete boolean ring is continuous (Example III.5.2). For a strongly regular ring A, the lattices $\mathsf{B}(A)$ and $\mathsf{Princ}(A.) = \mathsf{Princ}(.A)$ are isomorphic (Prop. III.4.7), and hence we have more generally:

Proposition 4.12. *The following properties of a strongly regular ring A are equivalent:*

(a) *A is continuous.*

(b) *A is complete.*

(c) *$\mathsf{B}(A)$ is complete.*

3. Full linear rings. Suppose A is a regular and right self-injective ring. The lattice $\mathsf{Princ}(A.)$ is locally atomic if and only if every principal right ideal contains a minimal right ideal, i.e. if and only if A is a product of endomorphism rings of vector spaces (Cor. 1.5).

4. Types of Baer rings. Recall that A is a Baer ring if every right (or left) annihilator ideal is generated by an idempotent (Chap. III, §8). A Baer ring is evidently right and left non-singular. On the other hand, every regular and right self-injective ring is a Baer ring. According to Kaplansky [1], every Baer ring can be decomposed uniquely into a direct product of Baer rings of five different types. An important (and vast) problem is to classify non-singular rings according to the types of their maximal rings of quotients. More particularly, it would be natural to investigate how the type of a Baer ring changes when taking its maximal ring of quotients. Partial results in this direction have been obtained by Roos [4] and by Cailleau and Renault [1].

§5. The Maximal Ring of Quotients of a Reduced Ring

A ring is said to be reduced if there are no non-zero nilpotent elements (Chap. I, §12). In this section we will look at the maximal ring of quotients of a reduced ring, and in particular we will find out when also $Q_{\max}$ is reduced.

Lemma 5.1. *If A is a reduced ring, then:*

(i) *$ab = 0$ implies $ba = 0$.*

(ii) *Every right annihilator ideal is a two-sided ideal.*

(iii) *$S \cap r(S) = 0$ for every $S \subset A$.*

(iv) *A is right and left non-singular.*

Proof. (i) $ab = 0$ implies $(ba)^2 = baba = 0$, so $ba = 0$.

(ii): For every $S \subset A$ we have $r(S) = l(S)$ by (*i*), so $r(S)$ is also a left ideal.

(iii): $(S \cap r(S))^2 \subset S\,r(S) = 0$, so $S \cap r(S) = 0$.

(iv): $r(a)$ cannot be essential in A since $aA \cap r(a) = aA \cap r(aA) = 0$ by (iii). ▯

It was proved in Chap. I, §12, that a regular ring is reduced if and only if it is strongly regular. It follows that if A is an arbitrary ring such that $Q_{\max}$ is strongly regular, then A is reduced (because there are no nilpotent elements $\neq 0$ in $Q_{\max}$). When A is commutative, the converse also holds. In fact, every commutative regular ring is reduced, and therefore a commutative ring is reduced if and only if it is non-singular. The non-commutative case is somewhat more complicated, as is shown by the following result (Renault [1], Utumi [4]):

Proposition 5.2. *The following properties of a reduced ring A are equivalent:*
(a) *$Q_{\max}$ is a strongly regular ring.*
(b) *The lattice $\mathrm{Cl}(A.)$ is distributive.*
(c) *A is a right Utumi ring.*
(d) *Every essentially closed right ideal of A is a two-sided ideal.*
(e) *$aA \cap bA = 0$ implies $ab = 0$, for all a and b in A.*
(f) *$\mathfrak{a} \cap \mathfrak{b} = 0$ implies $\mathfrak{a}\mathfrak{b} = 0$, for all right ideals $\mathfrak{a}$ and $\mathfrak{b}$ in A.*

Proof. (a)$\Leftrightarrow$(b): The lattice $\mathrm{Cl}(A.)$ is isomorphic to the lattice of principal right ideals of $Q_{\max}$, and this lattice is distributive if and only if $Q_{\max}$ is strongly regular (Chap. III, Props. 4.7 and 4.8).

(a)$\Rightarrow$(d): Every essentially closed right ideal $\mathfrak{a}$ may be written as $\mathfrak{a} = q Q_{\max} \cap A$ for some $q \in Q_{\max}$. Since $q Q_{\max}$ is a two-sided ideal of $Q_{\max}$, $\mathfrak{a}$ is two-sided in A.

(d)$\Rightarrow$(a): Suppose there is a nilpotent element $q \neq 0$ in $Q_{\max}$, say $q^2 = 0$. There exists $a \in A$ such that $qa \neq 0$ belongs to A. Then $qa \in q Q_{\max} \cap A$, which is essentially closed in A and hence is two-sided. Therefore also $aqa \in q Q_{\max}$, say $aqa = qq'$. Then

$$(qa)^2 = qaqa = q \cdot qq' = 0,$$

which is impossible.

(d)$\Rightarrow$(f): If $\mathfrak{a} \cap \mathfrak{b} = 0$, let $\mathfrak{c} \supset \mathfrak{b}$ be a pseudo-complement of $\mathfrak{a}$ in A. Then $\mathfrak{c}$ is essentially closed and thus two-sided, so $\mathfrak{a}\mathfrak{b} \subset \mathfrak{a}\mathfrak{c} \subset \mathfrak{a} \cap \mathfrak{c} = 0$.

(f)$\Leftrightarrow$(e) is clear.

(f)$\Rightarrow$(c): If $\mathfrak{a}$ is a right ideal which is not essential in A, then there exists a non-zero right ideal $\mathfrak{b}$ such that $\mathfrak{a} \cap \mathfrak{b} = 0$. Then $\mathfrak{b}\mathfrak{a} = 0$, which shows that $l(\mathfrak{a}) \neq 0$, so condition 4.7(c) is satisfied.

(c)$\Rightarrow$(d): If A is a right Utumi ring, then every essentially closed right ideal is a right annihilator and therefore a two-sided ideal by Lemma 5.1. ▯

Corollary 5.3. *If A is strongly regular, then also $Q_{\max}$ is strongly regular.*

Proof. If A is strongly regular, then every right ideal is two-sided, so 5.2(d) is satisfied, and hence $Q_{\max}$ is strongly regular. ▯

Proposition 5.4. *If A is strongly regular and right self-injective, then A is also left self-injective.*

Proof. The maximal left ring of quotients $Q^l_{\max}$ of A is also strongly regular, by Cor. 5.3. For each $q \in Q^l_{\max}$ there exists an idempotent e in A such that $0 \neq eq \in A$. Since every idempotent in $Q^l_{\max}$ is central, we get $0 \neq eq = qe$ in A, so A is an essential right A-submodule of $Q^l_{\max}$. Since A is right self-injective, $Q^l_{\max}$ must be equal to A. ▯

This result is a special case of the following more general result, due to Utumi [4] (which actually is a particular case of Prop. 4.9):

Proposition 5.5. *If both the maximal right and the maximal left rings of quotients of A are strongly regular, then these two rings coincide.*

Proof. It suffices to show that $Q^l_{\max}$ is an essential extension of A as a right A-module, because then there is a right A-linear imbedding $\mu\colon Q^l_{\max} \to Q^r_{\max}$ (which in fact is a ring homomorphism, by Exercise IX.6), and since $Q^l_{\max}$ is right self-injective by Prop. 5.4, μ must be an isomorphism.

So let $0 \neq q \in Q^l_{\max}$. Then there exists $a \in A$ such that $0 \neq a\,q = b \in A$. Since $Q^r_{\max}$ is strongly regular, we can find $x \in Q^r_{\max}$ such that $a = a^2\,x$ and $x = x^2\,a$ (cf. Exercise I.47). Then $a\,x\,b = a\,x\,a\,q = a\,q = b \neq 0$, so $x\,b$ is a non-zero element of $Q^r_{\max}$. Therefore there exists $c \in A$ such that $0 \neq x\,b\,c = d \in A$. We have

$$a^2\,d = a^2\,x b c = a b c = a^2\,q\,c,$$

and it follows from Lemma 5.1(i) that $q\,c\,a^2 = d\,a^2$. To conclude the proof, it remains to show that $d\,a^2 \neq 0$. Since $a\,x$ is a central idempotent of $Q^r_{\max}$, we have

$$d\,a\,x = a\,x\,d = a\,x \cdot x\,b\,c = x\,b\,c = d,$$

and therefore $d\,a^2\,x^2 = d\,a\,x = d \neq 0$ implies $d\,a^2 \neq 0$. □

The following result shows that a reduced Utumi ring has a kind of strongly regular "envelope":

Proposition 5.6. *Let A be a reduced, right Utumi ring. Let B denote the intersection of all regular subrings of $Q_{\max}$ containing A. Then B is a strongly regular ring, and $A \to B$ is a ring epimorphism.*

Proof. Since $Q_{\max}$ is strongly regular, also B is strongly regular by Exercise I.47(iii). In order to show that $A \to B$ is an epimorphism, it suffices to show that its dominion D is a regular ring. Suppose $a \in D$. Then there exists $b \in B$ such that $a = a^2\,b = b\,a^2$ and $b = b^2\,a = a\,b^2$. If $\alpha, \beta\colon D \to C$ are two ring homomorphisms which coincide on A, then

$$\begin{aligned}\alpha(b) &= \alpha(b^2\,a) = \alpha(b^2)\,\beta(a) = \alpha(b^2)\,\beta(a^3\,b^2) = \alpha(b^2\,a^3)\,\beta(b^2)\\ &= \alpha(a)\,\beta(b^2) = \beta(a\,b^2) = \beta(b).\end{aligned}$$

Hence $b \in D$, and it follows that D is regular. □

It is an interesting fact that every regular, right self-injective ring can be written as the product of a strongly regular ring and a regular ring with numerous nilpotent elements. More precisely (Utumi [2]):

Proposition 5.7. *Suppose A is a Baer ring such that every non-zero right ideal contains a non-zero idempotent. Then $A = B \times C$, where B is a reduced ring and every non-zero two-sided ideal of the ring C contains non-zero nilpotent elements.*

Proof. Let $\mathfrak{b}$ be the sum of all reduced ideals of A, i.e. two-sided ideals of A without nilpotent elements $\neq 0$. We will show that $r(l(\mathfrak{b}))$ is reduced, from which it will follow that $\mathfrak{b} = r(l(\mathfrak{b}))$ is the unique maximal reduced ideal of A.

So suppose there is a nilpotent element $a \neq 0$ in $r(l(\mathfrak{b}))$. Clearly we may as well assume $a^2=0$. The right ideal aA contains an idempotent $e \neq 0$. Suppose that $eA \cap \mathfrak{a}=0$ holds for all reduced ideals $\mathfrak{a}$. Then $eA\mathfrak{a} \subset eA \cap \mathfrak{a}=0$ implies that $e \in l(\mathfrak{b})$, and hence $e=e \cdot e \in l(\mathfrak{b}) \cdot r(l(\mathfrak{b}))=0$, which is impossible. Therefore we must have $eA \cap \mathfrak{a} \neq 0$ for a certain reduced ideal $\mathfrak{a}$. Let $f \neq 0$ be an idempotent belonging to $eA \cap \mathfrak{a}$. Since $e=ab$ and $f=ec$ for some $b, c \in A$, we have that $0 \neq f=f^2=fec$ implies $fe \neq 0$ and hence $fa \neq 0$. But on the other hand,

$$(fa)^2=fafa=fa \cdot abc \cdot a=0$$

since $a^2=0$. Since fa belongs to the reduced ideal $\mathfrak{a}$, this is a contradiction, and we may conclude that $\mathfrak{b}=r(l(\mathfrak{b}))$ is reduced.

Since A is a Baer ring, the right annihilator ideal $\mathfrak{b}$ is generated by an idempotent, which must be central in A by Lemma I.12.2. Therefore $A=\mathfrak{b} \oplus \mathfrak{c}$ for some two-sided ideal $\mathfrak{c}$. Clearly $\mathfrak{c}$ contains no reduced ideals $\neq 0$. □

Corollary 5.8. *If A is a complete regular ring, then $A=B \times C$, where B is a strongly regular ring and C is a regular ring such that every non-zero two-sided ideal contains non-zero nilpotent elements.*

Proof. The regularity of B and C follows e.g. from Exercise I.47(i). □

§6. Flatness of the Maximal Ring of Quotients

When the maximal right ring of quotients $Q_{\max}$ of A is semi-simple, then we have seen that $Q_{\max}$ is flat as a left A-module. But semi-simplicity is of course not a necessary condition for $Q_{\max}$ to be flat over A (Example 2.4). The following result states conditions equivalent to flatness:

Proposition 6.1. *The following properties of a right non-singular ring A are equivalent:*

(a) *$Q_{\max}$ is flat as a left A-module.*

(b) *$M \otimes_A Q_{\max}$ is a projective $Q_{\max}$-module for every finitely generated submodule M of a free right A-module.*

(c) *$\mathfrak{a} \otimes_A Q_{\max}$ is a projective $Q_{\max}$-module for every finitely generated right ideal $\mathfrak{a}$ of A.*

Proof. (a) ⇒ (b): If M is a finitely generated submodule of the free module F, then flatness of Q implies that $M \otimes Q$ is a submodule of the free Q-module $F \otimes Q$. Since Q is a regular ring, and $M \otimes Q$ is a finitely generated Q-module, it follows from Prop. I.6.10 that $M \otimes Q$ is projective over Q.

(b) ⇒ (c) is trivial.

(c) ⇒ (a): Let $\mathfrak{a}$ be a finitely generated right ideal of A. Since $\mathfrak{a} \otimes Q$ is projective over Q, $\mathfrak{a} \otimes Q$ is a submodule of a free Q-module and must therefore be a non-singular A-module. The flatness of Q now follows from Prop. I.10.6 with help of the following observation:

Lemma 6.2. *Let A be a right non-singular ring. If M is a right A-submodule of $Q_{\max}$, then*

$$Z(M \otimes_A Q) = \operatorname{Ker}(M \otimes_A Q \to Q).$$

Proof. The homomorphism μ: $M \otimes_A Q \to Q$ is of course the map induced by multiplication, i.e. $\sum x_i \otimes q_i \mapsto \sum x_i q_i$. Since Q_A is non-singular, it is clear that $Z(M \otimes Q) \subset \operatorname{Ker} \mu$. On the other hand, if $\sum x_i \otimes q_i \in \operatorname{Ker} \mu$, then we choose an essential right ideal $\mathfrak{a}$ such that $q_i \mathfrak{a} \subset A$ for all i. Then $(\sum x_i \otimes q_i)\mathfrak{a} = \sum x_i \otimes q_i \mathfrak{a} = \sum x_i(q_i \mathfrak{a}) \otimes 1 = 0$, so $\sum x_i \otimes q_i \in Z(M \otimes Q)$. □

Corollary 6.3. *Suppose A is a right non-singular ring such that $Q_{\max}$ is flat as a left A-module. Then every finitely generated flat submodule of a free right A-module is projective.*

Proof. If M is a finitely generated submodule of a free module, then $M \otimes Q$ is projective over Q by Prop. 6.1. If M also is flat, then its projectivity over A follows from Prop. I.11.6. □

We distinguish one class of rings for which $Q_{\max}$ always is flat:

Proposition 6.4 (Sandomierski [2]). *The following properties of a ring A are equivalent:*

(a) *A is right semi-hereditary.*

(b) *A is right non-singular, every right ideal of A is a flat module and $Q_{\max}$ is flat as a left A-module.*

Proof. (a) ⇒ (b): If $a \neq 0$ in A, then $A/r(a) \cong aA$ is a projective module, so $r(a)$ is a direct summand of A and therefore cannot be essential in A. Hence A is non-singular. Every right ideal of A is a direct union of finitely generated, hence projective, right ideals and is therefore a flat module. If $\mathfrak{a}$ is a finitely generated right ideal of A, then since $\mathfrak{a}$ is A-projective, it is clear that $\mathfrak{a} \otimes Q$ is Q-projective. It then follows from Prop. 6.1 that $Q_{\max}$ is flat over A. (This also follows from Example 2.4.)

(b) ⇒ (a): Every finitely generated right ideal of A is projective by Cor. 6.3. □

§7. Imbedding Non-Singular Modules in Free Modules

If A is an order in a semi-simple ring, then we know that every finitely generated torsion-free module admits a monomorphism into a free module (Prop. II.4.6). The following result by Goodearl [1] is a generalization of this basic fact to non-singular modules.

Theorem 7.1. *Let A be a right non-singular ring. The following statements are equivalent:*

(a) *Every finitely generated non-singular right module can be imbedded in a free module.*

(b) *$Q_{\max}$ is a perfect left localization of A.*

Proof. (a) $\Rightarrow$ (b): We first show that $Q_{\max}$ is flat as a right A-module. Let $\mathfrak{a}$ be a left ideal of A and suppose $x \in Q \otimes \mathfrak{a}$ maps to zero in $Q \otimes A \cong Q$. There exists a finitely generated right A-submodule M of Q and an element $y \in M \otimes \mathfrak{a}$ which maps to x. By hypothesis, we may imbed M in a free A-module F. Since Q_A is injective, there exists $F \to Q$ such that the diagram

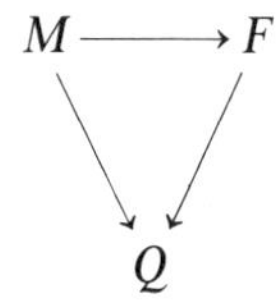

(1)

commutes. We get a big commutative diagram

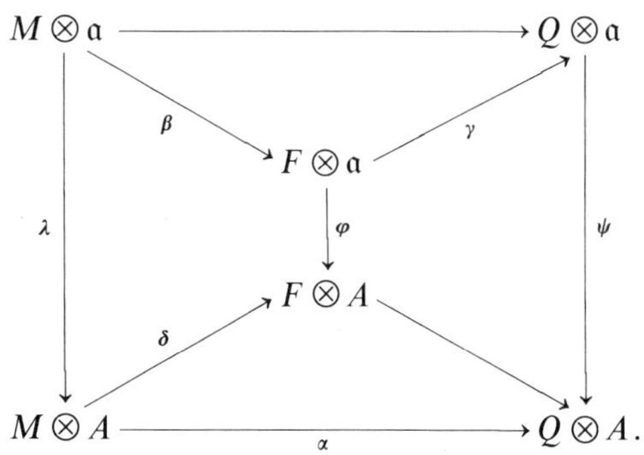

Now $\alpha\,\lambda(y) = \psi(x) = 0$, and since α is a monomorphism, this implies $\lambda(y) = 0$. Then it follows that $\varphi\,\beta(y) = \delta\,\lambda(y) = 0$, and since φ is a monomorphism, we get $\beta(y) = 0$. But then $x = \gamma\,\beta(y) = 0$. Hence $Q \otimes \mathfrak{a} \to Q$ is a monomorphism, so Q_A is flat.

Next we show that the multiplication map μ: $Q \otimes_A Q \to Q$ is an isomorphism, i.e. that $A \to Q$ is an epimorphism of rings. Suppose $x \in Q \otimes Q$ and $\mu(x) = 0$. As before we can find a finitely generated submodule M of Q such that x is the image of some $y \in M \otimes Q$, and we can imbed M in a free module F and construct a commutative diagram (1). There results a commutative diagram

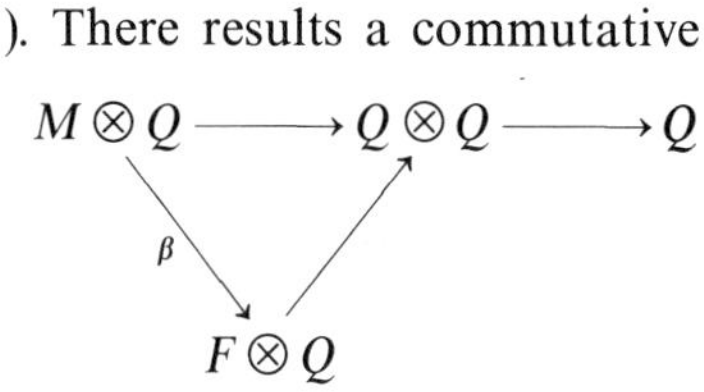

By Lemma 6.2 we have that $y \in Z(M \otimes Q)$, so $\beta(y) \in Z(F \otimes Q) = 0$, since $F \otimes Q$ is a free Q-module. Hence $x = 0$.

(b) $\Rightarrow$ (a): We first of all note:

Proposition 7.2. *If A is a right non-singular ring, then every finitely generated non-singular module can be imbedded in a finitely generated free Q-module.*

Proof. Let M be a finitely generated non-singular A-module. The canonical homomorphism $M \to M \otimes_A Q$ is a monomorphism (cf. the remark preceding Prop. 1.11 of Chap. IX). There is an exact sequence $F \to M \otimes Q \to 0$ where F is a finitely generated free Q-module. Since F is an injective module, it follows from Lemma VI.7.5 that $(M \otimes Q)/Z(M \otimes Q)$ is a direct summand of F. On the other hand, there is induced a monomorphism $M \to M \otimes Q/Z(M \otimes Q)$ since M is non-singular. □

It follows from this proposition that in order to prove the implication (b) ⇒ (a), it suffices to show that every finitely generated right A-submodule M of Q can be imbedded in a free A-module.

Since Q is a perfect left localization of A, the left ideals $\mathfrak{a}$ of A for which $Q\mathfrak{a} = Q$, form a perfect Gabriel topology $\mathfrak{F}$. Since M is finitely generated, there exists a finitely generated left ideal $\mathfrak{a} \in \mathfrak{F}$ such that $\mathfrak{a} M \subset A$. Let $a_1, \ldots, a_n$ generate $\mathfrak{a}$. Define $\varphi: M \to A^n$ as $\varphi(x) = (a_1 x, \ldots, a_n x)$. Then φ is a monomorphism, because ${}_A Q$ is $\mathfrak{F}$-torsion-free. □

Corollary 7.3. *Let A be a right non-singular ring of finite right rank. The following statements are equivalent:*

(a) *Every finitely generated non-singular right A-module can be imbedded in a free module.*

(b) *$Q_{\max}$ is flat as a right A-module.*

(c) *$Q_{\max}$ is also the left maximal ring of quotients of A.*

Proof. When A is non-singular and of finite rank, then $A \to Q_{\max}$ is an epimorphism by Cor. 2.6, so (a) ⇔ (b) is clear. The equivalence of (b) and (c) was proved in Prop. 2.9. □

Corollary 7.4 (Cateforis [2]). *The following properties of a right non-singular ring A are equivalent:*

(a) *Every finitely generated non-singular module is projective.*

(b) *A is right semi-hereditary and $Q_{\max}$ is a perfect left localization of A.*

When these conditions are satisfied, then A is also left semi-hereditary and $Q_{\max}$ is a perfect right localization of A.

Proof. The equivalence of the two conditions follows from Theorem 7.1 and Prop. I.6.9. It follows then from Prop. 6.4 that $Q_{\max}$ is a perfect right localization of A. It also follows that all non-singular modules are flat, and therefore A is left semi-hereditary (Exercise 18). □

Exercises

§1

1. Let C be an object of a spectral category **C**. Show that the endomorphism ring of C is strongly regular if and only if the lattice of subobjects of C is distributive. (Hint: cf. the proof of Prop. III.4.7.)

2. Let A be a right non-singular ring and $\mathfrak{E}$ the family of essential right ideals of A. Show that **Mod**-$(A, \mathfrak{E})$ is equivalent to a full subcategory of **Spec**(A).

§2

3. Suppose A is a right non-singular ring, and that $\varphi: A \to B$ is a ring monomorphism such that B is right self-injective and A is essential as a right A-submodule of B. Show that φ makes B into a maximal right ring of quotients of A. (Exercise 3 is useful when solving Exercise 4 7.)

4. Let K be a skew-field and put $T_2(K)=\begin{pmatrix} K & K \\ 0 & K\end{pmatrix}$. Show that:

(i) $T_2(K)$ is right and left artinian.

(ii) $\begin{pmatrix} 0 & K \\ 0 & K\end{pmatrix}$ is the minimal essential right ideal of $T_2(K)$.

(iii) The ring $M_2(K)$ of 2×2-matrices is the maximal right (and left) ring of quotients of $T_2(K)$.

(iv) $T_2(K)\to M_2(K)$ is a finite right (and left) localization.

5. Consider the ring $A=\begin{pmatrix} K & F \\ 0 & F\end{pmatrix}$, where K is a field and F is an extension field of K with $[F:K]>1$ (cf. Exercise VIII.7). Show that:

(i) The ring $M_2(F)$ is the maximal right ring of quotients of A, but it is not the maximal left ring of quotients of A.

(ii) $M_2(F)$ is a finite right localization of A.

(iii) $M_2(F)$ is not flat as a right A-module. (Hint: if it were flat, then $A\to M_2(F)$ would be a perfect left localization. Show that A is not essential as a left A-submodule of $M_2(F)$.)

6. Consider the ring $T_2(\mathbb{Z})=\begin{pmatrix} \mathbb{Z} & \mathbb{Z} \\ 0 & \mathbb{Z}\end{pmatrix}$. Show that:

(i) $M_2(\mathbb{Q})$ is the maximal right (and left) ring of quotients of $T_2(\mathbb{Z})$.

(ii) $T_2(\mathbb{Q})$ is a two-sided classical ring of quotients of A. (Note that A has nilpotent ideals $\neq 0$, so Q_{cl} is not semi-simple.)

7. Let $\mathfrak{F}$ be the set of right ideals of the form $\begin{pmatrix} \mathbb{Z} & \mathbb{Z} \\ 0 & n\mathbb{Z}\end{pmatrix}$ $(n\neq 0)$ in the ring $T_2(\mathbb{Z})$. Show that:

(i) $\mathfrak{F}$ is a Gabriel topology.

(ii) $A_{\mathfrak{F}}=\begin{pmatrix} \mathbb{Z} & \mathbb{Q} \\ 0 & \mathbb{Q}\end{pmatrix}$.

(iii) $A_{\mathfrak{F}}$ is a perfect right localization of A.

8. Let K be a field and $V=K\oplus K$ a two-dimensional vector space over K. Consider the ring $A=\begin{pmatrix} K & V \\ 0 & K\end{pmatrix}$. Show that:

(i) $\begin{pmatrix} 0 & V \\ 0 & K\end{pmatrix}$ is the minimal essential right ideal of A, and A is right (and left) non-singular.

(ii) $M_3(K)$ is the maximal right ring of quotients of A by means of the homomorphism

$$\varphi\colon \begin{pmatrix} K & K\oplus K \\ 0 & K\end{pmatrix}\to M_3(K)$$

defined as

$$\varphi\begin{pmatrix} a & (x,y) \\ 0 & b\end{pmatrix}=\begin{pmatrix} a & 0 & x \\ 0 & a & y \\ 0 & 0 & b\end{pmatrix}.$$

(iii) A is not essential as a left A-submodule of $M_3(K)$ by means of φ.

(iv) $M_3(K)$ is the maximal left ring of quotients of A by means of the homomorphism

$$\psi\colon \begin{pmatrix} K & K\oplus K \\ 0 & K\end{pmatrix}\to M_3(K)$$

defined as

$$\psi\begin{pmatrix} a & (x,y) \\ 0 & b\end{pmatrix}=\begin{pmatrix} a & x & y \\ 0 & b & 0 \\ 0 & 0 & b\end{pmatrix}.$$

(Hence $Q^l_{\max}(A)\cong Q^r_{\max}(A)$, but they do not coincide as overrings of A.)

(v) Generalize these results to the case $V=K^n$ $(n\geq 1)$.

9. Let A be a right non-singular ring. Show that:

(i) If P is a projective module containing a finitely generated essential submodule, then P is finitely generated. (Hint: show that P has finite projective coordinates.)

(ii) If A also has finite right rank, then every projective right ideal is finitely generated.

(iii) A right hereditary ring, which is of finite right rank, is right noetherian.

§3

10. Let A be a boolean ring such that $\mathsf{B}(A)$ is locally atomic. Show that:

(i) All finite sums of atoms form an ideal, which is the smallest essential ideal of A.

(ii) The completion of $\mathsf{B}(A)$ is isomorphic to the boolean algebra of all subsets of the set of atoms of $\mathsf{B}(A)$.

§4

11. Let A be a right non-singular ring with a non-singular module M. Show that if L is a submodule of M, then:

(i) $L^c = E(L) \cap M$.

(ii) If K is a pseudo-complement of L in M, then K is a complement of L^c in the lattice $\mathsf{Cl}(M)$.

12. Show that a right Utumi ring is a Baer ring if and only if every cyclic non-singular module is projective.

13. Let A be a right continuous regular ring and $\mathfrak{a}$ a two-sided ideal of A. Show that:

(i) Also $\mathfrak{a}^c$ is a two-sided ideal.

(ii) If $\mathfrak{a}$ is essentially closed as a right ideal, then $A = \mathfrak{a} \oplus r(\mathfrak{a})$.

(iii) If $\mathfrak{a}$ is a prime ideal, then $\mathfrak{a}$ is either essentially closed or essential as a right ideal. (Hint: note that $\mathfrak{a}^c \cdot r(\mathfrak{a}^c) = 0$ is contained in $\mathfrak{a}$.)

§5

14. Let K be a skew-field and I an arbitrary set. Let A be the subring of K^I consisting of (x_i) such that almost all (i.e. all except a finite number) x_i are equal. Show that:

(i) A is a strongly regular and continuous ring.

(ii) $K^{(I)}$ is a minimal essential ideal of A.

(iii) K^I is the maximal ring of quotients of A.

15. Show that a reduced prime ring has no zero-divisors $\neq 0$. (Hint: use the fact that annihilators are two-sided ideals.)

16. Show that if A is a right non-singular ring, then the center of A is a reduced ring. (Hint: show that if $c \in \operatorname{cen} A$ and $r(c) \cap \mathfrak{a} = 0$ for some non-zero right ideal $\mathfrak{a}$, then $c^n \mathfrak{a} \neq 0$ for all n.)

§6

17. Let A be right non-singular and of finite right rank. Show that every finitely generated flat right A-module is projective.

18. Show that the following properties of a right non-singular ring A are equivalent:

(a) All non-singular modules are flat.

(b) A is left semi-hereditary and $Q_{\max}$ is flat as a right A-module.

(c) Every right ideal is flat and $Q_{\max}$ is flat as a right A-module. (Hint: (a) $\Rightarrow$ (b): Use Exercise I.54; (c) $\Rightarrow$ (a): If M is non-singular, show that its module of quotients $M_{\mathfrak{E}}$ is flat, and that it follows that M is flat.)

§7

19. Show that the following properties of a regular ring A are equivalent:

(a) A is right self-injective.

(b) $Q_{\max}$ is a projective right A-module.

(c) Every finitely generated non-singular right A-module is projective.

Chapter XIII. Finiteness Conditions on **Mod**-$(A, \mathfrak{F})$

Let $\mathfrak{F}$ be a Gabriel topology on the ring A. In this chapter we will investigate how various finiteness properties of the quotient category **Mod**-$(A, \mathfrak{F})$ are reflected by properties of the ring A. The lattice of subobjects of $M_{\mathfrak{F}}$ in the category **Mod**-$(A, \mathfrak{F})$ is isomorphic to the lattice $\mathsf{Sat}_{\mathfrak{F}}(M)$ of $\mathfrak{F}$-saturated submodules of M, and the finiteness properties may therefore be formulated for the lattices of $\mathfrak{F}$-saturated submodules.

§ 1. Compact Lattices of Saturated Submodules

It is a natural question to ask for the precise conditions on the ring A which make the category **Mod**-$(A, \mathfrak{F})$ into e.g. a locally noetherian or a locally finitely generated category. It is clear that **Mod**-$(A, \mathfrak{F})$ is locally finitely generated if and only if the lattices $\mathsf{Sat}_{\mathfrak{F}}(M)$ are compactly generated for every A-module M. However, the concrete significance of this condition is not known, and so we specialize immediately to more particular cases.

Proposition 1.1. *The following properties of a module M are equivalent:*
(a) *$M_{\mathfrak{F}}$ is a finitely generated object in* **Mod**-$(A, \mathfrak{F})$.
(b) *The lattice $\mathsf{Sat}_{\mathfrak{F}}(M)$ is compact.*
(c) *The filter $\mathfrak{F}(M)$ has a basis consisting of finitely generated modules.*

Proof. (a) $\Leftrightarrow$ (b) is clear from Cor IX.4.4.

(b) $\Rightarrow$ (c): Each $L \in \mathfrak{F}(M)$ can be written as a direct union of finitely generated submodules L_i. In the lattice $\mathsf{Sat}_{\mathfrak{F}}(M)$ we then have

$$M = L^c = \Big(\sum_i L_i\Big)^c = \bigvee_i L_i^c,$$

and the compactness implies that $M = L_i^c$ for some i, and thus $L_i \in \mathfrak{F}(M)$.

(c) $\Rightarrow$ (b): Suppose (M_i) is a directed family of $\mathfrak{F}$-saturated submodules such that $M = \bigvee M_i$, i.e. $M = (\sum M_i)^c$. Then $\sum M_i \in \mathfrak{F}(M)$, so there exists a finitely generated module $L \in \mathfrak{F}(M)$ with $L \subset \sum M_i$. But then $L \subset M_i$ for some i, and so $M = L^c \subset M_i^c = M_i$, which gives $M_i = M$. □

Proposition 1.2. *The following properties of the Gabriel topology $\mathfrak{F}$ are equivalent:*
(a) *$A_{\mathfrak{F}}$ is a finitely generated object in* **Mod**-$(A, \mathfrak{F})$.
(b) *$\mathfrak{F}$ has a basis of finitely generated right ideals.*

(c) *Every direct union of $\mathfrak{F}$-closed modules is $\mathfrak{F}$-closed.*
(d) *The functor i: $\mathbf{Mod}$-$(A, \mathfrak{F}) \to \mathbf{Mod}$-$A$ preserves direct unions.*
(e) *The $\mathfrak{F}$-torsion radical preserves direct limits.*

Proof. (a) $\Leftrightarrow$ (b) has just been proved.

(b) $\Rightarrow$ (c): Let (M_α) be a directed family of $\mathfrak{F}$-closed submodules of some module. Then $\sum M_\alpha$ is $\mathfrak{F}$-torsion-free, so it remains to verify that it also is $\mathfrak{F}$-injective. Consider a homomorphism φ: $\mathfrak{a} \to \sum M_\alpha$ for some $\mathfrak{a} \in \mathfrak{F}$. By hypothesis, $\mathfrak{a}$ contains a finitely generated right ideal $\mathfrak{b} \in \mathfrak{F}$, and clearly $\varphi(\mathfrak{b}) \subset M_\alpha$ for some α. Since M_α is $\mathfrak{F}$-injective, there exists $x \in M_\alpha$ such that $\alpha(b) = x\,b$ for all $b \in \mathfrak{b}$. By Lemma IX.1.9 one has $\alpha(a) = x\,a$ also for $a \in \mathfrak{a}$.

(c) $\Rightarrow$ (d): Let (M_α) be a directed family of $\mathfrak{F}$-closed submodules of some module. The sum of the objects M_α in the category $\mathbf{Mod}$-$(A, \mathfrak{F})$ is $\sum M_\alpha = a(\sum i(M_\alpha))$, so if $\sum i(M_\alpha)$ is $\mathfrak{F}$-closed, then $i(\sum M_\alpha) = \sum i(M_\alpha)$.

(d) $\Rightarrow$ (a): If (M_α) is a directed family of subobjects of some object in $\mathbf{Mod}$-$(A, \mathfrak{F})$, then

$$\begin{aligned}\operatorname{Hom}_{\mathbf{C}}(A_{\mathfrak{F}}, \textstyle\sum M_\alpha) &= \operatorname{Hom}_A(A, i(\textstyle\sum M_\alpha)) = \operatorname{Hom}_A(A, \textstyle\sum i(M_\alpha)) \\ &= \textstyle\sum \operatorname{Hom}_A(A, i(M_\alpha)) = \textstyle\sum \operatorname{Hom}_{\mathbf{C}}(A_{\mathfrak{F}}, M_\alpha),\end{aligned}$$

where $\mathbf{C} = \mathbf{Mod}$-$(A, \mathfrak{F})$, and $A_{\mathfrak{F}}$ is finitely generated by Prop. V.3.2.

(b) $\Rightarrow$ (e): Let t denote the $\mathfrak{F}$-torsion radical. Suppose (M_α) is a direct system of modules. The inclusions $t(M_\alpha) \to M_\alpha$ induce in the limit an inclusion $\varinjlim t(M_\alpha) \to \varinjlim M_\alpha$. The class of $\mathfrak{F}$-torsion modules is closed under direct limits, since it is closed under direct sums and quotients, so $\varinjlim t(M_\alpha)$ is a submodule of $t(\varinjlim M_\alpha)$. To show that we actually have equality, suppose $x \in t(\varinjlim M_\alpha)$. Then $x\,\mathfrak{a} = 0$ for some finitely generated $\mathfrak{a} \in \mathfrak{F}$. Since $\mathfrak{a}$ is finitely generated, it is clear that we may represent x by some $x_\alpha \in M_\alpha$ such that still $x_\alpha\,\mathfrak{a} = 0$. Then $x_\alpha \in t(M_\alpha)$, and $x \in \varinjlim t(M_\alpha)$.

(e) $\Rightarrow$ (b): Write $\mathfrak{a} \in \mathfrak{F}$ as a direct union of finitely generated right ideals $\mathfrak{a}_\alpha$. Since $A/\mathfrak{a} = \varinjlim A/\mathfrak{a}_\alpha$ is a torsion module, we have $A/\mathfrak{a} = t(A/\mathfrak{a}) = \varinjlim t(A/\mathfrak{a}_\alpha)$. In particular, the generating element $\bar{1} \in A/\mathfrak{a}$ comes from some $t(A/\mathfrak{a}_\alpha)$, which means that there exist $a \in A$ and $\mathfrak{b} \in \mathfrak{F}$ such that $1 - a \in \mathfrak{a}$ and $a\,\mathfrak{b} \subset \mathfrak{a}_\alpha$. One may choose α so that $1 - a \in \mathfrak{a}_\alpha$, and then $\mathfrak{b} \subset \mathfrak{a}_\alpha$. Hence $\mathfrak{a}_\alpha \in \mathfrak{F}$. □

It should be remarked that if $\mathfrak{F}$ satisfies the equivalent conditions of Prop. 1.2, then in particular every direct sum of $\mathfrak{F}$-closed modules is $\mathfrak{F}$-closed.

§ 2. $\mathfrak{F}$-Noetherian Rings

A module M is *$\mathfrak{F}$-noetherian* if $M_{\mathfrak{F}}$ is a noetherian object in the category $\mathbf{Mod}$-$(A, \mathfrak{F})$.

Proposition 2.1. *The following properties of a module M are equivalent:*
(a) *M is $\mathfrak{F}$-noetherian.*
(b) *The lattice $\mathsf{Sat}_{\mathfrak{F}}(M)$ is noetherian.*
(c) *For every submodule L of M there exists a finitely generated module L' such that $L' \in \mathfrak{F}(L)$.*

Proof. (a) $\Leftrightarrow$ (b) is clear.

(a) $\Rightarrow$ (c): If $L \subset M$ and M is $\mathfrak{F}$-noetherian, then $L_{\mathfrak{F}}$ is finitely generated in **Mod**-$(A, \mathfrak{F})$, so the assertion follows from Prop. 1.1.

(c) $\Rightarrow$ (b): This is proved similarly to (c) $\Rightarrow$ (b) of Prop. 1.1. □

Corollary 2.2. *Every noetherian module is $\mathfrak{F}$-noetherian.*

Corollary 2.3. *If $\mathfrak{F}$ and $\mathfrak{F}'$ are Gabriel topologies with $\mathfrak{F} \subset \mathfrak{F}'$, then every $\mathfrak{F}$-noetherian module is $\mathfrak{F}'$-noetherian.*

The condition that A should be $\mathfrak{F}$-noetherian can be formulated quite explicitly in terms of a cogenerating injective module for the $\mathfrak{F}$-torsion theory. An injective module E is called *Σ-injective* if every direct sum $E^{(I)}$ of copies of E is an injective module.

Proposition 2.4. *Assume that the $\mathfrak{F}$-torsion theory is cogenerated by the injective module E. The following assertions are equivalent:*

(a) *A is $\mathfrak{F}$-noetherian.*

(b) *$\mathsf{Sat}_{\mathfrak{F}}(A)$ is a noetherian lattice.*

(c) *For every right ideal $\mathfrak{a}$ there exists a finitely generated right ideal $\mathfrak{b} \subset \mathfrak{a}$ such that $\mathfrak{a}/\mathfrak{b}$ is an $\mathfrak{F}$-torsion module.*

(d) *A satisfies* ACC *on annihilators of subsets of E.*

(e) *E is a Σ-injective module.*

(f) *Every direct sum of $\mathfrak{F}$-torsion-free injective modules is injective.*

Proof. (a) $\Leftrightarrow$ (b) $\Leftrightarrow$ (c) is clear from Prop. 2.1, and (b) $\Leftrightarrow$ (d) is an immediate consequence of Prop. IX.4.6.

(a) $\Rightarrow$ (f): If $A_{\mathfrak{F}}$ is noetherian, then **Mod**-$(A, \mathfrak{F})$ is a locally noetherian category. The inclusion functor i: **Mod**-$(A, \mathfrak{F}) \to$ **Mod**-A preserves direct unions by Prop. 1.1, and therefore in particular it preserves direct sums. The result therefore follows from Prop. V.4.3 and Prop. X.1.7.

(f) $\Rightarrow$ (e) is trivial.

(e) $\Rightarrow$ (d): Let $\mathfrak{a}_1 \subset \mathfrak{a}_2 \subset \cdots$ be a strictly ascending chain of annihilators of subsets of E. If this chain were not finite, we could choose for each n an element $x_n \in E$ such that $x_n \mathfrak{a}_n = 0$ but $x_n \mathfrak{a}_{n+1} \neq 0$. Put $\mathfrak{a} = \bigcup_1^\infty \mathfrak{a}_n$ and define α: $\mathfrak{a} \to \bigoplus_1^\infty E$ as $\alpha(a) = (x_1 a, x_2 a, \ldots)$. Note that α is well-defined! Since E is Σ-injective, α has the form $\alpha(a) = ya$ for some $y = (y_1, y_2, \ldots, y_m, 0, \ldots) \in \bigoplus_1^\infty E$, which contradicts the choice of the elements x_n. □

Corollary 2.5. *If A is $\mathfrak{F}$-noetherian, then $\mathfrak{F}$ has a basis of finitely generated right ideals.*

It should be noticed here that **Mod**-$(A, \mathfrak{F})$ may well be a locally noetherian category without having condition (f) satisfied, although the $\mathfrak{F}$-torsion-free injective modules are precisely the injective objects of **Mod**-$(A, \mathfrak{F})$; the reason for this is of course that the coproduct in **Mod**-$(A, \mathfrak{F})$ does not necessarily coincide with the direct sum in **Mod**-A. However, when A is $\mathfrak{F}$-noetherian, then coproducts in **Mod**-$(A, \mathfrak{F})$ are the same as in **Mod**-A, and since injective objects in **Mod**-$(A, \mathfrak{F})$ decompose into indecomposable objects (by Prop. V.4.5), we obtain:

Proposition 2.6. *If A is $\mathfrak{F}$-noetherian, then every $\mathfrak{F}$-torsion-free injective module is a direct sum of indecomposable submodules.*

It follows that if A is $\mathfrak{F}$-noetherian, then the $\mathfrak{F}$-torsion theory is cogenerated by the injective module $\bigoplus_I E_i$, where $\{E_i\}_I$ is the family of indecomposable injective $\mathfrak{F}$-torsion-free modules.

Example. Noetherian rings. From Prop. 2.4 follows that A is right noetherian if and only if there is a Σ-injective cogenerator for **Mod**-A.

§ 3. The Goldie Topology

We shall consider in more detail the particular case of the Goldie topology, i.e. the Gabriel topology $\mathfrak{G}$ generated by the family of essential right ideals of A.

Proposition 3.1. *The following properties of a module M are equivalent:*
(a) *M is $\mathfrak{G}$-noetherian.*
(b) *The lattice $\mathsf{Sat}_{\mathfrak{G}}(M)$ is noetherian.*
(c) *$\mathfrak{G}(M)$ has a basis of finitely generated modules.*
(d) *There is no infinite independent family of non-singular submodules of M.*

Proof. (a) $\Leftrightarrow$ (b) is clear as usual. Notice that $\mathsf{Sat}_{\mathfrak{G}}(M)$ consists of the submodules L of M such that M/L is non-singular.

(a) $\Rightarrow$ (c) is also clear from Prop. 2.1.

(c) $\Rightarrow$ (d): If $(M_i)_I$ is an independent family of non-singular submodules of M, then we get a monomorphism

$$\bigoplus_I a(M_i) \cong a(\bigoplus_I M_i) \to a(M)$$

in the category **Mod**-$(A, \mathfrak{G})$. Since each $a(M_i) \neq 0$ and $a(M)$ is noetherian, the family $(M_i)_I$ must be finite.

(d) $\Rightarrow$ (b): Suppose $M_1 \subset M_2 \subset \cdots$ is an ascending chain of $\mathfrak{G}$-saturated submodules of M. To each M_i we choose a pseudo-complement L_i in M_{i+1}. Then the family (L_i) is independent, and the modules L_i are non-singular since there are monomorphisms $L_i \to M_{i+1}/M_i$. So the family (L_i) is finite, and it follows that there exists an integer n such that M_i is essential in M_{i+1} for all $i \geq n$. But since M_{i+1}/M_i are non-singular, one must then have $M_i = M_n$ for $i \geq n$. $\square$

When $M = A$ we obtain (Teply [2]):

Proposition 3.2. *The following properties of the ring A are equivalent:*
(a) *A is $\mathfrak{G}$-noetherian.*
(b) *$\mathfrak{G}$ has a basis of finitely generated right ideals.*
(c) *There is no infinite independent family of non-singular right ideals.*
(d) *Every non-singular module has a unique maximal injective submodule.*
(e) *Every non-singular $E(A)$-divisible module is injective.*
(f) *Every direct sum of non-singular injective modules is injective.*

Proof. The equivalence of the conditions (a), (b) and (c) follows from Prop. 3.1, and they are equivalent to (f) by Prop. 2.4.

(f) ⇒ (d): Let M be a non-singular module, and let $(E_i)_I$ be the family of all injective submodules of M. Then $\bigoplus_I E_i$ is an injective module. It follows from Lemma VI.7.5 that also $\sum_I E_i$ is an injective module, and it is of course the maximal injective submodule of M.

(d) ⇒ (e): Recall that the $E(A)$-divisible modules are precisely the modules belonging to the torsion class generated by the injective modules (Example VI.2.3). If M is a non-singular $E(A)$-divisible module, let E be the maximal injective submodule of M. Then $M = E \oplus L$ for some module L. If $L \neq 0$, then $\mathrm{Hom}(E(A), L) \neq 0$ by the $E(A)$-divisibility. But the image of every homomorphism $E(A) \to L$ is an injective module (Lemma VI.7.5), so this contradicts the maximality of E.

(e) ⇒ (f) is clear, since direct sums of injective modules are $E(A)$-divisible. □

Examples

1. Modules of finite rank. Every module of finite rank is $\mathfrak{G}$-noetherian by Prop. 3.1.

2. Non-singular rings. When the ring A is right non-singular, the Prop. 3.2 reproduces some of the results proved in §2 of the preceding chapter. To summarize:

Proposition 3.3. *The following properties of a right non-singular ring A are equivalent:*

(a) *Every essential right ideal contains a finitely generated essential right ideal.*
(b) *A has finite right rank.*
(c) *Every non-singular module has a unique maximal injective submodule.*
(d) *Every non-singular $E(A)$-divisible module is injective.*
(e) *$E(A)$ is a Σ-injective module.*
(f) *Every direct sum of non-singular injective modules is injective.*

§4. $\mathscr{P}$-Noetherian Rings

Let A be a commutative ring, and let $\mathscr{P}$ be a set of prime ideals of A. In Example IX.6.6 we defined a Gabriel topology $\mathfrak{F}_{\mathscr{P}}$ as

$$\mathfrak{F}_{\mathscr{P}} = \{\mathfrak{a} \mid V(\mathfrak{a}) \cap \mathscr{P} = \emptyset\}.$$

It may as well be assumed that $\mathscr{P}$ is generically stable in the sense that if $\mathfrak{q} \subset \mathfrak{p}$ are prime ideals with $\mathfrak{p} \in \mathscr{P}$, then also $\mathfrak{q} \in \mathscr{P}$. We call a module *$\mathscr{P}$-noetherian* if it is $\mathfrak{F}_{\mathscr{P}}$-noetherian. One purpose of this section is to extend some of the results of Chap. VII to the case when A is $\mathscr{P}$-noetherian.

But first we make some general remarks. If M is a module, we know that the filter $\mathfrak{F}_{\mathscr{P}}(M)$ consists of all submodules L of M such that $\mathrm{Supp}(M/L) \cap \mathscr{P} = \emptyset$. For a submodule L of M we put

$$L^c = \{x \in M \mid (L : x) \in \mathfrak{F}_{\mathscr{P}}\}.$$

We say that L is a *$\mathscr{P}$-divisorial* submodule of M when $L^c = L$, i.e. when L is $\mathfrak{F}_{\mathscr{P}}$-saturated. A module M is *$\mathscr{P}$-codivisorial* if M is $\mathfrak{F}_{\mathscr{P}}$-torsion-free, i.e. if $\operatorname{Ann}(x)$ is a $\mathscr{P}$-divisorial ideal in A for each $x \in M$. We also put

$$L^{(\mathfrak{p})} = \{x \in M \mid xs \in L \text{ for some } s \notin \mathfrak{p}\}$$

for prime ideals $\mathfrak{p}$. Note that $L^{(\mathfrak{p})}$ is the inverse image of $L_{\mathfrak{p}}$ under the canonical homomorphism $M \to M_{\mathfrak{p}}$.

Lemma 4.1. *If L is a submodule of M, then*

$$L^c = \bigcap L^{(\mathfrak{p})} \quad \textit{with} \quad \mathfrak{p} \in \mathscr{P} \cap \operatorname{Supp}(M/L).$$

Proof. $x \in L^c$ if and only if $(L : x) \not\subset \mathfrak{p}$ holds for all $\mathfrak{p} \in \mathscr{P}$, i.e. if and only if for each $\mathfrak{p} \in \mathscr{P}$ there exists $s \notin \mathfrak{p}$ such that $xs \in L$. Since $L^{(\mathfrak{p})} = M$ for $\mathfrak{p} \notin \operatorname{Supp}(M/L)$, it suffices to consider prime ideals in $\operatorname{Supp}(M/L)$. □

Lemma 4.2. *A prime ideal $\mathfrak{p}$ is $\mathscr{P}$-divisorial in A if and only if $\mathfrak{p} \in \mathscr{P}$.*

Proof. Since $\operatorname{Ann}(x) = \mathfrak{p}$ for every non-zero $x \in A/\mathfrak{p}$, $\mathfrak{p}$ is $\mathscr{P}$-divisorial if and only if $\mathfrak{p} \notin \mathfrak{F}_{\mathscr{P}}$, i.e. $\mathfrak{p} \in \mathscr{P}$. □

We now consider $\mathscr{P}$-noetherian modules. The Prop. 2.1 translates into the following characterization of $\mathscr{P}$-noetherian modules:

Proposition 4.3. *The following properties of a module M are equivalent:*

(a) *M is $\mathscr{P}$-noetherian.*

(b) *The lattice of $\mathscr{P}$-divisorial submodules of M is noetherian.*

(c) *For every submodule L of M there exists a finitely generated module $L' \subset L$ such that* $\operatorname{Supp}(L/L') \cap \mathscr{P} = \emptyset$.

Proposition 4.4. *If M is a $\mathscr{P}$-noetherian module, then $M_{\mathfrak{p}}$ is a noetherian $A_{\mathfrak{p}}$-module for every $\mathfrak{p} \in \mathscr{P}$.*

Proof. Apply Cor. 2.3 to $\mathfrak{F}_{\mathscr{P}} \subset \mathfrak{F}_{\{\mathfrak{p}\}}$. □

In order to obtain a converse to this last result, we have to impose a further finiteness condition on M:

Proposition 4.5. *If M is an A-module such that $M_{\mathfrak{p}}$ is a noetherian $A_{\mathfrak{p}}$-module for each $\mathfrak{p} \in \mathscr{P}$, and such that* $\operatorname{Supp}(M/L) \cap \mathscr{P}$ *is finite for every submodule L of M, then M is $\mathscr{P}$-noetherian.*

Proof. Let $L_1 \subset L_2 \subset \cdots$ be an ascending chain of $\mathscr{P}$-divisorial submodules of M. Then we get a descending chain

$$\operatorname{Supp}(M/L_1) \supset \operatorname{Supp}(M/L_2) \supset \cdots,$$

and by the finiteness hypothesis there exists an integer k such that all $\mathscr{P} \cap \operatorname{Supp}(M/L_i)$ are equal for $i \geq k$. Therefore it follows from Lemma 4.1 that it suffices to show that the chain $L_1^{(\mathfrak{p})} \subset L_2^{(\mathfrak{p})} \subset \cdots$ is stationary for each $\mathfrak{p} \in \mathscr{P}$. But so is clearly the case when $M_{\mathfrak{p}}$ is a noetherian $A_{\mathfrak{p}}$-module. □

Corollary 4.6. *If $A_{\mathfrak{p}}$ is a noetherian ring for each $\mathfrak{p} \in \mathscr{P}$, and each $a \in A$ is contained in only finitely many prime ideals of $\mathscr{P}$, then A is $\mathscr{P}$-noetherian.*

Proof. Note that $\operatorname{Supp}(A/\mathfrak{a}) = V(\mathfrak{a})$. □

From Prop. 2.4 we obtain the following characterization of $\mathscr{P}$-noetherian rings:

Proposition 4.7. *The following properties of A are equivalent:*
(a) *A is $\mathscr{P}$-noetherian.*
(b) *A satisfies* ACC *on $\mathscr{P}$-divisorial ideals.*
(c) *Every direct sum of $\mathscr{P}$-codivisorial injective modules is injective.*

From Prop. 2.6 follows:

Proposition 4.8. *If A is $\mathscr{P}$-noetherian, then every $\mathscr{P}$-codivisorial injective module is a direct sum of indecomposable submodules.*

For the rest of this section we assume that A is $\mathscr{P}$-noetherian. If M is a module, then $\operatorname{Ass}(M)$ denotes the set of prime ideals $\mathfrak{p}$ for which there exist $x \in M$ such that $\operatorname{Ann}(x) = \mathfrak{p}$. As in the noetherian case (Chap. VII, § 1) one shows that every maximal member of the family of ideals $\operatorname{Ann}(x)$, for $0 \neq x \in M$, belongs to $\operatorname{Ass}(M)$. But it may well happen that $\operatorname{Ass}(M) = \emptyset$ for non-zero M. However:

Proposition 4.9. *If M is a non-zero $\mathscr{P}$-codivisorial module, then $\operatorname{Ass}(M) \neq \emptyset$ and $\operatorname{Ass}(M) \subset \mathscr{P}$.*

Proof. The family of ideals $\operatorname{Ann}(x)$, $0 \neq x \in M$, has a maximal member since the lattice of $\mathscr{P}$-divisorial ideals is noetherian. Hence $\operatorname{Ass}(M) \neq \emptyset$. It is clear from Lemma 4.2 that $\operatorname{Ass}(M) \subset \mathscr{P}$. □

In order to determine an injective module cogenerating the $\mathfrak{F}_{\mathscr{P}}$-torsion theory, it suffices (by the remark following Prop. 2.6) to describe the indecomposable $\mathscr{P}$-codivisorial injective modules.

From Lemma 4.2 and Prop. 4.9 follows:

Proposition 4.10. *An indecomposable injective module E is $\mathscr{P}$-codivisorial if and only if $E \cong E(A/\mathfrak{p})$ for some $\mathfrak{p} \in \mathscr{P}$.*

Corollary 4.11. *The $\mathfrak{F}_{\mathscr{P}}$-torsion theory is cogenerated by the injective module* $\bigoplus_{\mathfrak{p} \in \mathscr{P}} E(A/\mathfrak{p})$.

We would like to be able to characterize the $\mathfrak{F}_{\mathscr{P}}$-torsion modules in terms of their associated prime ideals. However, this can be done only in a rather weak sense, using an extension of the notion of associated prime ideal. If M is a module, then a prime ideal $\mathfrak{p}$ is *weakly associated* to M if there exists a non-zero $x \in M$ such that $\mathfrak{p}$ is a minimal element of $V(\operatorname{Ann}(x))$. Write $\operatorname{Ass}_f(M)$ for the set of weakly associated prime ideals of M.

Lemma 4.12. *If $\mathfrak{a} \neq A$ is an ideal, then there exists a minimal element of $V(\mathfrak{a})$.*

Proof. If $(\mathfrak{p}_\alpha)$ is a decreasing chain of prime ideals containing $\mathfrak{a}$, then $\mathfrak{p} = \bigcap_\alpha \mathfrak{p}_\alpha$ is a prime ideal containing $\mathfrak{a}$. In fact, if $a \notin \mathfrak{p}$, $b \notin \mathfrak{p}$, then there exists $\mathfrak{p}_\alpha$ with $a \notin \mathfrak{p}_\alpha$ and $b \notin \mathfrak{p}_\alpha$, and then $ab \notin \mathfrak{p}_\alpha$ implies $ab \notin \mathfrak{p}$. Zorn's lemma can therefore be used to establish the existence of a minimal element of $V(\mathfrak{a})$. □

Applying this result to $\mathfrak{a} = \operatorname{Ann}(x)$ for any non-zero $x \in M$, we get:

Lemma 4.13. *If M is a module, then $M = 0$ if and only if $\operatorname{Ass}_f(M) = \emptyset$.*

Lemma 4.14. *A prime ideal $\mathfrak{p}$ belongs to $\operatorname{Supp}(M)$ if and only if $\mathfrak{p}$ contains a prime ideal in $\operatorname{Ass}_f(M)$.*

Proof. This is obtained by a slight modification, using Lemma 4.12, of the proof of Prop. VII.5.3. □

Proposition 4.15. *A module M is an $\mathfrak{F}_{\mathscr{P}}$-torsion module if and only if*

$$\operatorname{Ass}_f(M) \cap \mathscr{P} = \emptyset.$$

Proof. Since M is an $\mathfrak{F}_{\mathscr{P}}$-torsion module if and only if $\operatorname{Supp}(M) \cap \mathscr{P} = \emptyset$, the assertion is a consequence of Lemma 4.14. □

§ 5. The Canonical Topology of a Krull Domain

As an illustration of the theory of $\mathscr{P}$-noetherian rings, we will make a study of the canonical topology of Krull domains, and thereby extend Prop. VII.6.10. Consider a commutative integral domain A with field of fractions Q. Let $\mathscr{P}$ denote the family of prime ideals of height ≤ 1 in A. The ring A is assumed to be a *Krull domain*, i.e. it satisfies:

$K1$. $A_{\mathfrak{p}}$ is a principal ideal domain for every non-zero $\mathfrak{p} \in \mathscr{P}$.

$K2$. $A = \bigcap_{\mathfrak{p} \in \mathscr{P}} A_{\mathfrak{p}}$.

$K3$. If $a \in A$ and $a \neq 0$, then there are only finitely many $\mathfrak{p} \in \mathscr{P}$ such that $a \in \mathfrak{p}$.

The rings $A_{\mathfrak{p}}$, for non-zero $\mathfrak{p} \in \mathscr{P}$, are then "discrete valuation rings", i.e. the non-zero ideals of $A_{\mathfrak{p}}$ are of the form $\pi^n A_{\mathfrak{p}}$ for $n \geq 0$. If $\mathfrak{a}$ is an ideal of A, then $\mathfrak{a}$ is $\mathscr{P}$-divisorial if and only if

$$\mathfrak{a} = \bigcap_{i=1}^{r} \mathfrak{p}_i^{(n_i)},$$

where $\mathfrak{a} A_{\mathfrak{p}_i} = \mathfrak{p}_i^{n_i} A_{\mathfrak{p}_i}$, $\mathfrak{p}_i^{(n_i)}$ denotes the inverse image of $\mathfrak{p}_i^{n_i} A_{\mathfrak{p}_i}$ under the map $A \to A_{\mathfrak{p}_i}$, and $\mathscr{P} \cap V(\mathfrak{a}) = \{\mathfrak{p}_1, \ldots, \mathfrak{p}_r\}$.

Such ideals have been called *divisorial* (Bourbaki [4]). From Cor. 4.6 we know that A is a $\mathscr{P}$-noetherian ring.

Proposition 5.1. *If M is a module with $\operatorname{Ann}(x) \neq 0$ for every $x \in M$, then $M_{\mathfrak{F}_{\mathscr{P}}} \cong \bigoplus_{\mathfrak{p} \in \mathscr{P}} M_{\mathfrak{p}}$.*

Proof. If $x \in M$, then by $K3$ there are only finitely many $\mathfrak{p} \in \mathscr{P}$ such that $\operatorname{Ann}(x) \subset \mathfrak{p}$. Hence the image of x in $M_{\mathfrak{p}}$ is zero for almost all $\mathfrak{p} \in \mathscr{P}$, and so there is a canonical homomorphism $\varphi: M \to \bigoplus_{\mathfrak{p} \in \mathscr{P}} M_{\mathfrak{p}}$. Since each $M_{\mathfrak{p}}$ is $\mathfrak{F}_{\mathscr{P}}$-closed, it follows from §1 that $\bigoplus M_{\mathfrak{p}}$ is $\mathfrak{F}_{\mathscr{P}}$-closed. It therefore only remains to show that $\operatorname{Ker} \varphi$ and $\operatorname{Coker} \varphi$ are $\mathfrak{F}_{\mathscr{P}}$-torsion modules. For this we need:

Lemma 5.2. *If $\mathfrak{p} \neq \mathfrak{q}$ are prime ideals of height 1, then $A_{\mathfrak{p}} \otimes_A A_{\mathfrak{q}} \cong Q$.*

Proof. Clearly $A_{\mathfrak{p}} \otimes_A A_{\mathfrak{q}} = A[S^{-1}]$, where S is the multiplicatively closed set $S = \{a\, b \mid a \notin \mathfrak{p},\ b \notin \mathfrak{q}\}$. The prime ideals in $A[S^{-1}]$ correspond to the prime ideals of A not intersecting S (Prop. VII.5.1), i.e. the prime ideals contained in $\mathfrak{p} \cap \mathfrak{q}$. But now the only prime ideal contained in $\mathfrak{p} \cap \mathfrak{q}$ is the zero ideal, so $A[S^{-1}] \cong Q$. $\square$

We return to the proof of Prop. 5.1. Localizing at the prime ideal $\mathfrak{q} \in \mathscr{P}$, we obtain an exact sequence

$$0 \to (\operatorname{Ker} \varphi)_{\mathfrak{q}} \to M_{\mathfrak{q}} \to \Big(\bigoplus_{\mathscr{P}} M_{\mathfrak{p}}\Big) \otimes A_{\mathfrak{q}} \to (\operatorname{Coker} \varphi)_{\mathfrak{q}} \to 0.$$

But for $\mathfrak{p} \neq \mathfrak{q}$ we have $M_{\mathfrak{p}} \otimes_A A_{\mathfrak{q}} \cong M \otimes_A A_{\mathfrak{p}} \otimes_A A_{\mathfrak{q}} \cong M \otimes_A Q = 0$, since M is a torsion module, and for $\mathfrak{p} = \mathfrak{q}$ we obtain $M_{\mathfrak{q}} \otimes_A A_{\mathfrak{q}} \cong M_{\mathfrak{q}}$. Hence the middle arrow of the sequence is an isomorphism, and $(\operatorname{Ker} \varphi)_{\mathfrak{q}} = (\operatorname{Coker} \varphi)_{\mathfrak{q}} = 0$ for all $\mathfrak{q} \in \mathscr{P}$. So $\operatorname{Ker} \varphi$ and $\operatorname{Coker} \varphi$ are $\mathfrak{F}_{\mathscr{P}}$-torsion modules. $\square$

Proposition 5.3 (Beck [1]). *The topology $\mathfrak{F}_{\mathscr{P}}$ coincides with the canonical topology of A.*

Proof. Since the canonical topology is cogenerated by $Q \oplus E(Q/A)$, it suffices by Cor. 4.11 to show that $E(Q/A) \cong \bigoplus E(A/\mathfrak{p})$ with the sum taken over non-zero $\mathfrak{p} \in \mathscr{P}$. We do this in a number of lemmas.

Lemma 5.4. *Q/A is a $\mathscr{P}$-codivisorial module.*

Proof. Suppose M is a submodule of Q such that $A \subset M$ and $(M/A)_{\mathfrak{p}} = 0$ for all $\mathfrak{p} \in \mathscr{P}$. Then $M \subset \bigcap_{\mathscr{P}} M_{\mathfrak{p}} = \bigcap_{\mathscr{P}} A_{\mathfrak{p}} = A$. So $M = A$, and the $\mathfrak{F}_{\mathscr{P}}$-torsion submodule of Q/A is zero. $\square$

Lemma 5.5. *$Q/A_{\mathfrak{p}} \cong E(A/\mathfrak{p})$ for every non-zero $\mathfrak{p} \in \mathscr{P}$.*

Proof. $Q/A_{\mathfrak{p}}$ is a divisible $A_{\mathfrak{p}}$-module, and is therefore an injective $A_{\mathfrak{p}}$-module (Prop. I.6.11). It is indecomposable, because all finitely generated $A_{\mathfrak{p}}$-submodules of Q are of the form $\pi^n A_{\mathfrak{p}}$ for a certain $\pi \in \mathfrak{p}$ and for integers n. On the other hand, also $E(A/\mathfrak{p})$ is an injective $A_{\mathfrak{p}}$-module (e.g. by Prop. II.3.7). Since $A/\mathfrak{p}$ may be imbedded in $Q/A_{\mathfrak{p}}$ by mapping $\bar{1}$ to $\overline{\pi^{-1}}$, we have $E(A/\mathfrak{p}) \cong Q/A_{\mathfrak{p}}$. $\square$

Proof of Prop. 5.3: Since Q/A is $\mathscr{P}$-codivisorial, it is an essential submodule of $(Q/A)_{\mathfrak{F}_{\mathscr{P}}} \cong \bigoplus_{\mathscr{P}} Q/A_{\mathfrak{p}}$ (Prop. 5.1). Hence $E(Q/A) \cong \bigoplus E(A/\mathfrak{p})$ since the second module is injective by Prop. 4.7. $\square$

As a particular consequence we have:

Corollary 5.6. $\operatorname{Ass}(Q/A)$ *consists of the prime ideals of height* 1.

Proposition 5.7. *A module M is an $\mathfrak{F}_{\mathscr{P}}$-torsion module if and only if* $\operatorname{Ass}(M) \cap \mathscr{P} = \emptyset$.

Proof. It suffices to show that if $M_{\mathfrak{p}} \neq 0$ for some $\mathfrak{p} \in \mathscr{P}$, then $\mathfrak{p} \in \operatorname{Ass}(M)$. When $\mathfrak{p} = (0)$, this is clear. Assume $\mathfrak{p}$ is of height 1. If $M_{\mathfrak{p}} \neq 0$, then there exists a non-zero $x \in M$ such that $\mathfrak{a} = \operatorname{Ann}(x) \subset \mathfrak{p}$, and we will show that $\mathfrak{p} \in \operatorname{Ass}(A/\mathfrak{a})$. By Cor. 5.6 there exists $q \in Q$ such that $\mathfrak{p} = \{a \in A \mid a\, q \in A\}$. Then we have $q\, \mathfrak{a} \not\subset \mathfrak{a}^{(\mathfrak{p})}$. For otherwise $q\, \mathfrak{a} A_{\mathfrak{p}} \subset \mathfrak{a} A_{\mathfrak{p}}$, which would imply $q \in A_{\mathfrak{p}}$ since $A_{\mathfrak{p}}$ is a discrete valuation ring, and thus $s\, q \in A$ for some $s \notin \mathfrak{p}$; a contradiction. Hence we can choose an element

$b \in q\,\mathfrak{a}$ such that $b \notin \mathfrak{a}^{(\mathfrak{p})}$. Then $b \in A$, since $q\,\mathfrak{a} \subset q\,\mathfrak{p} \subset A$, and $\mathfrak{p}\,b \subset \mathfrak{p}\,q\,\mathfrak{a} \subset \mathfrak{a}$. On the other hand, if $c\,b \in \mathfrak{a}$ for some $c \in A$, then we must have $c \in \mathfrak{p}$, for otherwise we would have $\frac{b}{1} \in \mathfrak{a}\,A_{\mathfrak{p}}$ and $b \in \mathfrak{a}^{(\mathfrak{p})}$. Thus $\mathfrak{p} \in \mathrm{Ass}(A/\mathfrak{a})$. □

Since $E(M)$ has the same associated prime ideals as M (Lemma VII.1.4), it follows:

Corollary 5.8. *The $\mathfrak{F}_{\mathscr{P}}$-torsion theory is stable.*

Exercises

Let $\mathfrak{F}$ denote a Gabriel topology on the ring A.

§1

1. Let E be an injective module cogenerating the $\mathfrak{F}$-torsion theory. Show that the following properties of $\mathfrak{F}$ are equivalent:
(a) $A_{\mathfrak{F}}$ is a Σ-generated object in **Mod**-$(A, \mathfrak{F})$ (cf. Exercise V.13).
(b) If $\mathfrak{a}_1 \subset \mathfrak{a}_2 \subset \cdots$ is a denumerable chain of right ideals such that $\bigcup_i \mathfrak{a}_i \in \mathfrak{F}$, then $\mathfrak{a}_n \in \mathfrak{F}$ for some n.
(c) Every direct sum of $\mathfrak{F}$-closed modules is $\mathfrak{F}$-closed.
(c′) Every denumerable direct sum of $\mathfrak{F}$-closed modules is $\mathfrak{F}$-closed.
(d) Every denumerable direct union of $\mathfrak{F}$-closed modules is $\mathfrak{F}$-closed.
(e) Every direct sum of copies of E is $\mathfrak{F}$-injective.
(f) The functor i: **Mod**-$(A, \mathfrak{F}) \to$ **Mod**-A preserves direct sums.

2. Show that $A_{\mathfrak{F}}$ is a finitely presented object in **Mod**-$(A, \mathfrak{F})$ if and only if every direct limit of $\mathfrak{F}$-closed modules is $\mathfrak{F}$-closed.

3. Show that the following properties of $\mathfrak{F}$ are equivalent:
(a) $A_{\mathfrak{F}}$ is a finitely generated object in **Mod**-$(A, \mathfrak{F})$.
(b) Every direct union of $\mathfrak{F}$-saturated submodules of an $\mathfrak{F}$-torsion-free module is $\mathfrak{F}$-saturated.
(c) Every direct limit of $\mathfrak{F}$-torsion-free modules is $\mathfrak{F}$-torsion-free.

§2

4. Show that the class of $\mathfrak{F}$-noetherian modules is closed under submodules, quotient modules and extensions.

5. Show that if M is a module such that $M_{\mathfrak{F}}$ is a finitely generated object in **Mod**-$(A, \mathfrak{F})$, then $M_{\mathfrak{F}} = L_{\mathfrak{F}}$ for some finitely generated submodule L of M. Show that the converse holds when $A_{\mathfrak{F}}$ is finitely generated in **Mod**-$(A, \mathfrak{F})$.

6. Show that a module M is $\mathfrak{F}$-noetherian if and only if for every ascending chain of submodules $M_1 \subset M_2 \subset \cdots$ of M, there exists n such that M_n/M_{n+1} is an $\mathfrak{F}$-torsion module.

7. Suppose A is $\mathfrak{F}$-noetherian. Show that if M is an $\mathfrak{F}$-torsion-free module, then $\mathrm{Ass}(M) \neq \emptyset$, (Hint: note that $\mathrm{Ass}(M) \subset \mathsf{Sat}_{\mathfrak{F}}(A)$.)

8. Show that A is $\mathfrak{F}$-noetherian if and only if every FP-injective (Exercises I.43 and V.18) and $\mathfrak{F}$-torsion-free module is injective.

§3

9. Let $\mathfrak{G}$ denote the Goldie topology on A. Show that the following assertions are equivalent:
(a) A is $\mathfrak{G}$-noetherian.
(b) Every direct limit of non-singular injective modules is non-singular injective (cf. Exercise 2).
(c) Every non-singular FP-injective module is injective (cf. Exercises I.43 and V.18).

10. Show that if A is $\mathfrak{D}$-noetherian, where $\mathfrak{D}$ consists of the dense right ideals of A, then A has finite right rank and satisfies ACC on right annihilators.

§4

Let A be a commutative ring with a saturated family $\mathscr{P}$ of prime ideals.

11. Show that if $\mathfrak{F}$ is a Gabriel topology on A, and M is a module, then:

(i) If M is $\mathfrak{F}$-torsion, then $\mathrm{Supp}(M) \subset \mathfrak{F}$.

(ii) If $\mathrm{Ass}_f(M) \cap \mathfrak{F} = \emptyset$, then M is $\mathfrak{F}$-torsion-free.

12. Show that if $\mathfrak{F}$ is a Gabriel topology on A such that A is $\mathfrak{F}$-noetherian, then $\mathfrak{F} = \mathfrak{F}_{\mathscr{P}}$ with $\mathscr{P} = \mathsf{Sat}_{\mathfrak{F}}(A) \cap \mathrm{Spec}(A)$.

13. Suppose $K \subset L$ are $\mathscr{P}$-divisorial submodules of a module M. Show that if $K_{\mathfrak{p}} = L_{\mathfrak{p}}$ for all $\mathfrak{p} \in \mathscr{P}$, then $K = L$.

14. Show that the lattice of $\mathscr{P}$-divisorial ideals is distributive if and only if $A_{\mathfrak{p}}$ is an arithmetical ring for each $\mathfrak{p} \in \mathscr{P}$.

15. Suppose A is an integral domain and let M be a torsion-free module (i.e. $\mathrm{Ann}(x) = 0$ for all $0 \neq x \in M$). Show that $M_{\mathfrak{F}_{\mathscr{P}}} = \bigcap_{\mathfrak{p} \in \mathscr{P}} M_{\mathfrak{p}}$.

§5

16. Let A be a Krull domain and $\mathscr{P} = \{$prime ideals of height $\leq 1\}$. Show that:

(i) For every ideal $\mathfrak{a}$ there exists an ideal $\mathfrak{b} \in \mathfrak{F}_{\mathscr{P}}$ such that $\mathfrak{a} = \mathfrak{a}^c \cap \mathfrak{b}$. (Hint: use the stability of $\mathfrak{F}_{\mathscr{P}}$ to decompose $E(A/\mathfrak{a})$.)

(ii) If $\mathfrak{a}$ is an irreducible ideal and $V(\mathfrak{a}) \cap \mathscr{P} \neq \emptyset$, then $\mathfrak{a}$ is divisorial.

Chapter XIV. Self-Injective Rings

There are very few general results known for the maximal ring of quotients for rings with non-zero singular ideal. Most of the results deal with the case when the maximal ring of quotients is a self-injective ring, a case which will be studied in this Chapter. For this we need to examine the properties of self-injective rings in some detail, and we devote the first three sections to that purpose.

§ 1. The Endomorphism Ring of an Injective Module

Let A be a ring, and consider the canonical functor P: **Mod**-$A \to$ **Spec**-A into the spectral category of **Mod**-A. For every module M there is induced a map

$$\mu_M\colon \operatorname{Hom}_A(M, M) \to \operatorname{Hom}_{\mathbf{Spec}\text{-}A}(P(M), P(M))$$

which is a ring homomorphism. When M is an injective module, this map is surjective, because any homomorphism $M' \to M$, where M' is an essential submodule of M, may be extended to $M \to M$. The kernel of μ_M is an ideal, consisting of those endomorphisms which vanish on an essential submodule of M. When M is injective, this ideal coincides with the Jacobson radical:

Proposition 1.1. *Let E be an injective module, and let H denote the endomorphism ring of E. The Jacobson radical of H consists of those endomorphisms whose kernels are essential submodules of E.*

Proof. Suppose $\alpha \in H$ and $\operatorname{Ker}\alpha$ is an essential submodule of E. Since $\operatorname{Ker}(1-\alpha) \cap \operatorname{Ker}\alpha = 0$, it follows that $\operatorname{Ker}(1-\alpha)=0$, and $1-\alpha$ is thus a monomorphism. Then $\operatorname{Im}(1-\alpha)$ is an injective module and hence a direct summand of E. But $\operatorname{Im}(1-\alpha)$ is also an essential submodule of E, because $\operatorname{Im}(1-\alpha) \supset \operatorname{Ker}\alpha$ for the simple reason that $\alpha(x)=0$ implies $x=(1-\alpha)(x)$. We conclude that $\operatorname{Im}(1-\alpha)$ coincides with E. Thus $1-\alpha$ is invertible for every α belonging to the two-sided ideal $N = \{\alpha \in H \mid \operatorname{Ker}\alpha \text{ is essential in } E\} = \operatorname{Ker}\mu_E$, and so $N \subset J(H)$ by Prop. VIII.1.5.

On the other hand, it is clear that $J(H)/N$ is the Jacobson radical of H/N, and H/N is isomorphic to the endomorphism ring of $P(E)$ in **Spec**-A, which is a regular ring by Prop. XII.1.2. It follows that $J(H)=N$. ☐

Combining this result with Prop. XII.1.2, we obtain (Osofsky [4], Renault [2], Roos [3], Utumi [7]):

Theorem 1.2. *Let E be an injective A-module and let H be its endomorphism ring. The ring $H/J(H)$ is regular and right self-injective.*

Corollary 1.3. *Let A be a right self-injective ring. Then:*
(i) *$J(A)$ is the right singular ideal of A.*
(ii) *$A/J(A)$ is a regular and right self-injective ring.*

The endomorphism ring of an injective module has one further important property:

Proposition 1.4. *Let H be the endomorphism ring of an injective A-module. Then idempotents may be lifted modulo $J(H)$.*

Proof. Assume $\alpha \in H$ and $\alpha - \alpha^2 \in J(H)$, which means that $L = \mathrm{Ker}(\alpha - \alpha^2)$ is an essential submodule of E. The injective envelope of $\alpha(L)$ is a direct summand of E and thus has the form $\mathrm{Im}\,\varepsilon$ for some idempotent ε of H. Then $\varepsilon\alpha = \alpha$ on the essential submodule L of E, so $\varepsilon\alpha - \alpha \in J(H)$. Put $\beta = \varepsilon + \varepsilon\alpha(1-\varepsilon)$, and note that $\beta = \beta^2$. Now $\beta - \varepsilon\alpha = \varepsilon - \varepsilon\alpha\varepsilon$ vanishes on $L' = \mathrm{Im}(1-\varepsilon) + \alpha(L)$, because we have $\varepsilon\alpha\varepsilon\alpha | L = \varepsilon\alpha^2 | L = \varepsilon\alpha | L$. But L' is an essential submodule of $\mathrm{Im}(1-\varepsilon) + \mathrm{Im}\,\varepsilon = E$, so it follows that $\beta - \varepsilon\alpha \in J(H)$. Since we already have shown $\varepsilon\alpha - \alpha \in J(H)$, we get $\beta - \alpha \in J(H)$, so β lifts the idempotent $\bar{\alpha}$. □

Corollary 1.5. *If A is a right self-injective ring, then idempotents can be lifted modulo $J(A)$.*

Since a regular ring is semi-simple if and only if it has no infinite family of orthogonal idempotents, we obtain from Cors. 1.3 and 1.5:

Proposition 1.6. *A right self-injective ring is semi-perfect if and only if it has no infinite family of orthogonal idempotents.*

More generally, since an injective module is of finite rank if and only if its endomorphism ring has no infinite family of orthogonal idempotents, we have:

Proposition 1.7. *The endomorphism ring of an injective module E is semi-perfect if and only if E has finite rank.*

Example. If H is the endomorphism ring of an indecomposable injective module, then $H/J(H)$ is a skew-field by Prop. V.4.1.

§ 2. Self-Injective Rings and Annihilator Conditions

For a right noetherian ring, the property of being right self-injective may be formulated as a condition on left annihilator ideals:

Proposition 2.1 (Ikeda and Nakayama [1]). *The following properties of A are equivalent:*

(a) *Every homomorphism $\alpha\colon \mathfrak{a} \to A$, where $\mathfrak{a}$ is a finitely generated right ideal, has the form $\alpha(a) = c\,a$ for some $c \in A$.*

(b) *A satisfies* (i) *$l(\mathfrak{a}_1 \cap \mathfrak{a}_2) = l(\mathfrak{a}_1) + l(\mathfrak{a}_2)$ for all finitely generated right ideals $\mathfrak{a}_1, \mathfrak{a}_2$.*

(ii) *$l(r(a)) = A\,a$ for every $a \in A$.*

Proof. (a) ⇒ (b): If $\mathfrak{a}_1$ and $\mathfrak{a}_2$ are finitely generated right ideals, then obviously $l(\mathfrak{a}_1)+l(\mathfrak{a}_2)\subset l(\mathfrak{a}_1\cap\mathfrak{a}_2)$. Suppose on the other hand that $a\in l(\mathfrak{a}_1\cap\mathfrak{a}_2)$. We can define a homomorphism $\alpha\colon \mathfrak{a}_1+\mathfrak{a}_2\to A$ as

$$\alpha(b)=\begin{cases} b & b\in\mathfrak{a}_1\\ (1+a)\,b & b\in\mathfrak{a}_2,\end{cases}$$

since these two expressions coincide on $\mathfrak{a}_1\cap\mathfrak{a}_2$. By (a) there exists $c\in A$ such that $\alpha(b)=c\,b$. For $b\in\mathfrak{a}_1$ we thus have $c\,b=b$, i.e. $(c-1)\,b=0$. As a consequence we can write $a=(c-1)+(1+a-c)\in l(\mathfrak{a}_1)+l(\mathfrak{a}_2)$, which proves (i).

For every $a\in A$ we have $A\,a\subset l(r(a))$. If $b\in l(r(a))$, we can define a homomorphism $a\,A\to b\,A$ as $a\,x\mapsto b\,x$. It must be given by left multiplication with some $c\in A$, so in particular $b=c\,a$ and $b\in A\,a$.

(b) ⇒ (a): Consider $\alpha\colon\mathfrak{a}\to A$ where $\mathfrak{a}$ is finitely generated, say $\mathfrak{a}=a_1\,A+\cdots+a_n\,A$. We use induction on n, the case $n=0$ being obvious. So we may assume there exist c and c' in A such that

$$\alpha(a)=\begin{cases} c\,a & \text{for}\quad a\in\mathfrak{a}'=a_1\,A+\cdots+a_{n-1}\,A\\ c'\,a & \text{for}\quad a\in a_n\,A.\end{cases}$$

For $a\in\mathfrak{a}'\cap a_n\,A$ we must have $(c-c')\,a=0$, so $c-c'\in l(\mathfrak{a}'\cap a_n\,A)=l(\mathfrak{a}')+l(a_n\,A)$ by (i). We accordingly write $c-c'=b-b'$ with $b\,\mathfrak{a}'=0$ and $b'\,a_n=0$. Then left multiplication by $c-b=c'-b'$ coincides with α on $\mathfrak{a}=\mathfrak{a}'+a_n\,A$. □

Proposition 2.2. *If A is right self-injective, then:*
(i) $l(\mathfrak{a}_1\cap\mathfrak{a}_2)=l(\mathfrak{a}_1)+l(\mathfrak{a}_2)$ *for all right ideals* $\mathfrak{a}_1$ *and* $\mathfrak{a}_2$.
(ii) $l(r(\mathfrak{b}))=\mathfrak{b}$ *for every finitely generated left ideal* $\mathfrak{b}$.

Proof. (i): The argument used in the preceding proof works equally well in the present situation.

(ii): Write $\mathfrak{b}=A\,a_1+\cdots+A\,a_n$. Then

$$\begin{aligned} l(r(A\,a_1+\cdots+A\,a_n))&=l(r(A\,a_1)\cap\cdots\cap r(A\,a_n))=l(r(A\,a_1))+\cdots+l(r(A\,a_n))\\ &=A\,a_1+\cdots+A\,a_n\end{aligned}$$

by (i) and Prop. 2.1(ii). □

We also briefly consider right self-injective rings satisfying the annihilator condition $r(l(\mathfrak{a}))=\mathfrak{a}$ for all right ideals $\mathfrak{a}$. Recall that a ring is a right Kasch ring if and only if it contains an isomorphic copy of each simple right module, or equivalently, $l(\mathfrak{a})\neq 0$ holds for every proper right ideal $\mathfrak{a}$ (Lemma XI.5.1).

Proposition 2.3. *When A is a right self-injective ring, the following conditions are equivalent:*
(a) *A is a right Kasch ring.*
(b) *A is an injective cogenerator for* **Mod**-*A*.
(c) $r(l(\mathfrak{a}))=\mathfrak{a}$ *for every right ideal* $\mathfrak{a}$.

Proof. (a) ⇒ (b) is clear from Lemma XI.5.1, and (c) ⇒ (a) is evident.
(b) ⇒ (c) follows from the more general observation:

Lemma 2.4. *If the ring A is a cogenerator for* **Mod**-*A, then* $r(l(\mathfrak{a}))=\mathfrak{a}$ *for every right ideal* $\mathfrak{a}$.

Proof. For every right ideal $\mathfrak{a}$ there is a monomorphism $\alpha: A/\mathfrak{a} \to A^I$ for some I. If $\alpha(\bar{1}) = (a_i)_I$, then clearly $\mathfrak{a} = \bigcap_I r(a_i) = r(\{a_i\})$, so $\mathfrak{a} = r(l(\mathfrak{a}))$. □

Proposition 2.5. *Let A be a semi-perfect and right self-injective ring, and assume that the right socle of A is an essential right ideal in A. Then A is a right Kasch ring.*

Proof. Write 1 as a sum of orthogonal primitive idempotents, $1 = e_1 + \cdots + e_n$, according to Prop. VIII.3.2. Since each $e_i A$ is an indecomposable injective module, it contains a unique minimal right ideal $\mathfrak{b}_i$. If now S is a simple A-module, then S is of course also a simple $A/J(A)$-module, and hence $S \cong \bar{e}_i A/J$ for some i. In particular, each $\mathfrak{b}_i$ is isomorphic to a certain $\bar{e}_{\pi(i)} A/J$. Since $e_i A$ is an injective envelope of $\mathfrak{b}_i$, one has $\mathfrak{b}_i \cong \mathfrak{b}_j$ if and only if $e_i A \cong e_j A$. It follows that the number of non-isomorphic $\bar{e}_i A/J$ is equal to the number of non-isomorphic $\mathfrak{b}_i$, and therefore each simple module is isomorphic to some minimal right ideal $\mathfrak{b}_i$. □

Examples

1. Noetherian self-injective rings. If A is a right noetherian ring, then A is right self-injective if and only if

(i) $l(\mathfrak{a}_1 \cap \mathfrak{a}_2) = l(\mathfrak{a}_1) + l(\mathfrak{a}_2)$ for all right ideals $\mathfrak{a}_1$ and $\mathfrak{a}_2$.

(ii) $l(r(\mathfrak{b})) = \mathfrak{b}$ for every finitely generated left ideal $\mathfrak{b}$.

2. Perfect self-injective rings. Every left perfect and right self-injective ring is a right Kasch ring, by Prop. 2.4.

§ 3. Quasi-Frobenius Rings

Definition. A right and left artinian ring A is a *quasi-Frobenius* ring (or shorter, a *QF-ring*) if it satisfies

$$r(l(\mathfrak{a})) = \mathfrak{a}, \quad l(r(\mathfrak{b})) = \mathfrak{b}$$

for all right ideals $\mathfrak{a}$ and left ideals $\mathfrak{b}$.

Proposition 3.1. *A right and left artinian ring is a QF-ring if and only if it is both right and left self-injective.*

Proof. If A is a QF-ring, then the Galois connection (r, l) defines an anti-isomorphism between the lattices of left, resp. right ideals (cf. Chap. III, §8). From this we may conclude that the annihilator conditions of Prop. 2.1 are satisfied, so A is right and left self-injective. The converse implication is an immediate consequence of Prop. 2.2. □

Corollary 3.2. *Every QF-ring is a right and left Kasch ring.*

Proof. Follows from Prop. 2.3. □

The sufficiency part of Prop. 3.1 may be improved by making the conditions one-sided:

Proposition 3.3. *If A is right or left artinian and is right or left self-injective, then A is a QF-ring.*

Proof. There are two cases to be considered:

(1) A is right artinian and right self-injective.

(2) A is left artinian and right self-injective.

The case (1) may be reduced to (2): If $\mathfrak{b}_1 \subset \mathfrak{b}_2 \subset \cdots$ is an ascending chain of finitely generated left ideals, then the descending chain $r(\mathfrak{b}_1) \supset r(\mathfrak{b}_2) \supset \cdots$ is stationary since A is right artinian. But since $\mathfrak{b}_i = l(r(\mathfrak{b}_i))$ by Prop. 2.2, also the ascending chain is then stationary. So A is left noetherian, and hence left artinian (Cor. VIII.2.4).

We next consider case (2). Since A is left artinian, it is right semi-artinian (Prop. VIII.5.4). It then follows from Prop. 2.5 that A is a right Kasch ring, which means that $r(l(\mathfrak{a})) = \mathfrak{a}$ holds for every right ideal $\mathfrak{a}$. By using this condition, and the fact that A is left artinian, we obtain that A is right noetherian. Then it follows as before that A actually is right artinian. Finally we have $l(r(\mathfrak{b})) = \mathfrak{b}$ for left ideals $\mathfrak{b}$ by Prop. 2.2. Thus we have established all the properties of a QF-ring. □

One may weaken the conditions for A to be QF quite a bit more:

Proposition 3.4. *If A is right or left noetherian and is right or left self-injective, then A is a QF-ring.*

Proof. Again there are two cases to be considered:

(1) A is left noetherian and right self-injective;

(2) A is right noetherian and right self-injective.

Case (1): To show that A is left artinian, it suffices by Cor. VIII.2.4 to show that A is semi-primary. $A/J(A)$ is left noetherian and regular, hence a semi-simple ring. It remains to show that $J(A)$ is nilpotent. The ascending chain of two-sided ideals $r(J) \subset r(J^2) \subset \cdots$ is stationary since A is left noetherian. Let $r(J^n) = r(J^{n+1})$. From Prop. 2.2 follows that $J^n = J^{n+1}$, which implies $J^n = 0$ by the Nakayama lemma (Prop. VIII.1.3).

Case (2): We prove a slightly stronger statement (and this is actually the result which will be needed for the applications to rings of quotients):

Theorem 3.5. *If A satisfies* ACC *on right or on left annihilators and is right or left self-injective, then A is a QF-ring.*

Proof. Case (1): Suppose A satisfies ACC on left annihilators and is right self-injective. Every finitely generated left ideal is a left annihilator by Prop. 2.2, so A is left noetherian. A is then QF by case (1) of the preceding proposition.

Case (2): Suppose A satisfies ACC on right annihilators and is right self-injective. If $\mathfrak{b}_1 \supset \mathfrak{b}_2 \supset \cdots$ is a descending chain of finitely generated left ideals, then the ascending chain of right annihilators $r(\mathfrak{b}_1) \subset r(\mathfrak{b}_2) \subset \cdots$ is stationary. From Prop. 2.2 follows that also the first chain is stationary, so A is a left perfect ring and is even semi-primary by Cor. VIII.6.4. We can apply Prop. 2.5 to conclude that A is a right Kasch ring, and therefore every right ideal is a right annihilator (Prop. 2.3). But then A is right noetherian, and since we have just seen that A is semi-primary, it follows that A is right artinian. Prop. 3.3 then gives that A is a QF-ring. □

It should be noticed that the annihilator conditions in the definition of QF-rings cannot be made one-sided (see Exercise 8). For a QF-ring one can describe the injective modules in a simple way:

Proposition 3.6. *The following properties of a module M over a QF-ring A are equivalent:*

(a) *M is an injective module.*

(b) *M is a projective module.*

(c) *$M \cong \bigoplus_I e_i A$ for a family $(e_i)_I$ of primitive idempotents.*

Proof. (a) $\Rightarrow$ (c): Since A is noetherian, every injective module can be decomposed as a direct sum of indecomposable injective modules. It thus suffices to consider an indecomposable injective module M. Since A is an artinian ring, M contains a simple submodule S, and clearly $M = E(S)$. But S is isomorphic to a minimal right ideal of A (Cor. 3.2), and since A is right self-injective, it follows that $E(S) \cong e_i A$ for some idempotent e_i.

(c) $\Rightarrow$ (b) is clear, since each $e_i A$ is a projective module.

(b) $\Rightarrow$ (a): Since A is self-injective and noetherian, every free module is injective (Prop. V.4.3), which implies that every projective module is injective. $\square$

We may now describe a class of right self-injective rings which are not left self-injective (Sandomierski [3]).

Proposition 3.7. *Let A be a QF-ring and F a free right A-module with endomorphism ring H. Then:*

(i) *H is a right self-injective ring.*

(ii) *If F is not finitely generated over A, then H is not left self-injective.*

Proof. (i) We have a natural equivalence of functors

$$\mathrm{Hom}_H(\cdot, H) = \mathrm{Hom}_H(\cdot, \mathrm{Hom}_A(F, F)) \cong \mathrm{Hom}_A(\cdot \otimes_H F, F).$$

Now F is an injective A-module by Prop. 3.6, and F is a finitely generated projective H-module by Prop. IV.6.7, so the last functor above is exact. Hence $\mathrm{Hom}_H(\cdot, H)$ is exact, and H is right self-injective.

(ii) Since F is a finitely generated projective H-module, we may equally well prove that F is not an injective H-module. But this is a consequence of:

Lemma 3.8. *Consider a module $M = \bigoplus_I M_i$ for an infinite index set I. The countermodule of M is not injective.*

Proof. Let H denote the endomorphism ring of M. For each $i \in I$, let $\varepsilon_i \in H$ be the projection onto the direct summand M_i, and also choose a non-zero $x_i \in M_i$. Consider the left ideal $\mathfrak{a}$ of H generated by the orthogonal idempotents ε_i ($i \in I$). Clearly one can define a homomorphism $f\colon \mathfrak{a} \to M$ of left H-modules as $f(\sum \alpha_i \varepsilon_i) = \sum \alpha_i(x_i)$. This homomorphism cannot be extended to H, for then it would be of the form $\alpha \mapsto \alpha(x)$ for some $x \in M$, but $\varepsilon_i(x) = 0$ for all but a finite number of $i \in I$. $\square$

Examples

1. Frobenius algebras. Let A be a finite-dimensional algebra over a field K. Then A is a *Frobenius algebra* if there exists a K-bilinear form $\Phi\colon A \times A \to K$, which is non-degenerated and is associative in the sense that

$$\Phi(a\,b, c) = \Phi(a, b\,c) \quad \text{for all} \quad a, b, c \in A.$$

Let $\mathfrak{a}$ be a left ideal of the Frobenius algebra A. Then for each $b \in r(\mathfrak{a})$ we get

$$\Phi(\mathfrak{a}, b) = \Phi(\mathfrak{a}\, b, 1) = \Phi(0, 1) = 0.$$

Conversely, if $\Phi(\mathfrak{a}, b) = 0$ holds, then $0 = \Phi(A\,\mathfrak{a}, b) = \Phi(A, \mathfrak{a}\, b)$ implies $\mathfrak{a}\, b = 0$. We thus have $r(\mathfrak{a}) = \{b \in A \mid \Phi(\mathfrak{a}, b) = 0\}$, and similarly $l(\mathfrak{b}) = \{a \in A \mid \Phi(a, \mathfrak{b}) = 0\}$ for a right ideal $\mathfrak{b}$. Using standard properties of bilinear forms, one finds that

$$\dim_K \mathfrak{a} + \dim_K r(\mathfrak{a}) = \dim_K A = \dim_K \mathfrak{b} + \dim_K l(\mathfrak{b}),$$

and we may conclude that $l(r(\mathfrak{a})) = \mathfrak{a}$, $r(l(\mathfrak{b})) = \mathfrak{b}$. Hence every Frobenius algebra is a QF-ring.

2. Group algebras. Let G be a finite group, and consider the group algebra $A = K[G]$ over a field K. Define a bilinear form $\Phi\colon A \times A \to K$ as $\Phi(\sum_g \alpha_g\, g, \sum_g \beta_h\, h) = \sum_g \alpha_g\, \beta_{g^{-1}}$, which is easily verified to be associative. It is also non-degenerate, because if $\Phi(\sum \alpha_g\, g, a) = 0$ for every $a \in K[G]$, then for each $h \in G$ we get $\alpha_h = \Phi(\sum \alpha_g\, g, h^{-1}) = 0$. $K[G]$ is thus a Frobenius algebra and in particular a QF-ring.

§ 4. Self-Injective Maximal Rings of Quotients

We recall that the maximal right ring of quotients $Q_{\max}$ of A is defined as $Q_{\max} = A_{\mathfrak{D}}$, where $\mathfrak{D}$ is the family of dense right ideals of A. In Theorem IX.3.3 we obtained a description of $Q_{\max}$ as the bicommutator of $E(A)$. Let H denote the endomorphism ring of the right A-module $E(A)$, and consider $E(A)$ as an H-A-bimodule. Let $\lambda\colon Q_{\max} \to \mathrm{Hom}_H(E(A), E(A))$ be the isomorphism in question. There is a commutative diagram

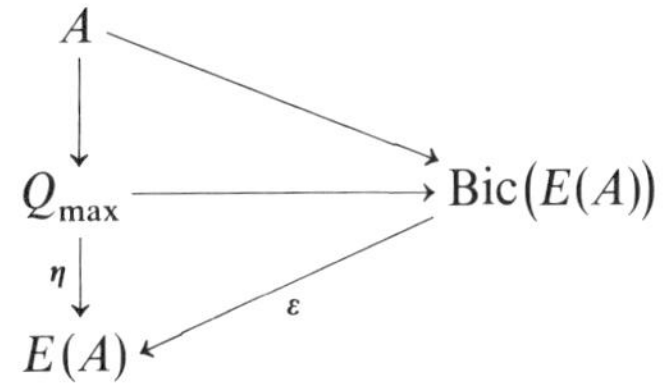

where $\varepsilon(f) = f(1)$, and the vertical arrows are the canonical imbeddings.

Proposition 4.1. *The following assertions are equivalent:*

(a) *$Q_{\max}$ is a right self-injective ring.*

(b) *$Q_{\max} \cong E(A)$.*

(c) *The right ideal $(A : x)$ is dense in A for every $x \in E(A)$.*

(d) *For every right ideal $\mathfrak{a}$ of A and homomorphism $\alpha\colon \mathfrak{a} \to A$, there exists a dense right ideal $\mathfrak{b}$ and a homomorphism $\beta\colon \mathfrak{b} \to A$ such that $\mathfrak{a} \subset \mathfrak{b}$ and $\beta \mid \mathfrak{a} = \alpha$.*

(e) *There is a ring isomorphism $\mu\colon H \to Q_{\max}$ such that $\eta\, \mu(h) = h(1)$.*

Proof. The equivalence of the four conditions (a)–(d) was obtained already in Chap. IX (Prop. IX.2.7).

(b) ⇒(e): We have $Q_{\max} \cong \mathrm{Hom}_Q(E(A), E(A)) \cong \mathrm{Hom}_A(E(A), E(A)) = H$, and it is clear that this isomorphism makes the diagram

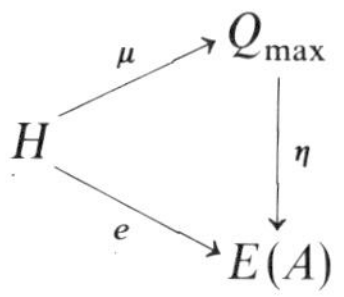

$$e(h) = h(1)$$

commutative.

(e) ⇒ (b): The fact that $E(A)$ is an injective module implies that the map e is surjective, so if μ exists, then also the imbedding $\eta: Q_{\max} \to E(A)$ is surjective. □

The condition (d) may be rephrased in a formulation which appears to be somewhat weaker, but which actually suffices to guarantee the self-injectivity of $Q_{\max}$. We introduce the family $\mathfrak{H}$ of right ideals $\mathfrak{a}$ of A for which there exists a homomorphism $\mathfrak{a} \to A$ which cannot be extended to a homomorphism $\mathfrak{b} \to A$ for any right ideal $\mathfrak{b}$ properly containing $\mathfrak{a}$. Every $\mathfrak{a} \in \mathfrak{H}$ is an essential right ideal, for if $\mathfrak{a} \cap \mathfrak{c} = 0$ and $\mathfrak{c} \neq 0$, then every $\alpha: \mathfrak{a} \to A$ can trivially be extended to $\mathfrak{a} \oplus \mathfrak{c} \to A$.

Proposition 4.2 (Masaike [2]). *$Q_{\max}$ is a right self-injective ring if and only if $l(\mathfrak{a}) = 0$ for every $\mathfrak{a} \in \mathfrak{H}$.*

Proof. If $Q_{\max}$ is right self-injective, then it follows from condition (d) of Prop. 4.1 that every $\mathfrak{a} \in \mathfrak{H}$ is a dense right ideal, and hence $l(\mathfrak{a}) = 0$.

Suppose conversely that $l(\mathfrak{a}) = 0$ for every $\mathfrak{a} \in \mathfrak{H}$. Let $\mathfrak{a}$ be an arbitrary right ideal of A, with a homomorphism $\alpha: \mathfrak{a} \to A$. By Zorn's lemma there is a homomorphism $\beta: \mathfrak{b} \to A$ such that $\mathfrak{a} \subset \mathfrak{b}$, β extends α, and β cannot be further extended to right ideals $\mathfrak{c}$ properly containing $\mathfrak{b}$. Thus $\mathfrak{b} \in \mathfrak{H}$ and $l(\mathfrak{b}) = 0$. By Prop. 4.1(d) it now suffices to show that $\mathfrak{b}$ is dense. Let a be an arbitrary element of A. We must show that $(\mathfrak{b} : a)$ has zero left annihilator, and we accomplish this by showing that $(\mathfrak{b} : a) \in \mathfrak{H}$. Define $\varphi: (\mathfrak{b} : a) \to A$ as $\varphi(c) = \beta(a\,c)$. We will show that φ cannot be further extended to bigger right ideals. Assume on the contrary that there exists a right ideal $\mathfrak{c} \supsetneqq (\mathfrak{b} : a)$ and $\psi: \mathfrak{c} \to A$ extending φ. We define $\gamma: \mathfrak{b} + a\,\mathfrak{c} \to A$ as $\gamma(b + a\,c) = \beta(b) + \psi(c)$; this is possible, because $b + a\,c = 0$ implies $c \in (\mathfrak{b} : a)$ and hence $\psi(c) = \varphi(c) = \beta(a\,c)$, so $\gamma(b + a\,c) = \beta(b) + \beta(a\,c) = 0$. Then γ is an extension of β. Since there exists $c \in \mathfrak{c}$ such that $a\,c \notin \mathfrak{b}$, we have $\mathfrak{b} \subsetneqq \mathfrak{b} + a\,\mathfrak{c}$, and this contradicts the maximality of β. □

When A is right non-singular, $Q_{\max}$ is semi-simple if and only if A is of finite right rank (Theorem XII.2.5). For the general case we have:

Proposition 4.3. *Assume $Q_{\max}$ is a right self-injective ring. Then $Q_{\max}$ is a semi-perfect ring if and only if A is of finite right rank.*

Proof. Since $Q_{\max} \cong \mathrm{Hom}_A(E(A), E(A))$ by Prop. 4.1(e), the assertion is a consequence of Prop. 1.7. □

By imposing further chain conditions on A we obtain a quasi-Frobenius maximal ring of quotients:

Theorem 4.4. *Assume $Q_{\max}$ is a right self-injective ring. The following assertions are then equivalent:*

(a) *Q_{max} is a quasi-Frobenius ring.*

(b) *The lattice $\mathrm{Sat}_{\mathfrak{D}}(A)$ is noetherian.*

(c) *For every right ideal $\mathfrak{a}$ of A there exists a finitely generated right ideal $\mathfrak{b} \subset \mathfrak{a}$ such that $l(\mathfrak{b} : a) = 0$ for all $a \in \mathfrak{a}$.*

(d) *A satisfies ACC on annihilators of subsets of $E(A)$.*

(e) *$E(A)$ is a Σ-injective module.*

(f) *Every direct sum of $\mathfrak{D}$-torsion-free injective modules is injective.*

Proof. The conditions (b)–(f) are all equivalent to each other, and they state that Q_{max} is a noetherian object of **Mod**-$(A, \mathfrak{D})$ (Prop. XIII.2.4). Since every QF-ring is noetherian, it is clear that (a) implies the other conditions.

(c) $\Rightarrow$ (a): By Theorem 3.5 it suffices to show that the ring Q_{max} satisfies ACC on right annihilator ideals. In view of the condition (d), it will be enough to show that if S_1 and S_2 are subsets of Q_{max}, then

$$\{q \in Q_{max} \mid S_1 q = 0\} \subset \{q \in Q_{max} \mid S_2 q = 0\}$$

holds if and only if

$$\{a \in A \mid S_1 a = 0\} \subset \{a \in A \mid S_2 a = 0\}.$$

The first inclusion trivially implies the second one. Suppose the second inclusion holds, and let $S_1 q = 0$ for some $q \in Q_{max}$. Choose $\mathfrak{a} \in \mathfrak{D}$ such that $q\,\mathfrak{a} \subset A$. Then $S_1 q\,\mathfrak{a} = 0$ implies $S_2 q\,\mathfrak{a} = 0$, which gives $S_2 q = 0$ since Q_{max} is $\mathfrak{D}$-torsion-free. $\square$

In Chapter XI we obtained sufficient conditions for $Q_{max}(A)$ to be equal to $Q_{tot}(A)$, the maximal flat epimorphic right ring of quotients of A. There it was proved that if $Q_{max}(A)$ is a right Kasch ring, then $Q_{max}(A) = Q_{tot}(A)$ (Prop. XI.5.3). Since every QF-ring is a right Kasch ring, we obtain in particular:

Proposition 4.5. *If $Q_{max}(A)$ is a QF-ring, then $Q_{max}(A) = Q_{tot}(A)$.*

Exercises

§1

1. Let M be an A-module, and let H be the endomorphism ring of $E(M)$. Show that

$$H/J(H) \cong \varinjlim \mathrm{Hom}_A(L, M),$$

where L runs through the down-wards directed family of essential submodules of M.

§2

2. Let A be a right self-injective ring. Show that every finite-dimensional non-singular module is semi-simple and injective. (Hint: use Lemma VI.4.5.)

3. Show that the following properties of A are equivalent:

(a) $r(l(\mathfrak{a})) = \mathfrak{a}$ for every right ideal $\mathfrak{a}$.

(b) Every cyclic module is torsionless.

4. Show that if A is right coherent and FP-injective (Exercise I.43), then $l(r(\mathfrak{b})) = \mathfrak{b}$ for every finitely generated left ideal $\mathfrak{b}$.

§3

5. If A is a commutative principal ideal domain, show that $A/\mathfrak{a}$ is a QF-ring for every ideal $\mathfrak{a} \neq 0$.

6. If A is a right noetherian ring satisfying

$$r(l(\mathfrak{a})) = \mathfrak{a} \text{ for right ideals } \mathfrak{a}, \qquad l(r(\mathfrak{b})) = \mathfrak{b} \text{ for left ideals } \mathfrak{b},$$

show that A is a QF-ring.

7. Show that A is a QF-ring if and only if every right ideal is the right annihilator of a finite subset of A and the analogous thing holds for left ideals.

8. Let K be the field $P(X)$ of rational functions over the prime field P with p elements. Let $\alpha: K \to K^p$ be the isomorphism given by $\alpha(f) = f^p$. Define a ring $A = eK \oplus yK$ with the multiplication rules

$$e\,a = a\,e = a \quad \text{for all } a \in A,$$

$$y^2 = 0,$$

$$\alpha(f)\,y = y\,f \qquad \text{for all } f \in K.$$

Show that:

(i) $Ay = Ky$ is the only non-trivial left ideal of A (so every left ideal is a left annihilator).

(ii) A is right and left artinian.

(iii) If $g_1, \ldots, g_r$ is a basis for K over K^p, then $g_1\,yA, \ldots, g_r\,yA$ are p distinct minimal right ideals of A. (Hence A cannot be a QF-ring.)

9. Let A be a semi-local ring.

(i) Show that $r(J)$ is the left socle $s_l(A)$ of A, while $l(J)$ is the right socle $s_r(A)$ of A.

(ii) If A is a QF-ring, show that $s_l(A) = s_r(A)$.

10. Define the *projective dimension* of a module M to be the smallest integer n (or ∞ if such a smallest integer does not exist) such that there is an exact sequence

$$0 \to P_n \to P_{n-1} \to \cdots \to P_0 \to M \to 0$$

with P_i projective. Show that if A is a QF-ring then every module has projective dimension 0 or ∞.

11. Let A be a commutative ring and G a group, and assume that $A[G]$ is self-injective. Show that if H is a subgroup of G, then also $A[H]$ is self-injective.

§4

12. Show that the following properties of a ring A are equivalent:

(a) A is a QF-ring.

(b) A is Σ-injective as a right A-module.

(c) Every projective module is injective.

13. If A is a ring without zero-divisors, show that A is a right Ore domain if and only if $E(A)$ is a Σ-injective module.

14. Let K be a field and consider the ring $R = K[X]/(X^4)$. Let A be the subring $K + K\bar{X}^2 + K\bar{X}^3$ of R. Show that:

(i) The ideal $K\bar{X}^2 + K\bar{X}^3$ of R belongs to $\mathfrak{H}$ (in the notation of Prop. 4.2), but is not dense.

(ii) $Q_{\max}(A)$ is not a self-injective ring.

Chapter XV. Classical Rings of Quotients

The major part of this chapter deals with the problem of finding necessary and sufficient conditions for a ring A to have a classical right ring of quotients of a certain kind (e.g. artinian or quasi-Frobenius). The basic result of this sort is the Goldie Theorem (Theorem II.2.2).

Let us record some terminology. An element $a \in A$ is *right (left) regular* if $r(a)=0$ (resp. $l(a)=0$), and a is *regular* if it is both right and left regular. A sub-module L of a module M is called *saturated* in M if $x\,s \in L$ for some regular $s \in A$ implies $x \in L$ (cf. Chap. IX, §4).

§1. Semi-Prime Ideals and the Prime Radical

Let A be an arbitrary ring. A two-sided ideal $\mathfrak{a}$ of A is called *semi-prime* if $A/\mathfrak{a}$ is a semi-prime ring.

Lemma 1.1. *An ideal $\mathfrak{a}$ is semi-prime if and only if it is an intersection of prime ideals.*

Proof. Since the prime ideals of A containing $\mathfrak{a}$ correspond bijectively to the prime ideals of $A/\mathfrak{a}$, it suffices to consider the case $\mathfrak{a}=(0)$. If the intersection of the prime ideals of A is (0), then A must be semi-prime, because a nilpotent ideal is contained in every prime ideal of A.

Suppose on the other hand that A is a semi-prime ring. It suffices to show that for every $a \neq 0$ there exists a prime ideal $\mathfrak{p}$ not containing a. Now since $A a A$ is not nilpotent, there exists $a_1 \in a A a$ such that $a_1 \neq 0$. Continuing in this way we obtain $a_i \neq 0$ $(i=1,2,\ldots)$ such that $a_0=a$ and $a_i \in a_{i-1} A_{i-1}$. With the use of Zorn's lemma we can find an ideal $\mathfrak{p}$ which is maximal with respect to $a_i \notin \mathfrak{p}$ for all i. It will be shown that $\mathfrak{p}$ is a prime ideal. Assume $\mathfrak{a}$ and $\mathfrak{b}$ are ideals not contained in $\mathfrak{p}$. Then by the maximality of $\mathfrak{p}$ we have $a_i \in \mathfrak{a}+\mathfrak{p}$ and $a_j \in \mathfrak{b}+\mathfrak{p}$ for some i,j. For $k=\max(i,j)$, this gives $a_{k+1} \in a_k\, A\, a_k \subset (\mathfrak{a}+\mathfrak{p})\cdot(\mathfrak{b}+\mathfrak{p}) \subset \mathfrak{a}\,\mathfrak{b}+\mathfrak{p}$. Since $a_{k+1} \notin \mathfrak{p}$, this implies $\mathfrak{a}\,\mathfrak{b} \not\subset \mathfrak{p}$, and so $\mathfrak{p}$ is a prime ideal. $\square$

The intersection of all prime ideals of A is called the *prime radical* of A, and is denoted by $N(A)$. From Lemma 1.1 it follows that $N(A)$ is the smallest ideal $\mathfrak{a}$ of A such that $A/\mathfrak{a}$ is a semi-prime ring. In particular, A is a semi-prime ring if and only if $N(A)=0$. Clearly $N(A)$ contains all nilpotent ideals of A.

Proposition 1.2. *$N(A)$ is a nil ideal.*

Proof. If $a \in A$ is not nilpotent, consider an ideal $\mathfrak{p}$ which is maximal with respect to $a^n \notin \mathfrak{p}$ for all n. As in the preceding lemma one shows that $\mathfrak{p}$ is a prime ideal. Hence $a \notin \mathfrak{p}$ implies $a \notin N(A)$. □

Corollary 1.3. $N(A) \subset J(A)$.

Proof. This was noted in Example VIII.1.1. □

Proposition 1.4. *If A satisfies* ACC *on two-sided ideals, then $N(A)$ is a nilpotent ideal.*

Proof. By hypothesis there is a maximal nilpotent ideal $\mathfrak{a}$. Then $\mathfrak{a} \subset N(A)$. But on the other hand it follows from the maximality of $\mathfrak{a}$ that $A/\mathfrak{a}$ is a semi-prime ring, so $\mathfrak{a} \supset N(A)$. Hence $\mathfrak{a} = N(A)$. □

Examples

1. Semi-primary rings. If $J(A)$ is a nilpotent ideal, e.g. if A is semi-primary, then $J(A) = N(A)$.

2. Commutative rings. When A is commutative, $N(A)$ consists of all nilpotent elements of A. For if a is nilpotent, then $A\,a$ is a nilpotent ideal, and hence $A\,a \subset N(A)$. Since every prime ideal contains a minimal prime ideal (cf. the proof of Lemma XIII.3.12), $N(A)$ is the intersection of the minimal prime ideals of A. So if A is noetherian, then $N(A)$ is the intersection of finitely many prime ideals (by the remark following Cor. VII.5.4).

3. The singular ideal. If A satisfies ACC on right annihilators, then the right singular ideal $Z(A)$ is nilpotent (Lemma II.2.5), so $Z(A) \subset N(A)$. In general one has $Z(A) \neq N(A)$ $\left(\text{consider e.g. } A = \begin{pmatrix} \mathbb{Z} & \mathbb{Z} \\ 0 & \mathbb{Z} \end{pmatrix}\right)$.

For an arbitrary commutative ring A one has $N(A) \subset Z(A)$. For suppose a is a nilpotent element; then for any ideal $\mathfrak{a} \neq 0$ there exists $n > 0$ such that $a^n\,\mathfrak{a} = 0$ but $a^{n-1}\,\mathfrak{a} \neq 0$, so $\operatorname{Ann}(a) \cap \mathfrak{a} \neq 0$, and $\operatorname{Ann}(a)$ is thus an essential ideal, i.e. $a \in Z(A)$. So for a commutative noetherian ring A one has $Z(A) = N(A)$.

§2. Rings of Quotients of Quotient Rings

In this section we make some useful observations concerning the relations between $Q_{\mathrm{cl}}(A)$ and $Q_{\mathrm{cl}}(A/\mathfrak{a})$ for a two-sided ideal $\mathfrak{a}$ of A. For each element $a \in A$ we let $\bar{a}$ denote its image in $A/\mathfrak{a}$. Following Feller and Swokowski ([1], [2]), we say that A is $\mathfrak{a}$-*reflective* if it satisfies the condition:

$a \in A$ is regular in A if and only if $\bar{a}$ is regular in $A/\mathfrak{a}$.

Clearly $\mathfrak{a}$-reflectivity has as a consequence that a right Ore condition for A is inherited by $A/\mathfrak{a}$. Thus we obtain statement (i) of:

Proposition 2.1. *If $Q_{\mathrm{cl}}(A)$ exists and A is $\mathfrak{a}$-reflective, then:*

(i) *$Q_{\mathrm{cl}}(A/\mathfrak{a})$ exists.*

(ii) *$\mathfrak{a}\,Q_{\mathrm{cl}}(A)$ is a two-sided ideal of $Q_{\mathrm{cl}}(A)$.*

(iii) $Q_{\mathrm{cl}}(A/\mathfrak{a}) \cong Q_{\mathrm{cl}}(A)/\mathfrak{a}\,Q_{\mathrm{cl}}(A)$.

Proof. It remains to show (ii) and (iii). Now it is easily verified that the map $a s^{-1} \mapsto \bar{a} \bar{s}^{-1}$ is a surjective ring homomorphism $Q_{cl}(A) \to Q_{cl}(A/\mathfrak{a})$, and that its kernel is $\mathfrak{a} Q_{cl}(A)$. This proves (ii) and (iii). □

In order to be able to lift an Ore condition from $A/\mathfrak{a}$ to A, we need to make a further assumption about $\mathfrak{a}$. We say that A is $\mathfrak{a}$-*quorite* if, given $a \in \mathfrak{a}$ and s regular in A, there exist $b \in \mathfrak{a}$ and t regular in A such that $a t = s b$.

Proposition 2.2. *If A is $\mathfrak{a}$-reflective and $\mathfrak{a}$-quorite, and if $Q_{cl}(A/\mathfrak{a})$ exists, then $Q_{cl}(A)$ exists.*

Proof. Consider elements a, s in A with s regular. Since A is $\mathfrak{a}$-quorite, it suffices to look at the case when $a \notin \mathfrak{a}$. Since $A/\mathfrak{a}$ satisfies the Ore condition, there exist b, t in A such that $a t - s b = c \in \mathfrak{a}$ and $\bar{t}$ is regular. By the $\mathfrak{a}$-reflectivity, t is regular in A. Since A is $\mathfrak{a}$-quorite, we can write $c u = s d$ for some $d \in \mathfrak{a}$ and u regular in A. This gives $a t u = s b u + c u = s(b u + d)$, so the Ore condition is satisfied by a and s. □

Examples

1. $J(A)$-reflectivity. It is of particular interest to determine conditions for A to be $J(A)$-reflective, where $J(A)$ is the Jacobson radical of A. E.g. we have:

Proposition 2.3. *Every left perfect ring A is $J(A)$-reflective.*

Proof. Every regular element of A is invertible by Lemma II.1.10. Hence a regular in A implies $\bar{a}$ regular in $A/J(A)$. Conversely, let $\bar{a}$ be regular in $A/J(A)$. Then $\bar{a}$ is invertible, so there exists $b \in A$ with $a b - 1 \in J(A)$. Since $J(A)$ is a nilideal, $(a b - 1)^n = 0$ for some n and this gives $a c = 1$ for some $c \in A$. Now $r(c) = 0$, so c is invertible by Lemma II.1.10, and hence a is invertible. □

2. Commutative noetherian rings

Proposition 2.4. *A commutative noetherian ring A is $N(A)$-reflective if and only if all prime ideals in $\operatorname{Ass}(A)$ are minimal prime ideals.*

Proof. Suppose $\operatorname{Ass}(A)$ consists of the minimal prime ideals $\mathfrak{p}_1, \ldots, \mathfrak{p}_n$, so $N(A) = \mathfrak{p}_1 \cap \cdots \cap \mathfrak{p}_n$. If a is regular in A, then $a \notin \mathfrak{p}_1 \ldots \mathfrak{p}_n$ by Lemma VIII.5.6. Therefore, if $a b \in N(A) = \mathfrak{p}_1 \cap \cdots \cap \mathfrak{p}_n$, then $b \in N(A)$. Hence $\bar{a}$ is regular in $A/N(A)$. On the other hand, if $a \in \mathfrak{p}_i$ and we choose $b \in \bigcap_{j \neq i} \mathfrak{p}_j$ such that $b \notin \mathfrak{p}_i$, then $a b \in N(A)$, so $\bar{a}$ is a zero-divisor. Hence if $\bar{a}$ is regular, then $a \notin \mathfrak{p}_1 \ldots \mathfrak{p}_n$, and a is regular in A.

Suppose conversely that there is a prime ideal $\mathfrak{q}$ in $\operatorname{Ass}(A)$ which is not a minimal prime ideal. Choose $a \in \mathfrak{q}$ such that a does not belong to any minimal prime ideal. Then a is not regular in A, but $\bar{a}$ is regular in $A/N(A)$, for if $a b \in N(A)$, then b belongs to every minimal prime ideal and hence $b \in N(A)$. Thus A is not $N(A)$-reflective. □

§3. Orders in a Semi-Primary Ring

The following result by Robson [1] characterizes those rings which are right orders in semi-primary rings:

Proposition 3.1. *A has a semi-primary classical right ring of quotients if and only if A satisfies the following conditions:*

(1) *A is $N(A)$-reflective and $N(A)$-quorite.*

(2) *$N(A)$ is a nilpotent ideal.*

(3) *The ring $A/N(A)$ is of finite right rank and satisfies* ACC *on right annihilators.*

Proof. Suppose A satisfies conditions (1)–(3). The condition (3) implies that $A/N(A)$ has a semi-simple classical right ring of quotients $\bar{Q}$, by Goldie's theorem (Theorem II.2.2). Then by Prop. 2.2, also $Q=Q_{\mathrm{cl}}(A)$ exists and $\bar{Q}\cong Q/N(A)\,Q$. Since $\bar{Q}$ is semi-simple, $N(A)\,Q$ is a semi-prime ideal, and hence $N(A)\,Q\supset N(Q)$. If we now can show that $N(A)\,Q$ is nilpotent, then it will follow that $N(A)\,Q=N(Q)$ and that $Q/N(Q)$ is semi-simple, which implies that Q is semi-primary.

Each element of $(N(A)\,Q)^2$ is a sum of elements of the form $a\,s^{-1}\,b\,t^{-1}$, with $a,b\in N(A)$ and s,t regular in A. Since A is $N(A)$-quorite, we can write $b\,u=s\,c$ for some $c\in N(A)$ and u regular in A. Then $a\,s^{-1}\,b\,t^{-1}=a\,c(t\,u)^{-1}\in N(A)^2\,Q$. Thus $(N(A)\,Q)^2\subset N(A)^2\,Q$, and by induction we more generally obtain $(N(A)\,Q)^n\subset N(A)^n\,Q$. So if $N(A)$ is nilpotent, then also $N(A)\,Q$ is nilpotent.

Suppose conversely that A has a semi-primary classical right ring of quotients Q. Then $N(Q)=J(Q)$ is a nilpotent ideal of Q, so $N(Q)\cap A\subset N(A)$ and Q is $N(Q)$-reflective (Prop. 2.3). We will prove that A is $(N(Q)\cap A)$-reflective. From the commutative diagram

$$\begin{array}{ccc} A & \longrightarrow & A/(N(Q)\cap A) \\ \downarrow & & \downarrow \\ Q & \longrightarrow & Q/N(Q) \end{array}$$

and the $N(Q)$-reflectivity of Q, we see that to do this, it suffices to show that $\bar{a}$ is regular in $A/(N(Q)\cap A)$ if and only if $\bar{a}$ is regular in the bigger ring $Q/N(Q)$. Now if $\bar{a}$ is regular in $A/(N(Q)\cap A)$, and $a\,q\in N(Q)$ for some $q=b\,s^{-1}\in Q$, then $a\,b\in N(Q)\cap A$, which implies $b\in N(Q)\cap A$ and hence $q\in N(Q)$; this shows that $r(\bar{a})=0$ in $Q/N(Q)$, which is equivalent to regularity (Lemma II.1.10). Conversely, it is clear that if $a\in A$ and $\bar{a}$ is regular in $Q/N(Q)$, then $\bar{a}$ is regular in the subring $A/(N(Q)\cap A)$.

The preceding argument shows in particular that the semi-simple ring $Q/N(Q)$ is a classical right ring of quotients of $A/(N(Q)\cap A)$. It then follows from the Goldie theorem that $A/(N(Q)\cap A)$ is of finite right rank, satisfies ACC on right annihilators and is semi-prime. The latter property implies that $N(Q)\cap A$ contains $N(A)$, and so we must have $N(Q)\cap N(A)=N(A)$. Finally we note that A is $N(A)$-quorite (Exercise 4). ▯

Corollary 3.2. *If A is a right order in a semi-primary ring Q, then $N(A)=A\cap N(Q)$.*

Proof. From the proof we have $N(A)=A\cap N(Q)$. ▯

One may prove a partial converse to this corollary:

Proposition 3.3 (Tachikawa [4]). *Assume the maximal right ring of quotients of A is a semi-primary, right Kasch ring. Then $Q_{\max}$ is a classical right ring of quotients of A if and only if $N(A)=A\cap N(Q_{\max})$.*

Proof. The sufficiency is clear from the corollary. Suppose conversely that $N(A)=A\cap N(Q_{\max})$. Then there is a ring monomorphism $A/N(A)\to Q_{\max}/N(Q_{\max})$. We first prove that $A/N(A)$ is a right order in the semi-simple ring $Q_{\max}/N(Q_{\max})$. To accomplish this it suffices to show that $Q_{\max}/N(Q_{\max})$ is the maximal right ring of quotients of $A/N(A)$, for then it follows that $A/N(A)$ is semi-prime of finite right rank (Theorem XII.2.5) and satisfies ACC on right annihilators (e.g. Theorem XIV.4.4), so the Goldie Theorem can be applied.

Thus we must show that the right ideal $\overline{(A:q)}$ of $A/N(A)$ is essential, for every $q\in Q_{\max}$. Suppose $a\notin N(A)$. Then $(A:q\,a)$ is a dense right ideal of A, and since $Q_{\max}$ is a Kasch ring, it follows that $(A:q\,a)\,Q_{\max}=Q_{\max}$ (Prop. XII.5.2). Therefore $a\notin N(A)$ clearly implies that $a(A:q\,a)\not\subset N(A)$. Hence there exists $b\in(A:q\,a)$ such that $a\,b\notin N(A)$. Thus for every non-zero $\bar{a}\in A/N(A)$ there exists $\bar{b}$ such that $0\neq\overline{a\,b}\in\overline{(A:q)}$. Thereby we have proved that $A/N(A)$ is a right order in $Q_{\max}/N(Q_{\max})$.

If $q\in Q_{\max}$, then the essential right ideal $\overline{(A:q)}$ contains a regular element $\bar{s}$, by the Goldie Theorem. Since $Q_{\max}$ is semi-primary, s is regular also in $Q_{\max}$ (Prop. 2.3). Thus there is a regular element $s\in A$ such that $q\,s\in A$, and it follows that $Q_{\max}$ is a classical right ring of quotients of A. □

Examples

1. Artinian classical rings of quotients. Since Q_{cl} is right noetherian if and only if the lattice of saturated right ideals of A is noetherian, it follows that a ring A has a right artinian classical right ring of quotients if and only if A satisfies conditions (1)–(3) and in addition the condition

(4) A satisfies ACC on saturated right ideals.

2. Noetherian rings. When A is right noetherian, the conditions (2)–(4) are automatically satisfied, so A has a right artinian classical right ring of quotients if and only if A is $N(A)$-reflective and $N(A)$-quorite. This result will be considerably improved in the next section, where it will be shown that only part of the $N(A)$-reflectivity is needed. In any case, we can already now conclude that if A is commutative and noetherian, then $Q_{\mathrm{cl}}(A)$ is artinian if and only if A is $N(A)$-reflective, i.e. all prime ideals in $\mathrm{Ass}(A)$ are minimal (Prop. 2.4).

§ 4. Noetherian Orders in an Artinian Ring

Throughout this section we assume A to be a *right noetherian* ring. Then $A/N(A)$ is a semi-prime, right noetherian ring, so it has a semi-simple classical right ring of quotients. For each subset S (or element s) of A, we denote by $\bar{S}$ (or $\bar{s}$) its image in $A/N(A)$.

Definition. A subset S of A is *preregular* if it satisfies:

(1) S is multiplicatively closed.

(2) $r(s)=0$ for every $s\in S$.

(3) $\bar{S}$ is equal to the set of regular elements of $A/N(A)$.

A preregular set satisfies axiom $S2$ of Prop. II.1.4, and we will show that it also satisfies $S1$.

Lemma 4.1. *Let S be a preregular subset of A. If $a \in l(N(A)) \cap N(A)$ and $s \in S$, then there exist $t \in S$ and $b \in A$ such that $a\,t = s\,b$.*

Proof. Since $a \in l(N(A))$, we trivially have $(sA : a) \supset N(A)$. To find $t \in (sA : a) \cap S$, it suffices by property (3) of preregular sets to show that $(sA : a)/N(A)$ contains regular elements. By the Goldie Theorem this is equivalent to showing that $(sA : a)/N(A)$ is an essential right ideal of $A/N(A)$. Thus let $\mathfrak{a}$ be a right ideal of A such that $\mathfrak{a} \not\subset N(A)$. We wish to show that $\mathfrak{a} \cap (sA : a) \not\subset N(A)$. If $a\,\mathfrak{a} = 0$, then $\mathfrak{a} \subset (sA : a)$ and hence $\mathfrak{a} \cap (sA : a) = \mathfrak{a} \not\subset N(A)$. On the other hand, if $a\,\mathfrak{a} \neq 0$, then $sA \cap a\,\mathfrak{a} \neq 0$ since $r(s) = 0$ implies sA essential in A (Lemma II.2.3). So there exists $c \in \mathfrak{a}$ such that $c \in \mathfrak{a} \cap (sA : a)$ and $a\,c \neq 0$. Since $a \in l(N(A))$, we must have $c \notin N(A)$, and it follows that $\mathfrak{a} \cap (sA : a) \not\subset N(A)$ also in this case. $\square$

We will denote the ring $A/(l(N(A)) \cap N(A))$ by A_1, and write a_1 for the image of $a \in A$ in A_1. Similarly we write S_1 for the image of $S \subset A$ in A_1. The prime radical of A_1 is $N(A_1) = N(A)/(l(N(A)) \cap N(A))$, and $A_1/N(A_1) \cong A/N(A)$.

Lemma 4.2. *If S is a preregular set in A, then S_1 is preregular in A_1.*

Proof. S_1 is clearly multiplicatively closed. The image of S_1 in $A_1/N(A_1) = A/N(A)$ is $\bar{S}$ and thus equal to the set of regular elements of $A_1/N(A_1)$. It remains to verify that $r(s_1) = 0$ for every $s \in S$. Suppose $s_1\,a_1 = 0$ for some $a \in A$. Then $s\,a \in l(N(A)) \cap N(A)$. But $\bar{s} \in A/N(A)$ is regular, so we must have $a \in N(A)$. $s\,a \in l(N(A))$ means that $s\,a\,N(A) = 0$, and $r(s) = 0$ implies $a\,N(A) = 0$. Hence $a \in l(N(A)) \cap N(A)$, and $a_1 = 0$. $\square$

Proposition 4.3. *Every preregular subset S of A is a right denominator set.*

Proof. We will prove this by induction on the index of nilpotence of $N(A)$, i.e. on the smallest integer k such that $N(A)^k = 0$. If this index is 1, the result is clear. Suppose the index is $n+1$. The index of nilpotence of $N(A_1)$ is then clearly n. Since S_1 is preregular in A_1 by Lemma 2.2, the induction hypothesis implies that S_1 satisfies the axiom $S1$. Suppose we are given $a \in A$ and $s \in S$. Then there exist $b \in A$ and $t \in S$ such that $a_1\,t_1 = s_1\,b_1$. Since $a\,t - s\,b \in l(N(A)) \cap N(A)$, we can apply Lemma 2.1 to get $u \in S$ and $c \in A$ such that $(a\,t - s\,b)\,u = s\,c$. From this we obtain $a\,t\,u = s(b\,u + c)$ with $t\,u \in S$, which shows that S satisfies $S1$. Since $S2$ is always satisfied, S is a right denominator set. $\square$

$A[S^{-1}]$ is clearly right noetherian, so it has a maximal nilpotent ideal $N(A[S^{-1}])$.

Lemma 4.4. $N(A[S^{-1}]) = N(A)\,A[S^{-1}]$.

Proof. Put $N(A[S^{-1}]) = J$. Then $J = (J \cap A)[S^{-1}]$ and clearly $J \cap A \subset N(A)$. Hence $N(A[S^{-1}]) \subset N(A)[S^{-1}]$. To prove equality, it suffices to show that $N(A)[S^{-1}]$ is a nilpotent ideal. Let $a\,s^{-1} \in N(A)[S^{-1}]$ and $b\,t^{-1} \in A[S^{-1}]$. Then $b\,t^{-1}\,a\,s^{-1} = b\,c\,u^{-1}\,s^{-1}$, where $a\,u = t\,c$ with $u \in S$ and $c \in A$. $a \in N(A)$ implies $t\,c \in N(A)$ and $\bar{t} \in A/N(A)$ regular implies $c \in N(A)$. This shows that $b\,t^{-1}\,a\,s^{-1} \in N(A)[S^{-1}]$, so $N(A)[S^{-1}]$ is a two-sided ideal. This argument also shows that $N(A)[S^{-1}]$ is a nilpotent ideal. $\square$

Proposition 4.5. *If S is preregular, then $A[S^{-1}]$ is a right artinian ring.*

Proof. We define a homomorphism

$$A[S^{-1}]/N(A[S^{-1}]) \to A/N(A)[\bar{S}^{-1}]$$

as $\overline{a\,s^{-1}} \mapsto \bar{a}\,\bar{s}^{-1}$. It is clear from the preceding lemma that this homomorphism is well-defined, and that it has a well-defined inverse $\bar{a}\,\bar{s}^{-1} \mapsto \overline{a\,s^{-1}}$. It is thus an isomorphism. Now $A/N(A)$ is a noetherian semi-prime ring and $\bar{S}$ is the set of all its regular elements, so $A/N(A)[\bar{S}^{-1}]$ is the classical right ring of quotients of $A/N(A)$, and it is semi-simple by the Goldie theorem. It follows that $N(A[S^{-1}])$ is equal to the Jacobson radical of $A[S^{-1}]$, and $A[S^{-1}]$ is thus a semi-primary right noetherian ring, hence right artinian. □

We will say that A satisfies the *regularity condition* if each element $a \in A$ for which $\bar{a} \in A/N(A)$ is regular, is itself regular (this is one half of $N(A)$-reflectivity).

Theorem 4.6 (Small [1]). *If A is a right noetherian ring satisfying the regularity condition, then A has a classical right ring of quotients which is right artinian.*

Proof. Put $S = \{a \in A \mid \bar{a} \in A/N(A) \text{ is regular}\}$. The regularity condition implies that S consists of regular elements. S is then preregular, so $A[S^{-1}]$ exists and is an artinian ring. It now suffices to show that $A[S^{-1}]$ is a classical right ring of quotients of A. Every element of $A[S^{-1}]$ has the form $a\,s^{-1}$ with $a \in A$, $s \in S$, and s is then regular. It remains to show that every regular element $a \in A$ is invertible in $A[S^{-1}]$. Suppose $a\,q = 0$ for some $q \in A[S^{-1}]$. Then $q = b\,s^{-1}$ with $b \in A$, $s \in S$. It follows that $a\,b = 0$ and hence $b = 0$. The right annihilator of a in $A[S^{-1}]$ is thus zero, and this implies that a is invertible in $A[S^{-1}]$ by Lemma II.1.10. □

Prop. 4.1 shows that the regularity condition is necessary for this result. An alternative proof of Small's theorem has been given by Lambek and Michler [3].

Examples

1. Hereditary orders

Proposition 4.7. *Every right hereditary and right noetherian ring is a right order in a right artinian ring.*

Proof. We verify the regularity condition. Suppose $a \in A$ and $\bar{a} \in A/N(A)$ is regular. The exact sequence $0 \to r(a) \to A \to a\,A \to 0$ splits, so $r(a) = eA$ for an idempotent e. But $\bar{a}$ regular implies that $e \in N(A)$, so $r(a) = e\,A = 0$. Since $a\,A$ is an essential right ideal of A (Lemma II.2.3) and A is right non-singular, we must also have $l(a) = 0$, so a is regular. □

2. Finitely generated algebras. Let R be a commutative ring which has an artinian classical ring of quotients $Q(R)$. Let A be an R-algebra, i.e. there is a ring homomorphism $\mu\colon R \to A$ with image contained in the center of A. Suppose A is finitely generated and torsion-free as an R-module. Then we may imbed A as a submodule of $A \otimes_R Q(R)$. $A \otimes_R Q(R)$ is a $Q(R)$-algebra, with multiplication given by $(a \otimes s^{-1})(b \otimes t^{-1}) = ab \otimes s^{-1} t^{-1}$. $A \otimes_R Q(R)$ is a finitely generated $Q(R)$-module, and is therefore an artinian ring. It is the classical ring of quotients of A, because for each $a \otimes s^{-1} \in A \otimes_R Q(R)$ we have

$$(a \otimes s^{-1})(\mu(s) \otimes 1) = a\,\mu(s) \otimes s^{-1} = a \otimes s\,s^{-1} = a \otimes 1$$

and hence

$$a\otimes s^{-1}=(a\otimes 1)(\mu(s)\otimes 1)^{-1}=a\mu(s)^{-1}$$

when identifying A with its image in $A\otimes_R Q(R)$, and $\mu(s)$ is a non-zero-divisor in A since A is torsion-free over R. Thus we have proved:

Proposition 4.8. *If R is a commutative ring with artinian classical ring of quotients $Q(R)$, and A is a finitely generated torsion-free R-algebra, then $A\otimes_R Q(R)$ is an artinian classical two-sided ring of quotients of A.*

As an illustration of this result we can take a finite group G and R a commutative ring with artinian classical ring of quotients Q. The group ring $R[G]$ is finitely generated and torsion-free over R. $R[G]$ is then an order in the artinian ring $Q[G]$.

§ 5. Orders in Self-Injective Rings

If A has a right self-injective classical right ring of quotients, then $Q_{\mathrm{cl}}(A)=Q_{\max}(A)$ and this ring coincides with the module $E(A)$. A necessary condition for this thing to happen is that for each $x\in E(A)$, the right ideal $(A:x)=\{a\in A\,|\,xa\in A\}$ contains a regular element. A technically useful restatement of this condition may be given by means of the family $\mathfrak{H}$ of right ideals $\mathfrak{a}$ for which there exists a homomorphism $\mathfrak{a}\to A$ that cannot be extended to right ideals properly containing $\mathfrak{a}$ (Chap. XIV, §4).

Lemma 5.1. *If $\mathfrak{a}\in\mathfrak{H}$, then there exists $x\in E(A)$ such that $\mathfrak{a}=(A:x)$.*

Proof. Let $\alpha\colon\mathfrak{a}\to A$ be a homomorphism not extendable to right ideals properly containing $\mathfrak{a}$. There exists $x\in E(A)$ such that $\alpha(a)=xa$ for every $a\in\mathfrak{a}$. By the maximality property of $\mathfrak{a}$ we must have $\mathfrak{a}=\{a\in A\,|\,xa\in A\}$. □

Proposition 5.2. *The following conditions are equivalent:*

(a) *For every $x\in E(A)$ there exists a regular element $s\in A$ such that $xs\in A$.*

(b) *Every $\mathfrak{a}\in\mathfrak{H}$ contains a regular element.*

Proof. (a) ⇒ (b) is immediate from Lemma 5.1.

(b) ⇒ (a): The condition (b) implies that $Q_{\max}$ is a right self-injective ring, by Prop. XIV.4.2. Hence for every $x\in E(A)$ we have that $(A:x)$ is a dense right ideal. Since every $\mathfrak{a}\in\mathfrak{H}$ contains regular elements, it suffices to show that $(A:x)\in\mathfrak{H}$. Now if $(A:x)\notin\mathfrak{H}$, then the homomorphism $\alpha\colon (A:x)\to A$, given by $\alpha(a)=xa$, can be extended to $\alpha'\colon\mathfrak{b}\to A$ for some right ideal $\mathfrak{b}\supsetneqq(A:x)$. There exists $y\in E(A)$ such that $\alpha'(b)=yb$ for all $b\in\mathfrak{b}$. But then $(x-y)a=0$ for all a in the dense right ideal $(A:x)$, which proves $x=y$. This contradicts $\mathfrak{b}\neq(A:x)$, and we must have $(A:x)\in\mathfrak{H}$. □

The following result is the "classical" analogue of Prop. XIV.4.3:

Proposition 5.3. *The following properties of A are equivalent:*

(a) *A is of finite right rank, and every $\mathfrak{a}\in\mathfrak{H}$ contains a regular element.*

(b) *Q_{cl} exists and is a semi-perfect, right self-injective ring.*

Proof. (b) ⇒ (a): This should be clear.

(a) ⇒ (b): We already know from the preceding Chapter that (a) implies that $Q_{\max}$ is semi-perfect and right self-injective. We will show that $Q_{\max}$ is a classical right ring of quotients of A. For each $q \in Q_{\max}$ we have $qa \in A$ for some regular $a \in A$, so it remains to show that every regular element a of A is invertible in $Q_{\max}$. Let $\varphi\colon Q_{\max} \to Q_{\max}$ be the map $\varphi(q)=aq$. Since $\operatorname{Ker}\varphi \cap A=0$ and A is essential in $Q_{\max}$, φ is a monomorphism. Thus $\operatorname{Im}\varphi$ is an injective submodule of $Q_{\max}$, and $Q_{\max}=\varphi(Q_{\max})\oplus K$ for some submodule K. Using the fact that φ is a monomorphism, one gets $\varphi(Q_{\max})=\varphi^2(Q_{\max})\oplus\varphi(K)$. Iteration of this gives that

$$Q_{\max}=\varphi^n(Q_{\max})\oplus\varphi^{n-1}(K)\oplus\cdots\oplus K$$

for every n. Since $E(A)$ has finite rank, we must therefore necessarily have $K=0$. Thus φ is an isomorphism, so a is right invertible in $Q_{\max}$. To conclude the proof, we use:

Lemma 5.4. *Suppose A has no infinite family of orthogonal idempotents. If $a \in A$ has a right inverse, then a is invertible.*

Proof. Suppose $ab=1$ but $ba \neq 1$. Define elements

$$e_i=b^i a^i - b^{i+1} a^{i+1} \quad \text{for} \quad i=1,2,\ldots.$$

Each e_i is different from zero, because $b^i a^i = b^{i+1} a^{i+1}$ would give

$$b^i a^i = a b^{i+1} a^{i+1} b = a b^i a^i b = b^{i-1} a^{i-1},$$

using repeatedly the fact that $ab=1$, and this would finally lead to $ba=1$. One verifies immediately that the e_i are orthogonal idempotents, contrary to our assumption. □

Proposition 5.5. *The following properties of A are equivalent:*

(a) *A is of finite right rank with $Z_r(A)$ nilpotent, and every $\mathfrak{a} \in \mathfrak{H}$ contains a regular element.*

(b) *$Q_{\max}$ is a semi-primary, right self-injective ring, and $Z_r(A)=N(A)$.*

(c) *Q_{cl} exists and is a semi-primary, right self-injective ring.*

Proof. First of all we note that $Z_r(A)=A \cap Z_r(Q_{\max})$ holds for an arbitrary ring A, as is easily verified. Furthermore, one has $Z_r(Q_{\max})=J(Q_{\max})$ whenever $Q_{\max}$ is right self-injective (Cor. XIV.1.3).

(a) ⇒ (c): In view of Prop. 5.3, it only remains to show that $J(Q_{\mathrm{cl}})$ is nilpotent. Assume $Z_r(A)^n=0$. If $q_1,\ldots,q_n \in J(Q_{\max})$, then one can write $q_i=a_i s_i^{-1}$ with $a_i \in Z_r(Q_{\max}) \cap A = Z_r(A)$ and s_i regular in A. By repeated use of the Ore condition, one obtains $q_1 \cdot \ldots \cdot q_n = b_1 \cdot \ldots \cdot b_n t^{-1}$ with $b_i,\ t \in A$ and t regular. Also $b_i \in Z_r(A)$, because a relation $at=sb$ with s regular implies $r(b)=(r(a):t)$. It follows that also $J(Q_{\max})^n=0$.

(c) ⇒ (b): We obtain

$$Z_r(A)=A \cap Z_r(Q_{\max})=A \cap N(Q_{\max})=N(A),$$

using Cor. 3.2.

(b) ⇒ (c): Q_{max} is a right Kasch ring by Prop. XIV.2.4. Furthermore we have

$$N(A) = Z_r(A) = A \cap Z_r(Q_{max}) = A \cap N(Q_{max}),$$

so Prop. 3.3 can be applied to give $Q_{cl} = Q_{max}$.

(c) ⇒ (a) follows from Prop. 5.3 and the remark that $Z_r(A) \subset Z_r(Q_{cl}) = J(Q_{cl})$. ▯

Recall that a sufficient condition for $Z_r(A)$ to be nilpotent is that A satisfies ACC on right annihilators (Lemma II.2.5).

The following characterization of orders in QF-rings is due to Masaike [2] and Tachikawa [4] (improving results by Jans [2] and Mewborn and Winton [1]):

Theorem 5.6. *The following properties of A are equivalent:*

(a) *A satisfies* ACC *on annihilators of subsets of $E(A)$, and every $\mathfrak{a} \in \mathfrak{H}$ contains a regular element.*

(a′) *A satisfies* ACC *on saturated right ideals, and every $\mathfrak{a} \in \mathfrak{H}$ contains a regular element.*

(b) *Q_{max} is a quasi-Frobenius ring, and $Z_r(A) = N(A)$.*

(c) *Q_{cl} exists and is a quasi-Frobenius ring.*

Proof. (a) ⇔ (a′) follows from Theorem XIV.4.4.

(a) ⇒ (c): We know from Theorem XIV.4.4 that Q_{max} is a QF-ring, and as in Prop. 5.3 it is then enough to prove that every regular element a of A is invertible in Q_{max}. By Lemma II.1.10 it even suffices to show that $r(a) = 0$ holds in Q_{max}. Suppose $aq = 0$ for some non-zero $q \in Q_{max}$. There exists $b \in A$ such that $0 \neq qb \in A$. But then $a(qb) = 0$ contradicts the regularity of a in A.

(c) ⇒ (a) should be clear, and (b) ⇔ (c) is a consequence of the corresponding implications in Prop. 5.5. ▯

§6. Rings of Quotients of Group Algebras

Let K be a field and G an arbitrary group. The group algebra of G over K will be denoted by $K[G]$. If G is a finite group, $K[G]$ is a finite-dimensional algebra, hence artinian and thus a ring of quotients. So the case of interest here is when G is infinite.

The *centralizer* of an element $g \in G$ is defined as $C(g) = \{h \in G \mid hg = gh\}$. $C(g)$ is a subgroup of G, and its index in G is equal to the number of conjugates of g in G, for if k and h are any elements of G, then $hgh^{-1} = kgk^{-1}$ is equivalent to $k^{-1}h \in C(g)$, i.e. k and h are in the same left coset of $C(g)$. Put

$$G^* = \{g \in G \mid C(g) \text{ has finite index in } G\}.$$

G^* is a normal subgroup of G, since every conjugate of hgh^{-1} is also conjugate of g.

Theorem 6.1 (Herstein and Small [1]). *If G^* has finite index in G, then $K[G]$ has a classical two-sided ring of quotients.*

Proof. By symmetry it suffices to verify the right Ore condition for $K[G]$. So let there be given elements a and s in $K[G]$ with s regular. a and s are linear combinations over K of a finite number of elements of G. So a and s are elements of a subring $K[H]$ where H is a finitely generated subgroup of G. Clearly

$H^* \supset G^* \cap H$, and it follows that H^* has finite index in H. Suppose we can show that there exist b and t in $K[H]$ such that t is regular in $K[H]$ and $at = sb$. Now t is regular also in $K[G]$, because $K[G]$ is a free module over $K[H]$. Thus the right Ore condition would be satisfied by a, s in $K[G]$.

From the preceding discussion it follows that we may assume without loss of generality that G is a finitely generated group. Since G^* is of finite index in G, also G^* is a finitely generated group by a classical theorem of Schreier (see e.g. Hall [1], Cor. 7.2.1). Let $g_1, \ldots, g_n$ be generators of G^*. Each $C(g_i)$ has finite index in G, and it is easy to see that this implies that also $\bigcap_1^n C(g_i)$ has finite index in G. But $\bigcap C(g_i)$ is the center of G^*, and is therefore a characteristic subgroup of G^* (i.e. invariant under all automorphisms of G^*) and since G^* is normal in G, it follows that $\bigcap C(g_i)$ is a normal subgroup of G. By this we have reduced the problem to showing:

Proposition 6.2. *If G is a group and contains an abelian normal subgroup N of finite index, then $K[G]$ has a classical two-sided ring of quotients.*

Proof. Let S be the set of regular elements in $K[N]$. S is a multiplicatively closed subset of $K[G]$, and we will first show that $K[G]$ has a right ring of fractions with respect to S. For this we have to verify properties $S1$ and $S2$ of Prop. II.1.4. $S2$ is clear, since we have noted above that every regular element of $K[N]$ remains regular in $K[G]$. It remains to verify $S1$. Let $a \in K[G]$ and $s \in S$. Choose a complete set $g_1, \ldots, g_r$ of coset representatives of N in G. The g_i's constitute a basis for the free $K[N]$-module $K[G]$. Write $a = g_1 c_1 + \cdots + g_r c_r$ with $c_r \in K[N]$. Put $s_i = g_i^{-1} s g_i$ for $i = 1, \ldots, r$. Since N is normal in G, each s_i belongs to $K[N]$, and in fact we have $s_i \in S$. Since $K[N]$ is a commutative ring, we can find $t \in S$ and $b_i \in K[N]$ such that $c_i t = s_i b_i$ for all i. If we put $b = g_1 b_1 + \cdots + g_r b_r$, we obtain

$$at = \sum g_i c_i t = \sum g_i s_i b_i = \sum g_i s_i g_i^{-1} g_i b_i = sb,$$

and this is condition $S1$.

Let $K[N]_S = (K[N])[S^{-1}]$ and $K[G]_S = (K[G])[S^{-1}]$. We assert that $K[G]_S$ is actually the classical right ring of quotients of $K[G]$. To prove this, we only have to show that each regular element u in $K[G]_S$ is invertible in $K[G]_S$. Consider $K[G]_S$ as a finitely generated free module over the commutative ring $K[N]_S$, and let $\varphi_u\colon K[G]_S \to K[G]_S$ be the $K[N]_S$-linear map given by $\varphi_u(x) = ux$. φ_u is a monomorphism. If we can show that φ_u is an epimorphism, then we know that there exists $y \in K[G]_S$ such that $uy = 1$, and since u is regular and $u(1 - yu) = 0$, we will have also $yu = 1$. We may therefore conclude the proof by showing:

Lemma 6.3. *Let Q be a commutative ring of quotients and let F be a finitely generated free right Q-module. Every monomorphic endomorphism of F is an isomorphism.*

Proof. Let $\varphi\colon F \to F$ be a monomorphism, and let $|\varphi|$ be its matrix with respect to some finite basis for F over Q. It is easy to show that the determinant of $|\varphi|$ must be a non-zero-divisor in Q because φ is a monomorphism. But then det $|\varphi|$ is invertible in Q, and φ is an isomorphism (cf. Lang [1], § 13.4). □

Exercises

§1

1. Let A be a semi-prime ring and e an idempotent of A. Show that eAe is a semi-prime ring.

2. Show that $N(M_n(A)) = M_n(N(A))$.

3. Let A be a right noetherian ring and $\mathfrak{q}$ a semi-prime ideal of A. Put $\mathscr{C}(\mathfrak{q}) = \{a \in A \mid \bar{a}$ is regular in $A/\mathfrak{q}\}$. Show that the Gabriel topology $\mathfrak{F}_{\mathscr{C}(\mathfrak{q})} = \{\mathfrak{a} \mid (\mathfrak{a}:a) \cap \mathscr{C}(\mathfrak{q}) \neq \emptyset$ for all $a \in A\}$ (cf. Exercise VI.18) corresponds to the torsion theory cogenerated by the injective A-module $E(A/\mathfrak{q})$. (Hint: note that if $\mathfrak{a}$ belongs to the topology defined from $E(A/\mathfrak{q})$, then $(\mathfrak{a}:a)$ maps to an essential right ideal in $A/\mathfrak{q}$ and therefore intersects $\mathscr{C}(\mathfrak{q})$ modulo $\mathfrak{q}$ by the Goldie theorem.)

§2

4. Let $\mathfrak{a}$ be a two-sided ideal of A, and assume that A is $\mathfrak{a}$-reflective. Show that if $Q_{\mathrm{cl}}(A)$ exists, then A is $\mathfrak{a}$-quorite.

§4

5. Consider the ring $\mathbb{Q}[[X, Y]]$ of formal power series over $\mathbb{Q}$, and put $A = \mathbb{Q}[[X, Y]]/(X^2, XY)$. Show that:

(i) A is its own classical ring of quotients.

(ii) Show that $A/N(A) \cong \mathbb{Q}[[Y]]$, so A is not artinian.

(iii) Verify explicitly that the regularity condition is not satisfied by A.

(iv) Show that both $\bar{X}A$ and $\bar{X}A + \bar{Y}A$ are associated prime ideals of A.

§5

6. Assume that A is an order in a ring Q. Show that the following conditions are equivalent:

(a) Every finitely generated torsion-free right A-module is a submodule of a free A-module.

(b) Every finitely generated right Q-module is a submodule of a free Q-module.

7. Let A be a right order in a QF-ring Q. Show that A is a two-sided order in Q if and only if every finitely generated torsion-free A-module is a submodule of a free A-module. (Hint: for the necessity, cf. Prop. II.4.6, and for the sufficiency, show that Q_A is flat (cf. Theorem XIII.7.1) and use Prop. 3.3.)

§6

8. Let A be a commutative ring and G a group. Show that if $A[G]$ has a classical right ring of quotients, then it has a classical two-sided ring of quotients. (Hint: use the involution of $A[G]$ given by $\sum a_g g \mapsto \sum a_g g^{-1}$).

Bibliography

Akiba, T.: 1. Remarks on generalized rings of quotients. Proc. Japan Acad. **40**, 801–806 (1964).
2. Remarks on generalized rings of quotients, II. J. Math. Kyoto Univ. **5**, 39–44 (1965).
3. Remarks on generalized rings of quotients, III. J. Math. Kyoto Univ. **9**, 205–212 (1969).

Albu, T.: 1. Asupra unor clase de module. Studii Cerc. Mat. **24**, 1329–1392, 1455–1501 (1972).

Albu, T., Năstăsescu, C.: 1. Décompositions primaires dans les catégories de Grothendieck commutatives. J. Reine Angew. Math. (to appear).

Alin, J.S.: 1. Structure of torsion modules. Ph. D. Thesis, Univ. of Nebraska 1967.
2. A primary decomposition theory for torsion modules. Proc. Amer. Math. Soc. **27**, 43–48 (1971).

Alin, J.S., Armendariz, E.P.: 1. TTF-classes over perfect rings. J. Austral. Math. Soc. **11**, 499–503 (1970).

Alin, J.S., Dickson, S.E.: 1. Goldie's torsion theory and its derived functor. Pacific J. Math. **24**, 195–203 (1968).

Almkvist, G.: 1. Fractional categories. Arkiv för Mat. **7**, 449–476 (1968).

Amitsur, S.A.: 1. General theory of radicals, II. Amer. J. Math. **76**, 100–125 (1954).
2. Prime rings having polynomial identities with arbitrary coefficients. Proc. London Math. Soc. **17**, 470–486 (1964).
3. Rings of quotients and Morita contexts. J. Algebra **17**, 273–298 (1971).
4. On rings of quotients. Symposia Math. **8**, 149–164 (1972).

Anderson, F.W.: 1. Lattice ordered rings of quotients. Canad. J. Math. **17**, 434–448 (1965).
2. Endomorphism rings of projective modules. Math. Z. **111**, 322–332 (1969).

Armendariz, E.P.: 1. Quasi-injective modules and stable torsion classes. Pacific J. Math. **31**, 277–280 (1969).
2. On finite-dimensional torsion-free modules and rings. Proc. Amer. Math. Soc. **24**, 566–571 (1970).
3. Q-divisible modules. Canad. Math. Bull. **14**, 491–494 (1971).

Armendariz, E.P., McDonald, G.R.: 1. Maximal quotient rings and S-rings. Canad. J. Math. **24**, 835–850 (1972).
2. Q-divisibility and injectivity. J. Algebra **27**, 1–10 (1973).

Artin, M., Grothendieck, A., Verdier, J.L.: 1. Théorie des topos et cohomologie étale des schémas. Lecture Notes in Math., Vol. 269. Berlin-Heidelberg-New York: Springer 1972.

Asano, K.: 1. Arithmetische Idealtheorie in nichtkommutativen Ringen. Jap. J. Math. **15**, 1–36 (1939).
2. Über die Quotientenbildung von Schiefringen. J. Math. Soc. Japan **1**, 73–78 (1949).

Auslander, M., Goldman, O.: 1. Maximal orders. Trans. Amer. Math. Soc. **97**, 1–24 (1960).
2. The Brauer group of a commutative ring. Trans. Amer. Math. Soc. **97**, 367–409 (1960).

Azumaya, G.: 1. Corrections and supplementations to my paper concerning Krull-Remak-Schmidt's theorem. Nagoya Math. J. **1**, 117–124 (1950).
2. Completely faithful modules and self-injective rings. Nagoya Math. J. **29**, 697–708 (1966).
3. Some properties of TTF-classes. Proceedings of the Conference on orders, group rings and related topics, pp. 72–83. Lecture Notes in Math., Vol. 353. Berlin-Heidelberg-New York: Springer 1973.

Ballet, B.: 1. Topologies linéaires et modules artiniens. Thèse, Univ. Paris-Sud (Orsay) 1971).

Bass, H.: 1. Finitistic dimension and a homological generalization of semi-primary rings. Trans. Amer. Math. Soc. **95**, 466–488 (1960).
2. Algebraic K-theory. New York: Benjamin 1968.

Beachy, J.A.: 1. Bicommutators of cofaithful, fully divisible modules. Canad. J. Math. **23**, 202–213 (1971).
2. Generating and cogenerating structures. Trans. Amer. Math. Soc. **158**, 75–92 (1971).
3. Cotorsion radicals and projective modules. Bull. Austr. Math. Soc. **5**, 241–253 (1971).

4. A characterization of torsion-free modules over rings of quotients. Proc. Amer. Math. Soc. **34**, 15–19 (1972).
5. A generalization of injectivity. Pacific J. Math. **41**, 313–327 (1972).
6. On maximal torsion radicals. Canad. J. Math. **25**, 712–726 (1973).

Beck, I.: 1. Injective modules over a Krull domain. J. Algebra **17**, 116–131 (1971).
2. Σ-injective modules. J. Algebra **21**, 232–249 (1972).

Behrens, E.A.: 1. Ring theory. New York: Academic Press 1972.

Bernhardt, R.L.: 1. Splitting torsion theories over semi-perfect rings. Proc. Amer. Math. Soc. **22**, 681–687 (1969).
2. On splitting in hereditary torsion theories. Pacific J. Math. **39**, 31–38 (1971).

Bican, L.: 1. QF-$3'$ modules and rings. Comm. Math. Univ. Carol. **14**, 295–303 (1973).

Bican, L., Jambor, P., Kepka, T., Němec, P.: 1. On rings with trivial torsion parts. Bull. Austr. Math. Soc. **9**, 275–290 (1973).
2. Preradicals. Comm. Math. Univ. Carol. **15**, 75–83 (1974).

Birkhoff, G.: 1. Lattice theory. AMS Colloquium Publ., Vol. 25, 3rd ed. Providence: American Math. Soc. 1967.

Björk, J.E.: 1. Rings satisfying a minimum condition on principal ideals. J. Reine Angew. Math. **236**, 112–119 (1969).

Bland, P.E.: 1. Perfect torsion theories. Proc. Amer. Math. Soc. **41**, 349–355 (1973).

Bourbaki, N.: 1. Elements of mathematics: Algebra. Chap. 1–3. Paris-Reading: Hermann and Addison-Wesley 1973.
2. Elements of mathematics: Commutative algebra. Paris-Reading: Hermann and Addison-Wesley 1972.

Brainerd, B., Lambek, J.: 1. On the ring of quotients of a boolean ring. Canad. Math. Bull. **2**, 25–29 (1959).

Bray, E.E., Byrd, K.A., Bernhardt, R.L.: 1. The injective envelope of the upper triangular matrix ring. Amer. Math. Monthly **78**, 883–886 (1971).

Brown, S.H.: 1. On rationally complete modules. Ph. D. Thesis, North Carolina State Univ. 1969.
2. Rings over which every simple module is rationally complete. Canad. J. Math. **25**, 693–701 (1973).

Bucur, I., Deleanu, A.: 1. Introduction to the theory of categories and functors. New York: J. Wiley 1968.

Budach, L.: 1. Quotientenfunktoren und Erweiterungstheorie. Berlin: Deutscher Verlag d. Wiss. 1967.

Burgess, W.D.: 1. Rings of quotients of group rings. Canad. J. Math. **21**, 865–875 (1969).

Butts, H.S., Spaht, C.G.: 1. Generalized quotient rings. Math. Nachr. **53**, 181–210 (1972).

Cahen, P.-J.: 1. Torsion theory and associated primes. Proc. Amer. Math. Soc. **38**, 471–476 (1973).
2. Commutative torsion theory. Trans. Amer. Math. Soc. **184**, 73–85 (1973).

Cailleau, A., Renault, G.: 1. Sur l'enveloppe injective des anneaux semi-premiers à l'idéal singulier nul. J. Algebra **15**, 133–141 (1970).

Cartan, H., Eilenberg, S.: 1. Homological algebra. Princeton: Princeton Univ. Press 1956.

Cateforis, V.C.: 1. Flat regular quotient rings. Trans. Amer. Math. Soc. **138**, 241–250 (1969).
2. On regular self-injective rings. Pacific J. Math. **30**, 39–45 (1969).
3. Two-sided semisimple maximal quotient rings. Trans. Amer. Math. Soc. **149**, 339–349 (1970).

Cateforis, V.C., Sandomierski, F.L.: 1. The singular submodule splits off. J. Algebra **10**, 149–165 (1968).
2. On modules of singular submodule zero. Canad. J. Math. **23**, 345–354 (1971).

Cauchon, G.: 1. Les T-anneaux et la condition de Gabriel. C. R. Acad. Sci. Paris **277**, 1153–1156 (1973).

Chamard, J.-Y.: 1. Anneaux semi-parfaits et presque-frobeniusiens. C. R. Acad. Sci. Paris **269**, 556–559 (1969).

Chase, S.U.: 1. Direct products of modules. Trans. Amer. Math. Soc. **97**, 457–473 (1960).

Chase, S.U., Faith, C.: 1. Quotient rings and direct products of full linear rings. Math. Z. **88**, 250–264 (1965).

Cheatham, T.J.: 1. Finite-dimensional torsion-free rings. Pacific J. Math. **39**, 113–118 (1971).

Chew, K.L.: 1. Closure operations in the study of rings of quotients. Bull. Math. Soc. Nanyang Univ. 1–20 (1965).

Claborn, L., Fossum, R.: 1. Class groups of n-noetherian rings. J. Algebra **10**, 263–285 (1968).

Cohn, P.M.: 1. On the free product of associative rings. Math. Z. **71**, 380–398 (1959).
2. Non-commutative unique factorization domains. Trans. Amer. Math. Soc. **109**, 313–331 (1963).
3. Universal algebra. New York: Harper and Row 1965.
4. Rings of fractions. Amer. Math. Monthly **78**, 596–615 (1971).

5. Inversive localization in noetherian rings. Comm. Pure Appl. Math. **26**, 679–691 (1973).

Colby, R.R., Rutter, E.A.: 1. Semi-primary *QF*-3 rings. Nagoya Math. J. **32**, 253–258 (1968).

2. *QF*-3 rings with zero singular ideal. Pacific J. Math. **28**, 303–308 (1969).

Courter, R.: 1. Finite direct sums of complete matrix rings over perfect completely primary rings. Canad. J. Math. **21**, 430–446 (1969).

Cunningham, R.S.: 1. On finite left localizations. Pacific J. Math. **51**, 407–415 (1974).

Cunningham, R.S., Rutter, E.A., Turnidge, D.R.: 1. Rings of quotients of endomorphism rings of projective modules. Pacific J. Math. **41**, 647–668 (1972).

Desq, R.: 1. Sur les anneaux de quotients. C. R. Acad. Sci. Paris **272**, 199–202 (1971).

Dickson, S.E.: 1. Decomposition of modules I (Classical rings). Math. Z. **90**, 9–13 (1965).

2. On torsion classes of abelian groups. J. Math. Soc. Japan **17**, 30–35 (1965).

3. A torsion theory for abelian categories. Trans. Amer. Math. Soc. **121**, 223–235 (1966).

4. Direct decomposition of radicals. Proceedings of the Conference on categorical algebra, La Jolla 1965. New York: Springer 1966.

5. Decomposition of modules II (Rings without chain conditions). Math. Z. **104**, 349–357 (1968).

Djabali, M.: 1. Anneau de fractions d'un *J*-anneau. Canad. J. Math. **20**, 182–202 (1968).

2. Anneaux de fractions "presque classiques" artiniens. J. Algebra **23**, 116–128 (1972).

Dlab, V.: 1. The concept of torsion module. Amer. Math. Monthly **75**, 973–976 (1968).

2. Distinguished submodules. J. Austral. Math. Soc. **8**, 661–670 (1968).

3. Distinguished sets of ideals of a ring. Czech. Math. J. **18 (93)**, 560–567 (1968).

4. A characterization of perfect rings. Pacific J. Math. **33**, 79–88 (1970).

5. On a class of perfect rings. Canad. J. Math. **22**, 822–826 (1970).

Eckmann, B., Schopf, A.: 1. Über injektive Moduln. Archiv d. Math. **4**, 75–78 (1953).

Eilenberg, S.: 1. Abstract description of some basic functors. J. Indian Math. Soc. **24**, 221–234 (1960).

Eilenberg, S., Nakayama, T.: 1. On the dimension of modules and algebras II. Nagoya Math. J. **9**, 1–16 (1955).

Elizarov, V.P.: 1. On quotient rings of associative rings. Izv. Akad. Nauk SSSR, Ser. Matem. **24**, 153–170 (1960) [Russian]. (Amer. Math. Soc. Transl. **52**, 1966).

2. Flat extensions of rings. Dokl. Akad. Nauk SSSR **175**, 759–761 (1967) [Russian].

3. Rings of quotients. Algebra i Logika **8**, 381–424 (1969) [Russian].

4. On the general theory of quotient rings. Sibirsk. Mat. Z. **11**, 526–546 (1970) [Russian].

5. Singular submodule and singular ideal. Sibirsk. Mat. Z. **12**, 465–468 (1971) [Russian].

Endler, O.: 1. Modules and rings of fractions. Summa Brasil. Math. **4**, 149–182 (1959).

Endo, S.: 1. On semi-hereditary rings. J. Math. Soc. Japan **13**, 109–119 (1961).

2. On flat modules over commutative rings. J. Math. Soc. Japan **14**, 284–291 (1962).

Evans, M.W.: 1. On commutative P. P. rings. Pacific J. Math. **41**, 687–697 (1972).

Faith, C.: 1. Orders in simple artinian rings. Trans. Amer. Math. Soc. **114**, 61–64 (1965).

2. Rings with ascending condition on annihilators. Nagoya Math. J. **27**, 179–191 (1966).

3. Lectures on injective modules and quotient rings. Lecture Notes in Math., Vol. 49. Berlin-Heidelberg-New York: Springer 1967.

4. Orders in semilocal rings. Bull. Amer. Math. Soc. **77**, 960–962 (1971).

5. Modules finite over endomorphism ring. Lectures on rings and modules, pp. 145–189. Lecture Notes in Math., Vol. 246. Berlin-Heidelberg-New York: Springer 1972.

6. Algebra: rings, modules and categories. Vol. I. Grundlehren Math. Wiss., Vol. 190. Berlin-Heidelberg-New York: Springer 1973.

Faith, C., Utumi, Y.: 1. On noetherian prime rings. Trans. Amer. Math. Soc. **114**, 53–60 (1965).

Faith, C., Walker, E.A.: 1. Direct-sum representations of injective modules. J. Algebra **5**, 203–221 (1967).

Farkas, D.: 1. Self-injective group algebras. J. Algebra **25**, 313–315 (1973).

Feller, E.H., Swokowski, E.W.: 1. Reflective *N*-prime rings with the ascending chain condition. Trans. Amer. Math. Soc. **99**, 264–271 (1961).

2. Correction to [1]. Trans. Amer. Math. Soc. **101**, 555 (1961).

3. Reflective rings with the ascending chain condition. Proc. Amer. Math. Soc. **12**, 651–653 (1961).

Findlay, G.D.: 1. Multiplicative ideal theory and rings of quotients. Proc. Roy. Soc. Edinburgh **A 68**, 30–53 (1968).

2. Flat epimorphic extensions of rings. Math. Z. **118**, 281–288 (1970).

3. A note on non-commutative localisation. J. London Math. Soc. **6**, 39–42 (1972).

Findlay, G.D., Lambek, J.: 1. A generalized ring of quotients. Canad. Math. Bull. **1**, 77–85, 155–167 (1958).

Fine, N.J., Gillman, L., Lambek, J.: 1. Rings of quotients of rings of functions. Montreal: McGill Univ. Press 1965.
Fischer, J.W.: 1. Structure of semi-prime P. I. rings. Proc. Amer. Math. Soc. **39**, 465–467 (1973).
Fuchs, L.: 1. Torsion preradicals and ascending Loewy series of modules. J. Reine Angew. Math. **239/240**, 168–179 (1969).
Fuller, K.R.: 1. Double centralizers of injectives and projectives over artinian rings. Illinois J. Math. **14**, 658–664 (1970).
Gabriel, P.: 1. La localisation dans les anneaux non commutatifs. Séminaire Dubreil-Pisot 1959/60.
2. Des catégories abéliennes. Bull. Soc. Math. France **90**, 323–448 (1962).
Gabriel, P., Oberst, U.: 1. Spektralkategorien und reguläre Ringe im Von-Neumannschen Sinn. Math. Z. **92**, 389–395 (1966).
Gabriel, P., Zisman, M.: 1. Calculus of fractions and homotopy theory. Ergebnisse der Math., Vol. 35. Berlin-Heidelberg-New York: Springer 1967.
Gentile, E.R.: 1. On rings with one-sided fields of quotients. Proc. Amer. Math. Soc. **11**, 380–384 (1960).
2. Singular submodule and injective hull. Indag. Math. **24**, 426–433 (1962).
Giraud, J.: 1. Analysis situs. Séminaire Bourbaki, Vol. 15, 1962–63, Exposé 256.
Goblot, R.: 1. Catégories modulaires. C. R. Acad. Sci. Paris **267**, 380–383 (1968).
2. Sur deux classes de catégories de Grothendieck. Thèse, Univ. de Lille 1971.
Godement, R.: 1. Théorie des faisceaux. Paris: Hermann 1958.
Golan, J., Teply, M.: 1. Finiteness conditions on filters of left ideals. J. Pure Appl. Alg. **3**, 251–260 (1973).
2. Torsion-free covers. Israel J. Math. **15**, 237–256 (1973).
Goldie, A.W.: 1. The structure of prime rings under ascending chain conditions. Proc. London Math. Soc. **8**, 589–608 (1958).
2. Semi-prime rings with maximum conditions. Proc. London Math. Soc. **10**, 201–220 (1960).
3. Non-commutative principal ideal rings. Arch. d. Math. **13**, 213–221 (1962).
4. Torsion-free modules and rings. J. Algebra **1**, 268–287 (1964).
5. Localization in non-commutative noetherian rings. J. Algebra **5**, 89–105 (1967).
6. A note on non-commutative localization. J. Algebra **8**, 41–44 (1968).
7. Some aspects of ring theory. Bull. London Math. Soc. **1**, 129–154 (1969).
8. The structure of noetherian rings. Lectures on rings and modules, pp. 213–321. Lecture Notes in Math., Vol. 246. Berlin-Heidelberg-New York: Springer 1972.
Goldman, O.: 1. Rings and modules of quotients. J. Algebra **13**, 10–47 (1969).
Goodearl, K.R.: 1. Embedding non-singular modules in free modules. J. Pure Appl. Algebra **1**, 275–279 (1971).
2. Singular torsion and the splitting properties. Mem. Amer. Math. Soc. **124** (1972).
3. Prime ideals in regular self-injective rings. Canad. J. Math. **25**, 829–839 (1973).
4. Triangular representations of splitting rings. Trans. Amer. Math. Soc. **185**, 271–285 (1973).
Gordon, R., 1. Semi-prime right Goldie rings which are direct sums of uniform right ideals. Bull. London Math. Soc. **3**, 277–282 (1971).
2. Classical quotient rings of PWD's. Proc. Amer. Math. Soc. **36**, 39–46 (1972).
Gordon, R., Robson, J.C.: 1. Krull dimension. Mem. Amer. Math. Soc. **133** (1973).
Gordon, R., Small, L.W.: 1. Piecewise domains. J. Algebra **23**, 553–564 (1972).
Gray, M.: 1. A radical approach to algebra. Reading: Addison-Wesley Publ. Co. 1970.
Grothendieck, A.: 1. Sur quelques points d'algèbre homologique. Tôhoku Math. J. **9**, 119–221 (1957).
2. Local cohomology. (Notes by R. Hartshorne.) Lecture Notes in Math., Vol. 41. Berlin-Heidelberg-New York: Springer 1967.
Grothendieck, A., Dieudonné, J.: 1. Eléments de géométrie algébrique. Partie IV: 1. Publ. Math. IHES **20** (1964).
Gupta, R.N.: 1. Self-injective quotient rings and injective quotient modules. Osaka J. Math. **5**, 69–87 (1968).
2. Characterization of rings whose classical quotient rings are perfect rings. Publ. Math. Debrecen **17**, 215–221 (1970).
Hacque, M.: 1. Mono-sous-catégories d'une catégorie de modules. Publ. Dép. Math. Lyon **6**, 13–48 (1969).
2. Localisations exactes et localisations plates. Publ. Dép. Math. Lyon **6**, 97–117 (1969).
3. Localisations commutatives stables. C. R. Acad. Sci. Paris **270**, 995–998 (1970).
4. Sur l'existence de l'épimorphisme injectif plat à gauche maximum. C. R. Acad. Sci. Paris **271**, 1218–1221 (1970).

Hajarnavis, C.R.: 1. Orders in QF and $QF2$ rings. J. Algebra **19**, 329–343 (1971).
2. On Small's theorem. J. London Math. Soc. **5**, 596–600 (1972).
Hall, M.: 1. The theory of groups. New York: MacMillan 1959.
Hart, R.: 1. Endomorphisms of modules over semi-prime rings. J. Algebra **4**, 46–51 (1966).
Harui, H.: 1. Modules over (qa)-rings. J. Sci. Hiroshima Univ., Ser. A-I **32**, 247–257 (1968).
Hattori, A.: 1. A foundation of torsion theory for modules over general rings. Nagoya Math. J. **17**, 147–158 (1960).
Heinicke, A.G.: 1. Triples and localizations. Canad. Math. Bull. **14**, 333–339 (1971).
Helzer, G.: 1. On divisibility and injectivity. Canad. J. Math. **18**, 900–919 (1966).
Herstein, I.N.: 1. Sul teorema di Goldie. Rend. Accad. Lincei **35**, 23–26 (1963).
2. Topics in algebra. Waltham: Blaisdell 1964.
3. Topics in ring theory. Chigaco: Univ. of Chicago Press (1969).
Herstein, I.N. and Small, L.W.: 1. Rings of quotients of group algebras. J. Algebra **19**, 153–155 (1971).
Hirata, K.: 1. Some types of separable extensions of rings. Nagoya Math. J. **33**, 107–115 (1968).
Hopkins, C.: 1. Rings with minimal conditions for left ideals. Ann. of Math. **40**, 712–730 (1939).
Hudry, A.: 1. Sur les anneux localement homogènes. C.R. Acad. Sci. Paris **271**, 1214–1217 (1970).
2. Sur les modules primaires de O. Goldman. C.R. Acad. Sci. Paris **274**, 1772–1774 (1972).
Ikeda, M. and Nakayama, T.: 1. On some characteristic properties of quasi-Frobenius and regular rings. Proc. Amer. Math. Soc. **5**, 15–18 (1954).
Jacobson, N.: 1. The theory of rings. AMS Surveys, Vol. 2. Providence: American Math. Soc. 1943.
2. Structure of rings. AMS Colloquium Publ., Vol. 37, 2nd ed. Providence: American Math. Soc. 1964.
Jans, J.P.: 1. Some aspects of torsion. Pacific J. Math. **15**, 1249–1259 (1965).
2. On orders in quasi-Frobenius rings. J. Algebra **7**, 35–43 (1967).
3. Torsion associated with duality. Tôhoku Math. J. **24**, 449–452 (1972).
Jategaonkar, A.V.: 1. Left principal ideal domains. J. Algebra **8**, 148–155 (1968).
2. Ore domains and free algebras. Bull. London Math. Soc. **1**, 45–46 (1969).
3. Left principal ideal rings. Lecture Notes in Math., Vol. 123. Berlin-Heidelberg-New York: Springer 1970.
4. Structure and classification of hereditary noetherian prime rings. In: Ring theory (Ed. R. Gordon), pp. 171–229. New York: Academic Press 1972.
Jensen, C.U.: 1. Arithmetical rings. Acta Math. Acad. Sci. Hung. **17**, 115–123 (1966).
Johnson, R.E.: 1. The extended centralizer of a ring over a module. Proc. Amer. Math. Soc. **2**, 891–895 (1951).
2. Prime rings. Duke Math. J. **18**, 799–809 (1951).
3. Structure theory of faithful rings I, II. Trans. Amer. Math. Soc. **84**, 508–522, 523–544 (1957).
4. Structure theory of faithful rings III. Proc. Amer. Math. Soc. **11**, 710–717 (1960).
5. Quotient rings of rings with zero singular ideal. Pacific J. Math. **11**, 1385–1392 (1961).
6. Rings with zero right and left singular ideals. Trans. Amer. Math. Soc. **118**, 150–157 (1965).
7. Remarks on a paper of Procesi. J. Algebra **2**, 38–41 (1965).
Johnson, R.E., Wong, E.T.: 1. Quasi-injective modules and irreducible rings. J. London Math. Soc. **36**, 260–268 (1961).
Jøndrup, S.: 1. On finitely generated flat modules II. Math. Scand. **27**, 105–112 (1970).
Kaplansky, I.: 1. Rings of operators. New York: W.A. Benjamin 1968.
Kasch, F., Mares, E.A.: 1. Eine Kennzeichnung semi-perfekter Moduln. Nagoya Math. J. **27**, 525–529 (1966).
Kašu, A.I.: 1. Closed classes of left Λ-modules and closed sets of left ideals of the ring Λ. Mat. Zametki **5**, 381–390 (1969) [Russian].
2. Radical closures in the category of modules. Mat. Issled. **5**, 91–104 (1970) [Russian].
Kato, T.: 1. Self-injective rings. Tôhoku Math. J. **19**, 485–495 (1967).
2. Torsionless modules. Tôhoku Math. J. **20**, 234–243 (1968).
3. Rings of U-dominant dimension $\geqq 1$. Tôhoku Math. J. **21**, 321–327 (1969).
4. Dominant modules. J. Algebra **14**, 341–349 (1970).
Kepka, T.: 1. Notes on radical filters of ideals. Comm. Math. Univ. Carol. **15**, 149–159 (1974).
Kiełpiński, R.: 1. Direct sums of torsions. Bull. Acad. Polon. Sci. **19**, 281–286 (1971).
Knight, J.T.: 1. On epimorphisms of non-commutative rings. Proc. Cambr. Phil. Soc. **68**, 589–601 (1970).
Koifman, L.A.: 1. Cogenerators of self-injective rings. Algebra i Logika **9**, 52–66 (1970) [Russian].

Krause, G.: 1. On fully left bounded left noetherian rings. J. Algebra **23**, 88–99 (1972).

Kurata, Y.: 1. On an n-fold torsion theory in the category ${}_R M$. J. Algebra **22**, 559–572 (1972).

Kuroš, A.G.: 1. Radicals in rings and algebras. Mat. Sb. **33**, 13–26 (1953) [Russian].

Lambek, J.: 1. On the structure of semi-prime rings and their rings of quotients. Canad. J. Math. **13**, 392–417 (1961).

2. On Utumi's ring of quotients. Canad. J. Math. **15**, 363–370 (1963).

3. On the ring of quotients of a noetherian ring. Canad. Math. Bull. **8**, 279–290 (1965).

4. Lectures on rings and modules. Waltham: Blaisdell 1966.

5. Torsion theories, additive semantics and rings of quotients. Lecture Notes in Math., Vol. 177. Berlin-Heidelberg-New York: Springer 1971.

6. Bicommutators of nice injectives. J. Algebra **21**, 60–73 (1972).

7. Localization and completion. J. Pure Appl. Algebra **2**, 343–370 (1972).

8. Noncommutative localization. Bull. Amer. Math. Soc. **79**, 857–872 (1973).

Lambek, J., Michler, G.: 1. The torsion theory at a prime ideal of a right noetherian ring. J. Algebra **25**, 364–389 (1973).

2. Completions and classical localizations of right noetherian rings. Pacific J. Math. **48**, 133–140 (1973).

3. Localization of right noetherian rings at semi-prime ideals. To appear.

Lambek, J., Rattray, B.A.: 1. Localization at injectives in complete categories. To appear.

2. Localization and sheaf reflectors. To appear.

Lang, S.: 1. Algebra. Reading: Addison-Wesley 1965.

Lazard, D.: 1. Autour de la platitude. Bull. Soc. Math. France **97**, 81–128 (1969).

Lesieur, L.: 1. Anneaux noethériens à gauche complètement primaires et anneaux co-irréductibles. J. Reine Angew. Math. **241**, 106–117 (1970).

2. Sur les T-anneaux. C.R. Acad. Sci. Paris **276**, 435–438 (1973).

Lesieur, L., Croisot, R.: 1. Sur la décomposition en idéaux primaires dans un anneau non necessairement commutatif. C.R. Acad. Sci. Paris **243**, 1988–1991 (1956).

2. Sur les anneaux premiers noethériens à gauche. Ann. Sci. Ecole Norm. Sup **76**, 161–183 (1959).

3. Extension au cas non-commutatif d'un théorème de Krull et d'un lemme d'Artin-Rees. J. Reine Angew. Math. **204**, 216–220 (1960).

4. Algèbre noethérienne non-commutative. Mém. Sci. Math. **154** (1963).

5. Cœur d'un module. J. Math. Pures Appl. **42**, 367–407 (1963).

Levy, L.: 1. Torsion-free and divisible modules over non-integral domains. Canad. J. Math. **15**, 132–151 (1963).

Luedeman, J.K.: 1. Σ-quasi-injective modules and Utumi's Σ-quotient ring. Bull. Soc. Math. Belgique **22**, 368–375 (1970).

2. A generalization of the concept of a ring of quotients. Canad. Math. Bull. **14**, 517–529 (1971).

MacLane, S.: 1. Categories for the working mathematician. Graduate Texts in Math., Vol. 5. New York-Heidelberg-Berlin: Springer 1971.

Maddox, B.H.: 1. Absolutely pure modules. Proc. Amer. Math. Soc. **18**, 155–158 (1967).

Maeda, F.: 1. Kontinuierliche Geometrien. Grundlehren der Math. Wiss., Vol. 95, Berlin-Heidelberg-New York: Springer 1958.

Maltsev, A.I.: 1. On the immersion of an algebraic ring into a field. Math. Ann. **113**, 686–691 (1936/37).

2. On the embedding of group algebras in algebras with division. Dokl. Akad. Nauk SSSR **60**, 1499–1501 (1948) [Russian].

Manocha, J.N.: 1. Finiteness conditions and torsion theories. To appear.

Maranda, J.-M.: 1. Injective structures. Trans. Amer. Math. Soc. **110**, 98–135 (1964).

Markov, V.T.: 1. Two remarks on the localization of Goldie. Sibirsk Mat. Z. **13**, 604–611 (1972) [Russian].

Masaike, K.: 1. Endomorphism rings of modules over orders in artinian rings. Proc. Japan Acad. **46**, 89–93 (1970).

2. Quasi-Frobenius maximal quotient rings. Sci. Rep. Tokyo Kyoiku D., Sec. A, **11**, 1–5 (1971).

3. On quotient rings and torsionless modules. Sci. Rep. Tokyo Kyoiku, D., Sec. A, **11**, 26–30 (1971).

Matlis, E.: 1. Injective modules over noetherian rings. Pacific J. Math. **8**, 511–528 (1958).

Maury, G.: 1. Localisations bilatères dans les ordres maximaux. C.R. Acad. Sci. Paris **270**, 1144–1146 (1970).

Megibben, C.: 1. Absolutely pure modules. Proc. Amer. Math. Soc. **26**, 561–566 (1970).

Mewborn, A.C.: 1. Some conditions on commutative semi-prime rings. J. Algebra **13**, 422–431 (1969).
2. Regular rings and Baer rings. Math. Z. **121**, 211–219 (1971).
Mewborn, A.C., Winton, C.N.: 1. Orders in self-injective semi-perfect rings. J. Algebra **13**, 5–9 (1969).
Mishina, A.P., Skornyakov, L.A.: 1. Abelian groups and modules. Moscow: Nauka 1969 [Russian].
Mitchell, B.: 1. Theory of categories. New York: Academic Press 1965.
Miyashita, Y.: 1. On quasi-injective modules. J. Fac. Sci. Hokkaido Univ., Ser. I, **18**, 158–187 (1965).
Morita, K.: 1. Duality for modules and its applications to the theory of rings with minimum condition. Sci. Reports Tokyo Kyoiku D. **6**, 83–142 (1958).
2. On S-rings in the sense of F. Kasch. Nagoya Math. J. **27**, 687–695 (1966).
3. Localization in categories of modules, I. Math. Z. **114**, 121–144 (1970).
4. Localization in categories of modules, II. J. Reine Angew. Math. **242**, 163–169 (1970).
5. Noetherian QF-3 rings and two-sided quasi-Frobenius maximal quotient rings. Proc. Japan Acad. **46**, 837–840 (1970).
6. Localization in categories of modules, III. Math. Z. **119**, 313–320 (1971).
7. Flat modules, injective modules and quotient rings. Math. Z. **120**, 25–40 (1971).
8. Quotient rings. In: Ring theory (Ed. R. Gordon), pp. 257–285. New York: Academic Press 1972.
Müller, B.J.: 1. The quotient category of a Morita context. J. Algebra **28**, 389–407 (1974).
Năstăsescu, C.: 1. Décomposition primaire dans les anneaux semi-artiniens. J. Algebra **14**, 170–181 (1970).
2. La structure des modules par rapport à une topologie additive. Tôhoku Math. J. **26**, 173–201 (1974).
Năstăsescu, C., Popescu, N.: 1. Quelques observations sur les topos abéliens. Rev. Roum. Math. Pures Appl. **12**, 553–563 (1967).
2. Anneaux semi-artiniens. Bull. Soc. Math. France **96**, 357–368 (1968).
3. On the localization ring of a ring. J. Algebra **15**, 41–56 (1970).
Nill, J.: 1. Torsionstheorien und Quotientenringe von rechtsnoetherschen Ringen. Mitteil. Math. Sem. Gießen **104** (1973).
Oberst, U.: 1. Duality theory for Grothendieck categories and linearly compact rings. J. Algebra **15**, 473–542 (1970).
Ore, Ø.: 1. Linear equations in non-commutative fields. Ann. of Math. **32**, 463–477 (1931).
2. Theory of non-commutative polynomials. Ann. of Math. **34**, 480–508 (1933).
Osofsky, B.L.: 1. On ring properties of injective hulls. Canad. Math. Bull. **7**, 405–413 (1964).
2. A generalization of quasi-Frobenius rings. J. Algebra **4**, 373–387 (1966).
3. A non-trivial ring with non-rational injective hull. Canad. Math. Bull. **10**, 275–282 (1967).
4. Endomorphism rings of quasi-injective modules. Canad. J. Math. **20**, 895–903 (1968).
Page, S.: 1. Properties of quotient rings. Canad. J. Math. **24**, 1122–1128 (1972).
Papp, Z.: 1. On algebraically closed modules. Publ. Math. Debrecen **6**, 311–327 (1959).
2. S-modules and torsion theories over artinian rings. Arch. d. Math. **23**, 598–602 (1972).
Pareigis, B.: 1. Categories and functors. New York: Academic Press 1970.
Pierce, R.S.: 1. Modules over commutative regular rings. Mem. Amer. Math. Soc. **70** (1967).
Pontryagin, L.: 1. Topological groups. Princeton: Princeton Univ. Press 1939.
Popescu, N.: 1. La localisation pour des sites. Revue Roum. Math. Pures Appl. **10**, 1031–1044 (1965).
2. Elemente de teoria fasciculelor, I–VI. Stud. Cerc. Math. **18**, 267–296, 407–456, 547–583, 645–669, 945–991 (1966); **19**, 205–240 (1967).
3. Le spectre à gauche d'un anneau. J. Algebra **18**, 213–228 (1971).
4. Categorii abeliene. Bucuresti: Ed. Acad. Rep. Soc. România 1971.
5. Sur les C.P.-anneaux. C.R. Acad. Sci. Paris **272**, 1493–1496 (1971).
6. Abelian categories with applications to rings and modules. London-New York: Academic Press 1973.
Popescu, N., Gabriel, P.: 1. Caractérisation des catégories abéliennes avec générateurs et limites inductives exactes. C.R. Acad. Sci. Paris **258**, 4188–4190 (1964).
Popescu, N., Spircu, T.: 1. Quelques observations sur les épimorphismes plats (à gauche) d'anneaux. J. Algebra **16**, 40–59 (1970).
Popescu, N., Spulber, D.: 1. Sur les quasi-ordres (à gauche) dans un anneau. J. Algebra **17**, 474–481 (1971).
Posner, E.C.: 1. Prime rings satisfying a polynomial identity. Proc. Amer. Math. Soc. **11**, 180–184 (1960).

Procesi, C.: 1. Sopra un teorema die Goldie riguardante la struttura degli anelli primi con condizione die massimo. Rend. Accad. Lincei **34**, 372–377 (1963).

2. Su un teorema di Faith ed Utumi. Rend. Math. e Appl. **24**, 346–347 (1965).

Procesi, C., Small, L.: 1. On a theorem of Goldie. J. Algebra **2**, 80–84 (1965).

Ramamurthi, V.S.: 1. On splitting cotorsion radicals. Proc. Amer. Math. Soc. **39**, 457–461 (1973).

Raphael, R.M.: 1. Rings of quotients and π-regularity. Pacific J. Math. **39**, 229–233 (1971).

Raynaud, J.: 1. Epimorphismes plats d'anneaux et propriétés de transfert. C.R. Acad. Sci. Paris **270**, 1714–1717 (1970).

2. Sur les épimorphismes plats d'anneaux. C.R. Acad. Sci. Paris **271**, 65–68 (1970).

3. Sur la théorie de la localisation. Thèse, Univ. de Lyon 1971.

4. Localisations et anneaux semi-noethériens à droite. Publ. Dép. Math. Lyon **8**, 77–112 (1971).

5. Localisations stables à droite et anneaux semi-noethériens à droite. C.R. Acad. Sci. Paris **275**, 13–16 (1972).

6. Quelques résultats sur les localisations. C.R. Acad. Sci. Paris **277**, 299–302 (1973).

Renault, G.: 1. Anneaux réduits noncommutatifs. J. Math. Pures Appl. **46**, 203–214 (1967).

2. Anneau associé à un module injectif. Bull. Sci. Math. **92**, 53–58 (1968).

3. Anneaux réguliers auto-injectifs à droite. Bull. Soc. Math. France **101**, 237–254 (1973).

Rentschler, R.: 1. Sur les modules M tels que $\operatorname{Hom}(M, -)$ commute avec les sommes directes. C.R. Acad. Sci. Paris **268**, 930–933 (1969).

Ribenboim, P.: 1. Rings and modules. New York: J. Wiley-Interscience 1969.

2. Extensions épi-plates de fractions. Symposia Math. **8**, 233–243 (1972).

Richman, F.: 1. Generalized quotient rings. Proc. Amer. Math. Soc. **16**, 794–799 (1965).

Riley, J.A.: 1. Axiomatic primary and tertiary decomposition theory. Trans. Amer. Math. Soc. **105**, 177–201 (1962).

2. Ideal theory in noetherian rings. Ph. D. Thesis, Brandeis Univ. 1964.

Rinehart, G.S.: 1. Note on the global dimension of a certain ring. Proc. Amer. Math. Soc. **13**, 341–346 (1962).

Rjabuhin, J.M.: 1. Radicals in categories. Bul. Akad. Stiince RSS Moldoven. **6**, 58–74 (1964) [Russian].

Robson, J.C.: 1. Artinian quotient rings. Proc. London Math. Soc. **17**, 600–616 (1967).

2. Non-commutative Dedekind rings. J. Algebra **9**, 249–265 (1968).

3. Idealizers and hereditary noetherian prime rings. J. Algebra **22**, 45–81 (1972).

Roos, J.-E.: 1. Caractérisation des catégories qui sont quotients de catégories de modules par des sous-catégories bilocalisantes. C.R. Acad. Sci. Paris **261**, 4954–4957 (1965).

2. Sur les catégories spectrales localement distributives. C.R. Acad. Sci. Paris **265**, 14–17 (1967).

3. Locally distributive spectral categories and strongly regular rings. Reports Midwest Category Seminar, pp. 156–181. Lecture Notes in Mathematics, Vol. 47. Berlin-Heidelberg-New York: Springer 1967.

4. Sur l'anneau maximal de fractions des AW^*-algèbres et des anneaux de Baer. C. R. Acad. Sci. Paris **266**, 449–452 (1968).

5. Locally noetherian categories and generalized strictly linearly compact rings. Proc. Battelle Conference II, pp. 197–277. Lecture Notes in Math., Vol. 92. Berlin-Heidelberg-New York: Springer 1969.

Samuel, Séminaire P.: 1. Les épimorphismes d'anneaux. Paris 1968.

Sanderson, D.F.: 1. A generalization of divisibility and injectivity in modules. Canad. Math. Bull. **8**, 505–513 (1965).

Sandomierski, F.L.: 1. Semisimple maximal quotient rings. Trans. Amer. Math. Soc. **128**, 112–120 (1967).

2. Non-singular rings. Proc. Amer. Math. Soc. **19**, 225–230 (1968).

3. Some examples of right self-injective rings which are not left self-injective. Proc. Amer. Math. Soc. **26**, 244–245 (1970).

Schelter, W.: 1. Products of torsion theories. Archiv der Math. **22**, 590–596 (1971).

2. Two-sided rings of quotients. Archiv der Math. **24**, 274–277 (1973).

Schelter, W., Roberts, P.: 1. Flat modules and torsion theories. Math. Z. **129**, 331–334 (1972).

Schelter, W., Small, L.: 1. Some pathological rings of quotients. To appear.

Schmidt, J.: 1. Zur Kennzeichnung der Dedekind-MacNeilleschen Hülle einer geordneten Menge. Archiv der Math. **7**, 241–249 (1956).

Schubert, H.: 1. Kategorien I–II. Berlin-Heidelberg: Springer 1970.

Sharpe, D.W., Vámos, P.: 1. Injective modules. London: Cambridge Univ. Press. 1972.

Shock, R.C.: 1. Injectivity, annihilators and orders. J. Algebra **19**, 96–103 (1971).
Silver, L.: 1. Non-commutative localization and applications. J. Algebra **7**, 44–76 (1967).
Skornyakov, L.A.: 1. Elizarov's quotient ring and the localization principle. Mat. Zametki **1**, 263–268 (1967) [Russian].
2. The decomposition of injective modules. Mat. Issled. **5**, 183–186 (1970) [Russian].
Small, L.W.: 1. Orders in artinian rings. J. Algebra **4**, 13–41 and 505–507 (1966).
2. On some questions in noetherian rings. Bull. Amer. Math. Soc. **72**, 853–857 (1966).
3. Orders in artinian rings II. J. Algebra **9**, 266–273 (1968).
Smith, P.F.: 1. Quotient rings of group rings. J. London Math. Soc. **3**, 645–660 (1971).
Spulber, D.: 1. Sur une classe des anneaux de Prüfer. C.R. Acad. Sci. Paris **271**, 365–367 (1970).
Stenström, B.: 1. Pure submodules. Arkiv f. Mat. **7**, 159–171 (1967).
2. On the completion of modules in an additive topology. J. Algebra **16**, 523–540 (1970).
3. Coherent rings and FP-injective modules. J. London Math. Soc. **2**, 323–329 (1970).
4. Flatness and localization over monoids. Math. Nachrichten **48**, 315–334 (1971).
5. Rings and modules of quotients. Lecture Notes in Math. Vol. 237. Berlin-Heidelberg-New York: Springer 1971.
6. The maximal ring of quotients of a triangular matrix ring. Math. Scand. **34**, 162–166 (1974).
Stephenson, W.: 1. Modules whose lattice of submodules is distributive. Proc. London Math. Soc. **28**, 291–310 (1974).
Storrer, H.H.: 1. Epimorphismen von kommutativen Ringen. Comm. Math. Helv. **43**, 378–401 (1968).
2. Rings of quotients of perfect rings. Math. Z. **122**, 151–165 (1971).
3. Rational extensions of modules. Pacific J. Math. **38**, 785–794 (1971).
4. Torsion theories and dominant dimension. Appendix to J. Lambek [5].
5. On Goldman's primary decomposition. Lectures on rings and modules, pp. 617–661. Lecture Notes in Math. Vol. 246. Berlin-Heidelberg-New York: Springer 1972.
6. Epimorphic extensions of non-commutative rings. Comm. Math. Helv. **48**, 72–86 (1973).
Tachikawa, H.: 1. On splitting of module categories. Math. Z. **111**, 145–150 (1969).
2. On left QF-3 rings. Pacific J. Math. **32**, 255–268 (1970).
3. Double centralizers and dominant dimensions. Math. Z. **116**, 79–88 (1970).
4. Localization and artinian quotient rings. Math. Z. **119**, 239–253 (1971).
5. Quasi-Frobenius rings and generalizations. Lecture Notes in Math., Vol. 351. Berlin-Heidelberg-New York: Springer 1973.
6. QF-3 rings and categories of projective modules. J. Algebra **28**, 408–413 (1974).
Takeuchi, M.: 1. A simple proof of Gabriel and Popescu's theorem. J. Algebra **18**, 112–113 (1971).
Talintyre, T.D.: 1. Quotient rings of rings with maximum condition for right ideals. J. London Math. Soc. **38**, 439–450 (1963).
2. Quotient rings with minimum conditions on right ideals. J. London Math. Soc. **41**, 141–144 (1966).
Teply, M.L.: 1. Torsion-free injective modules. Pacific J. Math. **28**, 441–453 (1969).
2. Some aspects of Goldie's torsion theory. Pacific J. Math. **29**, 447–459 (1969).
3. Homological dimension and splitting torsion theories. Pacific J. Math. **34**, 193–205 (1970).
4. The torsion submodule of a cyclic module splits off. Canad. J. Math. **24**, 450–464 (1972).
5. A class of divisible modules. To appear.
Turnidge, D.R.: 1. Torsion theories and semi-hereditary rings. Proc. Amer. Math. Soc. **24**, 137–143 (1970).
2. Torsion theories and rings of quotients of Morita equivalent rings. Pacific J. Math. **37**, 225–234 (1971).
Ulmer, F.: 1. On the existence and exactness of the associated sheaf functor. To appear.
Utumi, Y.: 1. On quotient rings. Osaka J. Math. **8**, 1–18 (1956).
2. On continuous regular rings and semisimple self-injective rings. Canad. J. Math. **12**, 597–605 (1960).
3. On continuous regular rings. Canad. Math. Bull. **4**, 63–69 (1961).
4. On rings of which any one-sided quotient rings are two-sided. Proc. Amer. Math. Soc. **14**, 141–147 (1963).
5. On continuous rings and self-injective rings. Trans. Amer. Math. Soc. **118**, 158–173 (1965).
6. On the continuity and self-injectivity of a complete regular ring. Canad. J. Math. **18**, 404–412 (1966).
7. Self-injective rings. J. Algebra **6**, 56–64 (1967).

Uzkov, A. I.: 1. On rings of quotients of commutative rings. Mat. Sb. **22 (64)**, 439–441 (1948) [Russian].

Verdier, J.-L.: 1. Topologies et faisceaux. Séminaire Artin-Grothendieck-Verdier 1963–64. Lecture Notes in Mathematics, Vol. 269. Berlin-Heidelberg-New York: Springer 1972.

Walker, C. L., Walker, E. A.: 1. Quotient categories and rings of quotients. Rocky Mountain J. Math. **2**, 513–555 (1972). (Mimeographed version appeared in 1963.)

2. Quotient categories of modules. Proc. Conf. Categorical Algebra, La Jolla 1965, p. 404–420. Berlin: Springer 1966.

Warfield, R. B.: 1. Purity and algebraic compactness for modules. Pacific J. Math. **28**, 699–719 (1969).

Watts, C.: 1. Intrinsic caracterizations of some additive functors. Proc. Amer. Math. Soc. **11**, 5–8 (1960).

Wei, D. Y.: 1. On the concept of torsion and divisibility for general rings. Illinois J. Math. **13**, 414–431 (1969).

Winton, C. N.: 1. On the derived quotient module. Trans. Amer. Math. Soc. **154**, 315–321 (1971).

Wong, E. T.: 1. Rings with nonzero singular ideals. J. Math. Kyoto Univ. **10**, 419–431 (1970).

Wong, E. T., Johnson, R. E.: 1. Self-injective rings. Canad. Math. Bull. **2**, 167–173 (1959).

Wu, L. E. T., Mochizuki, H. Y., Jans, J. P.: 1. A characterization of *QF*-3 rings. Nagoya Math. J. **27**, 7–13 (1966).

Zariski, O., Samuel, P.: 1. Commutative algebra, Vol. 1 and 2. Princeton: van Nostrand 1958 and 1960.

Zelmanowitz, J. M.: 1. A shorter proof of Goldie's theorem. Canad. Math. Bull. **12**, 597–602 (1969).

2. Orders in simple artinian rings are strongly equivalent to matrix rings. Pacific J. Math. **48**, 621–627 (1973).

Index

Die Grundlehren der mathematischen Wissenschaften in Einzeldarstellungen mit besonderer Berücksichtigung der Anwendungsgebiete

Eine Auswahl

---

23. Pasch: Vorlesungen über neuere Geometrie. In Vorbereitung
41. Steinitz: Vorlesungen über die Theorie der Polyeder. In Vorbereitung
45. Alexandroff/Hopf: Topologie. Band 1
46. Nevanlinna: Eindeutige analytische Funktionen
63. Eichler: Quadratische Formen und orthogonale Gruppen
102. Nevanlinna/Nevanlinna: Absolute Analysis
114. Mac Lane: Homology
123. Yosida: Functional Analysis
127. Hermes: Enumerability, Decidability, Computability
131. Hirzebruch: Topological Methods in Algebraic Geometry
135. Handbook for Automatic Computation. Vol. 1/Part a: Rutishauser: Description of ALGOL 60
136. Greub: Multilinear Algebra
137. Handbook for Automatic Computation. Vol. 1/Part b: Grau/Hill/Langmaack: Translation of ALGOL 60
138. Hahn: Stability of Motion
139. Mathematische Hilfsmittel des Ingenieurs. 1. Teil
140. Mathematische Hilfsmittel des Ingenieurs. 2. Teil
141. Mathematische Hilfsmittel des Ingenieurs. 3. Teil
142. Mathematische Hilfsmittel des Ingenieurs. 4. Teil
143. Schur/Grunsky: Vorlesungen über Invariantentheorie
144. Weil: Basic Number Theory
145. Butzer/Berens: Semi-Groups of Operators and Approximation
146. Treves: Locally Convex Spaces and Linear Partial Differential Equations
147. Lamotke: Semisimpliziale algebraische Topologie
148. Chandrasekharan: Introduction to Analytic Number Theory
149. Sario/Oikawa: Capacity Functions
150. Iosifescu/Theodorescu: Random Processes and Learning
151. Mandl: Analytical Treatment of One-dimensional Markov Processes
152. Hewitt/Ross: Abstract Harmonic Analysis. Vol. 2: Structure and Analysis for Compact Groups. Analysis on Locally Compact Abelian Groups
153. Federer: Geometric Measure Theory
154. Singer: Bases in Banach Spaces I
155. Müller: Foundations of the Mathematical Theory of Electromagnetic Waves
156. van der Waerden: Mathematical Statistics
157. Prohorov/Rozanov: Probability Theory. Basic Concepts. Limit Theorems. Random Processes
158. Constantinescu/Cornea: Potential Theory on Harmonic Spaces
159. Köthe: Topological Vector Spaces I
160. Agrest/Maksimov: Theory of Incomplete Cylindrical Functions and their Applications
161. Bhatia/Szegö: Stability Theory of Dynamical Systems
162. Nevanlinna: Analytic Functions
163. Stoer/Witzgall: Convexity and Optimization in Finite Dimensions I
164. Sario/Nakai: Classification Theory of Riemann Surfaces
165. Mitrinović/Vasić: Analytic Inequalities
166. Grothendieck/Dieudonné: Eléments de Géométrie Algébrique I
167. Chandrasekharan: Arithmetical Functions
168. Palamodov: Linear Differential Operators with Constant Coefficients
169. Rademacher: Topics in Analytic Number Theory
170. Lions: Optimal Control of Systems Governed by Partial Differential Equations
171. Singer: Best Approximation in Normed Linear Spaces by Elements of Linear Subspaces

172. Bühlmann: Mathematical Methods in Risk Theory
173. Maeda/Maeda: Theory of Symmetric Lattices
174. Stiefel/Scheifele: Linear and Regular Celestial Mechanics. Perturbed Two-body Motion—Numerical Methods—Canonical Theory
175. Larsen: An Introduction to the Theory of Multipliers
176. Grauert/Remmert: Analytische Stellenalgebren
177. Flügge: Practical Quantum Mechanics I
178. Flügge: Practical Quantum Mechanics II
179. Giraud: Cohomologie non abélienne
180. Landkof: Foundations of Modern Potential Theory
181. Lions/Magenes: Non-Homogeneous Boundary Value Problems and Applications I
182. Lions/Magenes: Non-Homogeneous Boundary Value Problems and Applications II
183. Lions/Magenes: Non-Homogeneous Boundary Value Problems and Applications III
184. Rosenblatt: Markov Processes. Structure and Asymptotic Behavior
185. Rubinowicz: Sommerfeldsche Polynommethode
186. Handbook for Automatic Computation. Vol. 2. Wilkinson/Reinsch: Linear Algebra
187. Siegel/Moser: Lectures on Celestial Mechanics
188. Warner: Harmonic Analysis on Semi-Simple Lie Groups I
189. Warner: Harmonic Analysis on Semi-Simple Lie Groups II
190. Faith: Algebra: Rings, Modules, and Categories I
192. Mal'cev: Algebraic Systems
193. Pólya/Szegö: Problems and Theorems in Analysis I
194. Igusa: Theta Functions
195. Berberian: Baer $*$-Rings
196. Athreya/Ney: Branching Processes
197. Benz: Vorlesungen über Geometrie der Algebren
198. Gaal: Linear Analysis and Representation Theory
199. Nitsche: Vorlesungen über Minimalflächen
200. Dold: Lectures on Algebraic Topology
201. Beck: Continuous Flows in the Plane
202. Schmetterer: Introduction to Mathematical Statistics
203. Schoeneberg: Elliptic Modular Functions
204. Popov: Hyperstability of Control Systems
205. Nikol'skii: Approximation of Functions of Several Variables and Imbedding Theorems
206. André: Homologie des Algèbres Commutatives
207. Donoghue: Monotone Matrix Functions and Analytic Continuation
208. Lacey: The Isometric Theory of Classical Banach Spaces
209. Ringel: Map Color Theorem
210. Gihman/Skorohod: The Theory of Stochastic Processes I
211. Comfort/Negrepontis: The Theory of Ultrafilters
212. Switzer: Algebraic Topology—Homotopy and Homology
213. Shafarevich: Basic Algebraic Geometry
214. van der Waerden: Group Theory and Quantum Mechanics
215. Schaefer: Banach Lattices and Positive Operators
216. Pólya/Szegő: Problems and Theorems in Analysis II. In preparation